Lecture Notes in Networks and Systems

Volume 127

The series “Lecture Notes in Networks and Systems” publishes the latest developments in Networks and Systems—quickly, informally and with high quality. Original research reported in proceedings and post-proceedings represents the core of LNNS.

Volumes published in LNNS embrace all aspects and subfields of, as well as new challenges in, Networks and Systems.

The series contains proceedings and edited volumes in systems and networks, spanning the areas of Cyber-Physical Systems, Autonomous Systems, Sensor Networks, Control Systems, Energy Systems, Automotive Systems, Biological Systems, Vehicular Networking and Connected Vehicles, Aerospace Systems, Automation, Manufacturing, Smart Grids, Nonlinear Systems, Power Systems, Robotics, Social Systems, Economic Systems and other. Of particular value to both the contributors and the readership are the short publication timeframe and the world-wide distribution and exposure which enable both a wide and rapid dissemination of research output.

The series covers the theory, applications, and perspectives on the state of the art and future developments relevant to systems and networks, decision making, control, complex processes and related areas, as embedded in the fields of interdisciplinary and applied sciences, engineering, computer science, physics, economics, social, and life sciences, as well as the paradigms and methodologies behind them.

**** Indexing: The books of this series are submitted to ISI Proceedings, SCOPUS, Google Scholar and Springerlink ****

More information about this series at http://www.springer.com/series/15179

Asis Kumar Tripathy · Mahasweta Sarkar ·
Jyoti Prakash Sahoo · Kuan-Ching Li ·
Suchismita Chinara
Editors

Advances in Distributed Computing and Machine Learning

Proceedings of ICADCML 2020

Springer

Editors
Asis Kumar Tripathy
Vellore Institute of Technology
Vellore, Tamil Nadu, India

Jyoti Prakash Sahoo
Institute of Technical Education
and Research (ITER)
Siksha 'O' Anusandhan (SOA)
Deemed to be University
Bhubaneswar, Odisha, India

Suchismita Chinara
National Institute of Technology, Rourkela
Rourkela, Odisha, India

Mahasweta Sarkar
San Diego State University
San Diego, CA, USA

Kuan-Ching Li
Providence University
Taichung, Taiwan

ISSN 2367-3370 ISSN 2367-3389 (electronic)
Lecture Notes in Networks and Systems
ISBN 978-981-15-4217-6 ISBN 978-981-15-4218-3 (eBook)
https://doi.org/10.1007/978-981-15-4218-3

This Springer imprint is published by the registered company Springer Nature Singapore Pte Ltd.
The registered company address is: 152 Beach Road, #21-01/04 Gateway East, Singapore 189721, Singapore

Organizing Committee

Chief Patron

G. Viswanathan, Chancellor, VIT

Patrons

Sankar Viswanathan, Vice-President, VIT
Sekar Viswanathan, Vice-President, VIT
G. V. Selvam, Vice-President, VIT
Kadhambari S. Viswanathan, Assistant Vice-President, VIT
Sandhya Pentareddy, Executive Director, VIT
Anand A. Samuel, Vice-Chancellor, VIT
S. Narayanan, Pro-Vice-Chancellor, VIT

General chairs

Suchismita Chinara, NIT, Rourkela
Mahasweta Sarkar, SDSU, CA, USA

International Advisory Chairs

Vincenzo Piuri, Università degli Studi di Milano, Italy
Kuan-ChingLi, Providence University, Taiwan
Kenji Suzuki, Illinois Institute of Technology, USA

Valentina Emilia Balas, Aurel Vlaicu University of Arad, Romania
João Manuel R. S. Tavares, Universidade do Porto (FEUP), Portugal
Florentina A. Pintea, Universitatea Tibiscus din Timişoara, Romania

Organizing Chair

Balakrushna Tripathy, Dean, SITE, VIT, Vellore

Organizing Co-chairs

Aswani Kumar Ch., Associate Dean, SITE, VIT, Vellore
Jasmine Norman, HOD, Department of IT, SITE, VIT, Vellore
Agilandeeswari L., HOD, Department of DC, SITE, VIT, Vellore
Ramkumar T., HOD, Department of CA, SITE, VIT, Vellore
Sree Dharinya S., HOD, Department of SSE, SITE, VIT, Vellore
Malaserene I., HOD, Department of Smart Computing, SITE, VIT, Vellore

Organizing Secretary

Asis Kumar Tripathy, VIT, Vellore

Co-organizing Secretaries

Tapan Kumar Das, VIT Vellore
Jyoti Prakash Sahoo, S 'O' A Deemed to be University
Alekha Kumar Mishra, NIT, Jamshedpur

Program Chairs

Debi Prasanna Acharjya, VIT, Vellore
Sathiyamoorthy E., VIT, Vellore

Executive Committee

Ajit Kumar Santra, VIT, Vellore
Ganesan K., VIT, Vellore
Hari Ram Vishwakarma, VIT, Vellore

Technical Program Committee

Xiao-Zhi Gao, University of Eastern Finland, Finland
Vinit Jakhetiya, IIT Jammu, India
Kenji Suzuki, Illinois Institute of Technology, USA
Ajit Kumar Nayak, S 'O' A Deemed to be University
Pritom Rajkhowa, CUHK, Hong Kong
Binayak Kar, National Taiwan University of Science and Technology, Taiwan
Deepak Puthal, Newcastle University, UK
Subrota Kumar Mondal, Macau University of Science and Technology
Dimitris Chatzopoulos, Hong Kong University of Science and Technology
ABM Rezbaul Islam, Sam Houston State University, USA
Biswa Mohan Acharya, S 'O' A Deemed to be University

Program Committee

Sushanta Karmakar, IIT Guwahati
Sanjaya Kumar Panda, IIITDM, Kurnool
Ashok Das, IIIT Hyderabad
Ansuman Mahapatra, NIT, Puducherry
Ramesh Mohapatra, NIT, Rourkela
Mukesh Kumar, NIT Patna
Rahul Dixit, IIIT, Pune
Umakanta Nanda, VIT AP
Ratnakar Dash, NIT, Rourkela
Ansuman Mohapatra, NIT, Puducherry
Arunashu Mahapatro, VSSUT, Odisha
Biswanath Sethi, IGIT, Sarang, Odisha
Deepak Ranjan Nayak, IIITDM Kanchipuram, India
Subasish Mohapatra, CET, Bhubaneswar
Soubhagya Shankar Barpanda, VIT AP
Munesh Singh, IIITDM Kancheepuram
Jitendra Kumar Rout, KIIT, Bhubaneswar
Debabrata Singh, S 'O' A Deemed to be University

Asish Kumar Dalai, VIT AP
Chandrasekar R., NIT, Puducherry
Jagadeesh Kakarla, IIITDM Kancheepuram
Suvendu Mohapatra, Volktek, Taiwan
Prakash Chandra Jena, Hi-Tech College of Engineering, Bhubaneswar
John Singh K., VIT, Vellore
Shashank Mouli Satapathy, VIT, Vellore
Mohd. Saifuzzaman, Daffodil International University, Dhaka, Bangladesh
Jeyanthi N., VIT, Vellore
Dharmveer Kumar Yadav, KEC, Katihar
Rahul Raman, VIT, Vellore

Publicity Chairs

Navaneethan C., VIT, Vellore
Manas Ranjan Prusty, VIT, Chennai
Kathiravan S., VIT, Vellore

Finance Chair

Chiranji Lal Chaudhury, VIT, Vellore

Advisory Committee

Banshidhar Majhi, IIITDM, Kancheepuram
Debasish Jena, IIIT, Bhubaneswar

Registration Chairs

Suganya P., VIT, Vellore
Meenatchi S., VIT, Vellore
Divya Udayan J., VIT, Vellore

Local Organizing Committee

Rajesh Kaluri, VIT, Vellore
Srinivas Koppu, VIT, Vellore
Praveen Kumar Reddy, VIT, Vellore
Priya M., VIT, Vellore
Thanapal P., VIT, Vellore
Prabhavathy P., VIT, Vellore
Jayaram Reddy A., VIT, Vellore
Sudha M., VIT, Vellore
Nadesh R. K., VIT, Vellore
Neelu Khare, VIT, Vellore
Srinivan P., VIT, Vellore
Vijayan R., VIT, Vellore
Pradeep Kumar Roy, VIT, Vellore
SwarnaPriya M., VIT, Vellore
P. J. Kumar, VIT, Vellore
Prabhu J., VIT, Vellore
Dharmendra Singh Rajput, VIT, Vellore
Vijay Anand R., VIT, Vellore
Parimala M., VIT, Vellore
Jagadeesh G., VIT, Vellore
Chemmalar Selvi G., VIT, Vellore

Keynotes, Workshops and Tutorials

I. KeyNotes

Overview of Resampling Methods
by Prof. D. V. L. N. Somayajulu, Director, IIITDM, Kurnool, India.
Research Roadmap of Offloading in Federated Cloud, Edge and Fog Systems
by Dr. Binayak Kar, NTUST, Taiwan.
Generating Pseudo-SQL Queries from Under-Specified Natural Language Questions
by Dr. Fuxiang Chen, Research Scientist, DeepSearchInc., Singapore.
Fog Computing: An Emerging Paradigm
by Prof. Prasanta K. Jana, IIT(ISM), Dhanbad, India.

II. Workshops

Demystifying DevOps with Docker and Kubernetes
by Prof. Jyoti Prakash Sahoo, Assistant Professor, S 'O' A Deemed to be University, India.

III. Tutorials

Energy-Efficient Cloud Computing
Dr. Sanjaya Kumar Panda, Asst. Prof., NIT Warangal, India.
Analysis of Keystroke Patterns with Machine Learning Approach
by Dr. Utpal Roy, Siksha Bhavana, Visva-Bharati, Santiniketan, India.

Preface

This issue of Lecture Notes in Networks and Systems is dedicated to the First International Conference on Advances in Distributed Computing and Machine Learning (ICADCML-2020). The ICADCML 2020 is an annual forum that aimed to bring together ideas, innovations, lessons, etc., associated with distributed computing and machine learning, and their application in diverse areas. Distributed computing performs an increasingly important role in modern data processing, information fusion, communication, network applications, real-time process control and parallel computing. Additionally, machine learning (ML) as a subset of artificial intelligence emerged with the scientific study of algorithms and statistical models to perform a specific task without using explicit instructions, relying on historical data, patterns and inference instead. This conference aspires to stimulate research in Advances in Distributed Computing and Machine Learning. The conference was organized by the School of Information Technology and Engineering (SITE), VIT, Vellore, from January 30–31, 2020.

VIT was founded in 1984 as Vellore Engineering College by the Chancellor Dr. G. Viswanathan. The VIT is the first institution in India to have the 4 STAR rating by QS world ranking. In addition to this, the consortium of industries, FICCI has adjudged VIT as the Excellence in Faculty. VIT has also completed three cycles of NAAC accreditation with an A grade, in addition to accreditation by coveted ABET, USA. Furthermore, Government of India has recognized VIT, Vellore, as an Institution of Eminence(IoE).

We would like to convey our earnest appreciation to all the authors for their contributions to this book. We would like to extend our gratitude to all the reviewers for their constructive comments on all papers, especially we would like to thank the Organizing Committee for their hard work. Finally, we would like to thank the Springer publications for producing this volume.

Vellore, India
January 2020

Asis Kumar Tripathy
Mahasweta Sarkar
Jyoti Prakash Sahoo
Kuan-Ching Li
Suchismita Chinara

Contents

About the Editors

Asis Kumar Tripathy is an Associate Professor in the School of Information Technology and Engineering, Vellore Institute of Technology, Vellore, India. He has more than ten years of teaching experience. He completed his Ph.D. from the National Institute of Technology, Rourkela, India, in 2016. His areas of research interests include wireless sensor networks, cloud computing, Internet of things and advanced network technologies. He has several publications in refereed journals, reputed conferences and book chapters to his credit. He has served as a program committee member in several conferences of repute. He has also been involved in many professional and editorial activities.

Mahasweta Sarkar is currently working as a Professor of the Department of Electrical and Computer Engineering in San Diego State University in 2006. Her M.S. and Ph.D. degrees were completed at the University of California, San Diego (UCSD), in 2003 and 2005, respectively. She received her B.S. degree in Computer Science & Engineering (Summa Cum Laude) in May 2000 from San Diego State University. Dr. Sarkar is a recipient of the "President's Leadership Award for Faculty Excellence" for the year 2010. She delivered invited lectures and keynotes in different universities spread all over the globe. The talks were on Wireless Body Area Networks and Brain–Computer Interface. Her research interest lies in the area of MAC layer power management algorithms and Quality-of-Service issues and protocols in WLANs, WMANs, WBANs, sensor networks and wireless adhoc networks. She has published over eighty research papers in these fields in various international journals and conferences of high repute.

Jyoti Prakash Sahoo is working as an Assistant Professor in the Institute of Technical Education and Research (ITER), Siksha 'O' Anusandhan (SOA) Deemed to be University, Bhubaneswar, India. He is having more than 12 years of academic and research experience in Computer science and engineering education. He has published several research papers in various international journals and conferences. He is also serving many journals and conferences as an editorial or reviewer board member. He is having expertise in the field of Cloud computing and Machine learning. He served as publicity chair and organizing member of technical program committees of many national and international conferences. Being a WIPRO Certified Faculty, he has also contributed to industry-academia collaboration, student enablement, and pedagogical learning. Furthermore, he is associated with various educational and research societies like IEEE, IET, IACSIT, IAENG, etc.

Kuan-Ching Li is currently appointed as Distinguished Professor at Providence University, Taiwan. He is a recipient of awards and funding support from several agencies and high-tech companies, as also received distinguished chair professorships from universities in several countries. He has been actively involved in many major conferences and workshops in program/general/steering conference chairman positions and as a program committee member, and has organized numerous conferences related to high-performance computing and computational science and engineering. Professor Li is the Editor-in-Chief of technical publications Connection Science (Taylor & Francis), International Journal of Computational Science and Engineering (Inderscience) and International Journal of Embedded Systems (Inderscience), and serves as associate editor, editorial board member and guest editor for several leading journals. Besides publication of journal and conference papers, he is the co-author/co-editor of several technical professional books published by CRC Press, Springer, McGraw-Hill, and IGI Global. His topics of interest include parallel and distributed computing, Big Data, and emerging technologies. He is a Member of the AAAS, a Senior Member of the IEEE, and a Fellow of the IET.

Suchismita Chinara is currently working as an Assistant Professor in the Department of Computer Science and Engineering, National Institute of Technology, India. She has received her M.E. degree in 2001 and Ph.D. in 2011 from the Department of Computer Science and Engineering, National Institute of Technology, Rourkela. She has authored and co-authored multiple peer-reviewed scientific papers and presented works at many national and international conferences. Her contributions have acclaimed recognition from honorable subject experts around the world. Her academic career is decorated with several reputed awards and funding. Her research interest includes wireless network, ad hoc network and MANET.

Distributed Computing Trends, Issues, and Applications

Customized Score-Based Security Threat Analysis in VANET

Alekha Kumar Mishra, Asis Kumar Tripathy, and Maitreyee Sinha

Abstract With the advancement of IoT, automated driving vehicles using VANET are the new trends of attraction. VANET is inherited from MANET and specifically used for communications between vehicles and between a vehicle and roadside infrastructures. VANET routing protocols are prone to security threats such as DoS, eavesdropping. In this paper, VANET routing protocols are analyzed to identify possible security threats, especially where and how cryptographic mechanisms could be implemented to achieve better security. A customized comparison of secured protocols is provided, followed by observations and conclusions.

Keywords VANET · Security threats · Routing protocols · Threat analysis · Threat level score

1 Introduction

In the era of IoT technologies, devices are being developed that can be installed in a vehicle in order to send and receive data related to vehicle movement, traffic, etc. Vehicular Adhoc Networks (VANETs) [1] are such networks that provide full-fledged communication between vehicles and the roadside established infrastructures. Since the technology is on the edge of manufacturing automated driving vehicle, it is required to standardize vehicle to vehicle (V2V) or vehicle to other agents communication. The common information that a vehicle requires to share includes

A. Kumar Mishra · M. Sinha (✉)
Department of Computer Applications, National Institute of Technology Jamshedpur, Jamshedpur, India
e-mail: maitreyeesinha2012@gmail.com

A. Kumar Mishra
e-mail: alekha.ca@nitjsr.ac.in

A. Kumar Tripathy
School of IT & Engineering, Vellore Institute of Technology, Vellore, India
e-mail: asistripathy@vit.ac.in

A. K. Tripathy et al. (eds.), *Advances in Distributed Computing and Machine Learning*, Lecture Notes in Networks and Systems 127, https://doi.org/10.1007/978-981-15-4218-3_1

velocity, position, fuel, and engine status, etc. When this information is available in the open air, this could lead to privacy concerns for vehicles under security. In an automated driving vehicle scenario, an agent completely depends on information of other vehicles to drive efficiently, and it is very much possible that a vehicle could simulate conditions and falsify the traffic information to induce accidents and jams. Thus, this needs to be assured that the communication between vehicles is secured.

This paper first enlists the possible attacks in a VANET communication scenario. Next, the analysis on the impact of each attack is performed highlighting their effect the safety. Specifically, we have discussed three protocols: (i) multi-hop vehicular broadcast, (ii) geo-based protocols , and (iii) clustering-based protocols and analyzed how these protocols have incorporated the security mechanisms. Finally, a customized study is made to compare the security strength of these protocols, with required improvements if any.

Rest of the paper has been organized as follows. The description of security threats in VANET is provided in Sect. 2. The methodology adapted to evaluate the severity of attacks is described in Sect. 3. The evaluation and scoring the threat level of various attacks are summarized in Sect. 4. At the end, the conclusion is provided in Sect. 6.

2 Security Threats in VANET

Due to network characteristics and features of VANET, it is exposed to a number of security threats [2, 3]. The most common threats are DoS, Sybil, message suppression, replay, eavesdropping, session hijacking, and location tracking [4]. The DoS attack [5] occurs when a malicious node in the network utilizes or blocks the resources of a victim such as communication channel bandwidth, and receiver buffer, thereby preventing other legitimate devices from accessing the resources. This attack is on the availability of the resources that brings down real-time communication and delay in the transfer of critical information between the vehicles. In Sybil attack, a malicious node illegally creates or claims to have a large number of identities and provide collaborative false information to legitimate vehicles about a traffic jam ahead, which force these vehicles to use an alternate route. This attack depends on three factors: (i) whether the network considers all the nodes to be identical, (ii) to what degree does a network entertain requests from nodes that cannot be trusted to be legitimate, and (iii) the minimum cost of manufacturing the VANET devices. In message suppression attack, a malicious node intuitively drops received data from the network that results in loss of critical traffic information. The attacker can also store this information and later re-inject into the network to broadcast false update. In various of these attacks called replay attack, an attacker resends a message that were sent earlier over the channel to mislead the receiver node. Here, the goal is to provide disguised information to the legitimate user to disrupt the network. The simplest form of passive attack is eavesdropping. In a various of these attacks, a malicious node gains unauthorized access to sensitive and confidential data which is not meant for the attacker. An eavesdropper collects these data through the communication channel without any consent and knowledge of the data owner. The attacker can misuse the data and may generate traffic issues for the other nodes. In session hijacking, a malicious node takes control of the session by getting the unique session ID for each session. It is mostly done in the beginning of communication during the process of connection establishment. Finally, using the location tracking attack, an attacker can mark the location of a victim node to eavesdrop or collect private information about the vehicle. This attack is a threat to privacy.

3 Methodology

A number of secured protocols have been proposed to handle various security issues in the literature. In this paper, three categories of protocols are selected for comparison and analyzed their performance under these attacks. These are multi-hop vehicular broadcast (MHVB) protocols, cluster-based protocols, and geo-based protocols. To provide a better comparison, these protocols are evaluated according to the issues that these protocols can solve. A score is assigned to each protocol type based on its ability. The score is decided depending on the possible attacks on VANETs discussed in Sect. 2. Each attack is assigned with some weight based on its severity level. In the end, the security measure for each of the attacks is presented followed by possible improvements on each categories of the protocol.

3.1 Measuring Security Attacks

It is necessary to analyze the severity of the attack before assigning a score to them. Based on the studies on these attacks as reported in the literature [2, 3], the following criteria are considered for evaluating the severity and assigning a score to each attack between the value 1 and 5.

- *Probability of an accident* (PA): This criterion checks whether a given attack can cause accidents among the vehicles.
- *Traffic jam time* (TJT): This criterion evaluates the time required by an attack to result in traffic jam.
- *Changing routes of vehicle* (CRV): It checks the ability of the attack to divert vehicles from their optimized route to deteriorate its performance. This task also involves checking the ability to trick the navigation system.
- *Threat to privacy* (TTP): This criterion tests the extent of an attack to launch a threat on privacy of a node in VANET.
- *After incident events* (AIE): These activities include blocking calls to fire departments or hospitals, blocking the roadside units, etc.

DoS is rated high for accident probability as it can clog up device's ability to receive a signal in a time of traffic jam and lead to an accident. An accident scenario can turn into traffic jams situation on the road. However, DoS has limited ability to manipulate vehicles in changing routes. DoS after incident events is also higher due to traffic jam and bringing down communication between vehicles. Sybil attack may use collaborated identities to provide false information about the vehicle routes. Therefore, the CRV criterion is highly affected by Sybil. Causing accident and after incidents such as jamming by a Sybil attacker is less probable. Dropping critical traffic information in crucial hours may lead to accident, traffic jam, change in vehicle routes, and blocking call to emergencies with a higher probability. However, similar to DoS and Sybil, it has very less impact on privacy. With replay attack, there is high chance of change in vehicle route which in turn may lead to traffic jam. Replay attack may also block communication service up to some extent. The attacks such as eavesdropping, session tracking, and location tracking mostly affect the privacy of crucial information that may lead to other attacks. It also has a limited impact on after incident events. Based on the above facts, it is observed that PA can be a most harmful situation in the VANET and TTP could also be dangerous but marginally lower than PA. Therefore, using the proposed evaluation criteria, PA is given the weight value of 2.0 , TTP is given weight value of 1.5, and rest of the criteria share weight value of 1 each. The overall threat value of an attack in VANET is given by

$$\text{Overall score} = 2 \times \text{PA} + \text{TJT} + \text{CRV} + 1.5 \times \text{TTP} + \text{AIE} \tag{1}$$

The normalized score is given by

$$\text{Norm. score} = \frac{\text{Overall score}}{32.5} \tag{2}$$

Using the above equation, the summary of evaluated measurement of threat levels based on the above five criteria is shown in Table 1. The effect ratio on criteria by each attack is also pictorially shown in Fig. 1.

Table 1 Attack measure evaluation

Attacks	PA	TJT	CRV	TTP	AIE	Overall	Norm. score
DoS	4	3	1	0	4	16	0.5
Sybil attack	2	2	4	0	1	11	0.34
Message suppression	4	3	4	0	4	19	0.58
Replay attack	2	4	4	0	3	15	0.46
Eavesdropping	0	0	0	4	4	10	0.31
Session attack	2	1	3	3	3	12	0.37
Location tracking	0	0	0	5	1	6	0.18

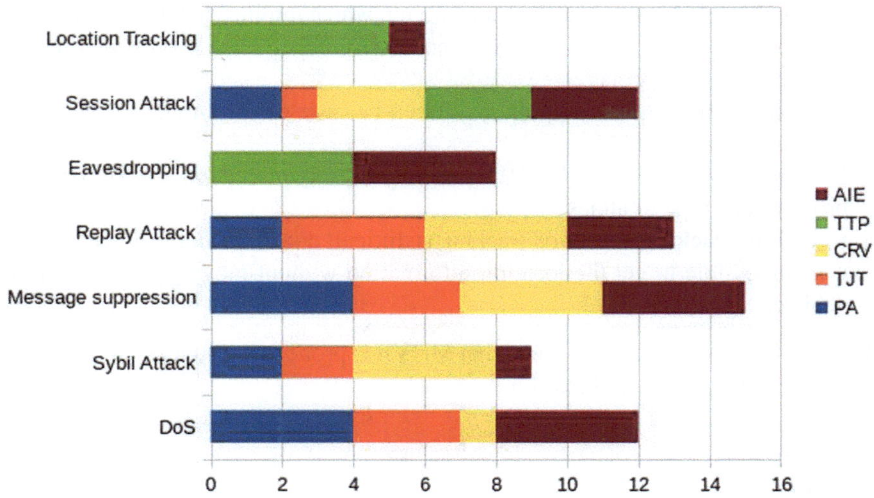

Fig. 1 Effect ratio on criteria of each attack

4 Scoring Various Protocols

In VANET, the correctness of the information is based on the authenticity of the received information. Each routing protocol of VANET must take some security measures to prevent the attacks. So, it is decided to calculate a final score of all security attacks by using the criteria discussed in Sect. 3 to know the effectiveness of each attacks in different scenarios.

4.1 Multi-hop Vehicular Broadcast

In [6], a secured routing protocol for multi-hop vehicular network is proposed. It uses elliptic curve cryptography (ECC) to encrypt the messages before it is transmitted. The challenge is associated with this scheme to achieve the same level of efficiency after using the ECC for encryption. In multi-hop vehicular broadcast (MHVB), it sends data from vehicle to vehicle in multi-hop fashion. Due to the multi-hop technology, it is very much prone to attacks as compared to one-hop networks.

1. **Denial of service attack**: Since there is no authentication used in MHVB, it will be difficult to stop from DOS attacks. An attacker can easily transmit the messages repeatedly to jam the entire road.
2. **Sybil attack**: It can create clones of captured vehicles again to jam the network. MHVB is prone to this type of attack.
3. **Message suppression**: MHVB is also prone to message suppression attack, because of its open nature of communication.

4. **Replay attack**: It uses wi-fi for transmission which implies that it is prone to this attack. Basic wi-fi does not contain sequence numbers/time stamps and has no protection against replay.
5. **Eavesdropping**: Eavesdropping is difficult because of short-range communication between the vehicles.
6. **Session Hijacking**: Since MHVB does not have any protection against it, so it is prone to this type of attack.
7. **Location Tracking**: Location tracking is difficult due to the multi-hop nature, so it is impossible to get the location information without moving as par with the vehicle.

Based on the observation, it is found that MHVB is prone to the attacks such as DOS, Sybil, message, replay, and session.

4.2 Trusted Variant of Multi-hop

In [7], a trusted multi-hop protocol is proposed that also encrypts the messages by using ECC.

1. **DoS**: It allows a vehicle to transmit a packet just once, which means probability of a DOS attack is very low and almost impossible.
2. **Sybil attack**: It is easy to authenticate the duplicates present in the network in this protocol. So, it is prone to Sybil attack.
3. **Message suppression**: This protocol is also open to message suppression because of its wireless medium of communication.
4. **Replay attack**: It is also prone to this attack because it is not using any time stamp to identify the messages.
5. **Eavesdropping**: There is a very less chance for the attacker to intercept the network due to the fixed transmission range of the vehicles.
6. **Session hijacking**: Another possible attack is session hijacking since it does not have any protection against it.
7. **Location tracking** This attack is almost impossible due to the frequent change of place of each vehicles.

Therefore, the possible attacks on a trusted variant of multi-hop network are message suppression, replay, and session.

4.3 Position-Based VANET

Position-based routing relies on geographic data for transmission purpose [8–10]. The information is exchanged to the vehicle which is nearest to the target, and it uses the greedy technique to find the route. Analysis of threats on the position-based VANET protocol is discussed below.

1. **False position advertisement**: Since it uses greedy routing, the attacker node may claim to be the only node between the source and destination by advertising a false position and tamper with the data received.
2. **Geographic Sybil attack**: Since this protocol uses greedy routing, the attacker node may create multiple IDs between the source and the destination which can result in a delay in delivery of data between the host and destination or result in the loss of packets.
3. **Message suppression**: Attackers selected as forwarders can simply drop packets, either all (black hole attack) or selectively (gray hole attack). Since this protocol uses greedy routing, the attacker node may simply drop the packets that are being transmitted which will result in loss of data transfer.
4. **Replay attack**: The attacker node may drop the packets being transmitted through its path, and instead, send previously received packets to the destination.
5. **Packet injection attack**: Since this protocol uses greedy routing, the attacker node may claim to be a forwarding node instead of a source node.

Therefore, this protocol is prone to DoS via packet injection, replay attack, message suppression, Sybil attack, and location tracking.

4.4 Clustering-Based Protocols in VANET

The VANET clustering methods [11] link nodes into group or collections known as clusters based on different rules. Then the cluster heads are selected for each cluster to pass information to other nodes. This clustering algorithm can be used as a secured routing protocol for VANET [12].

1. **DoS**: In clustering-based approaches, the DOS attack congests the data streamline which creates issues to transfer necessary information from one node to another.
2. **Sybil attack**: A trust-based clustering approach can be implemented to ensure that the cluster head of the nodes should be built on the trust of each other. If there is a communication failure between two nodes, then their trust index will decrease and it will detect that the node is defected.
3. **Message Suppression**: If in a clustering environment the cluster head turns out to be defective or compromised by an attacker, then it can drop some of the information which can cause anomaly and will cause loss of information in the network. It may be used later by the attacker. The critical information loss can cause errors and also may lead to accidents in real environment. It is a threat to the reliability of the network. However, recent cluster-based approaches overcome this issue by using the central authority approach. It ensures and authenticates the authenticity of the messages flowing in the network.

4. **Replay attack**: As such of now, no clustering method is able to solve this problem efficiently. Other routing protocols might have an optimal solution over clustering protocols.
5. **Eavesdropping**: While messages are flowing from one node to another, an eavesdropper may try to gain the information and might use it against them or may even try to alter the message. This may result in mismatch communication and may increase the distrust among corresponding nodes. Then, it collaboratively brings down the confidentiality level of the cluster. Many different and improved clustering approaches are introduced which reduced this effect although no method can completely stop this attack to occur.
6. **Session hijacking**: In this attack, the malicious node in the cluster can act as an actual member and can participate in the session establishment where no authentication process is required, which gives it the power to know all the necessary information in the beginning itself such as IP addresses, sequence number, and all, which later can be followed by DOS attack by that particular attacker. It can cause harm to trusted vehicles/ nodes and which can cause the unavailability of IP addresses of legitimate nodes.
7. **Location tracking**: This type of attack takes place mostly in dynamic clustering environments where the positions of the nodes or the vehicles keep changing. If the attacker is one of the nodes in the cluster, it can send wrong location information to other vehicles and cause disruption of the system. There are clustering techniques which can help in reducing such attacks using geo-based cluster mechanisms which locate the nodes directly with the help of location-based services.

It can be summarized that the clustering-based protocols are prone to DoS, Sybil attack, replay attack, eavesdropping, and session attack. Using the observation on the possible attacks on various routing protocols in VANET, Table 2 summarizes the total normalized score of each category of protocol.

Table 2 Normalized scores

	DoS	Sybil	Msg. supp.[a]	Replay attack	Eaves drop-ping	Session attack	Location tracking	Total
Multi-hop	✓	✓	✓	✓		✓		0.35
Trusted multi-hop			✓	✓		✓		0.23
Position-based	✓	✓	✓	✓			✓	0.31
Clustering-based	✓	✓		✓	✓	✓		0.31

[a]message suppression

5 Protocol Threat Analysis

The normalized score comparison of each attack is depicted in Fig. 2. It is observed from the figure that message suppression, DoS, and replay attack possess a higher threat level on VANET compared to other attacks. This is because of the impact it brings to the network in terms of various criteria such as PA, TJT, CRV, TTP, and AIE as discussed earlier.

The total threat score for each variants of protocol is shown in Fig. 3. It is found that multi-hop protocols have the highest level of threat while position-based and clustering-based protocols also have comparable level of threat. These threat levels are mostly due to open communication, and infrastructure-based pitfalls and protocol constraints.

6 Conclusion

In this paper, a score-based analysis of various threats on VANET is presented. Attacks on VANET are very common, and it is important to know the effect of each attack. We have calculated the impact of each attack for all the variants of routing protocols. The threat level of different types of protocols is also presented to clarify the importance of each attacks. All the attacks may not create the same type of issues

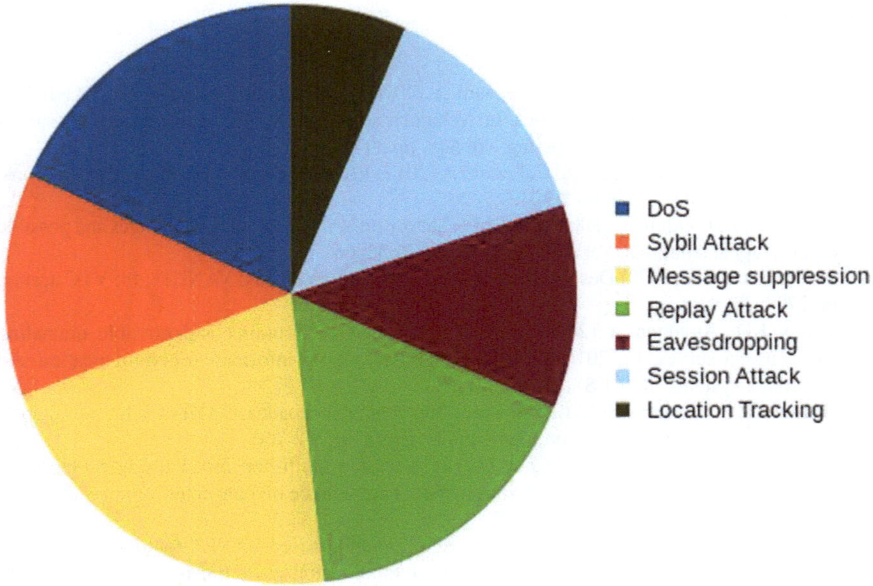

Fig. 2 Chart for normalized score comparison of the attacks

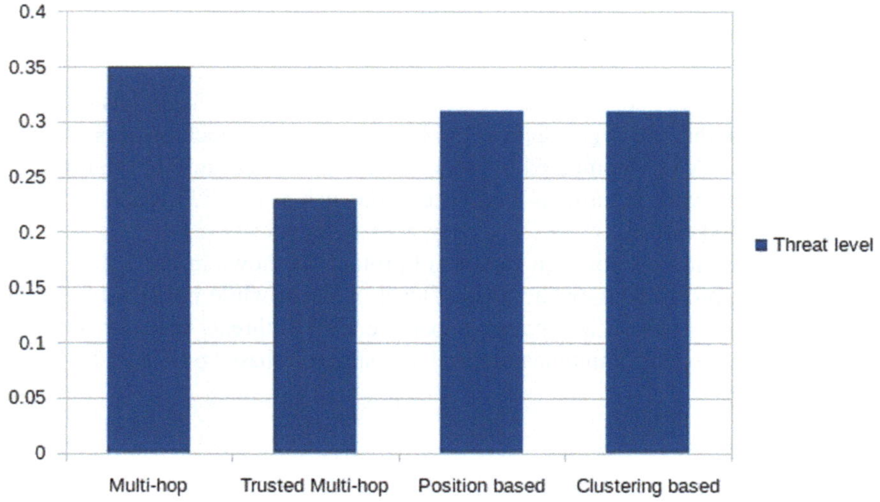

Fig. 3 Threat level of different protocol categories

to the network, so it is better to know the level of harmfulness before it has happened. This paper would help the researchers to be more alert and attentive before taking any security measures for the networks.

References

1. Rasheed A, Gillani S, Ajmal S, Qayyum A (2016) Vehicular ad hoc network (VANET): a survey, challenges, and applications. In: Vehicular ad-hoc networks for smart cities, advances in intelligent systems and computing, vol 548, pp 39–51
2. Hasrouny H, Samhat AE, Bassil C, Laouiti A (2017) VANET security challenges and solutions: a survey. Vehicular Commun 7(2017):7–20
3. Mejri MN, Ben-Othman J, Hamdi M (2014) Survey on VANET security challenges and possible cryptographic solutions. Vehicular Commun 1(2):53–66
4. Yi Q, Moayeri N (2008) Design of secure and application-oriented VANETs. In: VTC spring, pp 2794–2799
5. Thilak KD, Amuthan A (2016) DoS attack on VANET routing and possible defending solutions—a survey. In: 2016 international conference on information communication and embedded systems (ICICES). IEEE, pp 1–7
6. Osafune T, Lin L, Lenardi M (2006) Multi-hop vehicular broadcast (MHVB). In: 6th international conference on ITS telecommunications. IEEE, pp 757-760
7. Daxin T, Wang Y, Liu H, Zhang X (2012) A trusted multi-hop broadcasting protocol for vehicular ad hoc networks. In: 2012 international conference on connected vehicles and expo (ICCVE). IEEE, pp 18–22
8. Goel N, Sharma G, Dhyani I (2016) A study of position based VANET routing protocols. In 2016 international conference on computing, communication and automation (ICCCA). IEEE, pp 655–660
9. Lee KC, Hrri J, Lee U, Gerla M (2007) Enhanced perimeter routing for geographic forwarding protocols in urban vehicular scenarios. In: 2007 IEEE Globecom workshops. IEEE, pp 1-10

10. Moez J, Senouci S, Rasheed T, Ghamri-Doudane Y (2009) Towards efficient geographic routing in urban vehicular networks. IEEE Trans Veh Technol 58(9):5048–5059
11. Ahmed A, Hafid A (2012) A new stability based clustering algorithm (SBCA) for VANETs. In: 37th annual IEEE conference on local computer networks-workshops. IEEE, pp 843-847
12. Ahmad A, Kadoch M (2017) Performance improvement of cluster-based routing protocol in VANET. IEEE Access 5(2017):15354–15371
13. Slavik M, Mahgoub I (2013) Spatial distribution and channel quality adaptive protocol for multihop wireless broadcast routing in VANET. IEEE Trans Mob Comput 12(4):722–734
14. Tonguz OK, Wisitpongphan N, Parikh JS, Bai F, Mudalige P, Sadekar VK (2006) On the broadcast storm problem in ad hoc wireless networks. In: 2006 3rd international conference on broadband communications, networks and systems. IEEE, pp 1–11

FM—MANETs: A Novel Fuzzy Mobility on Multi-path Routing in Mobile Ad hoc Networks for Route Selection

T. Sudhakar and H. Hannah Inbarani

Abstract Mobile ad hoc networks (MANETs) are self-configured networks without any infrastructure-based network. The mobile nodes (MNs) are continuously deployed in an infrastructure-based network, whereas infrastructure-less network MNs are placed at random positions with mobility, i.e., movements of MNs. The major problem of ad hoc networks is routing and mobility of a node between the source and destination of a node. Due to the high movement in a wireless network, the MNs are not able to establish the connection between source and destination nodes. The network performances measures such as bandwidth and delay which are often changed due to MNs mobility, i.e., these network metrics include inconsistencies in a wireless ad hoc network. The uncertainties create a problem in the choice of the best route from the origin node to the recipient node unit. In this work, fuzzy mobility (FM) is used to resolve the issue of routing and mobility for better route selection in a wireless network. The performance results compared with the well-known existing wireless ad hoc routing protocols based on the packet delivery ratio (PDR), delay and throughput metrics. A fuzzy mobility-based selection gives better results when compared to other mobility models such as Random Way Point (RWP) and Random Direction (RD).

Keywords Ad hoc network · Fuzzy · Mobility · Routing first section

1 Introduction

In recent times [1], MANET is one of the major research areas to address many problems. The major problem of this network is that MNs are moving randomly, and they cannot predict route failure between source and destination nodes, due to the energy consumption problem while moving from another place. An ad hoc network

T. Sudhakar (✉) · H. Hannah Inbarani
Periyar University, Salem, Tamil Nadu, India
e-mail: sudhakarmecmca2013@gmail.com

H. Hannah Inbarani
e-mail: hhinba@periyaruniversity.ac.in

A. K. Tripathy et al. (eds.), *Advances in Distributed Computing and Machine Learning*, Lecture Notes in Networks and Systems 127, https://doi.org/10.1007/978-981-15-4218-3_2

is an autonomous mobile network, and each mobile node is acting as a router. The applications of the MANETs are a rescue operation, military purpose and battlefield operations.

In general, routing protocols are divided into three types. The first one is a proactive routing protocol; the second one is the reactive protocol and hybrid routing protocol. The proactive protocol is also known as a table-driven protocol in which each data packet movement is stored in a table format also known as a routing table. In the routing table, mobile node's id, source and destination IP addresses, etc., are considered as the header of the table. The reactive routing protocol is an on-demand based protocol that is when a route is needed, it forms a routing table on an ad hoc basis and periodically updates the routing information [2]. In this paper, fuzzy logic is used because of instability with high mobility of the mobile nodes from the sender to the receiver node group. It is a mathematical model to deal with fuzzy sets and a set of rule-based models [3].

1.1 The Motivation of the Paper

Mobility is a significant problem in wireless networks; due to this issue, packet delivery ratio, throughput and others are decreased. Due to the impact of this mobility, uncertainty issues, selection of an optimal path from the source node and set of receiver's nodes is needed. In this paper, the effort has been put to overcome and control the uncertainty issues using the fuzzy logic approach [4–5].

1.2 Contributions

The contributions of this work have been summarized as follows:

- The fuzzy mobility models are compared with existing proactive routing protocols such as Dynamic Source Routing (DSR) Protocol [6] and Ad hoc On-demand Vector (AODV) Routing Protocol [7].
- Performance of fuzzy mobility is compared with well-known temporal mobility models like Random Way Point (RWP) and Random Direction (RD) mobility models [8].

The rest of the paper formulation is as follows: Sect. 1 describes the introduction of the wireless ad hoc network and its challenges in real-world scenarios. Section 2 is about related works done so far in the fuzzy associated wireless network in Table 1. The proposed fuzzy mobility approach is explained in Sect. 3. Chapter 4 offers a simulation environment setup and subsequently discusses the comparative analysis of the proposed approach that has been done with the existing mobility models. Section 5 finally provides the conclusion of the paper.

Table 1 Summary of related works

S. no.	Year	Author name	Proposed model	Routing protocol	Parameters
1	2002	Gupta et al. [9]	Fuzzy metric approach to determine route lifetime	AODV	ART, node mobility and transmission power
2	2003	Alandjani and Eric Johnson [4]	Fuzzy routing in ad hoc networks	DSR, SMR	Resource allocation, network state and traffic importance of the network
3	2010	Thaw [10]	Fuzzy-based multi-constrained QoS distance vector routing protocol in MANETs	AODV	Hop count, mobility speed and bandwidth
4	2008	Naga Raju and Rmachandram [5]	Fuzzy cost-based multi-path routing for ad hoc mobile networks	IM-RMR	Hop count, mobility speed, computer efficiency, traffic load and bandwidth
5	2017	Vinoth Kumar et al. [11]	Efficient fuzzy logic-based multi-path routing	EMPR	Node mobility and stability to determine node reliability
6	2017	Yadav et al. [12]	Efficient fuzzy-based multi-cast routing protocol	MAODV, ODMRP	Residual energy, delay and bandwidth

2 Related Works

See Table 1.

3 Proposed Fuzzy Mobility Model

In this section, the proposed model discussed the fuzzy mobility model. Delay and bandwidth parameters have been considered as input variables. The delay and bandwidth of the path are determined using Eqs. 2 and 3 [12]

$$\text{Fuzzy Mobility Cost (FMC)} = (\text{Bandwidth, Delay}) \quad (1)$$

$$\text{Delay } (R(s,t)) = \sum_{i=1}^{n} \text{delay}[e_i, e \in R(s,t)] \quad (2)$$

$$\text{Bandwidth } (R(s,t)) = \sum_{i=1}^{n} \text{bandwidth}[e_i, e \in R(s,t)] \quad (3)$$

where R is a route from source s to destination t and e is channel transmission among mobile nodes.

3.1 *Fuzzy Mobility in Wireless Ad Hoc Networks*

Fuzzy logic [12] is a mathematical tool that deals with uncertainty issues. In this paper, two variables have considered as input in the fuzzy system, namely bandwidth and delay. The fuzzy system has four important components, i.e., fuzzification, fuzzy inference, fuzzy knowledge rule and defuzzification. The fuzzy mobility is compared with existing mobility models along with conventional reactive routing protocols such as DSR and AODV [13]. Each linguistic input variable converts into a single metric called fuzzy mobility cost (FMC).

(a) **Fuzzification-Input Phase**

In this phase [12], two input variables are fuzzified in the fuzzy logic system. Linguistic variables are decomposed into linguistic terms in a fuzzy logic system: three linguistic variables for each input parameter, such as bandwidth and delay, are low, medium and high. There are very small, low, medium and high-performance linguistic parameters. Table 2 shows the sample input variables of delay and bandwidth.

(b) **Fuzzy Mobility Membership Degree Phase**

In this phase [12], fuzzy membership functions are used in both fuzzification and defuzzification. It converts crisp input to fuzzy input, i.e., linguistic variables such as low medium and high in fuzzification step and vice versa [14]. The membership

Table 2 Membership function for delay and bandwidth

Linguistic input variables	Low	Medium	High
Delay	0	0	95
	9.65	95	175.6
	95.74	175.6	175.6
Bandwidth	0	30.5	118.4
	0	100.74	184.67
	78.4	184.67	190.45

degree has been used to describe a linguistic expression using a triangular membership function in the suggested routing protocol. Figures 1 and 2 demonstrate the membership function of delay and bandwidth. The final FCM is illustrated in Fig. 3.

Fig. 1 Membership function for the delay

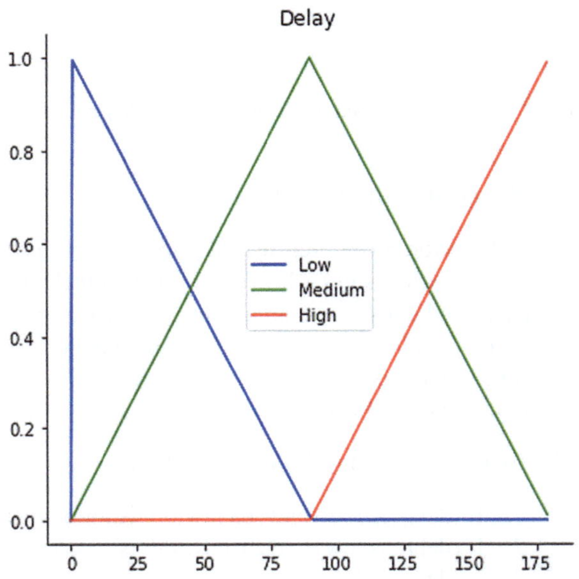

Fig. 2 Membership function for bandwidth

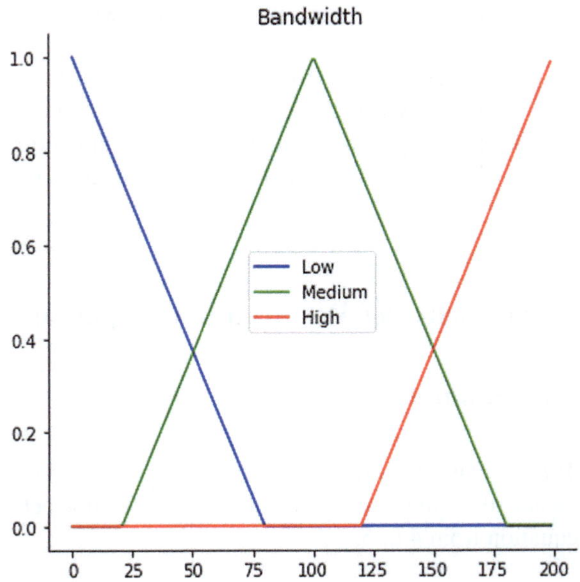

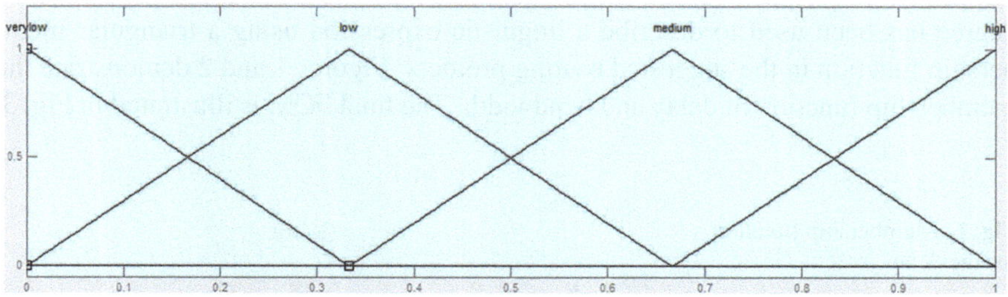

Fig. 3 Membership function for the **fuzzy mobility** model

(c) **Fuzzy Mobility Knowledge rule**

A rule base for controlling the decision variable is constructed in the fuzzy logic system (FLS). It is a basic IF-THEN rule that includes either AND or OR operator's precedent and consequent argument [12]. AND operator deals with the validity of numerous linguistic statements, and OR operator deals with the operator when at least one of them is correct. The following is an example of rules defining input and output mapping of FLS.

(d) **Fuzzy Knowledge Rule Generations**

1. If (delay is low) and (bandwidth is low), then (fuzzy_cost is very low)
2. If (delay is low) and (bandwidth is medium) then (fuzzy_cost is low)
3. If (delay is low) and (bandwidth is high) then (fuzzy_cost is medium)
4. If (delay is medium) and (bandwidth is low) then (fuzzy_cost is low)
5. If (delay is medium) and (Bandwidth is medium) then (fuzzy_cost is medium)
6. If (delay is medium) and (bandwidth is high) then (fuzzy_cost is high)
7. If (delay is high) and (bandwidth is low) then (fuzzy_cost is medium)
8. If (delay is high) and (bandwidth is medium) then (fuzzy_cost is high)
9. If (delay is high) and (bandwidth is high) then (fuzzy_cost is high).

4 Performance Results and Comparison

4.1 Simulation Setup

The simulation values and environment parameters are mentioned in Table 3. The results of routing protocols in the wireless ad hoc network were calculated using the equation from 4 to 6.

$$\text{Throughput} = \text{Number of packets sent successfully/total simulation time} \quad (4)$$

Table 3 Simulation setup configuration

Simulation environment parameters	Values of simulation
Simulation start time	0.5 s
Simulation end time	250.0 s
Propagation model	Two-ray ground and shadow model
Channel type	Wireless
Routing protocols	DSR and AODV
Mobility model	RD and RWP
Variation of pause time (In Seconds)	10, 20, 30, 40 and 50
Total no. of nodes	50
Mobility models	Random mobility model, random direction model
Network type	Wireless
MAC type	802.11

$$\text{Delay} = \text{Time spent on hop } 1 + \cdots + \text{time spent on hop } n \tag{5}$$

$$\text{PDR} = (\text{Number of packets sent} - \text{packet lost}) * 100/\text{Number of packets sent} \tag{6}$$

4.2 Simulation Comparison Results and Assessment

In this section, the performance results of the proposed work are discussed, along with the results of existing routing protocols. The comparison results prove that the FM model reveals better accuracy than conventional models. The pause time is used to describe the interval between node movements. In this scenario, while increasing pause time getting more PDR and throughput accuracy than existing model.

Figure 4 and Table 4 depicts the performance of PDR under variation of the mobile nodes. As the MNs pause time increases, the PDR decreases relatively because several nodes are present for communication. Network parameters can frequently change in a mobile ad hoc network due to high node flexibility, which causes instability challenges; the consequence is that the origin node cannot choose the ideal multicast routing route for information packet transmission. The proposed FM model's PDR is higher than the existing mobility models, such as RWP and RD routing protocols. The RWP and RD model does not consider a FLS for packet transmission [12].

Figure 5 and Table 5 show the performance of delay under the variation of the mobile nodes. The number of nodes presented in the network will lead to reducing

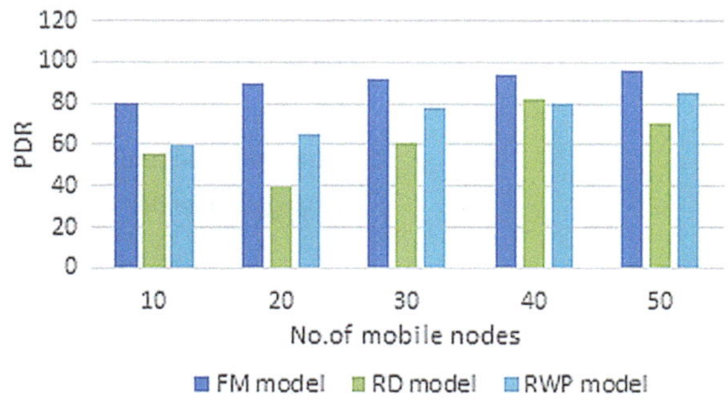

Fig. 4 PDR performance comparison

Table 4 Comparison of PDR under no. of mobile nodes

Mobility model/no. of mobile nodes	FM model RWP model RD model
10	60.5 55.4 53.4
20	86.5 60.5 65.8
30	84 80.5 72.6
40	89.5 85 79.8
50	92.4 86.8 80.4

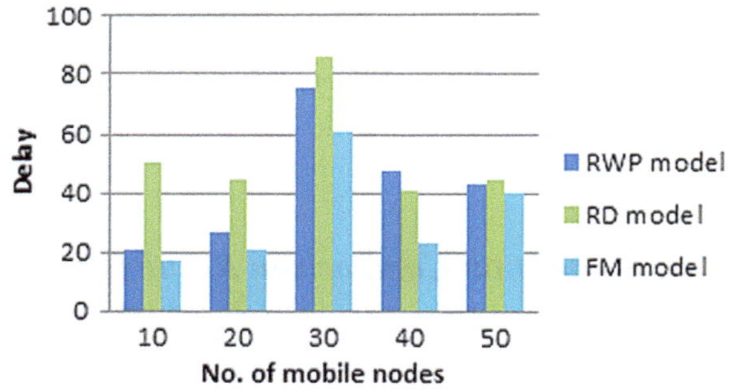

Fig. 5 Delay performance comparison

Table 5 Comparison of delay under no. of mobile nodes

Mobility model/no. of mobile nodes	FM model RWP model RD model
10	20.74 50.7 17.5
20	26.45 44.25 20.7
30	75.2 85.65 60.7
40	47.24 40.8 22.8
50	42.7 44.2 40.2

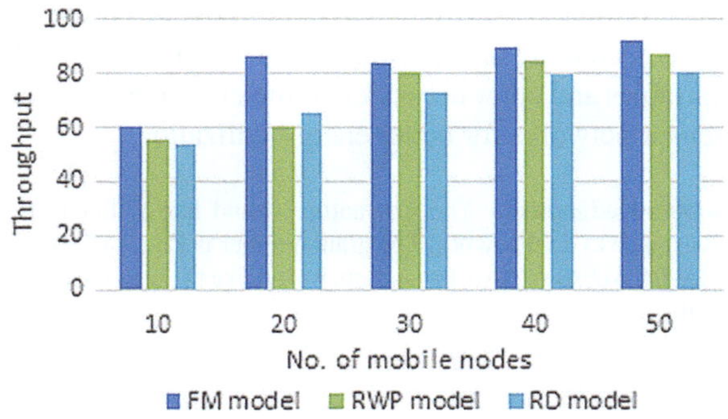

Fig. 6 Throughput performance comparison

Table 6 Comparison of throughput under no. of mobile nodes

Mobility model/no. of mobile nodes	FM model RWP model RD model
10	80.7 55.4 60.4
20	90 40 65.8
30	92.4 60.6 78.4
40	94.5 82 80.6
50	96.7 70.4 85.4

the performance of the network in terms of data loss and delay. Since the on-the-fly network has the uncertainty problem and to control such a problem, a fuzzy logic system is to be used [15]. The propounded FM model outperforms RWP and RD models in terms of delay, PDR and throughput.

Figure 6 and Table 6 show the performance of the throughput under the variation of several nodes. While increasing the number of nodes, pause time will decrease the throughput. The proposed FM model reveals better performance in terms of performance metrics, as discussed above. From Fig. 6, it can be observed that the throughput increases gradually as the number of mobile nodes increase [12]. The reason is that MNs increase the traffic among networks for the selection of multi-path routing.

5 Conclusion and Future Directions

In this work, a fuzzy mobility model is proposed for route selection in a mobile ad hoc network. The profound study shows that the fuzzy mobility model attains better outcomes as compared to other conventional mobility models, such as the random mobility model and a random direction model. Due to the increased mobility of the devices inside a wireless network, which causes instability issues, the network parameters change very often. The purpose of the work is to resolve the problem of uncertainty issues in a wireless network. This work has been proposed to select the minimum fuzzy mobility value for better route selection. The simulation works have been done on the NS2 and MATLAB tool. In the future, the number of input variables and other network parameters can be considered. Also, map the solution to sensor networks for better energy utilization.

Acknowledgements The first author would like to thank the UGC-RGNF-SC for the financial assistance to do this work. The grant number is F1-17.1/2014-15/RGNF-2014-15-SC-TAM-84052 (SA-III/Website). Also, the authors would like to thank the anonym's reviewers for the improvement of the paper.

References

1. Abbas NI, Ilkan M, Ozen E (2015) Fuzzy approach to improving route stability of the AODV routing protocol. EURASIP J Wirel Commun Netw 235:1–11
2. Pouyan AA, Yadollahzadeh Tabari M (2017) FPN-SAODV: using fuzzy petri nets for securing AODV routing protocol in mobile Ad hoc network. Int J Commun Syst 30(1):1–14
3. Liu H, Li J, Zhang YQ, Pan Y (2015) An adaptive genetic fuzzy multi-path routing protocol for wireless ad-hoc networks. In: 6th international conference on software engineering, artificial intelligence, networking and parallel/distributed computing and first ACIS international workshop on self-assembling wireless network. IEEE, USA, pp 468–475
4. Alandjani G, Eric Johnson EE (2003) Fuzzy routing in ad hoc networks. In: Conference proceedings of the 2003 IEEE international performance, computing, and communications conference. IEEE, USA, pp 525–530
5. Naga Raju A, Rmachandram S (2008) Fuzzy cost based multi-path routing for mobile ad-hoc networks. J Theor Appl Inf Technol 4(4):55–66
6. Pi S, Sun B (2012) Fuzzy controllers based multi-path routing algorithm in MANET. Phys Proc 24:1178–1185
7. Huang C-J, Chuang Y-T, Hu K-W (2009) Using particle swam optimization for QoS in ad-hoc multicast. Eng Appl Artif Intell 22(8):1188–1193
8. Doja MN, Alam B, Sharma V (2013) Analysis of reactive routing protocol using fuzzy inference system. AASRI Proc 5:164–169
9. Gupta S, Bharti PK, Choudhary V (2011) Fuzzy logic based routing algorithm for mobile ad hoc networks. In: Mantri A, Nandi S, Kumar G, Kumar S (eds) High performance architecture and grid computing, vol 169. HPAGC 2011. Springer, Berlin, Heidelberg, pp 574–579
10. Thaw MM (2010) Fuzzy-based multi-constrained quality of service distance vector routing protocol in mobile ad-hoc networks. In: 2nd international conference on computer and automation engineering (ICCAE), vol 3. IEEE (2010), pp 429–433
11. Vinoth Kumar K, Jayasankar T, Prabhakaran M, Srinivasan V (2017) Fuzzy logic based efficient multi-path routing for mobile ad hoc networks. Appl Math Inf Sci 11(2):449–455
12. Yadav AK, Das SK, Tripathi S (2017) EFMMRP: design of efficient fuzzy based multi-constraint multicast routing protocol for wireless ad-hoc network. Comput Netw 118:15–23
13. Pouyan AA, Yadollahzadeh Tabari M (2017) FPN-SAODV: using fuzzy petri nets for securing AODV routing protocol in mobile ad hoc network. Int J Commun Syst 30(1):29–35
14. Alkhodre AB (2016) Hybrid fuzzy social mobility model. Karbala Int J Modern Sci 2(1):29–40
15. Seethalakshmi P, Gomathi M, Rajendran G (2011) Path selection in wireless mobile ad hoc network using fuzzy and rough set theory. In: 2nd international conference on wireless communication, vehicular technology, information theory and aerospace and electronic systems technology (Wireless VITAE). IEEE (2011), pp 5–8

Black Hole Detection and Mitigation Using Active Trust in Wireless Sensor Networks

Venkata Abhishek Kanthuru and Kakelli Anil Kumar

Abstract The recent boom in wireless connectivity has seen a rapid deployment of wireless sensor networks (WSNs). This boom has led to new challenges in security due to the inherent resource and performance constraints in WSNs. The constraints mean that normal security models deployed on other types of networks are not feasible, and therefore a need for new models is present. One security issue present in these networks is the black hole attack. A black hole attack involves a malicious node, which tries to direct all traffic toward itself and drops all of it without forwarding it to the destination. To solve this issue, a hierarchical detection-based routing model is proposed, which along with the detection of black holes is also capable of actively re-routing the traffic during an attack to prevent any network downtime.

Keywords Black hole attack · Wireless sensor networks · Secure routing

1 Introduction

Wireless sensor networks (WSNs) are a collection of inexpensive, geographically distributed nodes used to monitor attributes related to environmental pollution, vehicle safety, building safety, warehouse inventories, among many others. Given the sensitive nature of some of these applications, it is essential to ensure that the WSN architecture is robust enough to run and be secure even in the presence of adversaries [1].

Nodes are required to be inexpensive and have limited computing power. The lack of resources is a challenge as traditional cryptographic solutions used for PCs and

V. A. Kanthuru (✉) · K. A. Kumar
Vellore Institute of Technology, Vellore, Tamil Nadu, India
e-mail: venkat.abhishek.1999@gmail.com

K. A. Kumar
e-mail: anilsekumar@gmail.com

A. K. Tripathy et al. (eds.), *Advances in Distributed Computing and Machine Learning*, Lecture Notes in Networks and Systems 127, https://doi.org/10.1007/978-981-15-4218-3_3

workstations are not feasible on these low-power sensor nodes [1, 2]. Additionally, wireless sensor nodes are deployed in dangerous or inaccessible areas for extended periods, and this adds to the complexity of securing the network since it is usually out of the physical reach of the operator. Adversaries, therefore, have the advantage of being able to steal nodes and reverse-engineer them to obtain cryptographic keys or other sensitive information. More worrying is the fact that adversaries can also introduce their malicious nodes into the network to disrupt the regular flow of network traffic or to wreak havoc, such as exhaustion, misdirection, greed, black holes, and collisions [3].

2 Related Work

Taylor et al. [2] proposed an attack detection scheme that uses a watchdog mechanism and an expert system on each node to detect anomalies in the behavior of neighboring nodes. Using this mechanism, all the nodes can sniff traffic of their nearby nodes. This strategy helps to detect the sender node of the abnormal data packet.

Liu et al. [3] proposed an active routing-based detection model where the protocol sends packets to convince the attacker to launch black holes in the network. By analyzing the behavior and modifying the trust of each node during this period, the network can effectively identify what nodes are malicious. The data is then routed only through nodes with high trust.

Priayoheswari et al. [4] have come up with a new model that primarily relies on the RSSI to compute the trust threshold. The thresholds are then utilized to plausibility-based algorithm to determine with high accuracy if a node is malicious or not. The algorithm was compared with the LEACH protocol and found that it performed a bit better during the delay analysis.

Shinde et al. [5] have focused on secure routing using a trustworthy pattern. The active trust routing scheme has been used to mitigate attacks like DOS, selective forwarding, and black hole during route detection and packet forwarding. Data hiding using the ECC mechanism has been used for security in WSN, which enhanced the network lifetime and network throughput and low energy conservation.

Mehetre et al. [6] proposed a trustable and secure routing scheme. The scheme uses a two-stage dual assurance security mechanism for securing the data packet in WSNs. Active trust is helpful in mitigating the security attacks during routing like selective forwarding, black hole. The scheme has identified the trusted routing paths using the cuckoo search algorithm and achieved energy efficiency with a better network lifetime.

3 Reactive Trust-Based Routing Protocol

The algorithms that make up the protocol are:

1. Route setup algorithm
2. Data forwarding algorithm
3. Active probing algorithm
4. Route resolution algorithm.

The primary goal of this protocol is to detect black holes in WSNs [7, 8]. The protocol starts by forming the routes using the route setup algorithm. Once a node is part of the network, it can start sending data using the data send algorithm. The algorithm relies on an ACKing mechanism where an ACK is sent every n minutes or for every m packets. If the source does not receive any ACK in this period, it starts the probing algorithm, which sends out a PROBE_REQUEST to the node's parent. The parent on receiving the PROBE_REQUEST sends back a PROBE_RESPONSE. Failure to do so leads to the trust of the parent being reduced. The PROBE_REQUESTs are sent until either a PROBE_RESPONSE is received or the trust of the parent falls below the threshold.

Once the parent sends the PROBE_RESPONSE, it starts the probing algorithm and repeats the steps described above. At any point, if a node's trust falls below the threshold, the child initiates the route resolution algorithm, which identifies a new parent and forms a new route to it. Once a new parent is found, the child notifies the sink about the malicious node.

Before we elaborate on the algorithms, the below subsections list some of the assumptions we make about the network.

3.1 Network Model

We assume that the wireless sensor network consists of wireless motes uniformly and randomly scattered around. The nodes are also assumed not to move around. There is only one sink in the entire network. All sensed data needs to be forwarded to the sink [9].

3.2 Adversary Model

We assume that the adversary can compromise some of the nodes. The compromised nodes become the black holes, and each black hole will unselectively discard all packets transmitted to it to prevent data from reaching the sink. We also assume that the adversary is unable to compromise the sink and its neighboring nodes [2].
The following subsections describe the individual algorithms in detail.

3.3 Route Setup Algorithm

The route setup algorithm (Algorithm 1) is run by the sink during the start of the network. The sink is initialized to be at level zero after which it starts broadcasting IDENT messages. Any node which is not yet part of the network and is close enough to the sink responds to it by sending an IDENT_REPLY. The sink on receiving the IDENT_REPLY adds the node to its routing table and sends an IDENT_ACK. The child adds the sink as its parent on receiving this message. The child also sets its node level to one level higher than the sink, which is one. This process is now repeated by the child. After a certain time, most of the nodes deployed will be part of the network. Any node that might not be a part of the network can add itself by sending out DISCOVER messages to which a nearby node can respond to and add to it as its child.

The network forms an N-ary tree-based structure. Every node has to forward the data, meant for the sink, to its parent. This algorithm also takes care of any black holes that might have been present during the network start-up, ensuring that only healthy nodes are part of the routing network.

3.4 Data Forwarding Algorithm

Once the level of a node is set, it can start transmitting data. It sets its ACK timer and sends the data to its parent. The parent then forwards it to its parent and so on till the data reaches the sink. The node then waits for the ACK, which is sent by the sink for every n packets or if a timer expires, whichever is earlier. Both the count of packets and timer are used to handle cases where parts of the data are received, but the rest is lost. In such cases, the source node might try sending the whole data again. By sending an ACK for the received packets, the source can synchronize accordingly. If, at any point, the source node does not receive the ACK, it starts the probing algorithm.

3.5 Active Probing Algorithm

This algorithm is called by the node that does not receive the ACK. The algorithm starts by sending a PROBE_REQUEST. The parent on receiving the PROBE_REQUEST responds with a PROBE_RESPONSE. Once the response is sent, it then starts the probing algorithm. This is because somewhere along the chain of nodes, there is a malicious node [10]. By running it serially, we are eliminating all possible nodes from the list of malicious nodes. Any node that does not respond to the PROBE_REQUEST will be sent a PROBE_REQUEST again. The request is sent until either the node responds or its trust falls below the threshold. Once the trust of

Algorithm 1: Route Setup Algorithm

```
Initialization
if node is SINK then
    Broadcast IDENT message
end
switch Received Message Type do
    case IDENT
        if node level is set then
            Discard IDENT
        else if distance of sender > γ (distance threshold) then
            Discard IDENT
        else
            Send IDENT_REPLY to node
        end
    case IDENT_REPLY
        if sender is already in routing table then
            Resend IDENT_SUCCESS
        else
            Add node to routing table
            Send IDENT_SUCCESS
        end
    case IDENT_SUCCESS
        Set sender as parent node in routing table
        Set parent's trust to MAX_TRUST
        Set Node Level = parent-level + 1
        Start Route Setup Algorithm
    case DISCOVER
        if sender already sent DISCOVER > n times then
            Notify SINK of possible malicous node
        end
        if sender is already in routing table then
            Resend IDENT_SUCCESS
        else
            Add node to routing table
            Send IDENT_SUCCESS
        end
    endsw
endsw
if node level is not set & time since startup > τ_startup then
    Broadcast DISCOVER message
end
```

Algorithm 2: Data Forwarding Algorithm

```
Initialize ACK timer τ_data
Send data message to parent node
switch Received Message Type do
    case DATA
        if node is not SINK then
            Add source IP address to routing table
            forward packet to parent
        else
            if number of packets sent by source exceeds threshold or ack timer exceeded then
                send DATA_ACK
            end
        end
    case DATA_ACK
        if node is destination then
            Reset τ_data
        else
            Forward DATA_ACK
        end
    endsw
endsw
if τ_data expired then
    if parent's trust > φ (trust threshold) then
        Reduce trust of parent node
        Start Probing Algorithm
    else
        Run Route-Resolution Algorithm
    end
end
```

a node falls below the threshold, the child marks the parent as a black hole and runs the route resolution algorithm.

3.6 Route Resolution Algorithm

If a parent's trust reduces below the threshold, then the suspicious node is removed from the routing table and is added to the black hole list. A DISCOVERY message is then broadcasted. Any node that receives the DISCOVERY message responds to it by sending an IDENT_ACK. The node sets the sender as its new parent and sends a message about the malicious node to the sink on receiving the IDENT_ACK.

Algorithm 3: Active Probing Algorithm

```
Initialize ACK timer τ_probe
Send PROBE_REQUEST to parent node
switch Received Message Type do
    case PROBE_REQUEST
        Send PROBE_RESPONSE to child
        Start Probing algorithm
    case PROBE_RESPONSE
        Stop τ_probe
    endsw
endsw
if τ_probe expired then
    if parent's trust > φ (trust threshold) then
        Reduce trust of parent node
        Send PROBE_REQUEST again
    else
        Run Route-Resolution Algorithm
    end
end
```

Algorithm 4: Route-Resolution Algorithm

```
Remove malicious node from routing table
Add malicious node to black hole list
Broadcast DISCOVERY message
switch Received Message Type do
    case DISCOVERY
        Add node to routing table
        Send IDENT_ACK
    endsw
endsw
Send malicious node info to sink
```

4 Results and Discussion

To simulate the protocol, we have used the Contiki-NG OS along with the COOJA network simulator [11]. Note that any trust calculation algorithm can be used to modify the parent's trust. For this paper, we have used a simple algorithm that linearly reduces the trust value. Another trust algorithm that can be used can be found in the paper [2].

The network is initialized with 20 normal nodes, one sink, and four nodes that become black holes after the route setup is done [12, 13].

To keep the results concise, we show only the logs of the section of the network shown in Fig. 1. We can see from Fig. 2 how node 8 sends an IDENT_REPLY to node 1 as it received an IDENT from node 1. Node 1, on receiving the IDENT_REPLY, replies with an IDENT_ACK and adds node 8 as its child. Node 8 now assigns a level

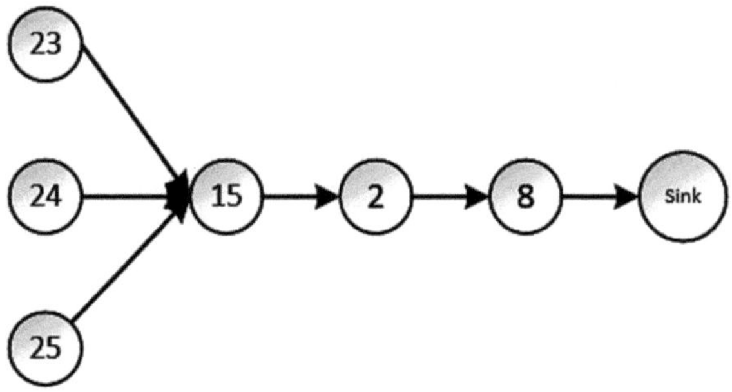

Fig. 1 The topology shown is a subsection of entire network created using the Cooja network simulator

Timestamp	Node ID	Log Level	Log Message
30:00.756	ID:8	[INFO: App]	Ident reply sent to fe80::212:7401:1:101
30:00.862	ID:1	[INFO: App]	Ident reply recieved from fe80::212:7408:8:808
30:00.869	ID:1	[INFO: App]	IDENT ACK sent to fe80::212:7408:8:808
30:00.929	ID:8	[INFO: App]	IDENT ACK received, status = 0 Node level: 1 fe80::212:7401:1:101
1:00:00.447	ID:2	[INFO: App]	Ident reply sent to fe80::212:7408:8:808
1:00:00.484	ID:8	[INFO: App]	Ident reply from fe80::212:7402:2:202
1:00:00.491	ID:8	[INFO: App]	Ident ack sent to fe80::212:7402:2:202
1:00:00.543	ID:2	[INFO: App]	IDENT ACK received, status = 1 Node level: 2 fe80::212:7408:8:808
1:00:00.565	ID:15	[INFO: App]	Ident reply sent to fe80::212:7402:2:202
1:00:00.586	ID:2	[INFO: App]	Ident reply from fe80::212:7415:15:1515
1:00:00.593	ID:2	[INFO: App]	Ident ack sent to fe80::212:7415:15:1515
1:00:00.642	ID:15	[INFO: App]	IDENT ACK received, status = 0 Node level: 3 fe80::212:7402:2:202

Fig. 2 Logs showing the route setup for few of the nodes

of 1 to itself and starts broadcasting IDENTS. In this manner, nodes 2 and 15 also become part of the network. Now, each of those nodes starts sending data to node 1. To simulate a black hole, we have assigned node 2 as the black hole. When node 15 does not receive any ACK from node 1, it starts the probing algorithm. We can see in Fig. 3 that node 15 sends a PROBE_REQUEST to node 2. But since node 2 does not respond, the timer expires, and the PROBE_REQUEST is sent again. This is repeated until node 2's trust falls below the threshold, which in our case was 3 tries. Once the trust falls below the threshold, node 15 looks for a new node. The new parent selected is node 14, which is also checked against the black hole list to see if it is also a malicious node. Since it is not, an IDENT_REPLY is sent to it. Node 14 adds node 15 as its child and sends back an IDENT_ACK. Once the new route is formed, node 15 notifies node 1 about the black hole through the new parent (node 14)

In Fig. 4, we can see the throughput calculated at node 1. During the interval of probing, node 1's throughput drops a bit. This drop can be seen clearly in Fig. 5.

Timestamp	Node ID	Log Level	Log Message
1:20:00.650	ID:15	[INFO: App]	Probe req sent to fe80::212:7402:2:202
1:23:00.659	ID:15	[INFO: App]	probe_rsp_timer_callback, probes_sent 1, max_probes 3
1:23:00.666	ID:15	[INFO: App]	Probe req sent to fe80::212:7402:2:202
1:26:00.677	ID:15	[INFO: App]	probe_rsp_timer_callback, probes_sent 2, max_probes 3
1:26:00.684	ID:15	[INFO: App]	Probe req sent to fe80::212:7402:2:202
1:29:00.689	ID:15	[INFO: App]	probe_rsp_timer_callback, probes_sent 3, max_probes 3
1:30:00.818	ID:15	[INFO: App]	NODE_NEW_ROUTE_RESOLUTION : new parent level:1 is_new_parent_blackhole: No (fe80::212:740e:e:e0e) Blackhole: fe80::212:7402:2:202
1:30:00.825	ID:15	[INFO: App]	Ident reply sent to fe80::212:740e:e:e0e
1:30:00.928	ID:15	[INFO: App]	IDENT ACK received, status = 3 Node level: 2 fe80::212:740e:e:e0e
1:30:00.936	ID:15	[INFO: App]	Blackhole node info sent to fe80::212:740e:e:e0e
1:30:01.178	ID:1	[INFO: App]	Blackhole node: fe80::212:7402:2:202

Fig. 3 Logs showing how a node probes the black hole and joins a new parent

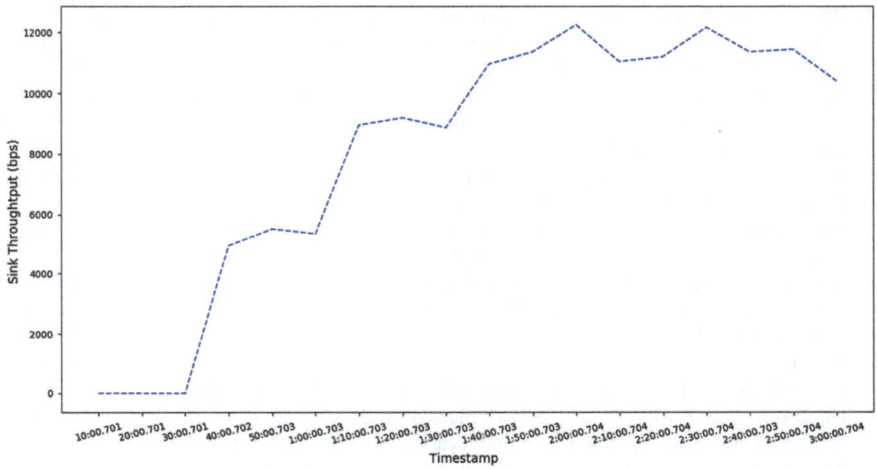

Fig. 4 Throughput at the sink

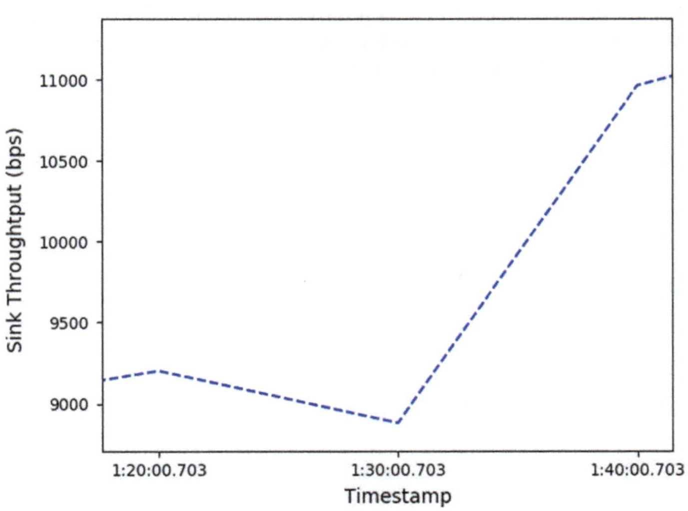

Fig. 5 Drop in the throughput at the sink

References

1. Taylor VF, Fokum DT (2014) Mitigating black hole attacks in wireless sensor networks using node-resident expert systems. In: Wireless telecommunications symposium, Washington, DC, pp 1–7. https://doi.org/10.1109/WTS.2014.6835013
2. Liu Y, Dong M, Ota K, Liu A (2016) ActiveTrust: secure and trustable routing in wireless sensor networks. IEEE Trans Inform Forens Secur
3. Taylor VF, Fokum DT (2014) Mitigating black hole attacks in wireless sensor networks using node-resident expert systems. In: Wireless telecommunications symposium
4. Priayoheswari B, Kulothungan K, Kannan A (2017) A novel trust based routing protocol to prevent the malicious nodes in wireless sensor networks. In: Proceedings—2017 2nd international conference on recent trends and challenges in computational models (ICRTCCM 2017)
5. Shinde M, Mehetre DC (2018) Black hole and selective forwarding attack detection and prevention in WSN. In: 2017 international conference on computing, communication, control and automation (ICCUBEA 2017)
6. Mehetre DC, Roslin SE, Wagh SJ (2018) Detection and prevention of black hole and selective forwarding attack in clustered WSN with Active Trust. Cluster Comput
7. Wazid M, Das AK (2017) A secure group-based blackhole node detection scheme for hierarchical wireless sensor networks. Wireless Pers Commun
8. Ali S, Khan MA, Ahmad J, Malik AW, Ur Rehman A (2018) Detection and prevention of black hole attacks in IOT & WSN. In: 2018 3rd international conference on fog and mobile edge computing (FMEC 2018)
9. Kumar KA, Krishna AVN, Chatrapati KS (2017) New secure routing protocol with elliptic curve cryptography for military heterogeneous wireless sensor networks. J Inform Optim Sci
10. Dhulipala VRS, Karthik N (2017) Trust management technique in wireless sensor networks: challenges and issues for reliable communication: a review. CSI Trans. ICT
11. Srivastava S, Singh M, Gupta S (2019) Wireless sensor network: a survey. In: 2018 international conference on automation and computational engineering (ICACE 2018)
12. Parganiha P, Kumar KA (2018) An energy-efficient clustering with hybrid coverage mechanism (EEC-HC) in wireless sensor network for precision agriculture. J Eng Sci Technol Rev
13. Kim KT, Youn HY (2017) A dynamic level-based routing protocol for energy effciency in wireless sensor networks. J Internet Technol

Power Allocation-Based QoS Guarantees in Millimeter-Wave-Enabled Vehicular Communications

Satyabrata Swain, Jyoti Prakash Sahoo, and Asis Kumar Tripathy

Abstract The next-generation vehicular network will see an unprecedented amount of data exchanged which is beyond the capacity of existing communication technologies for vehicular network. The much talked millimeter-wave (mmWave)-enabled communication technology is the potential candidate to this growing demand of ultra-high data transmission and related services. However, the unfavorable signal characteristics of mmWave bands make the quality of service guarantee more difficult when it is applied to user mobility. In this paper, we proposed a directional beam-based power allocation/re-allocation scheme to guarantee the quality of service (QoS) in a high user mobility scenario operating on mmWave band. The performed simulation results show that our proposed scheme outperformed the baseline scheme without power allocation.

Keywords mmWave · Vehicular network · Beam forming · QoS · Power allocation

1 Introduction

The next-generation vehicular network is envisioned to have numerous use cases which require high data rate and service continuity. Recently, popular millimeter-

The original version of the chapter has been revised: The reference 16 has been revised. The correction to this chapter can be found at https://doi.org/10.1007/978-981-15-4218-3_51

S. Swain (✉) · A. K. Tripathy
Vellore Institute of Technology, Vellore, Tamil Nadu, India
e-mail: satya.swain10@gmail.com

A. K. Tripathy
e-mail: asistripathy@gmail.com

J. P. Sahoo
Siksha 'O' Anusandhan (SOA) Deemed to be University, Bhubaneswar, India
e-mail: sahoo.jyotiprakash@gmail.com

A. K. Tripathy et al. (eds.), *Advances in Distributed Computing and Machine Learning*, Lecture Notes in Networks and Systems 127, https://doi.org/10.1007/978-981-15-4218-3_4

wave (mmWave) communication is a potential candidate to support these service requirementS. The mmWave spectrum loosely referred to the range above 30–300 GHz band. Recent measurement campaign [1–3, 26, 27] shows that the mmWave signal has several limitations. However, the use of mmWave band provides wide bandwidth channels, which enables multi-Gbps data rates among vehicles and infrastructure. This drives us to use mmWave band for vehicular communications. We describe the potential advantages and short fall of mmWave communications in the later part of this paper.

In this paper, we aim to provide multi-Gbps data connection from infrastructure to vehicle on a network operating on mmWave bands. We consider a mmWave-based beamforming system that uses narrow beams at both the transmitter and receiver. The key challenge in our system is how to change the serving beam of the receiver while the user in mobility, i.e., quickly aligning the transmitter and receiver beams. To solve this, we consider a power allocation-based serving beam switching/re-initialize method using the context information of vehicles. The key idea of this work is that with the estimated context information of the vehicle along with the channel condition, we compute the beam switching directions of the serving beam.

The paper is organized as follows. Section 2 provides an overview of the millimeter-wave communications. Section 3 provides the literature study. Section 4 presents the system overview and problem statement. Section 5 presents the performance evaluation. Finally, Sect. 6 concludes the paper.

2 Millimeter-Wave Communications

Recent study shows that there are many potential mmWave bands that could potentially be used for vehicular communications [4, 13–15]. The potential mmWave frequencies roughly range from 30 to 300GHz. The mmWave frequencies can be easily blocked by most solid materials such as concrete wall, metal, and even can also be blocked by human body [5]. Here in this section, we provide the key characteristics of mmWave spectrum.

Propagation: The high-frequency signals suffer from significantly higher attenuation as compared to the low-frequency microwave bands used for LTE networks or short-range WLAN wireless communications [6, 7]. The signal loss of mmWave channel is mainly due to its physical characteristics. However, recent studies for the mmWave propagation characteristics show that some of the sub-bands could be used for both indoor and outdoor environments [6]. The directional narrow beam can be used to overcome the higher path loss at the receiver side.

Beamforming: This technique is used to overcome the path attenuation by concentrating the transmitting power at certain direction over a narrow beam for high gain. The transmitter and receiver have to be capable of beamforming to communicate over certain direction. The transmission directivity will overcome the high attenuation problem; however, it will introduce deafness problem. To overcome this, the transmitter and the receiver have to be aligned over a direction before the actual communication begins. The challenges of finding a alignment point for communication between transmitter and receiver turn into an imperative purpose of design.

Due to its channel characteristics, the challenges in implementing the mmWave spectrum in a high mobility vehicular networks are numerous. Here, in this paper, we proposed a directional beam tracking mechanism to provide uninterrupted service connection in mmWave-enabled vehicular network.

3 Related Works

In paper [6], they proposed a technique to establish mmWave link at sub-6 GHz using spatial information. They have modeled a compressed beam selection as a weighted signal problem where it gets the sight of the signal information from sub-channel of 6GHz band. In [5] paper, researcher validated channel model that is proposed by 3GPP for NR-V2X system. Urban and highway are considered for testing the validity of the proposed work. This scenario has been considered for the line-of-sight probability due to shawdowing, and attenuation due to static and dynamic blockage. In this paper [7], for pairing vahicle efficiently and dynammically, they have proposed a novel technique using swarm intelligence and matching theory. Transmissiona and reception beamwidth optimization also been done by considering both channel queue information and channel state information. In [4] paper, they have studied by simulating mmWave-related strategies which would support vehicle-to-network communication below 6GHz. They have also compared the result with the state-of-the-art LTE network. Choi et al. claimed that only by using mmWave communication chhannel, high-bandwidth vehicle can be communicated. They have highlighted both pros and cons of mmWave while using V2V and V2I communications. He proposed a novel technique to get rid of the mmWave beam training overheads. In this paper [8], Kato et al. explained all the existing solutions for mmWave inter-vehicle communication system. They have specifically characterized propagation of 60 GHz band. In paper [9], Lu et al. have explained pros and cons of existing wireless solution for V2V, V2I, vehicle-to-road infrastructure, and vehicle-to-sensor connectivity. They also identified the future research challenges on vehicular communication. Yamamoto et al. [10] have carried out experiment in order to calculate the path loss that affects inter-vehicle communication. They have considered line-of-sight (LOS) scenario which is at fixed distance on a plane surface and non-LOS (NLOS) on which three intermediate vehicles cause obstruction. In NLOS scenario, using other intermediate vehicle, wave propagation is calculated based on uniform theory of diffraction. In this paper [11], propagation channel characteristics such as shadowing effect, number of paths, and root mean square delay speed are investigated. They have estimated linear relationship between the number of paths and the delay spread, negative cross-correlation between the shadow fading and the delay spread, and an upper bound exponential model of the delay spread and the path loss. Ali et al. proposed a technique to formulate beam selection as a weighted sparse signal recovery problem and obtain the weights using sub-6 GHz angular information.

4 System Overview

Software-defined network-enabled mmWave 5G network architecture is proposed. Propagation characteristics of the mmWave bands have been described in the next section.

4.1 Network Model

We consider a highway scenario where mmWave-enabled BSs, installed at road-side poll, provides lines-of-sight (LoS) network coverage to highly mobile vehicles as illustrated in Fig. 1. The primary goal of our evaluation is to characterize the service availability of a user while moving fast in a highway scenario, where intra-cell beam switching and inter-cell beam switching occur very frequently which led to disruption in service continuity. Without loss of generality, for the sake of simplicity of evaluation, we consider the scenario where BS serves only those vehicles which drive on its side of the road. In order to capture the mmWave beamforming system characteristics, we assume that both the transmitter and receiver are operating on directional beamforming mechanism.

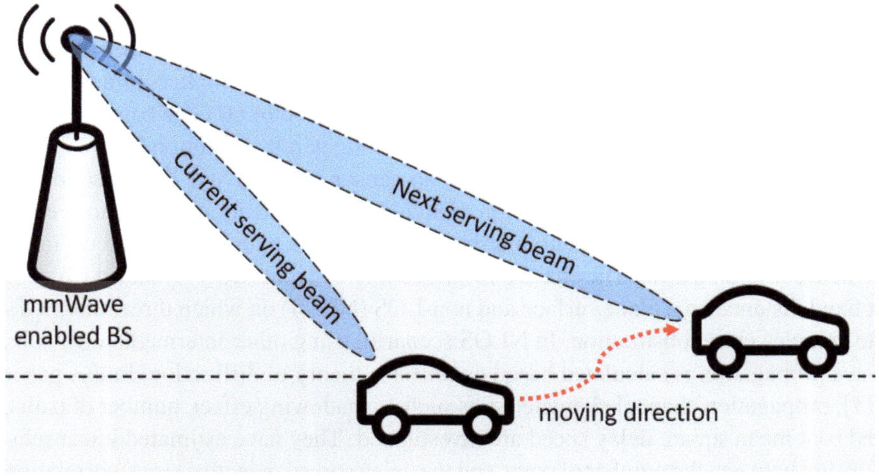

Fig. 1 A pictorial representation of the network model used in this work

4.2 Millimeter-Wave Channel Models

The performance evaluation of the beamforming mmWave networks requires an appropriate channel models. In this paper, we adopt a codebook-based beamforming channel from the mmWave-enabled gNBs to vehicle or UE. The time-varying channel condition is stored in a matrix $\mathbf{H}$. For the small-scale fading model, each of gNB B is synthesized to $L = 16$ paths. For simplicity, here we have only considered the elevation angle. However, the horizontal variations are straightforward. The vertical AoAs, θ_{rx}^{bl}, and vertical AoDs, θ_{tx}^{bl}, where $b = 1, \ldots, B$ represent the index of the gNB, and $l = 1, \ldots, L$ represents the index of path between gNB to UE. We assume the receiver with α_{rx} antenna elements and the transmitter with α_{tx} antenna elements. The channel gain matrix can be represented as follows.

$$\mathbf{H}(t) = \frac{1}{\sqrt{L}} \sum_{b=1}^{B} \sum_{l=1}^{L} z_{bl} \cdot \mathbf{u}_{rx}\left(\theta_{bl}^{rx}\right) \cdot \mathbf{u}_{tx}\left(\theta_{bl}^{tx}\right) \tag{1}$$

where $z_b(t)$ is the complex small-scale fading gain on the lth sub-path from the bth gNB and $\mathbf{u}_{rx} \in \mathbb{C}^{\alpha_{rx}}$ and $\mathbf{u}_{tx} \in \mathbb{C}^{\alpha_{tx}}$ are the precode beamforming vector for the RX and TX antenna arrays to the angular arrivals and departures.

In radio communications, the signal power at the receiver is affected by attenuation characterized by path loss (PL). The received power P_{rx} at the receiver is given by:

$$P_{rx} = P_t \cdot \psi \cdot \sigma^{-1} \cdot PL^{-1} \tag{2}$$

where P_{tx} is a reference power or transmitted power, ψ is the combined antenna gain of transmitter and receiver, σ is the sub-path attenuation, and PL^{-1} denotes the associated line-of-sight (LOS) path loss in dB and can be derived as:

$$PL(d)\ [dB] = \alpha + \beta \cdot 10 \log_{10}(d) + \eta \tag{3}$$

where $PL(d)$ is the mean path loss over a reference distance d, in dB, α is the floating intercept in dB, β is the path loss exponent, $\eta \sim N(0, \sigma^2)$.

5 Protocol Overview

5.1 Main Idea

At a particular time, a vehicle will be connected to a base station. As the vehicle is moving, there is a chance that it may not be connected after certain time. There will be a variation of signal strength when the vehicle moves. So it is necessary to increase the signal strength when it falls below certain threshold. Base station has no idea whether the current signal strength is sufficient for the UE or not. To fulfill this requirement, all the vehicles are connected using LTE network. The purpose of LTE network is to transmit control information to road-side unit, infrastructure unit, and other vehicles. It is assumed that LTE network is always connected without interruption. We have proposed a power allocation scheme on which base station can increase signal strength by using X power allocation scheme when needed. From the inherent characteristics of mmWave, serving beam will change when the vehicle changes it position.

5.2 Working Procedure

Beam updating is needed for uninterrupted data transmission. When vehicle goes out of range of a base station, it is necessary to connect to another base station as soon as possible. As a vehicle know that it is going out of range of a base station, it will inform this information to the base station. Then, the base station can transmit all the information of that vehicle to other nearby base station. Therefore, handoff can be completed with very less time.

5.3 Power Allocation

In this paper, we investigate power allocation schemes in a multi-user systems for QoS-based scheduling. We deploy a proportional-fair power allocation scheme to satisfy a target throughput. We assume a UE required minimum of power $P_{\min}$ to satisfy its throughput requirement. Initially, the scheduled beam is transmitted with a power $P_b < P_{\text{tot}}$, where P_{tot} is the total power of the BS. The maximum power $P_{\max}$ can be allocated to one transmitting beam, where $P_{\max} \leq P_{\text{tot}}$. For instance, during mobility if the channel condition falls, at receiver, beyond a threshold level, BS reallocates its power to meet the target of QoS performance at the receiver. However, when this proportional power allocation cannot solve the deteriorating channel condition, the BS has to change its current serving beam to next serving beam. As mentioned earlier, the procedure of changing the serving beam is based on the periodic geo-location and channel state feedback from UE.

5.4 Performance Evaluation

In this section, we provide some performance results to compare the throughput of our power allocation-based beam switching mechanism and baseline scheme which do not re-allocate power during beam switch. We evaluated the system performance

Table 1 System Parameters

Parameter	Value
Operating frequency	38 GHz
Operating bandwidth	1 GHz
Transmit power	33 dBm
Fading	8.2 dB
SINR requirement	−10 dB
Background noise	−174 dBm/Hz

in terms of maximum achievable data rate. The system parameters used here are summarized in Table 1.

In Fig. 2, we compare the two schemes in terms of maximum achievable throughput at the receiver end. Due to user mobility, the serving beam changes very frequently which leads to beam misalignment. Here, we calculate the data rate using Eqs. (2) and (3). We observe that our proposed scheme provides much higher throughput than the baseline scheme. In addition, we observe with increasing UE speed the data rate starts decreasing. However, the proposed scheme always outperforms the baseline scheme at even higher speed which makes the proposed scheme more practical and solution oriented.

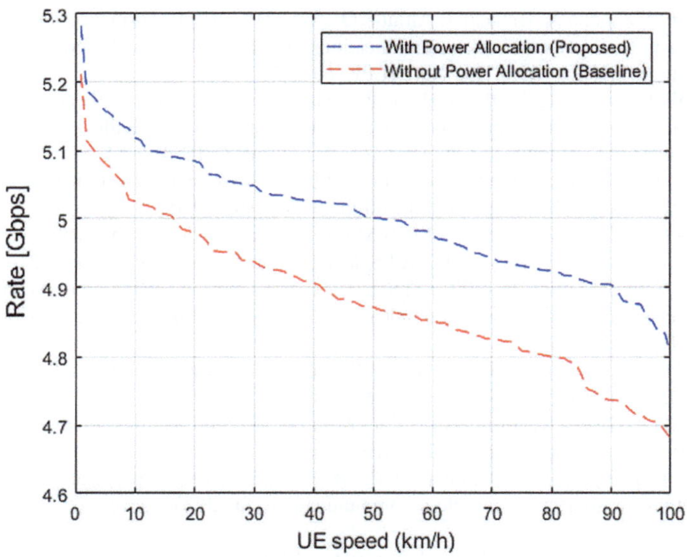

Fig. 2 Connection stability with varying UE speed considering mmWave propagation and beam switching

6 Conclusion

In recent years, the mmWave-based vehicular communications are becoming popular among academia and industry due to its huge potential in future V2X applications. Highly directional communications in mmWave system make mobility applications in vehicular application more challenging. In this paper, we explored the transmitter receiver beam misalignment problem and proposed a power allocation-based mechanism to provide stability in the system. We evaluate the system performance through MATLAB simulations. Our proposed scheme outperforms the baseline scheme in terms of user throughput. We believe our proposed scheme can be applied to support data-hungry applications in QoS system.

References

1. 3GPP, Technical Specification Group Services and System Aspects (2017) Enhancement of 3GPP support for V2X scenarios; stage 1 (Release 15), TS 22.186
2. Weng C-W, Lin K-H, Sahoo BPS, Wei H-Y (2018) Beam-aware dormant and scheduling mechanism for 5G millimeter-wave cellular systems. IEEE Trans Veh Technol 67(11):1–1 Nov
3. Sahoo BPS, Weng C-W, Wei H-Y (2018) SDN—architectural enabler for reliable communication over millimeter-wave 5G network. IEEE Globecom Workshops (GC Wkshps), p 1-1, December 2018
4. Giordani M, Zanella A, Higuchi T, Altintas O, Zorzi M (2018) Performance study of LTE and mmWave in vehicle-to-network communications. In: 2018 17th annual Mediterranean ad hoc networking workshop (Med-Hoc-Net). IEEE, pp 1–7
5. Giordani M, Shimizu T, Zanella A, Higuchi T, Al-tintas O, Zorzi M (2019) Path loss models for V2V mmWave communication: performance evaluation and open challenges. In: 2019 IEEE 2nd connected and automated vehicles symposium (CAVS). IEEE, pp 1–5
6. Ali A, Gonzlez-Prelcic N, Heath RW (2017) Millimeter wave beam-selection using out-of-band spatial information. IEEE Trans Wirel Commun 17(2):1038–1052
7. Perfecto C, Del Ser J, Bennis M (2017) Millimeter-wave V2V communications: distributed association and beam alignment. IEEE J Selected Areas Commun 35(9):2148–2162
8. Kato A, Sato K, Fujise M (2001) ITS wireless transmission technology. Technologies of millimeter-wave inter-vehicle communications: propagation characteristics. J Commun Res Lab 48:99–110 March
9. Lu N, Cheng N, Zhang N, Shen X, Mark JW (2014) Connected vehicles: solutions and challenges. IEEE Internet Things J 1(4):289–299 Aug
10. Yamamoto A, Ogawa K, Horimatsu T, Kato A, Fujise M (2008) Pathloss prediction models for intervehicle communication at 60 GHz. IEEE Trans Veh Technol 57(1):65–78
11. Geng S, Kivinen J, Zhao X, Vainikainen P (2009) Millimeter-wave propagation channel characterization for short-range wireless communications. IEEE Trans Veh Technol 58(1):3–13 Jan
12. Va V, Shimizu T, Bansal G, Heath RW (2016) Millimeter wave vehicular communications: a survey. Found Trends R Network 10(1):1113 [Online]. Available: https://doi.org/10.1561/1300000054
13. Choi J, Va V, Gonzalez-Prelcic N, Daniels R, Bhat CR, Heath RW (2016) Milli-meter-wave vehicular communication to support massive automotive sensing. IEEE Commun Mag 54(12):160167 December

14. Rappaport TS, Heath Jr RW, Daniels RC, Murdock JN (2014) Millimeter wave wireless communications. Pearson Education, London
15. Shah S, Ahmed E, Imran M, Zeadally S (2018) 5G for vehicular communications. IEEE Commun Mag 56(1):111–117 Jan
16. Mason F, Giordani M, Chiariotti F, Zanella A, Zorzi Z (2020) An adaptive broadcasting strategy for efficient dynamic mapping in vehicular networks. IEEE Trans Wirel Commun (TWC) 19(8):5605–5620 Aug
17. Hartenstein H, Laberteaux L (2008) A tutorial survey on vehicular ad hoc networks. IEEE Commun Mag 46(6):164–171 Jun
18. Alam N, Dempster AG (2013) Cooperative positioning for vehicular networks: facts and future. IEEE Trans Intell Transport Syst 14(4):1708–1717 Dec
19. Higuchi T, Giordani M, Zanella A, Zorzi M, Altintas O (2019) Value-anticipating V2V communications for cooperative perceptio. In: 30th IEEE intelligent vehicles symposium (IV)
20. Polese M, Giordani M, Mezzavilla M, Rangan S, Zorzi M (2017) Improved handover through dual connectivity in 5G mmWave mobile networks. IEEE J Selected Areas Commun 35(9):2069–2084 Sept
21. Balico LN, Loureiro AAF, Nakamura EF, Barreto RS, Pazzi RW, Oliveira HABF (2018) Localization prediction in vehicular ad hoc networks. IEEE Commun Surv Tutor 20(4):2784–2803 May
22. Weng C-W, Sahoo BPS, Wei H-Y, Yu C-H (2018) Directional reference signal design for 5G millimeter wave cellular systems. IEEE Trans Veh Technol 67(11):1–1 Nov
23. Ali A, Heath Jr. RW (2017) Compressed beam-selection in millimeter wave systems with out-of-band partial support information. In: Proceedings of IEEE international conference on acoustics, speech, and signal processing (ICASSP), Mar 2017, pp 3499–3503
24. Seo J, Sung Y, Lee G, Kim D (2016) Training beam sequence design for millimeter-wave MIMO systems: a POMDP framework. IEEE Trans Signal Process 64(5):1228–1242 Mar
25. Va V, Shimizu T, Bansal G, Heath Jr. RW (2016) Millimeter wave vehicular communications: a survey. Found Trends Network 10(1):1–113
26. Sahoo BPS, Chou C-C, Weng C-W, Wei H-Y (2018) Enabling millimeter-wave 5G networks for massive IoT applications: a closer look at the issues impacting millimeter-waves in consumer devices under the 5G framework. IEEE Consum Electron Mag 8(1):49–54
27. Sahoo BPS, Yao CH, Wei HY (2017) Millimeter-wave multi-hop wireless back-hauling for 5G cellular networks. In: IEEE 85th vehicular technology conference (VTC Spring), Sydney, NSW, pp 1–5
28. Yao CH, Chen YY, Sahoo BPS, Wei HY (2017) Outage reduction with joint scheduling and power allocation in 5G mmWave cellular networks. IEEE 28th annual international symposium on personal, indoor, and mobile radio communications (PIMRC), Montreal, QC, pp 1–6

Renewable Energy-Based Resource Management in Cloud Computing: A Review

Sanjib Kumar Nayak, Sanjaya Kumar Panda, and Satyabrata Das

Abstract Energy conservation is one of the most challenging problems in cloud datacenters. It is produced using fossil fuels, especially coal, gas, orimulsion and petroleum, and the cost of these fuels is increasing rapidly due to high demand of electricity. Moreover, fuels emit a huge amount of carbon dioxide and heat, which make adverse effect on the environment. Therefore, the cloud service providers are planning to use renewable energy sources, such as biomass, hydro, solar and wind to run their datacenters. It will reduce the usage of fossil fuels, the carbon dioxide emission and the energy cost of the datacenters to some extent. In this paper, we present a short review on renewable energy-based resource management in cloud computing. Here, the resources (like datacenters) use renewable energy sources to provide the services to the users and use non-renewable energy sources in case of scarcity of renewable energy. Furthermore, we discuss a load balancing problem in which the aim is to distribute the user requests to the datacenters and provide possible solutions to show the impact of non-renewable (brown) and renewable (green) energy in this context. These solutions are simulated using MATLAB R2014a, and their performances are tested using twenty instances of four different datasets in the form of two metrics, namely the overall cost and the number of used renewable energy resources.

Keywords Cloud computing · Resource management · Load balancing · Renewable energy · Green energy · Brown energy

S. K. Nayak · S. Das
Veer Surendra Sai University of Technology, Burla, Odisha 768018, India
e-mail: fortunatesanjib@gmail.com

S. Das
e-mail: teacher.satya@gmail.com

S. K. Panda (✉)
National Institute of Technology, Warangal, Warangal, Telangana 506004, India
e-mail: sanjayauce@gmail.com

A. K. Tripathy et al. (eds.), *Advances in Distributed Computing and Machine Learning*, Lecture Notes in Networks and Systems 127, https://doi.org/10.1007/978-981-15-4218-3_5

1 Introduction

Datacenter is a massive pool of networked servers, residing in a centralized location, for processing, storing, managing and distributing its data. The number of datacenters is estimated globally as 8.6 million, including 3 million in the USA [1]. Each datacenter typically hosts 50,000–80,000 servers, which require a power capacity of 25–30 MW [2]. According to the National Resources Defense Council (NRDC) report [3], the USA requires approximately 140 billion kilowatt-hours of electricity by 2020 to run its datacenters, whereas global datacenters require approximately 416 TW of electricity per year [4]. Moreover, the amount of electricity is doubled in every quadrennial period as per the British Petroleum plc [5]. Therefore, energy conservation is a challenging problem in cloud datacenters [6, 7].

Energy is produced using non-renewable energy sources like fossil fuels [8]. The cost of these fuels is increasing rapidly due to the high demand of electricity. Moreover, burning these fuels generates carbon dioxide, particle and heat, which are harmful for the environment and leads to the greenhouse effect. The released particles pollute water, land and air. Therefore, the cloud service providers (CSPs) are looking for eco-friendly solutions in order to protect the environment and reduce the usage of these fuels. One of the possible solutions is to use renewable energy sources, such as biomass, hydro, solar and wind to run the datacenters [8]. The CSPs can invest in on-site power sources without relying on non-renewable energy only. For instance, they can use photovoltaics to convert sunlight into electricity or wind energy converter to convert wind energy into electricity or flowing water to generate electricity or burn biomass to create steam in order to generate electricity. On the other hand, the energy cost and the carbon dioxide emission are also reduced to some extent. It is noteworthy to mention that renewable energy sources are not available round the clock. Therefore, many researchers [8–11] have suggested to use both non-renewable and renewable energy sources to fulfill the demands of users. More specifically, the renewable energy sources run the datacenters to provide the services to the users. In case of scarcity of renewable energy, the non-renewable energy sources are used to run the datacenters.

In this paper, we present a short review on renewable energy-based resource management in cloud computing. The review focuses on energy sources (i.e., non-renewable and renewable), environment, future knowledge about energy sources and others. Moreover, we discuss a load balancing problem in this context. The aim of this problem is to distribute the user requests to the datacenters, such that the overall cost is minimized and the number of used renewable energy resources is maximized. For this problem, we present various possible solutions and show the impact of energy sources. These solutions are simulated and rigorously tested using twenty instances of four different datasets in terms of two performance metrics.

The rest of this paper is structured as follows. Section 2 discusses the related work. Section 3 presents the mathematical modeling of the load balancing problem. In Sect. 4, we illustrate some possible solutions and discuss the step-by-step process. The simulation results of these solutions are presented in Sect. 5. Finally, we conclude in Sect. 6.

2 Related Work

Many energy-based algorithms [6–16] have been developed to improve energy efficiency of the datacenter(s). Beloglazov et al. [12] have studied various causes of high energy consumption and presented a taxonomy for future advancements/directions. The taxonomy focuses on various levels of computing systems, such as hardware, firmware, operating system, virtualization and datacenter. Rahman et al. [16] have presented a state of the art on power management of datacenters. They have categorized the research works based on their objectives, such as maximizing/minimizing carbon dioxide emission, electricity cost, quality of service and renewable energy sources. They have suggested various factors, especially environmental factors for future site construction of datacenters.

Le et al. [14] have proposed a dynamic load balancing policy, which saves electricity-related costs and provides cooling effects. This policy takes care of both load placement and migration in the datacenters. Panda and Jana [6] have presented an algorithm to minimize the energy consumption of the datacenters. They have mapped the user requests to the virtual machines and further mapped the virtual machines to the physical machines of the datacenters. Further, Panda and Jana [7] have addressed the problems associated with task scheduling and task consolidation and proposed an algorithm to balance energy consumption and makespan. However, the above works were not considered any renewable energy sources in their models.

Khosravi et al. [13] have proposed three algorithms in which one algorithm is offline and the other two algorithms are online. In the offline algorithm, they have assumed that the knowledge of renewable energy sources is known. In the online algorithms, they have assumed no future knowledge and limited knowledge, respectively, about the renewable energy sources. Toosi and Buyya [8] have studied various renewable energy-based algorithms and proposed a load balancing algorithm. This algorithm is based on fuzzy logic, which does not require any future knowledge of energy sources, cost and loads. They have achieved better cost performance in five-hour window size, and suggested to regulate the window size as per the load and environment. Pierson et al. [15] have developed a project, called DATAZERO to provide high availability of services. For this, it makes a trade-off between the electrical side and IT management side by introducing a negotiation process. They have suggested to further tune the negotiation algorithm using various approaches like game theory as a part of future work. In summary, we present a comparative study of some energy-based algorithms in Table 1.

Table 1 Comparative study of energy-based algorithms

Article	Algorithm	Objective	Energy resources		Environment	Future knowledge	Experimentation platform	Drawback(s)
			Renewable	Non-renewable				
Toosi and Buyya (2015) [8]	Fuzzy logic-based load balancing	Maximize the renewable energy consumption and minimize the total incurred cost	Yes	Yes	Homogeneous	No	CloudSim	The outcomes are based on the window size, and it is regulated as per the workload and environment
	Future-aware best fit							It distributes the user requests to the datacenters by considering the lowest total incurred cost only
	Round robin							It distributes the user requests to the datacenters evenly without considering renewable energy consumption and total incurred cost
	Highest available renewable first							It distributes the user requests by considering the availability of renewable energy resource slots only
Xu et al. (2018) [11]	Reinforcement learning-based job scheduling	Minimize the total cost	Yes	Yes	Homogeneous	No	Not specified	The learning process is slower due to usage of single processor
Chen et al. (2012) [9]	MinBrown	Minimize the brown energy consumption	Yes	Yes	Homogeneous	No	Own simulator	The cost of energy is not considered
Le et al. (2011) [14]	Worst fit	Minimize the electricity cost	No	Yes	Homogeneous	No	Own simulator	It considers three electricity-based costs without focusing renewable energy sources
	Static cost-aware ordering							It follows static datacenter ordering and selects cheapest datacenter first
Li et al. (2011) [10]	iSwitch	Trade-off between renewable energy utilization and switching frequency	Yes	Yes	Not applicable	Not applicable	Own simulator	The over-provisioned power is not absorbed

3 Problem Statement

Consider a set of datacenters, $D = \{D_1, D_2, D_3, \ldots, D_m\}$, where each datacenter D_j, $1 \leq j \leq m$ contains a set of resources (nodes), $R_j = \{R_{j1}, R_{j2}, R_{j3}, \ldots, R_{j|R_j|}\}$. A resource $R_{jk} \in R_j$ contains a set of slots, $R_{jk} = \{R_{jk1}, R_{jk2}, R_{jk3}, \ldots, R_{jko}\}$, where $1 \leq k \leq |R_j|$ and $1 \leq o \leq |R_{jk}|$. In general, resources are of two types, non-renewable or brown (NRE) and renewable or green (RE), with respect to time. More specifically, a resource slot $R_{jkl} \in R_{jk}$, $1 \leq j \leq m$, $1 \leq k \leq |R_j|$, $1 \leq l \leq o$ may be non-renewable or renewable with respect to time l, where o is the future knowledge about energy sources (i.e., future time window). However, it cannot be both non-renewable or renewable at the same time. If a resource $R_{jk} \in RE$, then the cost of using this resource is fixed and predetermined within the future time window. On the other hand, if a resource $R_{jk} \in NRE$, then the cost of using this resource is dynamic and predetermined within the future time window.

Consider a set of user requests, $U = \{U_1, U_2, U_3, \ldots, U_n\}$, where each user request U_i, $1 \leq i \leq n$ is represented in the form of 3-tuple. Mathematically, $U_i = <ST_i, N_i, D_i>$, where ST_i is the start time, N_i is the number of nodes, and D_i is the duration of user request i. A global queue Q keeps the user requests in the ascending order of their start time.

The problem is to map the user requests U to the datacenters D (i.e., $f: U \rightarrow D$, where f is a mapping function), such that the following objectives are fulfilled. (1) The overall cost is minimized. (2) The number of used renewable energy resources is maximized.

4 Illustration

Let us assume that there are nine tasks, U_1–U_9 (i.e., three assigned tasks, U_1–U_3 and six unassigned tasks, U_4–U_9) and two datacenters, D_1 and D_2 (each with five nodes) as shown in Table 2 [8]. The initial Gantt chart, after assigning three tasks, is shown in Table 3. Note that gray color (□) indicates the non-renewable energy (brown) resource slots, white color (□) indicates the renewable energy (green) resource slots, and numeric values, in the top of each datacenter, indicate the cost. Without the loss of generality, we can assume that the cost of using renewable energy resource slots is zero. Therefore, the initial cost of datacenters, D_1 and D_2, is zero as first three

Table 2 Set of nine tasks with their start time, nodes and duration

User request	U_1	U_2	U_3	U_4	U_5	U_6	U_7	U_8	U_9
Start time	1	1	1	3	4	5	5	7	8
Nodes	1	1	1	2	1	1	1	2	3
Duration	4	1	4	5	3	2	3	2	2

Table 3 Initial Gantt chart (after assigning three tasks)

	0.1	0.1	0.3	0.4	0.4	0.2	0.2	0.1	0.1
Data center D_1									
	U_2								
	U_1	U_1	U_1	U_1					
	0.1	0.1	0.5	0.3	0.2	0.3	0.4	0.5	0.1
Data center D_2									
	U_3	U_3	U_3	U_3					
	$t = 1$	$t = 2$	$t = 3$	$t = 4$	$t = 5$	$t = 6$	$t = 7$	$t = 8$	$t = 9$

tasks are executed in renewable energy resource slots only. We assume that the future window size is 7 [8].

Let us discuss the future-aware best fit algorithm. At time $t = 3$, user request U_4 is mapped with datacenters, D_1 and D_2, and it requires two nodes for five units of time. The cost of these datacenters is $0.3 + 0.2 = 0.5$ and $0.3 + 0.3 + 0.3 = 0.9$ for executing user request U_4. As datacenter D_1 takes lowest cost, user request U_4 is assigned to datacenter D_1. At time $t = 4$, user request U_5 is mapped with two datacenters, and it requires one node for three units of time. The cost of these datacenters is $0.4 + 0.4 + 0.2 = 1.0$ and 0.3 for executing it. As datacenter D_2 takes lowest cost, user request U_5 is assigned to datacenter D_2. Similarly, other user requests (i.e., U_6 to U_9) are assigned to datacenters D_2, D_2, D_1 and D_2, respectively. The total cost of two datacenters is 0.8 and 1.4 for executing all the user requests. The final Gantt chart is shown in Table 4.

Now, we discuss the round robin algorithm using the same illustration. At time $t = 3$, user request U_4 is assigned to datacenter D_1 in a circular fashion. The cost of datacenter D_1 is $0.3 + 0.2 = 0.5$ for executing it. At time $t = 4$, user request U_5 is assigned to datacenter D_2, and the cost of executing the same is 0.3. Similarly, other user requests (i.e., U_6–U_9) are assigned to datacenters D_1, D_2, D_1 and D_2, respectively. The total cost of two datacenters is 1.4 and 0.9 for executing all the user requests. The final Gantt chart is shown in Table 5.

Let us now discuss the highest available renewable first algorithm. At time $t = 3$, user request U_4 is mapped with datacenters, D_1 and D_2. The number of available renewable energy resource slots of these datacenters is 8 and 7 for exe-

Table 4 Gantt chart for future-aware best fit algorithm

	0.1	0.1	0.3	0.4	0.4	0.2	0.2	0.1	0.1
Data center D_1									
							U_8		
			U_4	U_4			U_8		
	U_2		U_4	U_4	U_4	U_4	U_4	U_8	
	U_1	U_1	U_1	U_1	U_4	U_4	U_4	U_8	
	0.1	0.1	0.5	0.3	0.2	0.3	0.4	0.5	0.1
Data center D_2									
					U_7	U_7		U_9	U_9
				U_5	U_6	U_6		U_9	U_9
	U_3	U_3	U_3	U_3	U_5	U_5	U_7	U_9	U_9
	$t = 1$	$t = 2$	$t = 3$	$t = 4$	$t = 5$	$t = 6$	$t = 7$	$t = 8$	$t = 9$

Table 5 Gantt chart for round robin algorithm

	0.1	0.1	0.3	0.4	0.4	0.2	0.2	0.1	0.1
Data center D_1							U_8		
			U_4	U_4	U_6	U_6	U_8		
	U_2		U_4	U_4	U_4	U_4	U_4	U_8	
	U_1	U_1	U_1	U_1	U_4	U_4	U_4	U_8	
	0.1	0.1	0.5	0.3	0.2	0.3	0.4	0.5	0.1
Data center D_2									
								U_9	U_9
				U_5	U_7	U_7		U_9	U_9
	U_3	U_3	U_3	U_3	U_5	U_5	U_7	U_9	U_9
	$t = 1$	$t = 2$	$t = 3$	$t = 4$	$t = 5$	$t = 6$	$t = 7$	$t = 8$	$t = 9$

cuting user request U_4. As datacenter D_1 contains the highest number of available renewable energy resource slots, user request U_4 is assigned to datacenter D_1. At time $t = 4$, user request U_5 is mapped with two datacenters. The number of available renewable energy resource slots of these datacenters is 0 and 2 for executing it. As datacenter D_2 contains the highest number of available renewable energy resource slots, user request U_5 is assigned to datacenter D_2. Similarly, other user requests (i.e., U_6–U_9) are assigned to datacenters D_2, D_2, D_2 and D_1, respectively. The total cost of two datacenters is 0.8 and 1.5 for executing all the user requests. The final Gantt chart is shown in Table 6. We also present the Gantt chart for optimal geographical load balancing algorithm in Table 7 [8]. Here, the total cost of two datacenters is 0.3 and 1.7 for executing all the user requests.

Table 6 Gantt chart for the highest available renewable first algorithm

	0.1	0.1	0.3	0.4	0.4	0.2	0.2	0.1	0.1
Data center D_1									
			U_4	U_4				U_9	U_9
	U_2		U_4	U_4	U_4	U_4	U_4	U_9	U_9
	U_1	U_1	U_1	U_1	U_4	U_4	U_4	U_9	U_9
	0.1	0.1	0.5	0.3	0.2	0.3	0.4	0.5	0.1
Data center D_2									
					U_7	U_7	U_8		
				U_5	U_6	U_6	U_8	U_8	
	U_3	U_3	U_3	U_3	U_5	U_5	U_7	U_8	
	$t = 1$	$t = 2$	$t = 3$	$t = 4$	$t = 5$	$t = 6$	$t = 7$	$t = 8$	$t = 9$

Table 7 Gantt chart for optimal geographical load balancing algorithm

	0.1	0.1	0.3	0.4	0.4	0.2	0.2	0.1	0.1
Data center D_1									
							U_8		
	U_2			U_5	U_7	U_7	U_8	U_8	
	U_1	U_1	U_1	U_1	U_5	U_5	U_7	U_8	
	0.1	0.1	0.5	0.3	0.2	0.3	0.4	0.5	0.1
Data center D_2									
			U_4	U_4	U_6	U_6		U_9	U_9
			U_4	U_4	U_4	U_4	U_4	U_9	U_9
	U_3	U_3	U_3	U_3	U_4	U_4	U_4	U_9	U_9
	$t = 1$	$t = 2$	$t = 3$	$t = 4$	$t = 5$	$t = 6$	$t = 7$	$t = 8$	$t = 9$

Table 8 Gantt chart for worst fit algorithm

Data center	$t=1$	$t=2$	$t=3$	$t=4$	$t=5$	$t=6$	$t=7$	$t=8$	$t=9$
D_1 cost	0.1	0.1	0.3	0.4	0.4	0.2	0.2	0.1	0.1
D_1					U_7	U_7	U_8		
D_1	U_2			U_5	U_6	U_6	U_8	U_8	
D_1	U_1	U_1	U_1	U_1	U_5	U_5	U_7	U_8	
D_2 cost	0.1	0.1	0.5	0.3	0.2	0.3	0.4	0.5	0.1
D_2									
D_2			U_4	U_4				U_9	U_9
D_2			U_4	U_4	U_4	U_4	U_4	U_9	U_9
D_2	U_3	U_3	U_3	U_3	U_4	U_4	U_4	U_9	U_9

Let us now discuss the MinBrown algorithm. It distributes the user requests to the datacenters with minimum required non-renewable energy resource slots. Note that we have not considered all the parameters of the MinBrown algorithm, such as bandwidth, deadline, epoch, migration and network latency. At time t = 3, user request U_4 is mapped with datacenters, D_1 and D_2. The number of required non-renewable energy resource slots of these datacenters is 2 and 3 for executing user request U_4. As datacenter D_1 contains the minimum required non-renewable energy resource slots, user request U_4 is assigned to datacenter D_1. Similarly, other user requests are assigned to the datacenters. The final Gantt chart is shown in Table 6, which is same as the highest available renewable first algorithm. Now, we discuss the static cost-aware ordering algorithm. It follows static datacenter ordering and selects the cheapest datacenter first. Alternatively, it performs the ordering of datacenters based on cost (i.e., cheapest to expensive). The final Gantt chart is shown in Table 4, which is same as the future-aware best fit algorithm. Let us now discuss the worst fit algorithm. It distributes the user requests to the datacenters with maximum available resource slots. At time t = 3, user request U_4 is mapped with datacenters, D_1 and D_2. As datacenter D_2 contains the maximum available resource slots, user request U_4 is assigned to datacenter D_2. At time t = 4, user request U_5 is mapped with two datacenters. As datacenter D_1 contains the maximum available resource slots, user request U_5 is assigned to datacenter D_1. Similarly, other user requests (i.e., U_6–U_9) are assigned to datacenters D_1, D_1, D_1 and D_2, respectively. The total cost of two datacenters is 0.9 and 1.2 for executing all the user requests. The final Gantt chart is shown in Table 8.

5 Simulation Results

We utilize MATLAB R2014a (running on Intel(R) Core(TM) $i5$-4210M CPU @ 1.70 GHz 1.70 GHz processor with 8 GB installed memory and Windows 7 64-bit operating system) to simulate some of the existing algorithms, namely future-aware best fit (FABEF) or static cost-aware ordering (SCA), round robin (RR) and highest available renewable first (HAREF) or MinBrown (MB). We generated four different datasets, namely 400 × 50 (i.e., 400 indicates the number of user requests and 50 indicates the number of datacenters), 800 × 100, 1200 × 150 and 1600 × 200 using MATLAB R2014a inbuilt function. Each dataset contains five instances, namely $i1$–$i5$ of the same size. The simulation parameters and their values are shown in Table 9.

The overall cost and the number of used renewable energy resources of the FABEF/SCA, RR and HAREF/MB algorithms are calculated for all the instances of four datasets and compared as shown in Table 10, Figs. 1 and 2. The comparison results show that the FABEF/SCA outperforms than the RR and HAREF/MB algorithms in terms of overall cost, whereas the HAREF/MB outperforms than the FABEF/SCA and RR algorithms in terms of number of used renewable energy resources. Note that we have not shown the simulation results of the worst fit algorithm due to space limitations.

Table 9 Simulation parameters and their values

Parameter	Value
Number of user requests	400, 800, 1200, 1600
Number of datacenters	50, 100, 150, 200
Structure of datasets	Number of user requests × number of datacenters
Range of datasets	[10–2000]
Instances	$i1, i2, i3, i4, i5$
Distribution	Uniform
Simulation method	Monte Carlo
Cost of renewable resources	5
Cost of non-renewable resources	[20–40]

Table 10 Comparison of overall cost and number of used renewable energy resources for FABEF or SCA, RR and HAREF or MB algorithms

Dataset	Instance	Overall cost			Number of used renewable energy resources		
		FABEF or SCA	RR	HAREF or MB	FABEF or SCA	RR	HAREF or MB
400 × 50	$i1$	262,450	0,618,574	0,270,188	203,036	148131	204,263
	$i2$	254,274	0,639,999	0,262,899	192,624	124,705	193,759
	$i3$	253,488	0,772,139	0,261,954	191,990	148,136	193,454
	$i4$	247,760	0,736,032	0,257,347	190,859	124705	192514
	$i5$	261,052	0,772,139	0,267,561	198,467	138,232	200,004

(continued)

Table 10 (continued)

Dataset	Instance	Overall cost			Number of used renewable energy resources		
		FABEF or SCA	RR	HAREF or MB	FABEF or SCA	RR	HAREF or MB
800 × 100	$i1$	509,146	1,209,273	0,529,687	402,754	307,123	407,322
	$i2$	499,424	1,249,969	0,518,239	396,594	279,442	400,813
	$i3$	498,041	1,285,652	0,518,788	394,539	265,099	398,854
	$i4$	481,993	1,255,054	0,500,812	388,434	252,372	392,302
	$i5$	513,160	1,251,225	0,539,511	404,211	297,363	408,917
1200 × 150	$i1$	714,208	1,894,150	0,743,036	581,537	423,480	587,951
	$i2$	708,719	1,792,889	0,739,374	578,916	423,191	585,139
	$i3$	732,902	1,841,103	0,763,980	595,799	409,378	601,791
	$i4$	718,685	1,804,653	0,746,495	584,977	439,528	590,988
	$i5$	732,324	1,822,021	0,764,081	600,193	419,658	606,293
1600 × 200	$i1$	966,843	2,545,480	1,008,371	793,906	527,842	803,455
	$i2$	971,400	2,508,121	1,014,267	800,359	541,380	809,565
	$i3$	962,787	2,368,714	1,004,627	785,672	566,335	794,821
	$i4$	979,457	2,317,138	1,022,388	798,617	557,535	809,113
	$i5$	982,562	2,362,988	1,027,227	803,185	558,533	812,180

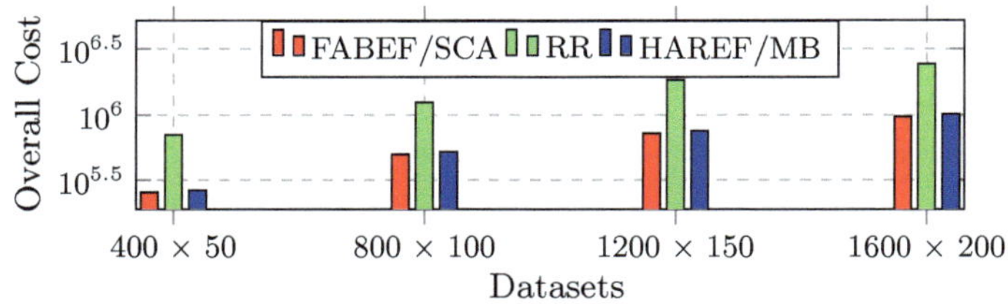

Fig. 1 Comparison of overall cost for FABEF/SCA, RR and HAREF/MB algorithms

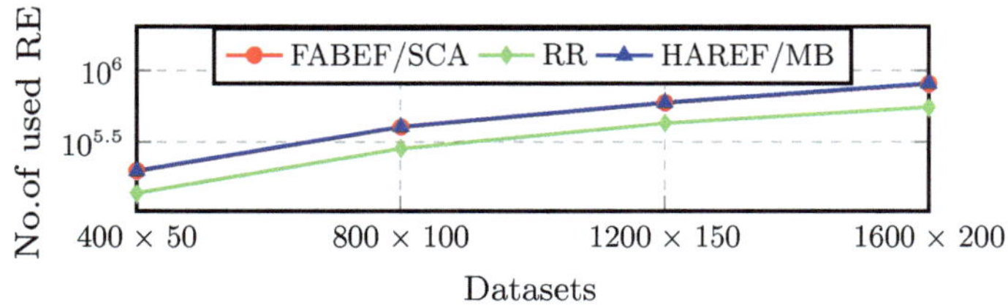

Fig. 2 Comparison of number of used renewable energy resources for FABEF/SCA, RR and HAREF/MB algorithms

6 Conclusion

In this paper, we have presented a short review on renewable energy-based resource management in cloud computing. We have also presented a load balancing problem and discussed possible solutions to show the impact of non-renewable and renewable energy resources. We have simulated some of the possible solutions and tested their performance using four different datasets in two performance metrics. The simulation results have shown that the FABEF/SCA algorithm outperforms in terms of overall cost, whereas the HAREF/MB algorithm outperforms in terms of number of used renewable energy resources. As future work, we suggest to develop an efficient algorithm, which takes care of the overall cost as well as the number of used renewable energy resources.

References

1. Harvey C Data center. https://www.datamation.com/data-center/what-is-data-center.html/. Accessed 10 Jan 2020
2. Miller R Inside amazon's cloud computing infrastructure. https://datacenterfrontier.com/inside-amazon-cloud-computing-infrastructure/. Accessed 10 Jan 2020
3. Delforge P (2014) America's data centers are wasting huge amounts of energy. In: Natural resources defense council (NRDC), pp 1–5
4. Danilak R (2017) Why energy is a big and rapidly growing problem for data centers. Forbes 15:12–17
5. Dudley B et al (2018) Bp statistical review of world energy. BP Stat Rev London 6
6. Panda SK, Jana PK (2017) An efficient request-based virtual machine placement algorithm for cloud computing. In: International conference on distributed computing and internet technology, pp 129–143. Springer
7. Panda SK, Jana PK (2019) An energy-efficient task scheduling algorithm for heterogeneous cloud computing systems. Cluster Comput 22(2):509–527
8. Nadjaran Toosi A, Buyya R (2015) A fuzzy logic-based controller for cost and energy efficient load balancing in geo-distributed data centers. In: Proceedings of the 8th international conference on utility and cloud computing, pp 186–194. IEEE Press
9. Chen C, He B, Tang X (2012) Green-aware workload scheduling in geographically distributed data centers. In: 4th IEEE international conference on cloud computing technology and science proceedings, pp 82–89. IEEE
10. Li C, Qouneh A, Li T (2011) Characterizing and analyzing renewable energy driven data centers. In: Proceedings of the ACM SIGMETRICS joint international conference on measurement and modeling of computer systems, pp 131–132. ACM
11. Xu C, Wang K, Li P, Xia R, Guo S, Guo M (2018) Renewable energy-aware big data analytics in geo-distributed data centers with reinforcement learning. IEEE Trans Netw Sci Eng
12. Beloglazov A, Buyya R, Lee YC, Zomaya A (2011) A taxonomy and survey of energy-efficient data centers and cloud computing systems. In: Adv Comput 82:47–111 (Elsevier)
13. Khosravi A, Toosi N, Buyya R (2017) Online virtual machine migration for renewable energy usage maximization in geographically distributed cloud data centers. Concurr Comput Pract Exp 29(18):e4125
14. Le K, Bianchini R, Zhang J, Jaluria Y, Meng J, Nguyen TD (2011) Reducing electricity cost through virtual machine placement in high performance computing clouds. In: Proceedings of 2011 international conference for high performance computing, networking, storage and analysis, p 22. ACM

15. Pierson JM, Baudic G, Caux S, Celik B, Da Costa G, Grange L, Haddad M, Lecuivre J, Nicod JM, Philippe L et al (2019) Datazero: Datacenter with zero emission and robust management using renewable energy. IEEE Access 7:103209–103230
16. Rahman A, Liu X, Kong F (2013) A survey on geographic load balancing based data center power management in the smart grid environment. IEEE Commun Surv Tutor 16(1):214–233

A Multi-objective Optimization Scheduling Algorithm in Cloud Computing

Madhu Bala Myneni and Siva Abhishek Sirivella

Abstract Task scheduling plays a major role in cloud computing that creates a direct impact on performance issues and reduces the system load. In this paper, a novel task scheduling algorithm has proposed for the optimization of multi-objective problem in the cloud environment. It addresses a model to define the demand of resources by a job. It gives a relationship between the resources and costs within a project. The scheduling of multi-objective problem is optimized with the use of ant colony optimization algorithm. The evaluation of the cost and performance of the task has two major constraints considered as makespan and budget's cost. The two considered constraints will make the algorithm to achieve the optimal result within time and enhance the quality of performance of the system considered. This method is very powerful than other methods with single objectives considered such as makespan, utilization of resources, violation of deadline rate and cost.

Keywords Cloud computing · Task · Resources · Ant colony algorithm · Pheromone · Fitness function

1 Introduction

The task scheduling creates a direct impact on the system's load and performance. Hence, to decrease the load on computer, we can use cloud computing. An effective task scheduling method should meet the needs of user, and the system's efficiency is improved. As we know that task scheduling problems come under NP-hard problems, we can use the ant colony algorithm because it is a global optimization algorithm as well as probabilistic approach.

M. B. Myneni (✉) · S. A. Sirivella
Institute of Aeronautical Engineering, Dundigal, Hyderabad, Telangana 500043, India
e-mail: baladandamudi@gmail.com

S. A. Sirivella
e-mail: sivaabhishek3@gmail.com

A. K. Tripathy et al. (eds.), *Advances in Distributed Computing and Machine Learning*, Lecture Notes in Networks and Systems 127, https://doi.org/10.1007/978-981-15-4218-3_6

Without the management by user, cloud computing gives the system the required resources with more data storage and a faster computing ability. To achieve consistency and economies at different scales, cloud computing relies majorly on sharing of resources. The existing approaches to handle the task scheduling include various single objective methods such as operator non-availability period [1] for optimizing task completion time, reduce start time for self adopting scheduling [4], deadline guaranteed scheduling [5, 7] for improvement in resource utilization, energy-aware real-time scheduling [6] and minimum task length [10]. A heuristic approach for cloud [2] and then multi-objective optimization function [8] is developed by using ant colony [9] optimization. Later, the task scheduling focuses on the performance and effectiveness of heterogeneous computing environments [3]. Various cloud environments with data hosting service are evaluated for effective cost and high availability [11].

Based on the behavior of ants that search for their foods, the ant colony algorithm was proposed to find the optimal paths. The ants wander around in search of food and then walk back to the colony leaving pheromones which are referred here as markers that will tell the other ants that the path has food. Other ants get across these pheromones (markers), they tend to follow the markers and get to their food, and a time comes where all the ants will follow the same path so as to obtain their food.

2 Definition and Description of Problem

Symbol	Definitions
T_i	The task i, $1 < j < K$
U_j	The resource j, $1 < i < N$
N, K	The amount of resources and tasks
C_i, M_i	CPU and memory of T_i
D_i	The deadline of task T_i
Bi	The budget cost of task T_i
C_j, M_j	CPU and memory of R_j
$C_{\text{cost}}(j), M_{\text{cost}}(j)$	The costs of CPU and memory of R_j
C_{base}	The base cost of CPU under the lowest usage
M_{base}	The base cost of memory under 1 GB memory
t_{ij}	The duration time of task T_i in resource R_j
$C_{\text{Trans}}, M_{\text{Trans}}$	The transmission cost associated with CPU and memory
α, β	The weight factors of the heuristic information and pheromone
γ, δ	The weight factors of the performance and cost
ρ	The pheromone evaporation factor

(continued)

(continued)

Symbol	Definitions
D_v	The deadline violation rate
n_d	The number violating the deadline time in K tasks

2.1 *Definition of Tasks and Resources*

Let us consider that there are K number of tasks $T = \{T_1, T_2 \ldots, T_i, \ldots, T_K\}$ and N number of resources $R = \left\{R_1, R_2 \ldots, R_j, \ldots, R_N\right\}$ present in a cloud computing system.

- **TASK**: The parameters which the user has specified for the cloud computing system are as follows:
  - C_i CPU utilization
  - M_i memory use
  - D_i deadlines of the considered tasks
  - B_i budget cost of task.

  The task is defined as follows:

$$T_i = (C_i, M_i, D_i, L_i)$$

- **RESOURCE**: The resources defined are virtual in nature and will include the CPU and memory and are defined as follows:

$$R_j = (C_j, M_j)$$

In order to continue the research, we need to assume the resources based on the above-defined definitions. The user provides the information of the resource demanded by a task monitoring of resource usage in the cloud which is managed by the virtualization concept. The system will cut off task performance if the required resources exceed the number of users; therefore, the task tends to fail.

3 Resource Cost and Task Scheduling Mechanism

The multi-objective scheduling is attained in cloud by considering resource cost model.

3.1 Resource Cost Model

In general, the tasks are categorized based on resources used and storage space required. In cloud, tasks and resources are distinct, i.e., some tasks require more storage space and other tasks need more CPU resources. To address the said problem, the proposed model builds on resource cost as the multi-objective which divides the resources as CPU and memory.

Resource 1 cost: The CPU (processor) cost is explained in Formula 1:

$$C_{\text{cost}}(j) = C_{\text{base}} * C_j * t_{ij} + C_{\text{Trans}} \tag{1}$$

Resource 2 cost: The memory cost is explained in Formula 2:

$$M_{\text{cost}}(j) = M_{\text{base}} * M_j * t_{ij} + M_{\text{Trans}} \tag{2}$$

Therefore, the cost functions can be obtained from Formulas 3 and 4 as follows:

$$C_j = \sum_{j=1}^{n} C_{\text{cost}} j \tag{3}$$

$$M_j = \sum_{j=1}^{n} M_{\text{cost}} j \tag{4}$$

3.2 Task Scheduling in Cloud

The task scheduling in cloud is built by considering performance and budget as major constraints. Let us assume that there are K number of tasks $T = \{T_1, T_2, \ldots, T_i, \ldots, T_k\}$ and N number of resources $R = \{R_1, R_2, \ldots, R_j, \ldots, R_n\}$ in the cloud computing system. Now, this can be modeled by considering two major constraints such as the deadline cost and budget cost constraints of the task. The minimum cost of the above said constraints is optimization need to be considered for cloud model. This multi-objective optimization function to this problem is defined in Formula (5) as follows:

Minimize

$$\sum_{x} \text{OPT}(x) = \text{PER}(x), \text{BUD}(x) \tag{5}$$

subject to two identified constraints, which are given in Formulas 6, 7, 8

$$T(x) = C1(x) + C2(x) \tag{6}$$

$$\text{BUD}(x) \leq \sum_{i=1}^{k} \text{BUD}_i \tag{7}$$

$$\text{PER}(x) \leq \sum_{i=1}^{k} \text{PER}_i \tag{8}$$

Here, x acts as feasible solution.

- $T(x)$ refers to total cost.
- PER(x) refers to makespan of performance objective.
- BUD(x) is the user budget cost for the task demand for CPU.

Hence, to solve the optimization of multi-objective problem in cloud, the ant colony algorithm is used.

4 Multi-objective Optimization

The optimization of multi-objective scheduling is implemented by using the optimization algorithm as ant colony. This algorithm premises its own advantages when it comes to solve a multi-objective problem. The existing researches promise that this can used to solve many scheduling problems and also prove they assured results. This can also avoid the solution to fall into local optima.

4.1 Proposed Scheduling Optimization

The task scheduling optimization of multi-objective in cloud is implemented by using the ant colony algorithm. The process includes the following method. The ants choose a path randomly and attain their desired targets and calculate the "fitness of path." So, in order to focus on fitness of ants and also to achieve the optimal solution, we need to update the choices and pheromone.

4.2 Transition Behaviors

The transition behavior of ants is based on probability of more pheromone level in a path. First, we need to input the no. of tasks, their deadlines, no. of resources, budget costs, ability and other relevant parameters. Since we are using the ant colony algorithm, we need to assign a task to each ant. We need to allocate a task T_i to resource R_i. Next, continue the above process until all the tasks are allocated to resources. So, the process of assigning a task to the resource is similar to the formation of path by an ant.

The behavioral choice process for each task allocation to resources has a relationship with heuristic information and pheromone. Let us assume that $g_k(T_i, R_j)$ is a set that indicates a resource and it should meet the budget and deadline constraints in the Kth iteration of a task T_i. Therefore, getting the occurrences of the same task T_i is defined in Formula 9.

$$P_k(T_i, R_j) = \begin{cases} \frac{[\tau(T_i, R_j)]^\alpha [\eta(T_i, R_j)]^\beta}{\sum_{h \in g_k(T_i, R_j)} [\tau(T_i, h)]^\alpha [\eta(T_i, h)]^\beta}, & R_j \in g_k(T_i, R_j) \\ 0, & \text{otherwise.} \end{cases} \tag{9}$$

Here,

- τ (T_i, R_j) = The task associated pheromone T_i and the assigned resource R_j in the same path.
- $\eta(T_i, R_j)$ = Heuristic information, which is taken as inverse of starting time of task T_i.

The attributes α represents weight factor of heuristic information and β represents weight factors of pheromone.

4.3 Fitness Function

The quality of feasible solutions is evaluated with the specified fitness function that is used. Based on the optimization required, the fitness function is arranged. In the taken problem, there are two scheduling targets such as costs and makespan. These are to be minimized in scheduling through optimization. Therefore, the fitness function is named here as evaluation function in Formula (10) as follows:

$$\mathrm{Fit}(x) = \gamma \mathrm{e}^{(-\mathrm{PER}(x))} + \delta \mathrm{e}^{(-\mathrm{BUD}(x))} \tag{10}$$

In this, γ is factor of performance and δ is a factor of cost, where $\gamma > \delta$ and γ, δ $\in (0, 1)$.

Here, two objectives functions are PER(x) which is the performance and BUD(x) which is the cost. The fitness value is higher when the values of cost and makespan are smaller.

4.4 Updating the Pheromone

To find a path, more ants are to be allowed, the pheromone of that particular path should be more strengthened, and the fitness value of that path should be high. So, for each side of the path, in each iteration, the pheromone should be updated. The pheromone updating rule is given in Formula (11).

$$\tau(T_i, R_j) = (1 - \rho) \cdot \tau(T_i, R_j) + \Delta\tau(T_i, R_j) \tag{11}$$

where

ρ pheromone evaporation factor
$\Delta\tau(T_i, R_j)$ pheromone level incrementation.

The incremental amount is high when the fitness of path is high.

$$\Delta\tau(T_i, R_j) = \begin{cases} Q\left(\gamma e^{(-F(x))} + \delta e^{(-B(x))}\right), & (T_i, R_j) \in \text{path}_l \\ 0, & \text{Otherwise.} \end{cases} \tag{12}$$

where Q is a constant and considered as 100.

For every iteration, the pheromone level increments when PER(x) and BUG(x) are getting small. By the pheromone update, the accurate result is enhanced, and the local solution is minimized. Hence, more ants will move toward the optimal path after several iterations, and the local optimum can be prevented using pheromone evaporation factor ρ (Fig. 1).

Here, $\text{iter}_{\max}$ is the maximum no. of iterations, and its value is considered as 100.

4.5 Complexity Analysis

To find out the optimal path, initially the algorithm takes $O(k)$ time complexity. In the optimization judgement process, to meet the performance constraints and cost, the algorithm takes $O(nk)$ time complexity. And the space complexity taken by the algorithm will be $O(1)$.

5 Conclusion

A multi-objective task scheduling optimization model is discussed with the ant colony algorithm. The demand of resource for individual task is estimated using the proposed resource cost optimization model and projects the relation between resource and task as well. Since this algorithm was developed to optimize the scheduling of tasks, preventing solution falling into local optima, this algorithm used the constraint

Fig. 1 Algorithm for multi-objective ant colony optimization scheduling method

Algorithm 1 Multi-Objective Ant Colony Optimization Scheduling Method

Input:
$T_1, T_2, \ldots, T_i, \ldots, T_K, R_1, R_2, \ldots, R_j, \ldots, R_N, iter_{max}$
Output:
map (T_i, R_j)
BEGIN
  Initialize ants distribution among R_j;
  DO
    FOR each ant do
      FOR each T_i do
        Select next route;;
      END FOR
      Evaluate fitness of individual path by formula 10;
      IF r_j meets the optimization problem Then
        Output the map (T_i, R_j);
        Update pheromone along its path by formula 11 and 12;
      END IF
      END FOR
    UNTIL $iter_{max}$
END

functions such as budget and performance to calculate the costs and finding the quality of the solution. This method is performed better than other existing methods when considering individual (single) constraints such as the cost, utilization of resources and rate of deadline violation thus proving the effectiveness.

References

1. Chen Y, Zhang A, Tan Z (2013) Complexity and approximation of single machine scheduling with an operator non-availability period to minimize total completion time. Inf Sci 25(1):150–163
2. Tsai CW, Huang W-C, Chiang M-H, Chiang M-C, Yang C-S (2014) A hyper-heuristic scheduling algorithm for cloud. IEEE Trans Cloud Comput 2(2):236–250
3. Topcuoglu H, Hariri S, Wu M-Y (2002) Performance-effective and lowcomplexity task scheduling for heterogeneous computing. IEEE Trans Parallel Distrib Syst 13(3):260–274
4. Tang Z, Jiang L, Zhou J, Li K, Li K (2015) A self-adaptive scheduling algorithm for reduce start time. Future Gen Comput Syst 4344(3):51–60
5. Shin S, Kim Y, Lee S (2015) Deadline-guaranteed scheduling algorithm with improved resource utilization for cloud computing. In: Proceedings of 12th annual IEEE consumer communication networking conference (CCNC), pp 814–819
6. Zhu X, Yang LT, Chen H, Wang J, Yin S, Liu X (2014) Real-time tasks oriented energy-aware scheduling in virtualized clouds. IEEE Trans Cloud Comput 2(2):168–180
7. Van den Bossche R, Vanmechelen K, Broeckhove J (2011) Costefficient scheduling heuristics for deadline constrained workloads on hybrid clouds. In: Proceedings of IEEE 3rd international conference on cloud computing technology science (CloudCom), pp 320–327

8. Zuo L, Shu L, Dong S, Zhu C, Hara T (2015) A multi-objective optimization scheduling method based on the ant colony algorithm in cloud computing. IEEE Access 3:2687–2699
9. Farahnakian F et al (2015) Using ant colony system to consolidate VMs for green cloud computing. IEEE Trans Serv Comput 8(2):187–198
10. Di S, Wang C-L, Cappello F (2014) Adaptive algorithm for minimizing cloud task length with prediction errors. IEEE Trans Cloud Comput 2(2):194–207
11. Myneni MB, Narasimha Prasad LV, Naveen Kumar D (2017) Intelligent hybrid cloud data hosting services with effective cost and high availability. Int J Electr Comput Eng 7(4):2176–2189

Addressing Security and Computation Challenges in IoT Using Machine Learning

Bhabendu Kumar Mohanta, Utkalika Satapathy, and Debasish Jena

Abstract The Internet of things (IoT) is widely used to implement different applications like smart home, smart health care, smart city, and smart farming system. The development of large smart devices/sensors enables smart technologies making it possible to implement the smart application in real time. The IoT system has security challenges like authentication, data privacy, access control, and intrusion detection system. Similarly, the computation of the sensed information from the environment is a challenging task. The computation must perform using distributed or decentralized architecture to overcome the centralized system difficulty. In a distributed/decentralized system when multiple nodes participate in a computational process, there is the risk of mutual consensus problem, malicious node detection, or data modification attacks. In this paper, the authors have identified machine learning as a solution to address some of the existing security and computational challenges. The paper also explained the implementation platform available for the integration of IoT with machine learning.

Keywords Security · Computation · Machine learning · Secure communication · IoT

1 Introduction

The IoT is one of the main research topics in both academic and industry. The lots of smart devices are manufactured by the industry to sense and act in intelligent way. IoT devices are connected in wireless or through the wire to the network layer

B. K. Mohanta (✉) · U. Satapathy · D. Jena
Information Security Laboratory, IIIT Bhubaneswar, Odisha 751003, India
e-mail: C116004@iiit-bh.ac.in

U. Satapathy
e-mail: a117010@iiit-bh.ac.in

D. Jena
e-mail: debasish@iiit-bh.ac.in

A. K. Tripathy et al. (eds.), *Advances in Distributed Computing and Machine Learning*, Lecture Notes in Networks and Systems 127, https://doi.org/10.1007/978-981-15-4218-3_7

and next to the application layer. Basically, it follows a three-layer architecture. The security vulnerability of smart devices and future challenges are explained in papers [1, 2]. The patient data collection using IoT cloud framework in smart hospital application is one of the applications where privacy of personal user is important [3]. The IoT application needs a security issue to be addressed like the confidentiality and integrity of the user must be maintained which is described by authors in [4]. The basis of any security system is addressing the confidentiality, integrity, and availability. Before developing the security solution of the IoT system, it is essential to know the difference between the traditional security framework and the current system. To ensure the security in IoT framework design and challenges, the potential threads [5]. The IoT needs some enabling technologies like fog computing, software-defined network, to integrate with the model to address the security issue [6, 7]. In a smart home system, lots of home appliances are embedded with smart technology. Those devices are connected to the home network and communicate with the home user through mobile devices. The real-time monitoring security of the system, even it is easy to monitor the fire in case of home from the kitchen or any unusual activity can be detected. Similarly in the case of smart farming, using IoT makes the farmer job easy by sensing the environment and processing this information. The IoT-based smart farming makes the system more efficient and reduces the excess use of material. Because of resource constraint smart devices, different security and privacy issues exist in IoT applications. Some of the issues are already addressed by the research group. Still a lot of work needs to be done to make the system more efficient and more trustful to the end-user.

1.1 Motivation and Organization of the Paper

Internet of things is improving human living standards, accessing home appliances from a remote location, monitoring the environment in real time using IoT techniques. From automatically ordering fruits by the smart fridge to monitoring the soil condition in the agriculture system, IoT system has come a long way. But the security and performing the computation to take a decision are done by fog devices or at the cloud end. The demand is to build a lightweight security protocol that will make the system robust and tamperproof. Some of the work regarding is already done by the research community but till lot more need to be done. The machine learning is one of the intelligent algorithms to make decisions automatically. So in this work, authors motivate to integrate machine learning with IoT to address the security issues. The paper is organized in this way.

Related work of machine learning with IoT is explained in Sect. 2. In Sect. 3, security issues of IoT system are discussed, especially secure computation using multiple nodes. In Sect. 4, machine learning algorithms and related work are already done in the security domain of IoT and are described in detail. Section 6 addresses the implementation platform available and the solution approach for a security issue in IoT in terms of machine learning. Section 5 concludes the paper using the summary of the paper and addressing future work.

2 Related Work

IoT is one of the promising technologies in the recent past. The growth of IoT is exponential, as estimated 50 billion devices are connected to the smart IoT network. It generated a huge volume of data for processing and computing. So, some related technique like fog computing and cloud computing is used to process and store the data generated from the IoT applications. Fog computing provides computation at the edge of the network. The security issues are the most challenging part to implement IoT applications. Machine learning techniques like supervised learning, unsupervised learning, and reinforcement learning are widely used for many domains for classification and regression purposes. The technique such as machine learning and an artificial neural network can address the security issue in IoT application. As shown in Table 1, related research done on IoT security and machine learning is explained.

3 Security Challenges in IoT System

The Internet of things (IoT) reference model is already described in different research works. The general architecture of IoT consists of the physical/perception layer, network layer, transport layer, and application layer [13]. Each layer uses some standard protocol of their respective work. The IoT system needs two things to make the system tamperproof: One is securing smart devices, and the second is preventing different security attacks. Confidentiality, integrity, and availability are the fundamental requirements of any security system. The IoT application carries critical information like personal health records in hospital cases and user privacy in the smart home system. For example, IoT device authentication [14] in a smart home system is important to maintain confidentiality among all smart home users [15]. In all these cases, the authorized user and the unauthorized user must distinguish by the network. The information must properly share among users which means proper access control must be designed. Similarly, in the case of integrity, the applications must provide a reliable service to the end-user which means sender and receiver must be legitimate node. To monitor the smart environment devices are must available and connected to the others devices through internet. Some of the security issue and attack cases in IoT are:

- Denial-of-service attacks
- Eavesdropping

Table 1 Related work on security issue in IoT and machine learning

Papers	Year	Contribution
Canedo and Skjellum [8]	2016	Machine learning is used in IoT gateway to secure the devices. The artificial neural network (ANN) is used to detect the anomalies present in the edge of the IoT application. The paper explained the IoT scenario where the temperature sensor was used to collect data and ANN technique is used to predict the anomalies present in the network. By doing so, some of the security issues of IoT applications are solved
Xiao et al. [9]	2018	In this article, machine learning techniques such as supervised, unsupervised, and reinforcement learning are used to address the security issues in the IoT system. The privacy issues like secure authentication, data protection, malware detection, and secure offloading are investigated by the authors in this work. Finally, the paper also explained the challenges of implementing machine learning in IoT applications
Fernandez Molanes et al. [10]	2018	The safety and security of IoT are explained in this paper using deep learning and neural network. The integration of machine learning and artificial network addressed the security issue in the IoT platform
Hussain et al. [11]	2019	In this paper, in-depth layer-wise (physical, network, and application) security analysis is done by the authors. The different techniques and algorithms used for IoT security are explained in detail in this paper
Zantalis et al. [12]	2019	The IoT and machine learning are integral to solve many security issues in IoT application. A lot of algorithms are designed and implemented to address different security issues in the IoT system. In this paper, an intelligent transportation system is considered as the application and explained the security issue associated with this infrastructure how machine learning can solve these issues

- Spoofing
- Privacy leakage
- Side channel
- Impersonation
- Sybil attack
- Routing information
- Man-in-the-middle attack
- Repudiation
- Distributed attack
- SQL injection
- Cross-site scripting
- Malicious script
- Phishing
- Access control

- Jamming attack
- Physical attack.

The computing techniques associated with IoT applications are fog computing, edge computing, and cloud computing. The authors in [16] explained all the computing technique integration with IoT. The communication technologies like nearest field communication (NFC), radio frequency identification (RFID), Zigbee, WiFi, 6LoWPAN, Z-Wave, Sigfox, LoRaWAN challenges, and security attacks are possible with these communication protocols.

4 Machine Learning for IoT

The IoT application needs to compute the collected information in real time. The IoT cloud-based infrastructure case processing and computation are done on cloud computing which also provides storage and security. But using cloud computing has some issues like latency, bandwidth, and delay. The critical application like detecting road surface [17] and fire detection in the smart home requiring the computation and processing must be done in a quick time, and the corresponding event is a trigger. The recent development of fog computing and edge computing provides computation and processing in real time, as the fog computing or edge computing devices are low-end, less storage space capability. The computation is done using multiple fog/edge nodes present in the network. During computation, multiple nodes perform computation in a distributed way. There are several security issues and attack possible as an intermediate node, as well as sensor node, is vulnerable to the outside world. The machine learning has different learnings like supervised learning, unsupervised learning, and reinforcement learning. Figure 1 shows IoT infrastructure and machine

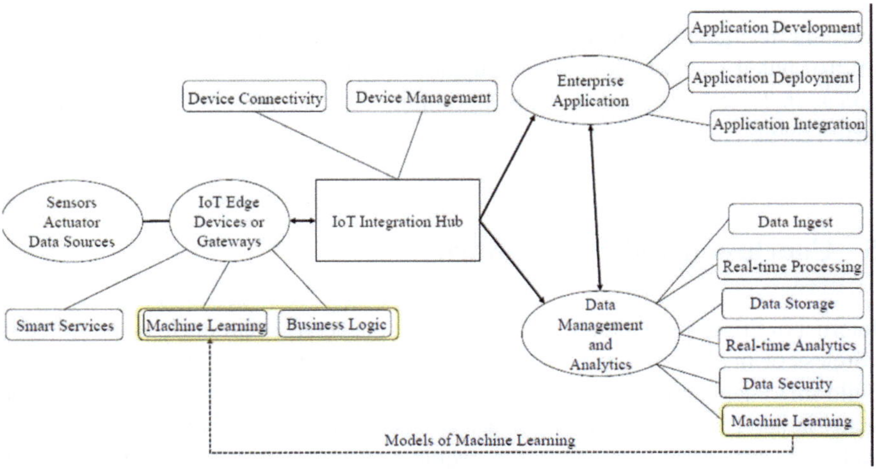

Fig. 1 Basic functionality in IoT and machine learning

Table 2 Machine learning addresses IoT security

Security threats	ML solutions
Denial-of-service (DoS) attack	• Multilayer perceptron (MLP)-based protocol [18] • Particle swarm optimization and back propagation algorithm [19] • Deep learning with dense random neural network (RNN) [31]
Eavesdropping	• Q-learning-based offloading strategy [20] • Nonparametric Bayesian technique [21] • Q-learning and Dyna-Q technique [22]
Spoofing	• Q-learning and Dyna-Q technique [22] • Support vector machines (SVMs) [23] • Incremental aggregated gradient (IAG) [24] • Distributed Frank–Wolfe (dFW) technique [25]
Privacy leakage	• Privacy-preserving scientific computations (PPSC) [26] • Commodity integrity detection algorithm (CIDA) [27]
Digital fingerprinting	• Support vector machines [28] • Artificial neural network [29]
Real-time distributed attack	• ELM-based semi-supervised fuzzy C-means (ESFCM) method [30]
Network attack (TCP SYN attack)	• Deep learning with dense random neural network (RNN) [31] • A darknet traffic analysis using association rule learning [32]

learning functionality. The multiple edge devices can perform computation mutually in distributed way which give computation results reliable and build trust among nodes. In Table 2, machine learning algorithm which addresses the security issue in IoT is mentioned.

5 Conclusion and Future Work

Internet of things is an interconnection of heterogeneous smart devices connected in wireless or in a wired way. The development of smart devices made it possible to build smart applications like smart home systems, smart cities, smart agriculture, and many more. The application like smart patient monitoring and earthquake detection and even smart fire detection systems requires the collected data computation faster and accurate. The IoT system is vulnerable to different types of security attacks. In this paper, the authors have identified security challenges and their corresponding imparts of the IoT system. Then, the authors reviewed some of the work already that suggests the integration of machine learning with IoT. Finally, the paper explained that machine learning can address some of the security challenges including secure computation in an IoT network. In the future, we want to take smart home as an IoT-based application to implement the smart environment and apply the machine learning algorithm which can perform the computation in distributed way like blockchain technology and give reliable solution.

References

1. Liu X, Zhao M, Li S, Zhang F, Trappe W (2017) A security framework for the internet of things in the future internet architecture. Future Internet 9(3):27
2. Andrea I, Chrysostomou C, Hadjichristofi G (2015) Internet of things: security vulnerabilities and challenges. In: 2015 IEEE symposium on computers and communication (ISCC), pp 180–187. IEEE
3. Jaiswal K, Sobhanayak S, Mohanta BK, Jena D (2017) IoT-cloud based framework for patient's data collection in smart healthcare system using raspberry-pi. In: 2017 International conference on electrical and computing technologies and applications (ICECTA), pp 1–4. IEEE
4. Hassan WH (2019) Current research on Internet of things (iot) security: a survey. Comput Netw 148:283–294
5. Ammar M, Russello G, Crispo B (2018) Internet of things: a survey on the security of IoT frameworks. J Inf Sec Appl 38:8–27
6. Salman O, Elhajj I, Chehab A, Kayssi A (2018) IoT survey: an SDN and fog computing perspective. Comput Netw 143:221–246
7. Čolaković, A., & Hadžialić, M. (2018) Internet of Things (IoT): A review of enabling technologies, challenges, and open research issues. Computer Networks 144:17–39
8. Canedo J, Skjellum A (2016) Using machine learning to secure IoT systems. In: 2016 14th annual conference on privacy, security and trust (PST), pp 219–222. IEEE
9. Xiao L, Wan X, Lu X, Zhang Y, Wu D (2018) IoT security techniques based on machine learning. arXiv:1801.06275
10. Fernandez Molanes R, Amarasinghe K, Rodriguez-Andina J, Manic M (2018) Deep learning and reconfigurable platforms in the internet of things: challenges and opportunities in algorithms and hardware. IEEE Ind Electron Mag 12(2)
11. Hussain F, Hussain R, Hassan SA, Hossain E (2019) Machine learning in IoT security: current solutions and future challenges. arXiv:1904.05735
12. Zantalis F, Koulouras G, Karabetsos S, Kandris D (2019) A review of machine learning and IoT in smart transportation. Future Internet 11(4):94
13. Satapathy U, Mohanta BK, Panda SS, Sobhanayak S, Jena D (2019) A secure framework for communication in internet of things application using hyperledger based blockchain. In: 2019 10th international conference on computing, communication and networking technologies (ICCCNT), pp. 1–7. IEEE
14. Mohanta BK, Sahoo A, Patel S, Panda SS, Jena D, Gountia D (2019) DecAuth: decentralized authentication scheme for IoT device using ethereum blockchain. In: TENCON 2019-2019 IEEE region 10 conference (TENCON), pp 558–563. IEEE
15. Satapathy U, Mohanta BK, Jena D, Sobhanayak S (2018) An ECC based lightweight authentication protocol for mobile phone in smart home. In: 2018 IEEE 13th international conference on industrial and information systems (ICIIS), pp 303–308. IEEE

16. Santhadevi D, Janet B (2018) Security challenges in computing system, communication technology and protocols in IoT system. In: 2018 international conference on circuits and systems in digital enterprise technology (ICCSDET), pp 1–7. IEEE
17. Dey MR, Satapathy U, Bhanse P, Mohanta BK, Jena D (2019) MagTrack: detecting road surface condition using smartphone sensors and machine learning. In: TENCON 2019-2019 IEEE region 10 conference (TENCON), pp 2485–2489. IEEE
18. Pavani K, Damodaram A (2013) Intrusion detection using MLP for MANETs, pp 440–444
19. Kulkarni RV, Venayagamoorthy GK (2009) Neural network based secure media access control protocol for wireless sensor networks. In: 2009 international joint conference on neural networks, pp 1680–1687. IEEE
20. Xiao L, Xie C, Chen T, Dai H, Poor HV (2016) A mobile offloading game against smart attacks. IEEE Access 4:2281–2291
21. Xiao L, Yan Q, Lou W, Chen G, Hou YT (2013) Proximity-based security techniques for mobile users in wireless networks. IEEE Trans Inf Foren Sec 8(12):2089–2100
22. Xiao L, Li Y, Han G, Liu G, Zhuang W (2016) PHY-layer spoofing detection with reinforcement learning in wireless networks. IEEE Trans Veh Technol 65(12):10037–10047
23. Ozay M, Esnaola I, Vural FTY, Kulkarni SR, Poor HV (2015) Machine learning methods for attack detection in the smart grid. IEEE Trans Neural Netw Learn Syst 27(8):1773–1786
24. Xiao L, Wan X, Lu X, Zhang Y, Wu D (2018) IoT security techniques based on machine learning: how do IoT devices use AI to enhance security? IEEE Signal Process Mag 35(5):41–49
25. Xiao L, Wan X, Han Z (2017) PHY-layer authentication with multiple landmarks with reduced overhead. IEEE Trans Wirel Commun 17(3):1676–1687
26. Yan Z, Zhang P, Vasilakos AV (2014) A survey on trust management for internet of things. J Netw Comput Appl 42:120–134
27. Li C, Wang G (2012) A light-weight commodity integrity detection algorithm based on Chinese remainder theorem. In: 2012 IEEE 11th international conference on trust, security and privacy in computing and communications, pp 1018–1023. IEEE
28. Hassija V, Chamola V, Saxena V, Jain D, Goyal P, Sikdar B (2019) A survey on IoT security: application areas, security threats, and solution architectures. IEEE Access 7:82721–82743
29. Oulhiq R, Ibntahir S, Sebgui M, Guennoun Z (2015) A fingerprint recognition framework using artificial neural network. In: 2015 10th international conference on intelligent systems: theories and applications (SITA), pp 1–6. IEEE
30. Rathore S, Park JH (2018) Semi-supervised learning based distributed attack detection framework for IoT. Appl Soft Comput 72:79–89
31. Brun O, Yin Y, Gelenbe E (2018) Deep learning with dense random neural network for detecting attacks against iot-connected home environments. Proc Comput Sci 134:458–463
32. Hashimoto N, Ozawa S, Ban T, Nakazato J, Shimamura J (2018) A darknet traffic analysis for IoT malwares using association rule learning. Proc Comput Sci 144:118–123

A Survey: Security Issues and Challenges in Internet of Things

Balaji Yedle, Gunjan Shrivastava, Arun Kumar, Alekha Kumar Mishra, and Tapas Kumar Mishra

Abstract Internet of Things (IoT) is an emerging technology, in which sensible objects are connected to form a network and are in constant touch with one another. Today's IoT applications are developed using different architecture approaches like centralized, distributed, and cooperative approaches. In the IoT applications, security is a main concern which needs to be studied thoroughly and deduce feasible solutions. This paper presents a brief survey of existing security challenges at different layers of IoT protocols and the initial simulation results of the work.

Keywords Internet of things · Protocols · Security · IoT architecture

1 Introduction

In last decade, IoT applications have increased in fast pace and IoT is providing trendy development in radio-frequency identification (RFID), sensible sensors, communication technologies, and web protocols. The objective of IoT is to form the network of billions of wireless identifiable objects, which can communicate with each other

B. Yedle · G. Shrivastava · A. Kumar (✉) · T. K. Mishra
Department of Computer Science and Engineering, National Institute of Technology, Rourkela, Rourkela, India
e-mail: kumararun@nitrkl.ac.in

B. Yedle
e-mail: 218CS2436@nitrkl.ac.in

G. Shrivastava
e-mail: 218CS2316@nitrkl.ac.in

T. K. Mishra
e-mail: mishrat@nitrkl.ac.in

A. K. Mishra
Department of Computer Applications, National Institute of Technology, Jamshedpur, Jamshedpur, India
e-mail: alekha.ca@nitjsr.ac.in

A. K. Tripathy et al. (eds.), *Advances in Distributed Computing and Machine Learning*, Lecture Notes in Networks and Systems 127, https://doi.org/10.1007/978-981-15-4218-3_8

anywhere, anytime using any service [1]. The communication is not only restricted to devices, but it can also happen between people and their environment. IoT is a technology which enables new applications by connecting data acquisition devices together in support of intelligent decision making. The devices are embedded with sensors and actuators to collect surrounding information. The collected raw data is then processed by smart devices, and the decisions are made accordingly [2]. The popular IoT applications include transportation, health care, industrial automation, and animal tracking.

IoT devices communicate using different channels like Bluetooth, Wi-Fi, RFID, NFC, etc. In IoT, conventional protocols of different layers cannot be used due to limited memory space and battery lifetime; therefore, several protocols are proposed for different layers. As a large number of uniquely identifiable devices, IoT devices are interacting with one another, and it creates complex network where large amount of data is exchanged [3]. Therefore, it increases the risk of several potential attacks. So, there must be proper security infrastructure with new systems and protocols which can bound the threats related to confidentiality, integrity, and availability [4].

The remainder of this paper is organized as follows: Sect. 2 describes the three main approaches of the IoT architecture. The protocols used at different layers of IoT are presented in Sect. 3. Section 4 briefly discusses the security goals in IoT, and various security threats at each layer are presented in Sect. 5. Section 6 verifies the proposed authentication protocol, and finally, Sect. 7 concludes the paper.

2 IoT Architecture

The IoT approaches are based on two principles: edge intelligence and collaboration. The authors [5] have discussed different approaches which are used to build IoT network. The approaches are depicted in Fig. 1.

- Centralized IoT: This approach does not provide any of the two mentioned principles. In this approach, the network of things is passive, all the tasks, such as retrieval of data, processing data, combining, and providing it to the users, is done by a single central entity. If a user wants to use the services, it has to connect to the central entity.
- Collaborative IoT: In this approach, the main load is within the central entity, the only difference is with the collaboration principle. Therefore, various central entities are used to exchange data. The users combine various service providers to complete a given task.
- Connected Intranets of Things: In this approach, the data acquisition networks process information locally and provide it to central entities as well as to local and remote end users. Due to the limitation of these entities, the information is mainly processed by central authority. However, if the central authority fails to provide information, the local information can still be accessed.

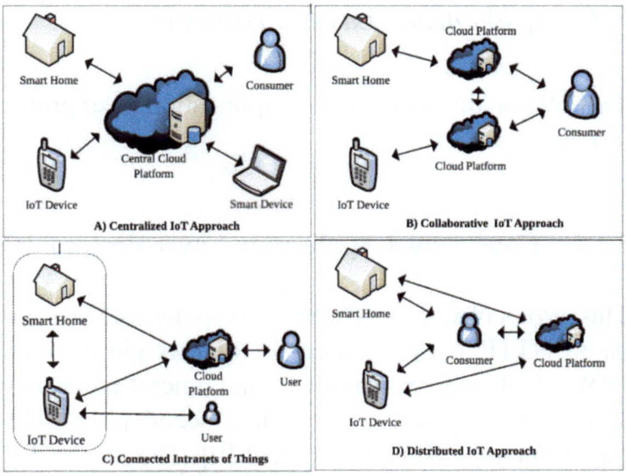

Fig. 1 Overview of different IoT approaches [6]

- Distributed IoT: In distributed IoT approach, physical objects process, retrieve, combine, and provide information and services to other entities. In this approach, nodes are distributed geographically. Nodes collaborate with other nodes to form an IoT network to provide real-time application to user.

3 IoT Protocols

The different protocols at different layers are discussed below, and the protocol stack is shown in Fig. 2.

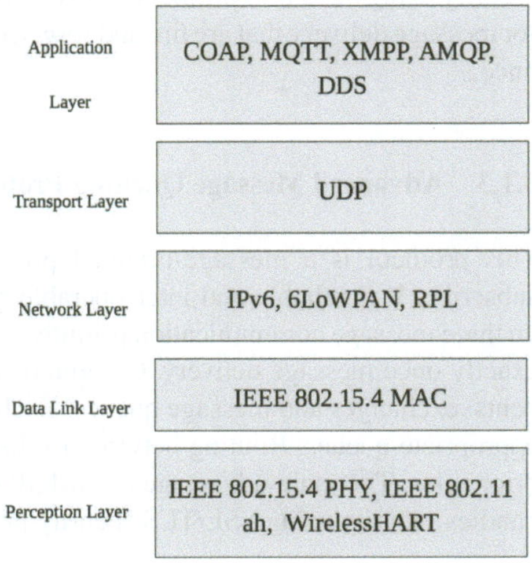

Fig. 2 IoT protocol stack [7]

3.1 Application Layer Protocols

This section discusses various application layer protocols which are preferably used in IoT.

3.1.1 Constrained Application Layer Protocol (CoAP)

This protocol is a synchronous response/request application layer protocol similar to HTTP. CoAP is used for lightweight devices [8]. This protocol works on UDP which supports multicast and unicast requests with low header overhead and less complexity and also provides security with the help of Datagram Transport Layer Security (DTLS) protocol [9]. The protocol stack of CoAP is divided into two parts: Request/Response layer and Message layer. Message layer is to exchange messages between two end users over UDP [10]. Request/Response layer is responsible to store request/response method code which helps to check if the messages are arrived in order. The different types of messages used in CoAP are conformable, non-conformable, acknowledgment, and reset.

3.1.2 Message Queue Telemetry Transport (MQTT)

MQTT protocol is for lightweight devices and constrained network. This protocol works on publish/subscribe mechanism. MQTT protocol is divided into three components: publisher, broker (server), and subscriber. Each user can act as a publisher by registering to the broker as well as a subscriber by subscribing to a topic. The publisher is generator of data, and it publishes specific topics and sends to the subscribed users through the brokers. MQTT protocol works on TCP and provides the security using SSL/TLS protocol. MQTT provides three levels of Quality of Service for message delivery that are fire and forget, deliver at least once, and delivery exactly once.

3.1.3 Advanced Message Queuing Protocol (AMQP)

This protocol is a message-oriented protocol in IoT. AMQP supports publish/subscribe. It is reliable and inter-operable protocol which works on TCP. It works on three message communication primitives such as at-most-once, at-least-once, and exactly once message delivery. Communications are handled by two main components: exchanges and message queues. Exchanges are used to route the messages to appropriate queues. Routing between exchanges and message queues are based on prior rules. The main advantage of AMQP is its store and forward feature. AMQP handles security using SSL/TLS security protocol.

3.2 *Network Layer Protocols*

Network layer is divided into routing layer and encapsulation layer. The role of routing layer is to transfer packets between source and destination, and encapsulation layer helps in forming these packets.

3.2.1 IPv6

With the growing number of devices in the IoT network and limited size of IPv4, the addressing scheme has been shifted to IPv6. IPv6 addresses are 128-bit-long fixed length address given to every device. Since in IPv4 the number of addresses is limited, NAT is used to map multiple devices with the same IP address. Therefore, these devices can easily access Internet but they cannot be accessed through Internet. Ipv6 helps in solving NAT issue and is more suitable in IoT environment. It provide multicast communication in contrast to IPv4 which only supports broadcast communication, which saves a lot of battery usage and hence less power is consumed.

3.2.2 6LoWPAN

6LoWPAN is referred as IPv6 over low-power wireless personal area network. This protocol helps in transmission of IP version 6 packets consuming low power. The protocol aims to connect the entire system at low data rates. It supports several IPv6 operations and specifications such as mapping IPv6 addresses with the system identifiers and identifying the neighboring devices. Security at this layer can be imposed using IPSec [11].

3.2.3 RPL

RPL stands for routing protocol for low-power and lossy networks. It works on distance vectors and creates destination oriented directed acyclic graph (DODAG). The authors [12] have briefly described how RPL is beneficial for IoT. RPL protocols provide three types of communication: point-to-multipoint, multipoint-to-point, and point-to-point [13]. The IoT environment is dynamic so routing protocols should fulfill several requirements, such as providing routing topology to the moving objects [14]. The authors [15] have discussed several location-based routing protocols which can be used in dynamic IoT environment.

3.3 Perception Layer Protocols

This section discusses physical and MAC layer protocols. A novel MAC layer protocol is briefly discussed in [16] for safely broadcasting messages.

3.3.1 IEEE 802.15.4

IEEE 802.15.4 is a protocol for lightweight devices and personal area network. This protocol has a fixed-frame format [17]. The protocol works as the base for many other standard technologies. This protocol provides security, encryption, and authentication [18]. IEEE 802.15.4 contains two types of network nodes, full-function devices (FFD) and reduced function devices (RFD) [19]. The hardware in IEEE 802.15.4 is supported by symmetric cryptography such as Advanced Encryption Standard (AES). This protocol supports several security modes at MAC layer such as AES-CTR, AES-CBC-MAC-32, and AES-CCM-32. There are several limitations of this protocol such as unbounded delay, limited communication reliability, and no protection against interference.

4 Security Goals in IoT

The security goals that should be considered in IoT environment are as follows:

- **Confidentiality**: Confidentiality means to prevent sensitive information from being accessed by unauthorized users. There are several ways to provide confidentiality such as data encryption, managing data access, and authentication of the user [20].
- **Integrity**: Data integrity means to maintain accuracy, consistency, and trustworthiness of information. Data should not be altered during communication such as modification of data by a third party or affected due to other factors that are human uncontrolled including server crash [21].
- **Availability**: Data availability means information should be available to users whenever required. It ensures immediate access of information to authorized users.

5 Security Challenges in IoT

The categorization of security threats at different levels is discussed below.

5.1 Perception Layer Challenges

Perception layer is also called sensor layer because it consists of different sensors like RFID, 2D bar codes, and sensor networks, which act as data acquisition devices [22]. The threats of perception layer are discussed below.

- Eavesdropping: In this attack, the attacker can steal information such as passwords by continuously listening the communication channel.
- Tag Cloning: In this attack, the attacker copies the tag of legitimate node, so user cannot differentiate between real and compromised node.
- Spoofing: In this attack, the attacker broadcasts fake information to the nodes and make them believe it is coming from real source.
- RF Jamming: In this attack, the attacker attacks on wireless devices by disrupting the network with excess noise signals.

5.2 Network Layer Challenges

The purpose of this layer is to connect smart devices, devices in network, and two or more networks to each other [22]. Network layer threats are discussed below:

- Denial of Service (DOS): In this attack, attacker prevents authentic users from accessing physical objects or network resources. The attacker floods the network with unnecessary requests, thus increasing the network traffic and exhausting the resources.
- Man-in-the-middle Attack: In this attack, attacker secretly intercepts the communication between sender and receiver and spoofs both the ends, convincing them that they are communicating with each other directly. Attacker gains full control over communication and can alter messages based on requirements.
- Sleep Deprivation Attack: Since the sensor nodes are powered with batteries, they follow a routine to sleep in between to extend battery life. So in this attack, attacker keeps the sensor nodes awake and reduces the battery lifetime which causes shutdown [4].

5.3 Application Layer Challenges

This layer provides services to the user so that both parties can communicate with each other efficiently. In this layer, the attacker mainly attacks on privacy of each user. Application layer threats are discussed below:

- Malicious Code Injection: It is a piece of code which attacker injects into the system to steal the information. It also causes damage to the system and cannot be controlled or blocked by the anti-virus tools.

- Spear-Phishing Attack: In this attack, attacker sends personalized message which contains malicious link of malware to a particular victim.
- Sniffing Attack: In this attack, attacker forces sniffer application to intercept the network traffic. If any data packets are transmitted without encryption, it can be easily read by attacker.
- Privacy Leak: Since most of the IoT applications are executed on the common operating system, there are chances that attacker might steal users' data.

6 Simulations and Results

There are several security requirements which should be kept in mind before deploying the protocol in IoT network. Some of which are data privacy, trust management, authentication, identity management, and access control. The authors [23] have proposed a method to register a new device in the IoT environment by authenticating it first with the server. In the proposed protocol, there are four entities: device, controller, authenticating server, and manufacturing server as shown in Fig. 3. Since the devices added to the IoT environment are lightweight, it does not have much resources; hence, this protocol uses NFC as a channel to communicate with the controller. Controller, authenticating server, and manufacturing server communicate with each other using Internet. Since the device added to the IoT environment is lightweight, it does not have much resources; hence, this protocol uses NFC as a channel to communicate with the controller. NFC is a secure communication channel which works well upto a limited range. Controller, authenticating server, and manufacturing server communicate with each other using Internet. The protocol works with some assumptions; they are as follows:

- There is no trust relationship between device and the controller until the device is registered.
- There is no trust relationship between authenticating and manufacturing server until the certificate verification.
- The controller and the authenticating server trust each other from advance and also share a secret key, SK_{cs}.
- The device and the manufacturing server have an initial key, IK_D (Table 1).

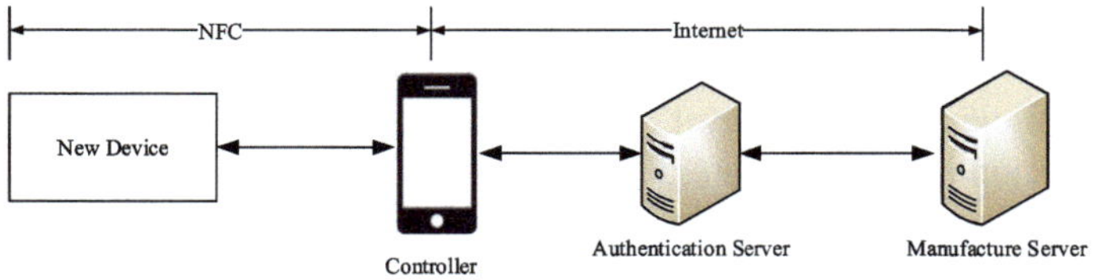

Fig. 3 Proposed system model

Table 1 Abbreviations used in the protocol

Notations	Description
ID_d	Identifier of device
IK_d	Initial key for device
FW_d	Firmware of device
PSK	Preshared key between device and authentication server
RN_x	Nonce of entity 'x'
SK_{cs}	Secret key between controller and authentication server
TID	Transaction identifier of device
TS_i	Timestamp
H()	Hash function

Before any process steps begin, each entity has some prior knowledge. The new device has knowledge about the initial key, IK_d, and the controller has manufacturing server port, hash function $H(FW_d||ID)$, and IP details which are obtained by scanning the QR code attached with the device by the manufacturing server. The controller and the authenticating server also share a secret key, SK_{cs}. The manufacturing server has prior knowledge about the initial key, IK_d. The operational steps of the proposed protocol are depicted in Fig. 4. In this paper, the proposed protocol is verified using scyther tool. To verify the protocol, Scyther tool has several security claims or attributes. If the security claims are satisfied, the status of each claim is displayed OK; otherwise it displays fail. In Fig. 5, the status of each security attributes

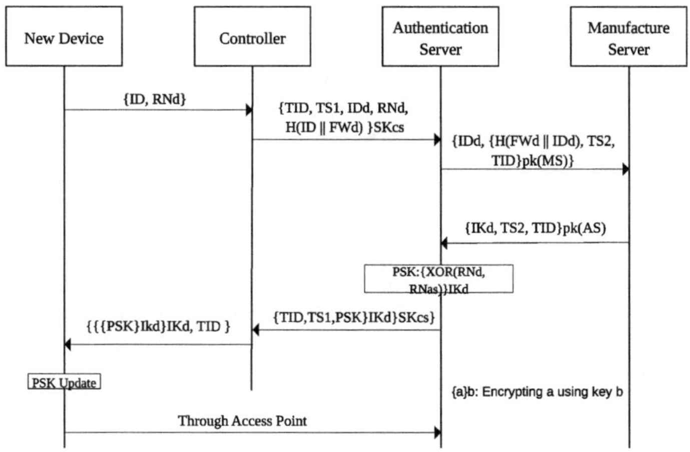

Fig. 4 Proposed operation model

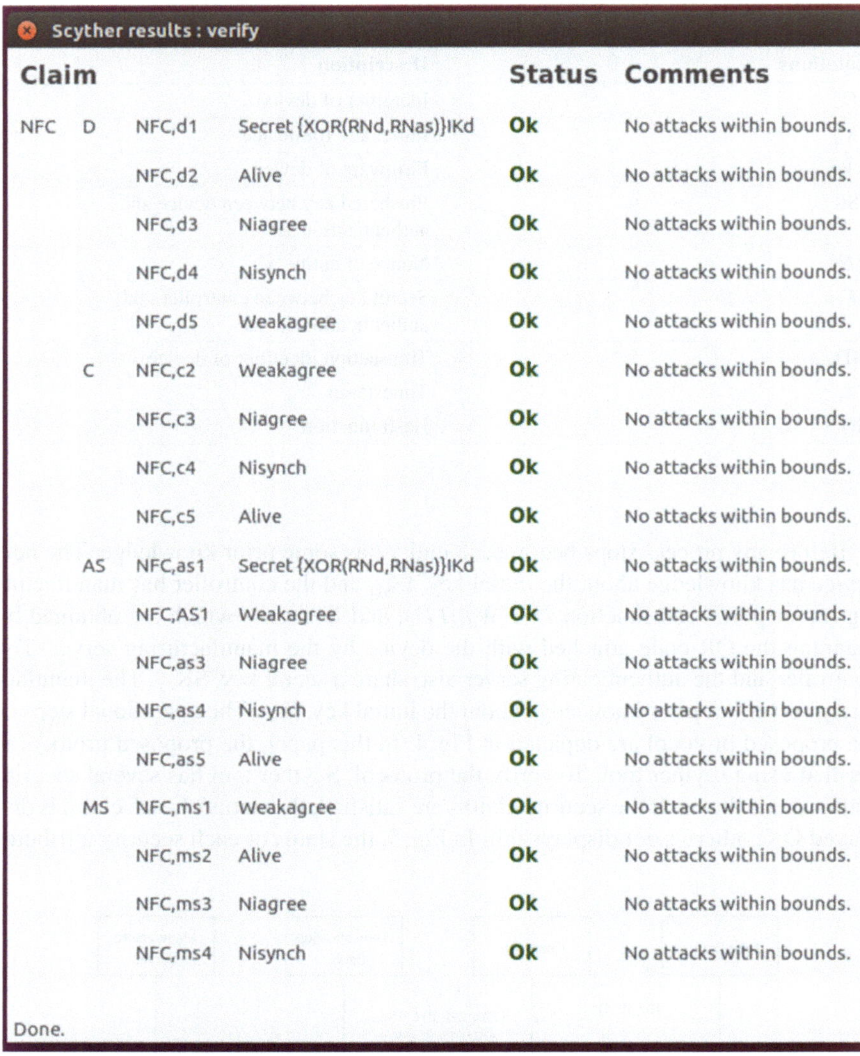

Claim				Status	Comments
NFC	D	NFC,d1	Secret {XOR(RNd,RNas)}IKd	Ok	No attacks within bounds.
		NFC,d2	Alive	Ok	No attacks within bounds.
		NFC,d3	Niagree	Ok	No attacks within bounds.
		NFC,d4	Nisynch	Ok	No attacks within bounds.
		NFC,d5	Weakagree	Ok	No attacks within bounds.
	C	NFC,c2	Weakagree	Ok	No attacks within bounds.
		NFC,c3	Niagree	Ok	No attacks within bounds.
		NFC,c4	Nisynch	Ok	No attacks within bounds.
		NFC,c5	Alive	Ok	No attacks within bounds.
	AS	NFC,as1	Secret {XOR(RNd,RNas)}IKd	Ok	No attacks within bounds.
		NFC,AS1	Weakagree	Ok	No attacks within bounds.
		NFC,as3	Niagree	Ok	No attacks within bounds.
		NFC,as4	Nisynch	Ok	No attacks within bounds.
		NFC,as5	Alive	Ok	No attacks within bounds.
	MS	NFC,ms1	Weakagree	Ok	No attacks within bounds.
		NFC,ms2	Alive	Ok	No attacks within bounds.
		NFC,ms3	Niagree	Ok	No attacks within bounds.
		NFC,ms4	Nisynch	Ok	No attacks within bounds.

Done.

Fig. 5 Verification result using Scyther tool

like Alive, Weakagree, Niagree, and Nisynch is OK. Therefore,the proposed method is safe from all attacks like man-in-the-middle attack, replay attack, and impersonation attack. Also it shows that intial key (IK_d) of new device and preshared key (PSK) is secure, and it will not be intercepted by any attacker.

7 Conclusion

This paper has presented a brief survey of existing security challenges in IoT and the initial results of the work. The different architectures of the IoT infrastructure like centralized, collaborative, and connected Intranets are discussed in detail. The paper also discusses the security goals like confidentiality, availability, and integrity. At last, the paper presents the security issues at each layer of IoT protocols and initial simulation results. In future, we plan to simulate different security threats at each layer of the protocols and analyze the obtained results. Also, solutions to these security threats will be presented.

References

1. Yaqoob I, Ahmed E, Hashem IAT, Ahmed AIA, Gani A, Imran M, Guizani Mohsen (2017) Internet of things architecture: recent advances, taxonomy, requirements, and open challenges. IEEE Wirel Commun 24(3):10–16
2. Hossain M, Riazul Islam SM, Ali F, Kwak K-S, Hasan R (2018) An internet of things-based health prescription assistant and its security system design. Future Gener Comput Syst 82:422–439
3. Chasaki D, Mansour C (2015) Security challenges in the internet of things. Int J Space Situated Compu 5(3):141–149
4. Farooq Muhammad MU, Anjum W, Sadia Mazhar K (2015) A critical analysis on the security concerns of internet of things (IoT). Int J Comput Appl (0975 8887) 111(7)
5. Riahi A, Natalizio E, Challal Y, Mitton N, Iera A (2014) A systemic and cognitive approach for IoT security. In: 2014 International conference on computing, networking and communications (ICNC), pp 183–188. IEEE
6. Roman R, Zhou J, Lopez J (2013) On the features and challenges of security and privacy in distributed internet of things. Comput Netw 57(10):2266–2279
7. Makkad A (2017) A survey on application layer protocols for internet of things (IoT). Int J Adv Res Comput Sci 8(3)
8. Abdul Rahman R, Shah B (2016) Security analysis of IoT protocols: a focus in coap. In: 2016 3rd MEC international conference on big data and smart city (ICBDSC), pp 1–7. IEEE
9. Makkad A (2017) Security in application layer protocols for IoT: a focus on coap. Int J Adv Res Comput Sci 8(5)
10. Zhao M, Kumar A, Ristaniemi T, Chong PHJ (2017) Machine-to-machine communication and research challenges: a survey. Wirel Pers Commun 97(3):3569–3585
11. Granjal J, Monteiro E, Silva J (2015) Security for the internet of things: a survey of existing protocols and open research issues. IEEE Commun Surv Tutor 17(3):1294–1312
12. Iova O, Picco P, Istomin T, Kiraly C (2016) Rpl: the routing standard for the internet of things... or is it? IEEE Commun Mag 54(12):16–22
13. Zhao M, Kumar A, Chong PHJ, Rongxing L (2017) A comprehensive study of rpl and p2p-rpl routing protocols: implementation, challenges and opportunities. Peer-to-Peer Netw Appl 10(5):1232–1256
14. Zhao M, Kumar A, Chong PHJ, Rongxing L (2016) A reliable and energy-efficient opportunistic routing protocol for dense lossy networks. IEEE Wirel Commun Lett 6(1):26–29
15. Kumar A, Yu Shwe H, Juan Wong K, Chong PH (2017) Location-based routing protocols for wireless sensor networks: a survey. Wirel Sensor Netw 9(1):25–72
16. Zhang M, Md Nawaz Ali GG, Joo Chong PH, Seet B-C, Kumar A (2019) A novel hybrid mac protocol for basic safety message broadcasting in vehicular networks. IEEE Trans Intell Transp Syst

17. Tara S, Raj J (2015) Networking protocols and standards for internet of things. Internet Things Data Anal Handb 7
18. Al-Fuqaha A, Guizani M, Mohammadi M, Aledhari M, Ayyash M (2015) Internet of things: a survey on enabling technologies, protocols, and applications. IEEE Commun Surv Tutor 17(4):2347–2376
19. Kumar A, Zhao M, Wong K-J, Liang Guan Y, Joo Chong PH (2018) A comprehensive study of IoT and wsn mac protocols: research issues, challenges and opportunities. IEEE Access 6:76228–76262
20. Miorandi D, Sicari S, De Pellegrini F, Chlamtac I (2012) Internet of things: vision, applications and research challenges. Ad hoc Netw 10(7):1497–1516
21. Atzori L, Iera A, Morabito G (2010) The internet of things: a survey. Comput Netw 54(15):2787–2805
22. Burhan M, Rehman R, Khan B, Kim B-S (2018) Iot elements, layered architectures and security issues: a comprehensive survey. Sensors 18(9):2796
23. Kang N, Kim J (2018) Entity authentication and secure registration for lightweight devices in internet of things. Int J Control Autom 11(5):37–48

Low-Cost Real-Time Implementation of Malicious Packet Dropping Detection in Agricultural IoT Platform

J. Sebastian Terence and Geethanjali Purushothaman

Abstract Internet of Things (IoT) enables to connect various devices to Internet. It also gives access to various devices from remote place at anytime. IoT applied in various areas such as smart city, health care, agriculture, waste management and food supply. A major drawback of IoT is lack of protection against security issues. One of the security problems in wireless network is packet dropping attacks. In packet dropping attacks, malicious node drops data packet intensively to disturb the network traffic. We studied different agricultural IoT systems and found that most the systems are defenseless against malicious packet dropping attack. In this paper, we proposed novel technique to detect malicious packet dropping attack in IoT platform. The proposed technique is implemented in real-time agriculture application with low-cost IoT devices. The result shows that the proposed technique is able to detect malicious packet dropping effectively with less false positive and false negative. Also it helps to increase the packet delivery rate and throughput of the network.

Keywords Internet of Things · Smart agriculture · Packet dropping attacks · Malicious node detection

1 Introduction

Internet of Things (IoT) is one of the emerging trends in scientific and research society. IoT is applied in various real-world applications. Through improved operational efficiencies as well as new revenue-creating products and services, by 2025, the potential economic impact of the IoT is estimated to be $2.7 to $6.2 trillion per year [1]. IoT is considered as growth of Internet techniques that include wireless sensor network, smart objects, actuators, gateways, etc. IoT enables objects and things to be smarter, so that these devices can be accessed from anywhere and anytime. IoT

J. S. Terence
Department of CSE, Karunya Institute of Technology and Sciences, Coimbatore, India

G. Purushothaman (✉)
School of Electrical Engineering, VIT, Vellore, India
e-mail: pgeethanjali@vit.ac.in

A. K. Tripathy et al. (eds.), *Advances in Distributed Computing and Machine Learning*, Lecture Notes in Networks and Systems 127, https://doi.org/10.1007/978-981-15-4218-3_9

is used multiple application such as agriculture, smart city, smart home, industrial application and food supply [2].

Even though IoT is used in multiple application, it has security limitations. Various researches reveal that IoT applications are suffered due to numerous security issues [3]. Some traditional wireless network such as mobile ad hoc network, sensor networks is suffered due to many attacks such as blackhole attack (packet dropping attack), grayhole attack (selective packet dropping attack), sinkhole attack, warmhole attack, selfish node, etc. One of the most dangerous attack is the packet dropping attack, in this attack, wireless node drops all packets or selective data packets. This kind of packet drops has happened due to the malicious activity of adversaries. The adversaries compromise wireless node with false data then wireless node start packet dropping to disturb the network flow [4].

Since IoT also comprises various wireless devices such as sensors and gateway, there are a lot of possibilities of launching packet dropping attacks on IoT [5]. In this paper, we launched packet dropping attacks in IoT platform and proposed novel technique to detect the packet dropping attacks in IoT platform. To implement and identify the packet dropping attacks in IoT, we used agricultural IoT system. The soil moisture of various plants is observed through soil moisture sensors. The sensor sends data to the gateway. Gateway is a powerful device which collects data from many sensors. The gateway sends collected data to the cloud platform through the Internet. From the cloud, users can access data from anywhere and anytime. In this process, we compromised one of the gateways and compromised gateway drops' data packet selectively. Our objective is to detect the malicious gateway which drops data packets by a novel packet dropping detection algorithm. To detect the malicious packet dropping in IoT platform, we choose agricultural IoT. The results show that the proposed algorithm is able to detect malicious packet dropping with a less false positive and false negative.

The rest of the paper is organized as follows, description of agriculture IoT system is given in Sect. 2. The malicious packet dropping is explained in Sect. 3. Section 4 describes the proposed algorithm for malicious packet dropping detection. Performance of the proposed technique is discussed in Sect. 5. The conclusion is given in Sect. 6.

2 Agricultural IoT Systems

In this section, the role of IoT techniques in an agricultural area is discussed. The yield of agriculture depends on many parameters such as plant type, water quantity, environment temperature, soil type, soil temperature, nutrients in the soil, sunlight and so on. To measure these parameters from different environment medium low-cost sensors are used. These sensors are light weighted, they observe environment conditions, but many of such sensors are unable to connect to the Internet and send data to store in database or cloud. To receive data from sensors and do initial processing of received raw data, the small powered computer called gateways are used.

Gateway is a device, which collects data from a variety of devices and transfers data to the Internet which is then stored in the server or cloud platform. To build communication between sensors and gateway, a variety of protocols can be used such as ZigBee and message queuing telemetry transport (MQTT) protocol. These protocols were mainly developed for low-cost devices. Gateway uses Wi-fi modules (or) general packet radio service (GPRS) to connect to the Internet. The collected data are stored in any database (MYSQL, NoSQL, etc.) or cloud platform. The stored data are processed using machine learning or big data analytics for decision-making purpose. The state of art of various IoT agricultural systems is given below.

In [6], the authors analyzed water quality using IoT components in agricultural pounds located at Jiangsu Providence, China. In [7], the authors demonstrated IoT-based farmland monitoring system. The authors [8] analyzed performance of ZigBee communication in greenhouse environment. In [9], IoT-based automatic hydroponic agriculture technique is demonstrated successfully. In [10], greenhouse data are collected by IoT technologies, and the collected data are analyzed by Hadoop. In [11], the authors used &Cube (installed in Raspberry—Pi) as IoT gateway to collect various farm-related data. In [12], Lingzhi mushroom farms were monitored by IoT techniques, and mushroom growth increased with help of IoT technologies. In [13], greenhouse parameters were controlled using IoT communication. In [14], IoT-based precision-based agriculture is demonstrated successfully. In [15], the authors customized IoT-based smart device which consists of temperature sensors, humidity sensors and luminosity sensors to analyze various environment variables. In [16], the authors measured soil parameters such as moisture, temperature and pH value using IoT components. In [17], the authors projected smart vegetable storage using IoT. The system was developed to monitor the environment atmosphere of potato seed warehouse. In [18], the authors proposed IoT-based vineyard monitoring system. In [19], the authors used MQTT protocol to monitor agriculture land. In [20], the authors used IoT components to monitors different environment features in the citrus orchard of the Three Georges Reservoir in China. In [21], the authors used ADS 1115 (analog-to-digital converter) which has I2C protocol for plant monitoring. In [22], the authors designed customized IoT device called SmartFarmNet which consists of sensors, camera and so on. SmartFarmNet helps farmers to monitor the plants. In [23], the authors utilized IoT components to monitor and control the greenhouse parameters. In [24], the authors used IoT technique to observe nutrient film technique (NFT) farm which utilizes the hydroponic technique. In [25], the authors used IoT components to measure environment data, and the comfort level of the crop was calculated from the environment data. In [26], the authors used IoT technologies with big data mechanism to supervise and manage greenhouse. In [27], IoT techniques were utilized to maintain pH level (bicarbonates, salts and nitric acid) in nutrient solutions of hydroponic agriculture system. In [28], the authors utilized various sensors with IoT components and cloud techniques to measure the soil parameters like phosphorus and other nutrients. In [29], the authors combined IoT technology with neural network technique to predict the weather details to help farmers. In [30], the authors designed the IoT-based monitoring system to monitor greenhouse parameters such as temperature, humidity and different gases using IoT technology. In [31], the authors use IoT-based system to monitor remote places. The system was implemented in the desert and coastal area of China (Kubuqi desert, Ulanbuh, Taklimakan, etc.), and data were collected for one year period. In [32], the authors integrated IoT techniques with deep learning methodology for plant monitoring. In [33], the authors monitored and analyzed tomato seedling growth in Anhui Agricultural University, China, and also the authors in [34] used IoT techniques to monitor tomatoes seeds. In [35], IoT techniques were used to monitor mango plants which were cultivated in UniMAP Agro-tech greenhouse, Malaysia. In [36], IoT-based system helped to monitor and control environmental parameters of hydroponic crops. In [37], the authors applied MicaZ motes to monitor various environmental parameters in greenhouses.

In general, the IoT system uses different sensors to read or observe the environment data, and sensors sends observed data to gateway. Gateway devices act as small computer. Gateway collects data from different sensors then sends collected data to server or cloud platform. The user used to access these details (from server/cloud) through Web services (or) mobile services. The major problem of existing plant monitoring systems suffers due to security threats. These systems do not use any special mechanism (or) technique to prevent security threats. The adversary can launch malicious attacks on IoT devices. The detailed description of malicious packet dropping is given in the next section.

Form the state of art, it is clear that many smart farming techniques use Wi-fi communications. To analyze the security aspects of Wi-fi, we implemented packet dropping attacks in IoT systems, which uses Wi-fi as communication module. We proposed novel technique to detect malicious packet dropping attack in IoT systems.

3 Malicious Packet Dropping Attacks

A wireless network, such as MANET, sensor networks, delay-tolerant network, VANET, is suffering due to packet dropping attacks. In these attacks, the adversary capture wireless node and compromised wireless node with fault data. The compromised node act like genuine node but to disturb the network traffic, it drops data packet selectively or drops all data packets. Most of the time attackers compromise the wireless node which is an open environment [38, 39]. In the same way, packet dropping attacks can be launched in IoT platform. The attackers can capture IoT gateway which is placed in an open environment and compromised captured gateway with malicious code, so that the compromised gateway may involve in a malicious activity such as selectively drops data which is received from sensor's actuators or drops all received data. Most of the times, IoT devices are kept in an open environment to observe the physical phenomena. This provides a way for attackers to capture the IoT devices and compromise them for various malicious activities. Many approaches used for detecting malicious activities in a wireless network such as MANET, sensor network and IoT. In [5], the authors used sequence number and time stamp to detect selective forwarding attack and blackhole attack in medical IoT system. The instruction detection system for IoT [40] is developed to deal with the attacks, but it lacks in the autonomic detection process. In [41], the authors integrated fingerprint verification mechanism to prevent the blackhole attack in IoT. In [42], the nodes overhear neighbor packet transferring the behavior to mitigate blackhole in IoT. Even though many approaches used to malicious attacks in wireless network, these methods have its own advantages and disadvantages. We proposed novel lightweight mechanism to detect malicious packet dropping attacks. Our mechanism does not require any special hardware, and result shows that it increases packet delivery rate and throughput and decreases false detection ratio. The detailed description of proposed mechanism is given in next section.

4 Malicious Packet Dropping Detection in Agriculture IoT System

IoT net network contains a group of sensors, actuators and gateway. As mentioned in the previous section, researchers applied IoT technologies in various real-time agriculture-related applications. But most of these techniques do not provide defense against malicious packet dropping attack. To solve this problem, we projected malicious packet dropping detection in agriculture IoT platform. In this work, we have used a soil moisture sensor to measure the moisture of the soil, and measured data are converted to digital form using ADC and then transformed to gateway called NodeMCU. NodeMCU sends data to a public cloud platform called ThingSpeak. IoT platform is intended to measure the soil moisture of the plants. The objective of this work is to identify the malicious packet dropping in IoT platform. In this work,

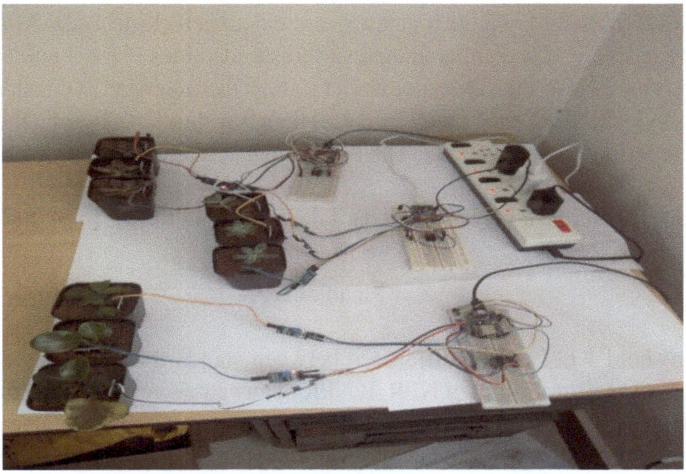

Fig. 1 Plant monitoring using IoT

as shown in Fig. 1, totally nine plants were taken for the experiment which can be kept as indoor plants.

Plants are grouped into three categories based on its family, namely black prince, peperomia and ghost succulent. Each category has three plants, and every plant is equipped with a soil moisture sensor. Three sensors are connected to one NodeMCU, so that three NodeMCU (namely gateway1, gateway2 and gateway3) are used to fetch the details from the nine sensor nodes. NodeMCU sends the data to open-source cloud platform called ThingSpeak. The users can get plant moisture details from the cloud platform anytime. In this experiment, gateway2 is a malicious node. It drops data packet selectively (or) it may drop all received data packets. To detect the malicious node, we implemented detection algorithm in Java platform. Apache POI was used with the Java platform to read the data, which was collected from ThingSpeak cloud platform.

Algorithm to Detect Malicious Gateway
// $i = 1$ to n; where n is number of gateway

1. **if** moisture reading found in gateway i **then**
2. increase NPR of gateway i
3. **end if**
4. **if** NPR of gateway i < $\lambda 1$ **then**
5. gateway i is a malicious node
6. **if** NPR of gateway $i < \lambda 2$ **then**
7. gateway i is a suspicious node
8. **end if**
9. **end if**.

This algorithm is executed in the Java platform. In this algorithm, a number of packet (NPR) received from each gateway are calculated (Line 1–2). After finding NPR of the gateway, it is compared against threshold values (Line 3 and 5). If NPR of the gateway is less than threshold1 ($\lambda 1$), then the gateway is considered as a malicious node. If NPR of the gateway is less than threshold2 ($\lambda 2$), then the gateway is considered as a suspicious node. The threshold calculation is given below.

$$\lambda 1 = \text{Difference between standard deviation and the mean of NPR} \quad (1)$$

$$\lambda 2 = \text{Mean of NPR} \quad (2)$$

The threshold 1 ($\lambda 1$) is the difference between standard deviation and mean of NPR, i.e., it gives a lower limit of NPR. If NPR of any gateway is less than threshold 1 ($\lambda 1$), then it means that gateway drops more packets. Then respective gateway is considered as a malicious node. The threshold 2 ($\lambda 2$) is the mean of NPR. If NPR of any gateway is less than threshold 2 ($\lambda 2$), then it means that gateway drops fewer packets. Then respective gateway is considered as a suspicious node. The optimal value of $\lambda 1$ and $\lambda 2$ is depends on number of packet received from the gateways. Some existing methods [4, 39, 43] used constant (C) value in threshold value $[0 < C \leq 1]$. The C value helps to differentiate genuine node and malicious node, but it leads to false positive and false negative. As shown in Eqs. 1 and 2, the proposed technique does not use any constant value in threshold calculation. Instead, it identifies suspicious node (line 4) and confirms the malicious node (line 6) based on its packet dropping nature. It helps to reduce false positive and false negative value. The result shows that the detection technique detects malicious node with high accuracy. Performance evaluation of the implemented technique is discussed in the next section.

5 Performance Analysis

The proposed detection technique is evaluated in a real-time scenario. Various network parameters are discussed to the estimated performance of the detection algorithm. The packet delivery ratio (pdr) of IoT devices is shown in Fig. 2. PDR is defined as the ratio of the number of data packets successfully received by the destination node to the number of packets sent by the source node. Here, gateway2 is a malicious node, and pdr of gateway2 is reduced when packet drop rate increased. Since other gateway (such as gateway1 and gateway3) does not drop data, pdr of other gateway is high. But when detection algorithm is implemented, it finds the malicious node (gateway2) and gateway2 is reconstructed. Since malicious activity of gateway2 is removed, the pdr of gateway2 is increased more than 95%.

Throughputs of three gateway devices are shown in Fig. 3. Throughput is the ratio of the packets successfully received with respect to the time. From the figure, it is clear that gateway devices give maximum threshold when there is no malicious activity. But in gateway2, throughout is decreased due to malicious packet dropping. The gateway2's throughput is increased after applying a detection algorithm in the

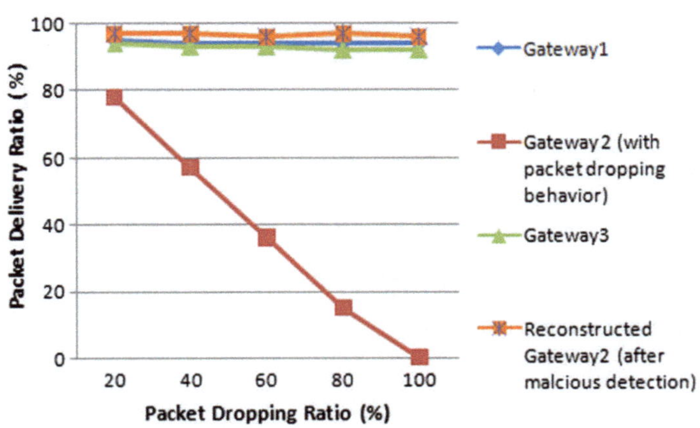

Fig. 2 Packet delivery ratio (PDR) of gateway devices

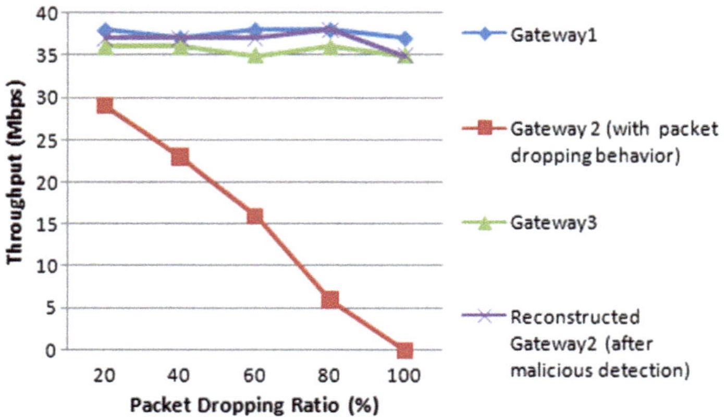

Fig. 3 Throughput of gateway devices

network. Similarly, Fig. 4 shows the false positive and false negative of the proposed algorithm. False positive is defined as some good nodes that are erroneously detected as compromised node, and false negative is nothing but the malicious node determined as a good node. Figure 4 shows that when the packet drop rate is less, it increases false positive and false negative. But, false negative and false positive are decreased when the packet drop is increased. The reason is the detection algorithm which is unable to detect malicious node when it drops less number of the data packet.

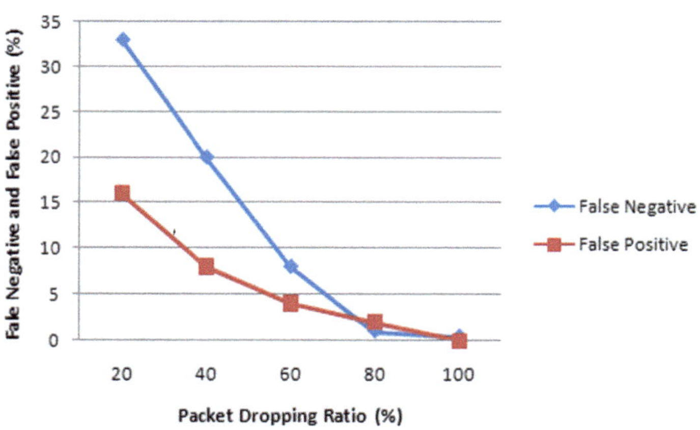

Fig. 4 False negative and false positive ratio

6 Conclusion and Future Scope

The application of IoT is increased rapidly. Many wireless attacks can launch in IoT platform to disturb the network traffic. To investigate the impact of malicious packet dropping attacks in IoT platform, we implemented malicious packet dropping attacks in agricultural IoT devices and also novel technique proposed to detect malicious packet dropping attack in IoT devices. The result shows that the proposed detection technique is successful in detecting malicious IoT devices, and also, it gives good impact on packet delivery ratio and throughput. The proposed technique yields a less false positive and false negative. In future, the current research is extended by using more number of sensors and gateways in malicious packet dropping detection, and also, the performance of proposed technique would be compared with existing methods.

References

1. International Communication Union (ITU) Fact an figures for ICT revolution and remaining gaps. Available at www.itu.int/ict
2. Asghari P, Rahmani AM, Javadi HHS (2019) Internet of things applications: a systematic review. Comput Netw 148:241–261
3. Tzounis A, Katsoulas N, Bartzanas T, Kittas C (2017) Internet of things in agriculture, recent advances and future challenges. Biosys Eng 164:31–48
4. Terence JS, Geethanjali P (2019) A novel technique to detect malicious packet dropping attacks in wireless sensor networks. J Inf Process Syst 15(1)
5. Mathur A, Newe T, Rao M (2016) Defence against blackhole and selective forwarding attacks for medical WSNs in the IoT. Sensors 16:118
6. Ma D, Ding Q, Li Z, Li D, Wei Y (2012) Prototype of an aquacultural information system based on internet of things E-Nose. Intell Autom Soft Comput 18(5):569–579
7. Liu J (2016) Design and implementation of an intelligent environmental-control system: perception, network, and application with fused data collected from multiple sensors in a Greenhouse at Jiangsu, China. Int J Distrib Sens Netw 12(7):5056460
8. Lamprinos I, Charalambides M (2015) Experimental assessment of ZigBee as the communication technology of a wireless sensor network for greenhouse monitoring. Int J Adv Smart Sens Netw Syst 6
9. Palande V, Zaheer A, George K (2018) Fully automated hydroponic system for indoor plant growth. Proc Comput Sci 129:482–488
10. Yang J, Liu M, Lu J, Miao Y, Hossain MA, Alhamid MF (2018) Botanical internet of things: toward smart indoor farming by connecting people, plant, data and clouds. Mobile Netw Appl 23(2):188–202
11. Ryu M, Yun J, Miao T, Ahn I, Choi S, Kim J (2015) Design and implementation of a connected farm for smart farming system. In SENSORS, IEEE, 2015, pp 1–4
12. Chieochan O, Saokaew A, Boonchieng E (2017) IOT for smart farm: a case study of the Lingzhi mushroom farm at Maejo University. In: 2017 14th international joint conference on computer science and software engineering (JCSSE). IEEE, pp 1–6
13. Kodali RK, Vishal J, Karagwal S (2016) IoT based smart greenhouse. In: Humanitarian technology conference (R10-HTC), IEEE Region 10, pp 1–6
14. Ferrández-Pastor FJ, García-Chamizo JM, Nieto-Hidalgo M, Mora-Martínez J (2018) Precision agriculture design method using a distributed computing architecture on internet of things context. Sensors 18(6):1731

15. Maia RF, Netto I, Ho Tran AL (2017) Precision agriculture using remote monitoring systems in Brazil. In: Global humanitarian technology conference (GHTC). IEEE, pp 1–6
16. Na A, Isaac W, Varshney S, Khan E (2016) An IoT based system for remote monitoring of soil characteristics. In: Connect your worlds, international conference on information technology (InCITe)-the next generation IT summit on the theme-internet of things. IEEE, pp 316–320
17. Tervonen J (2018) Experiment of the quality control of vegetable storage based on the internet-of-things. Proc Comput Sci 130:440–447
18. Pérez-Expósito JP, Fernández-Caramés TM, Lamas PF, Castedo L (2017) An IoT monitoring system for precision viticulture. In: 2017 IEEE international conference on internet of things (iThings) and IEEE green computing and communications (GreenCom) and IEEE cyber, physical and social computing (CPSCom) and IEEE smart data (SmartData). IEEE, 2017, pp 662–669
19. Pooja S, Uday DV, Nagesh UB, Talekar SG (2017) Application of MQTT protocol for real time weather monitoring and precision farming. In: 2017 international conference on electrical, electronics, communication, computer, and optimization techniques (ICEECCOT). IEEE, 2017, pp 1–6
20. Zhang X, Zhang J, Li L, Zhang Y, Yang G (2017) Monitoring citrus soil moisture and nutrients using an IoT based system. Sensors 17(3):447
21. Bachuwar VD, Shligram AD, Deshmukh LP (2018) Monitoring the soil parameters using IoT and Android based application for smart agriculture. In: AIP conference proceedings, 1989(1):020003. AIP Publishing
22. Jayaraman PP, Yavari A, Georgakopoulos D, Morshed A, Zaslavsky A (2016) Internet of things platform for smart farming: experiences and lessons learnt. Sensors 16(11):1884
23. J. Pitakphongmetha, N. Boonnam, S. Wongkoon, T. Horanont, D. Somkiadcharoen, and J. Prapakornpilai, "Internet of things for planting in smart farm hydroponics style," In Computer Science and Engineering Conference (ICSEC), pp. 1–5. IEEE, 2016
24. Crisnapati PN, Wardana INK, Aryanto IKAA, Hermawan A (2017) Hommons: Hydroponic management and monitoring system for an IOT based NFT farm using web technology. In: 2017 5th international conference on cyber and IT service management (CITSM). IEEE, 2017, pp 1–6
25. Mekala MS, Viswanathan P (2019) CLAY-MIST: IoT-cloud enabled CMM index for smart agriculture monitoring system. Measurement 134:236–244
26. Lee M, Kim H, Yoe H (2018) ICBM-based smart farm environment management system. In: International conference on software engineering, artificial intelligence, networking and parallel/distributed computing. Springer, Cham, pp 42–56
27. Cambra C, Sendra S, Lloret J, Lacuesta R (2018) Smart system for bicarbonate control in irrigation for hydroponic precision farming. Sensors (Basel, Switzerland) 18(5)
28. Estrada-López J, Castillo-Atoche AA, Vázquez-Castillo J, Sánchez-Sinencio E (2018) Smart soil parameters estimation system using an autonomous wireless sensor network with dynamic power management strategy. IEEE Sens J 18(21):8913–8923
29. Aliev K, Jawaid MM, Narejo S, Pasero E, Pulatov A (2018) Internet of plants application for smart agriculture. Int J Adv Comput Sci Appl 9(4)
30. Singh TA, Chandra J (2018) IOT based Green House Monitoring System. J Comput Sci 14(5):639–644
31. Yan M, Liu P, Zhao R, Liu L, Chen W, Yu X, Zhang J (2018) Field microclimate monitoring system based on wireless sensor network. J Intell Fuzzy Syst 35:1–13
32. Geng L, Dong T (2017) An agricultural monitoring system based on wireless sensor and depth learning algorithm. Int J Online Eng (iJOE) 13(12):127–137
33. Jiao J, Ma H, Qiao Y, Du Y, Kong W, Wu Z (2014) Design of farm environmental monitoring system based on the internet of things. Adv J Food Sci Technol 6(3):368–373
34. Kalathas J, Bandekas DV, Kosmidis A, Kanakaris V (2016) Seedbed based on IoT: a case study. J Eng Sci Technol Rev 9(2):1–6
35. Halim AAA, Hassan NM, Zakaria A, Kamarudi LM, Bakar Abu (2016) Internet of things technology for greenhouse monitoring and management system based on wireless sensor network. ARPN J Eng Appl Sci 11(22):13169–13175

36. Ferrández-Pastor FJ, García-Chamizo JM, Nieto-Hidalgo M, Mora-Pascual J, Mora-Martínez J (2016) Developing ubiquitous sensor network platform using internet of things: application in precision agriculture. Sensors 16(7):1141
37. Akkaş MA, Sokullu R (2017) An IoT-based greenhouse monitoring system with Micaz motes. Proc Comput Sci 113:603–608
38. Mathew A, Terence JS (2017) A survey on various detection techniques of sinkhole attacks in WSN. In: 2017 international conference on communication and signal processing (ICCSP). IEEE, 2017, pp 1115–1119
39. Terence Sebastian, Purushothaman Geethanjali (2019) Behavior based routing misbehavior detection in wireless sensor networks. KSII Trans Internet Inf Syst 13(11):5354–5469
40. de AC Mello, Ribeiro RL, Almedia, Moreno ED (2017) Mitigating attacks in the internet of things with a self-protecting architecture. AICT2017: the thirteen advanced international conference on telecommunications, pp 14–19
41. Kumar R., "Internet of Things for the Prevention of Black Hole Using Fingerprint Authentication and Genetic Algorithm Optimization," International Journal of Computer Network and Information Security, Aug 1;9(8):17, 2018
42. Ahmed F, Ko YB (2016) Mitigation of black hole attacks in routing protocol for low power and lossy networks. Secur Commun Netw 9(18):5143–5154
43. Madria S, Jian Y (2009) SeRWA: a secure routing protocol against wormhole attacks in sensor networks. Ad Hoc Netw 7(6):1051–1063

IoT-Based Air Pollution Controlling System for Garments Industry: Bangladesh Perspective

Marzan Tasnim Oyshi, Moushumi Zaman Bonny, Syed Ahmed Zaki, and Susmita Saha

Abstract Environment is changing dramatically due to global warming and environment pollution and leading the next generation toward a vulnerable future. Air pollution is one of the key factors of the unstable environment around the globe. Garment industry is playing one of the major roles causing air pollution. Internet of Things (IoT) is enabling the communication of real-life objects and engendering incipient opportunities for the betterment of human. In this research, a novel method of air pollution controlling system for garments industry has been proposed using IoT technology in the context of Bangladesh. The research will avail to assist the government of Bangladesh for the identification of industries who are emitting extortionate toxic gases, fine them, and use the collected money for a green and development world.

Keywords Air pollution · IoT · Garments industry · Gas sensor

1 Introduction

Industry 4.0 came up with the concept of fourth technological revolution, combining basic concepts of Cyber-Physical Systems (CPS), Internet of Things (IoT), Information and Communications Technology (ICT), Enterprise Architecture (EA), and

M. T. Oyshi (✉) · M. Z. Bonny
Department of Computer Science & Engineering, Daffodil International University, Dhaka, Bangladesh
e-mail: marzan@diu.edu.bd

M. Z. Bonny
e-mail: bonny.cse@diu.edu.bd

S. A. Zaki
Frankfurt University of Applied Sciences, Frankfurt, Germany
e-mail: zaki.cse.daffodil@gmail.com

S. Saha
University of Lodz, Lodz, Poland
e-mail: Susmitasaha2458@gmail.com

A. K. Tripathy et al. (eds.), *Advances in Distributed Computing and Machine Learning*, Lecture Notes in Networks and Systems 127, https://doi.org/10.1007/978-981-15-4218-3_10

Enterprise Integration (EI). IoT represents a fundamental part in the revelation of Industry 4.0. In recent days, IoT has been considered as the technology and economic wave in the global industry. IoT enables an intelligent network that connects all real-life objects to the Internet for the purpose of exchanging information with a view to communicate through the information sensing devices in using mutually agreed protocols. IoT is the connectivity of physical objects to the Internet with a view to monitor and control their behavioral pattern to gain efficiencies and create ultimate new capabilities.

Bangladesh is a highly populated country in South Asia. The industry-based economy of Bangladesh is 46th largest in the world in nominal terms and 33rd largest by purchasing power parity. According to CIA World Factbook, the economy of Bangladesh has grown roughly 6% per year since 1996 [1]. The half of total GDP is generated from the agriculture and industrial sectors. Garment industry is the backbone of industrial structure of Bangladesh.

The industrial production growth rate of Bangladesh is 8.4%, and it ranked seventh in country comparison to the world [2]. But this industry is being a great threat to the environment and leading Bangladesh to a very uncertain future. Industrial Revolution has upgraded Bangladesh from poor to developing country but this specific unplanned industry has become the ultimate threat to the people by violating the environment ecosystem. The global climate is extremely affected by the revolution of industries. Industries are helping us to rise the economic condition but nature is paying the cost. Bangladesh is one of the world's highest overly populated country.
There are thousands of factories located in Dhaka and many more around the country. Garment is the backbone of the industry of Bangladesh. The garment factories emit toxic elements that get mixed with the environment and affect the environment badly. It is the high time to take proper steps against this national hazard and protect our future. This research focuses on the air pollution caused by the garments industry and proposes a solution to control air pollution using IoT-based solution.

2 Primary Documentary

According to the size of the population, Bangladesh is extremely limited. Dhaka has been termed as a fourth polluted city in the world according to Daily Star 2014, and the condition is getting worse each and every day. Air pollution and water pollution are serious health hazards among all the pollutions. According to the Department of Environment (DOE), the National Ambient Air Quality Standards is given in Table 1.

But the current status of Dhaka city has already exceeded the threshold limit. The current status of Dhaka city is given below where ∗P denotes as permissible and ∗C denotes as concentration (Table 2).

According to the research work on "Industrial Pollution in Bangladesh," the main sources of air pollution have been identified as

Table 1 Air quality standards according to Department of Environment (DOE) and National Ambient

Land use category	8 h average concentration in μg/m^3			
	CO	NO_2	SPM	SO_2
Industrial/mixed use	5000	100	500	120
Commercial/mixed use	5000	100	400	100
Residential/rural Use	2000	80	200	80
Sensitive use[a]	1000	30	100	30

Source Department of Environment (DOE) (1997)
[a]Sensitive areas include national monuments, health resorts, hospitals, archeological spots, and educational institutions

Table 2 Current status of Dhaka city

Location at Dhaka City	CO		NO_2		SPM		SO_2	
	P	C	P	C	P	C	P	C
Gulistan	5000	33200	100	500	400	1332	100	800
Jatrabari		67,000		500		4667		1300
Pantho path		85,100		500		2666		900
Mohakhali		69,300		500		2111		1200

CO—carbon monoxide
NO_2—nitrogen dioxide
SPM—suspended particulate matter
SO_2—sulfur dioxide

- Burning of fossil fuel
- Industrial discharge
- Emission from vehicles

The research work results industries emission as an alarming source of environment pollution. If the specific industry gets identified, it is possible to take proper steps to force industries to emit toxic disposals in a threshold point.

3 Literature Review

IoT is capable of building an ecosystem for the considered problem area and providing solutions by communicating. According to Saifuzzaman et al. [3], real-life IoT application can enhance the quality of life. As per Ray [4], IoT is approving users to acquire the finest worth by connecting devices to Internet in health sector and care [5], agriculture [6], manufacturing of industrial material [7], transportation [8], commercial business [9], e-education [10], logistic [11], retail and marketing

[12], empowerment of e-governance [13], smart city [14], assisted living [15], and many other sectors. Gartner [16] indicated that by 2020 about 25 billion things will be connected to Internet and will be in practical use. This specifies that there will be connected device, numbering in more than double of the world population [17]. Broring et al. [16] introduced an ecosystem of IoT architecture with five interoperability patterns. Roblek et al. [18] presented a conceptual study of "Industry 4.0" by focusing Internet of Things. Tapashetti et al. [19] presented a prototype for monitoring indoor air quality by measuring the concentration of CO and HCHO gases and notified the user when the gases reach to the threshold.

This research intends to control air pollution by identifying the level of emitted toxic gases, identifying the industry (garments will be focused mainly), and assisting the government of Bangladesh environment friendly by accomplishing the Internet of Things.

4 Proposed Methodology

IoT-based air pollution controlling system for garments industry plans to introduce a service that will come up with an IoT device that will be placed in the exit points of garments, a dashboard for the ministry to keep track of the emission of toxic gases by garments and a user manual. The device will contain

- Microcontroller
- GSM module
- Battery
- Gas sensor—MQ135
- Alarm

The proposed system architecture of IoT-based air pollution controlling system for garments industry is shown in Fig. 1. The device will be powered using battery. MQ135 is basically a gas sensor, and it can detect NH3,NOx, alcohol, benzene, smoke, CO_2 ,etc. MQ135 will collect data from the environment and compare the toxic levels with the permitted level of emission. Data will be published to the broker, and MongoDB will be in the subscribing mode and store the data. Sensor data will be plotted in the dashboard. Here the dashboard will be the user interface, and it will be accessible by the ministry itself. Analyzing the data system will take actions accordingly.

If the emission is below threshold, the system will do nothing but if the emission crosses threshold, it will turn on the alarm. The ministry will be notified immediately via dashboard. Government will be able to send warning and fix certain amount of bill depending on the emission of the toxic gases by the garments. Location of the specific garments will be easily identified using GSM module. The alarm will not be turned off until the industry clears the bill generated by the government. The block diagram of IoT-based air pollution controlling system for garments industry is shown in Fig. 2.

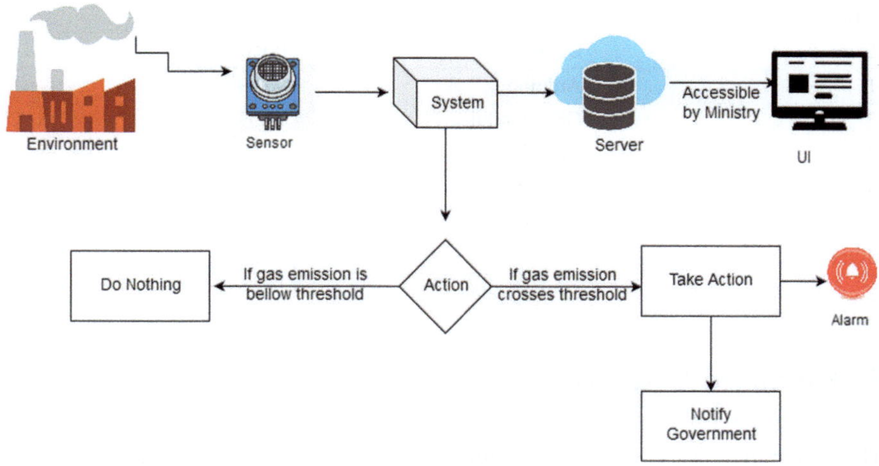

Fig. 1 System architecture

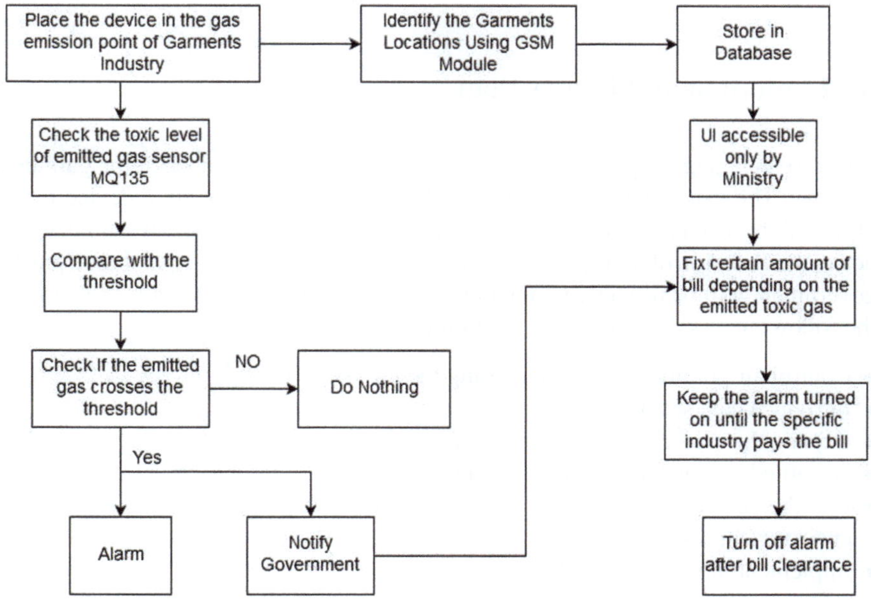

Fig. 2 Block diagram

The device will be powered on using pencil battery. Specially designed casing will be used to protect the device from heat, smoke, and dust. End user will be able to use the service via Web application and mobile application. To use the service, the user will need to have a specific account by sharing the official details. Without logging in the system, no one will be able to get the service.

Table 3 Cost structure

S. no.	Device	Price
1	Arduino	6.09
2	GSM module	14.59
3	Battery	0.99
4	MQ135	4.12
5	Alarm	5.79
Total		31.58

5 Estimated Cost Structure

This research focuses on the cost optimization. It is possible to implement the project having a low budget. The estimated cost structure for IoT-based air pollution controlling system for garments industry is shown in Table 3.

6 Limitation and Future Plan

This research plans to place the device in the exit point of the pipes of the garments industry. The exit points are sophisticated and vulnerable. Placing a device in the pipes will be a challenge here. The industries will not easily allow a system that will detect the emission of toxic gases and make them to pay bills. Entire support from the government will be needed to implement the project. This research aims to perform the following tasks to perform in the future:

- Control air pollution by implementing the project
- Introduce data security
- Expand the research to other industries
- Rise fund by collecting bills from industries
- Use the funds to contribute to the development of people who are affected by environment pollution
- Implement the solution globally.

7 Conclusion

Environment pollution is a great threat for global climate change. The climate is transmuting significantly around the world, and the entire world is moving ahead to a very uncertain future. It is the high time to take felicitous steps to control environmental pollution and save us from extinction. We are already paying the price of global warming and witnessing tsunami, tornado, earthquake, and many more devastating natural disaster. It is our responsibility to rehabilitate our environment, rather than eradicating it.

References

1. Bangladesh Economy (2019) https://theodora.com/wfbcurrent/bangladesh/bangladesheconomy.html. Last accessed 8 Feb 2019
2. Environmental Pollution of Bangladesh—it's effect and control. http://www.bangladeshenvironment.com/index.php/polution-s/294-environmental-pollution-of-bangladesh-it-s-effect-and-control. Last accessed 24 Oct 2019
3. Saifuzzaman M, Moon NN, Nur FN (2017) IoT based street lighting and traffic management system. In: 2017 IEEE region 10 humanitarian technology conference (R10-HTC). https://doi.org/10.1109/r10-htc.2017.8288921
4. Ray PP (2018) A survey on Internet of Things architectures. J King Saud Univ Comput Inform Sci 30(03):291–319
5. Carlos C, Sandra S, Jaime L, Laura G (2017) An IoT service-oriented system for agriculture monitoring. In: ICC 2017—2017 IEEE international conference on communications. https://doi.org/10.1109/ICC.2017.7996640
6. Gabriel N, Stefan P, Alexandru S, Vladimir F (2017) A cloud-IoT based sensing service for health monitoring. In: 2017 E-Health and bioengineering conference (EHB). https://doi.org/10.1109/EHB.2017.7995359
7. Andrea M, Marco P, Laura S, Franco C (2016) Public transportation, IoT, trust and urban habits. In: International conference on internet science. https://doi.org/10.10007/978-3-319-45982-0-27
8. Kleanthis T, Theodoros F (2016) From mechatronic components to industrial automation things: an IoT model for cyber-physical manufacturing systems. J Softw Eng Appl 10(08). https://doi.org/10.4236/jsea.2017.108040
9. Sok PC, Kim JG (2016) E-Learning Based on Internet of Things. J Comput Theor Nanosci 22(11):3294–3298
10. Martin F, Stefan M (2013) IoT in practice: examples: IoT in logistics and health. Enabling Things Talk. https://doi.org/10.1007/978-3-642-40403-0-4
11. Naoshi U, Hirokazu I, Keisuke I (2016) IoT service business ecosystem design in a global, competitive, and collaborative environment. In: 2016 Portland international conference on Management of Engineering and Technology (PICMET). https://doi.org/10.1109/PICMET.2016.7806694
12. Wei Z, Frederique AB, Selwyn P (2016)Dynamic organizational learning with IoT and retail social network data. In: 2016 49th Hawaii international conference on system sciences (HICSS). https://doi.org/10.1109/HICSS.2016.476
13. Paul B, Marijn J (2015) Advancing e-government using the internet of things: a systematic review of benefits. In: International conference on electronic government
14. Hamidreza A, Vahid H, Vincenzo L, Aurelio T, Orlando T, Miadreza SK, Pierluigi S (2016) IoT-based smart cities: a survey. In: The 16th IEEE international conference on environment and electrical engineering (IEEE-EEEIC'16). https://doi.org/10.1109/EEEIC.2016.7555867
15. Angelika D, Robert MO, Mario D, Dieter H, Günter S (2010) The internet of things for ambient assisted living. In: Seventh international conference on information technology: new generations. https://doi.org/10.1109/ITNG.2010.104
16. Broring A, Schmid S, Schindhelm CK, Khelil A, Kabisch S, Kramer D, Teniente E (2017) Enabling IoT ecosystems through platform interoperability. IEEE Software 34(1):54–61. https://doi.org/10.1109/ms.2017.2

17. World population: past, present, and future. https://www.worldometers.info. Last accessed 8 July 2018
18. Roblek V, Meško M, Krapež A (2016) A complex view of industry 4.0. SAGE Open 6(2):215824401665398. https://doi.org/10.1177/2158244016653987
19. Tapashetti A, Vegiraju D,Ogunfunmi T (2016) IoT-enabled air quality monitoring device: a low cost smart health solution. In: 2016 IEEE global humanitarian technology conference (GHTC). https://doi.org/10.1109/ghtc.2016.7857352
20. Gartner Press Release. 4.9 billion connected "things" will be in use in 2015, Barcelona, Spain, 11 Nov. Last accessed 8 July 2018

Automated Plant Robot

U. Nanda, A. Biswas, K. L. G. Prathyusha, S. Gaurav, V. S. L. Samhita, S. S. Mane, S. Chatterjee, and J. Kumar

Abstract This work is a Raspberry Pi-based prototype that has been built while keeping both the Indian agricultural scenario and hobbyist backyard gardeners in mind. It aims not only to act as a solution to conventional method of growing plants but also to approach along the lines of sustainable development. With our project, we are also aiming to optimize the amount of resources required to grow a plant. The model acts as the brain of the plant by providing directions to the technical implementations of the setup such as an attached pair of wheels and a water pump to act according to the requirement. This prototype could be a possible solution to majority of the urban population planning to have potted plants and greenery in their apartment but with space, time and effort constraints. Further details like implementation, advantages and future are mentioned in the next sections.

Keywords Automated plant · Robotics · Water resource constraint · Raspberry Pi · IOT

1 Introduction

Agriculture and farming sector is one of the leading sectors in India and provides the greatest number of jobs in the country. So, it is inevitable that the livelihoods of farmers and other people involved in agriculture be the most affected by a slightest drop in the crop yield. The agricultural output in turn heavily depends on factors like sunlight, water and soil quality. All of these factors, if not adequately available, render the seed unable to sprout. Even after the seed sprouts, adequate conditions are required for the plant to survive.

U. Nanda (✉) · J. Kumar
School of Electronics Engineering, VIT-AP University, Andhra Pradesh, Amaravati 522237, India
e-mail: uk_nanda@yahoo.co.in

A. Biswas · K. L. G. Prathyusha · S. Gaurav · V. S. L. Samhita · S. S. Mane · S. Chatterjee
School of Computer Science Engineering, VIT-AP University, Andhra Pradesh, Amaravati 522237, India

A. K. Tripathy et al. (eds.), *Advances in Distributed Computing and Machine Learning*, Lecture Notes in Networks and Systems 127, https://doi.org/10.1007/978-981-15-4218-3_11

This work deals with providing a smart platform for plants to grow. Our aim is not only to get better yield but also to proceed along the lines of sustainable development. In conventional methods, water is not used efficiently enough, leading to a significant contribution to water wastage. Overwatering of plants leads to waterlogging, which often renders the soil unfavorable for the healthy growth of plants. Overwatering plants is also one of the prime reasons of water wastage. Our model consists of various interconnected sensors which provide the best possible environment for plants to grow. Here, we are aiming not only to ensure better growth and survival rates of plants but also to optimize the use of energy and water resources required to grow the plant. This prototype can be used by a wide range of users to solve real-world problems—be it as a hobbyist backyard gardening tool or when applied on a larger scale, a potential solution to global water scarcity.
The work presented here has the following novel aspects.

- Sustainable growth and development of plants.
- Optimization of water and energy resources required for the growth of a plant.
- Minimizing human effort and attention required to grow a plant.
- Motivate the urban population to have a backyard garden by minimizing the time and effort required to maintain one.

2 Problem Description

Water scarcity is increasing in India at an alarming rate. This water crisis is attributed to increasing demand due to growing population and inefficient use of water resources. Using conventional methods, water is not used efficiently enough, leading to a significant contribution to water wastage. According to a research paper by NABARD and ICRIER [1], approximately 78% of the freshwater available in the country is used up by the agriculture and farming sector. A closer look at the cropping and irrigation patterns within the country reveals a daunting inefficiency that causes the most water-related issues in India, as well as depletion of the groundwater tables at an unsettling rate.

Lack of substantial investments in the infrastructure for canals and reservoirs, inefficient control and preservation of surface water and rainwater, combined with misaligned water usage policies can lead to heavy usage of groundwater for agricultural purposes. Such a situation is seen in the states of Punjab and Haryana. The Nature study [2], published in the year 2009, warned that, if counteractions are not taken soon, this region would not only face an acute shortage of drinking water, but the agricultural yield would also diminish. This could lead to extensive socioeconomic stresses.

The evidence so far is in accordance with this prediction. Along with potable water shortage and water pollution, the agricultural yield growth in Punjab which was once the top-performing state has reduced considerably in recent years, creating a variety of problems in the state [3].

The existing solutions to water scarcity for agriculture are not only financially infeasible for the majority of farmers but also they are not applicable for any small-scale usage for a backyard garden hobbyist. This work aims to tackle both of these problems while aiming to minimize human effort.

3 Proposed Idea

The proper, coordinated working of all the mentioned components together leads to the efficient functioning of the model as a whole. This section intends to define the overall working of the project as a whole and how each component helps in controlling the plant, more so, how the plant desires to move. The potted plant is placed on a thin, light wooden platform on wheels as shown in Fig. 1, equipped with two 9V DC motors driving the rear wheels. The LDRs attached at the two sides of the platform measure the light intensity and send the electric signal output to the Raspberry Pi. The Pi checks if the current light energy is sufficient for the plant. If the received light energy is more than the ideal amount for the plant, the platform moves toward the direction where lesser light is registered. In case of deficiency of light, the platform moves toward the direction where more light is available, thus making the optimum availability of sunlight.

As the platform moves, there certainly arises the question regarding obstacles. This is solved using ultrasonic sensors attached to the platform. The model has been designed in such a way that the direction in which the platform moves depends on the output of the LDRs. While the platform is in motion, the ultrasonic sensor checks for any obstacles present in the path. If the ultrasonic sensors detect an obstacle within 15 cm of the platform, it automatically stops the motors so that no collision takes place.

Fig. 1 Photograph of our developed prototype

To consider the water levels in the pot, we have included a water level sensor in the pot. This sensor effectively measures the water level present in the soil (in mm) and sends the output to the Pi. The Pi decides if the received input from the water level sensor is in the range of the ideal water level or not depending on the datasheet of ideal moisture requirements of more than 350 crops. In case the water level is detected to be excess for the particular plant in the pot, a two-way water pump is switched on through a 5 V relay module to drain out the excess water. This excess water is then stored in a water reservoir built in the platform for later use. In case the water level sensor detects a scarcity of moisture in the soil, the water pump pumps back from the reservoir to the pot. The main intention behind this module is to avoid water logging in the pot that might hinder the growth of the plant and to provide better usage of water resources.

The temperature and the humidity sensor, on the other hand, are the physical variables that cannot be altered by any human means. This requires human manipulation like moving the plant to a different location with ideal temperature and humidity conditions. For this, the user needs to be notified about the current temperature and humidity and whether it is a threat to the survival of the plant. Figure 1 is that of the prototype of our project. The DC motors are attached to the rear wheels of the model to move it forward and backward. The ultrasonic sensors and LDR sensors are placed at the front and rear side of the model for obstacle avoidance and to measure and compare light intensities, respectively. The water level sensor and the pump are placed in the pot. A DHT11 sensor is placed beside the pot. The DHT11 sensor onboard measures the temperature and humidity levels of the surroundings and sends a mail to the user if the same is beyond the plant's tolerable limit.

The ultrasonic sensors (HC-SR04) are connected to the Raspberry Pi through GPIO pins as shown in Fig. 2. The LDRs are also connected to the Raspberry Pi

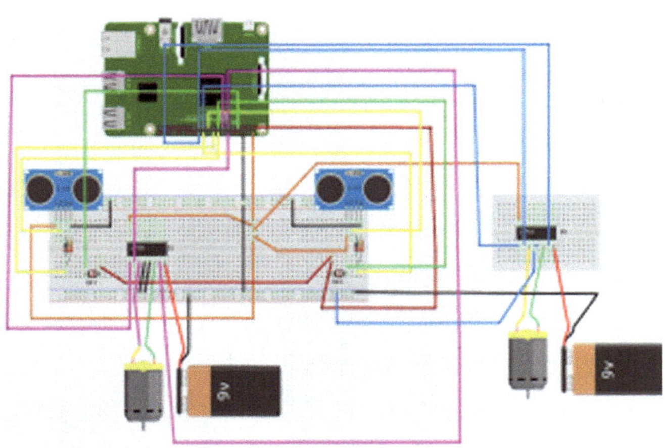

Fig. 2 Schematic circuit diagram and wiring of the LDR-ultrasonic-DC Motor module of the prototype

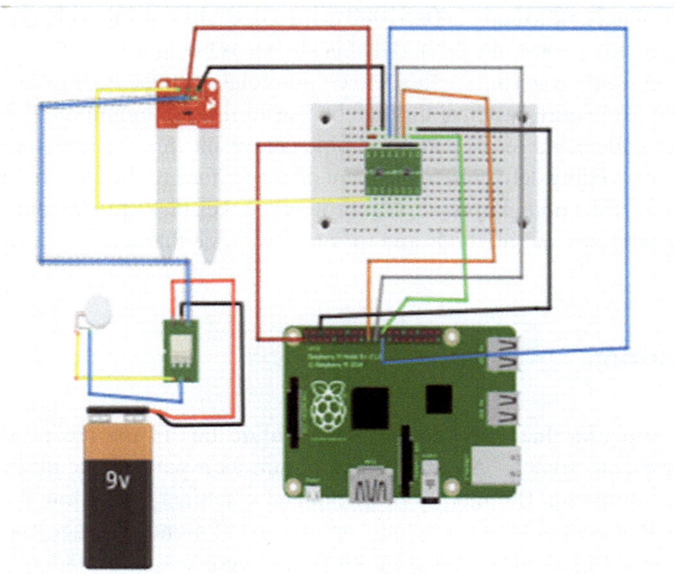

Fig. 3 Schematic circuit diagram and wiring of the water level sensor-water pump-water reservoir of the prototype

through GPIO pins. Two DC motors used to move the platform are powered by two 9V batteries and connected to Raspberry Pi GPIO pins through L293D motor driver ICs.

The water level detection sensor is connected to the Raspberry Pi GPIO pins through an 8-channel 10-bit ADC as shown in Fig. 3. There is a water pump present in the white pot to drain the water. It is powered by a 5V battery and connected to the Raspberry Pi GPIO pins through a 5V relay switch.

4 Related Work

A few commercial products and equipment are present in contemporary market that aim to provide a sustainable environment for plants to grow and there are other products and equipment that aim to tackle overwatering of plants as well. However, the reason why we decided to work on a prototype is that there are no products that achieve both simultaneously while being easily accessible to consumers. We have tried to minimize the time and effort that is required on the part of a consumer to have greener surroundings.

Another area where our prototype shines is that how cost efficient and affordable it is for Indian consumers. Classic forms of saving water in backyard-gardens include automatic rain shutoff device and smart controller for watering system both of which cost approximately $200 and $300–$1000 to install, excluding labor charges [4]. Drip irrigation is another watering method that is known as being a more efficient alternative to conventional watering methods in commercial settings. Although convenient and efficient than spray systems, this can be wasteful if not monitored properly. For instance, an undetected leak or a sliced pipe can result in excessive wastage. And because water is being delivered to the plant on the surface of the soil, though slower, evaporation is still a potential issue. Our prototype takes these fundamental issues of preexisting products and methods and tries to overcome them.

5 Conclusion

This work suggests that a more sustainable future lies in the creation of a new "cybernetic-plant" society. An augmented nature as a part of our design process is only the beginning. The merger of plants and machines will allow us to realize capabilities that we could never achieve in this world alone. The agenda is turning the project into a marketable tinkering kit so that people start to understand nature much more than what they used to and perhaps then give a second thought to the functional lifeforms or they will end up turfing in the trash the next time they forget to water their plants or provide the proper amount of sunlight.
The following features can be embedded to further improve our model.

- Integrate everything into a user-friendly app.
- A database for storing the temperatures recorded, amount of water and number of watering required.
- Use the above-mentioned database to predict the amount of watering required each morning.

References

1. Gulati A, Mohan G Towards sustainable, productive and profitable agriculture: case of rice and sugarcane. Working paper no. 358, Indian Council for Research on International Economic Relations (ICRIER), New Delhi (2018)
2. Rodell M, Velicogna I, Famiglietti J (2009) Satellite-based estimates of groundwater depletion in India. Nature 460:999–1002
3. Saranga H, Ananda Kumar S (2018) Misaligned agriculture: a major source of India's water problems, Forbes India, July
4. Rudd H, Vala J, Schaefer V (2002) Importance of backyard habitat in a comprehensive biodiversity conservation strategy: a connectivity analysis of urban green spaces. Restor Ecol 10(2):368–375

IoT Security Issues and Possible Solution Using Blockchain Technology

Marzan Tasnim Oyshi, Moushumi Zaman Bonny, Susmita Saha, and Zerin Nasrin Tumpa

Abstract The world is moving along with the technical advancement where smart home and smart cities are creating smart lifestyle and Internet of things (IoT) is being one of the most popular technologies in this decade. IoT is making life better by connecting devices and producing billions of data. But for the heterogeneous connectivity, smaller size, memory, storage and less security measurement, it is easy to hack and manipulate the IoT devices. To have a safe and secured connection, it is a must to ensure the data security in IoT too. This research proposes a novel method to secure the IoT devices using blockchain technology.

Keywords Internet of things · Security · Threat · Blockchain

1 Introduction

Internet of things (IoT) has become one of the greatest hype in this decade. IoT is being publicized and reaching out to people as this technology is acquiescing sensor nodes equipped with diverse monitoring capacities to have added facilities in conventional communication. Along with the expeditious augmentation of high-speed networks and smart devices, IoT has attained wide approval and recognition as the classical standard for low-power lossy networks (LLNs) included constrained

M. T. Oyshi (✉) · M. Z. Bonny · Z. N. Tumpa
Department of Computer Science & Engineering, Daffodil International University, Dhaka, Bangladesh
e-mail: marzan@diu.edu.bd

M. Z. Bonny
e-mail: bonny.cse@diu.edu.bd

Z. N. Tumpa
e-mail: zerinnasrin@gmail.com

S. Saha
Department of Information and Mathematics, University of Lodz, Lodz, Poland
e-mail: Susmitasaha2458@gmail.com

A. K. Tripathy et al. (eds.), *Advances in Distributed Computing and Machine Learning*, Lecture Notes in Networks and Systems 127, https://doi.org/10.1007/978-981-15-4218-3_12

resources. According to Zaslavsky et al. [1], Internet of things is allowing Internet to be expanded into the physical world and to interconnect physical objects along with the Internet. IoT devices are approximately narrow in storage, compute and network capacity. Most of the time, designers keep the security apart, hoping that it might be added later on. Hence, they are more likely to face attacks comparing to other edged devices like smartphones, computers, or tablets. In today's world, IoT is being dedicated to enhance the quality of life by connecting things to the Internet and formulating life better. But less security control might lead this amazing technology to a threat. According to Cisco Inc., it is being predicted to have about 50 billion connected devices up to 2020. This connectivity will be able to gain access of all types of personal data and sophisticated information. Without using accurate security measurements, it will allow intruders to the get the control over enormous data. This major problem has worked as a motivation of this research.

Blockchain is an undeniably innovation invention by group of people which is mainly known as Satoshi Nakamoto. Since the beginning, blockchain has been used for cryptocurrency and bitcoin. Later on, blockchain has expanded its branches in finance, accounting, agriculture, industry, medicine, supply chain, education, architecture, digital IDs, food safety, tax regulation, weapon tracking, security and many more. In recent days, blockchain is taking part in security issues. There is a huge possibility to fix security issues raised in IoT technology by integrating blockchain with it. This research discusses IoT security issues, focuses on IoT data security and proposes a method to resolve security issues using blockchain technology. The IoT security issues and their ongoing research status have been described in Sect. 2

2 Literature Review

According to the research study by Hewlett Packard Enterprise [2], around 80% of things in IoT decline the requirement of having passwords with satisfactory complexity and length, 70% facilitate an intruder to identify authentic user accounts through account listing, network services used by 70% which is unencrypted and around 60% boost security concerns along with the user interfaces. Khan et al. [3] described briefly about major security issues in IoT. Authors categorized the security issues into three segments. The segments include low-level security issues, intermediate-level security issues and high-level security issues. They outlined security requirements for IoT in conjunction with the current threats, attacks and presented the state-of-the-art solutions. Authors discussed blockchain and its possibility to solve IoT security issues. Alaba et al. [4] discussed the IoT security circumstances by providing an analysis of the possible attacks. They categorized security issues in the terms of data, communication, architecture and application. Authors prospected taxonomy for IoT security which is different than the other conventional layered architecture. Granjal et al. [5] analyzed about how current approaches ensured elementary security requirements and protected communications in IoT, together along with the strategies and open challenges for further research work in the context IoT. They mainly discussed and analyzed the security issues for the protocols which are defined for IoT. The security investigation represented in [6–8] discussed and compared cryptographic algorithms and diverse key management systems. Sicari et al. [9] discussed security, contribution providing confidentiality, access control and privacy for IoT including the security for middleware. They discussed privacy issues, trust management, data security, authentication, intrusion detection systems and network security.

Zhang et al. [10] discussed current challenges and research opportunities of IoT based on security. Authors discussed major IoT security issues in terms of exclusive identification of objects, authentication and authorization, privacy, malware, the call for software vulnerabilities and lightweight cryptographic procedures. Huh et al. [11] proposed a system to maintain IoT devices using blockchain technology. They have used RSA public key cryptosystems for managing keys. They have stored public keys in Ethereum and saved private keys in individual devices. They have used Ethereum as it allows to write Turing-complete codes with a view to run on the top of Ethereum. They have expressed their desire to work in large scale with the IoT devices in their upcoming work. Kosba et al. [12] presented a model named "Hawk" which offers a smart contract system which is decentralized and does not store any kind of financial transaction with a purpose to retain transactional privacy. Azaria et al. [13] used blockchain technology for distributed record management system on the context of medical to handle EMRs. Biswas et al. [14] proposed a security framework by incorporating blockchain technology along with the smart devices. They aimed to provide security communication platform on the context of smart city.

Kshetri [15] stated blockchain as a data structure which creates a temper proof digital ledger regarding transaction and sharing them. According to him, blockchain technology is using public key cryptography with a view to sign transactions among the responsible parties. All the transaction remains stored in the distributed ledger. Apparently, it is quite difficult to change or remove any blocks from the blockchain. The combination of blockchain and IoT is possibly to be very powerful. The combination of big data and artificial intelligence (AI) might produce more significant impact. For IoT application and platform, the main drawback is the dependency of centralized cloud in security purpose. Decentralized blockchain is capable enough of solving the centralized cloud issue to strengthen the security concerns.

Kiayias et al. [16] proposed the first blockchain protocol "Ouroboros" basing on proof of stake with precise security guarantees. They have established security properties for protocol comparable which is achieved by bitcoin blockchain protocol. They have also proposed method to neutralize attacks like selfish mining. Pass et al. [17] stated that Nakamoto's most popular blockchain protocol is capable of preventing "Sybil attacks" depending on computational puzzles which are introduced by Naor and Dwork. They have proved that blockchain consensus mechanism is capable enough of satisfying a powerful form of consistency and liveness in asynchronized network along with adversarial delays for priority bounded. Puthal et al. [18] described blockchain as a decentralized security framework. They represented a detailed overview of blockchain technology on the context of security across distributed parties in an invincible and transparent way. Analyzing above ideas, it is clear that blockchain is possible to enable security for IoT. Security for IoT devices using blockchain technology has been discussed in Sect. 4.

3 IoT Security Issues

IoT security issues mostly rise up because of spending less time and resource in security of IoT devices and services by the manufacturers. IoT is a wonderful technology but this technology is failing to be matured as this technology is failing to provide complete security services. The major security problems are given below:

- Virus
- Hacker
- Malware
- Trojan Horses
- Password Cracking
- SQL Injection
- Cross-site Scripting
- Denial of Service Attack or DDoS
- Phishing
- Data Diddling
- Sade Chinnel Attack
- Web Jacking
- Identify Theft
- Botnets
- Logic Bomb
- Email Bomb.

In terms of IoT, security might be vulnerable to both cyber and physical layer. The security issues of IoT in both layers are shown in Table 1.

Table 1 Security in terms of IoT

	Cyber	Physical
Cyber	Cyber Crimes (SQL injection,Cross-Site scripting, Logic Bombs, Phishing, Web jacking and many more)	DOS attack Data Integrity AttackForming botnet
Physical	Side Channel Analysis (Hacking what a 3D printer prints by listening to sound made by the printer)	Physical Sabotage

4 Proposed Methodology

For IoT-based solution, it is important to identify the real-life problem to get the proper solution. Field work determines if the problem identification is appropriate or not. The proper field work leads to the sensor identification. The sensor collects data from the environment. The data is stored in the database and finally represented to the end user using Web application or mobile application. But here, most of the time, data remains floating and stored in the cloud. If the data is not protected, an intruder can access and alter the data. The proposed workflow for IoT solution is shown in Fig. 1.

To secure the data, blockchain technology has been used here. Blockchain is a chain of blocks that contains information. Blockchain is widely used for cryptocurrency. However, blockchain can be used to data security as well. This technology is mainly a distributed ledger and opens to everyone. Whenever a record is entered in blockchain, it becomes very difficult to change as the blocks are connected with each other like widely shared network. Each block has three components - data, hash and hash of previous block. Block structure is shown in Fig. 2.

Data stored inside the block depends on the type of data. Hash of the block defines a block which contains all the information, and the hash is always unique. Hash is calculated by the time of creating a block itself. Any kind of change inside the block changes the hash of that correspondent block. Having the hash of previous block creates the chain of blocks and introduces blockchain. Initially, the hash of previous data block is 0 and the rest of the hash depends on the data and hash. There are number of algorithms to generate hash. In this research, SHA-256 has been used to create a unique hash. SHA-256 is a cryptographic hash algorithm. SHA-256 generates unique

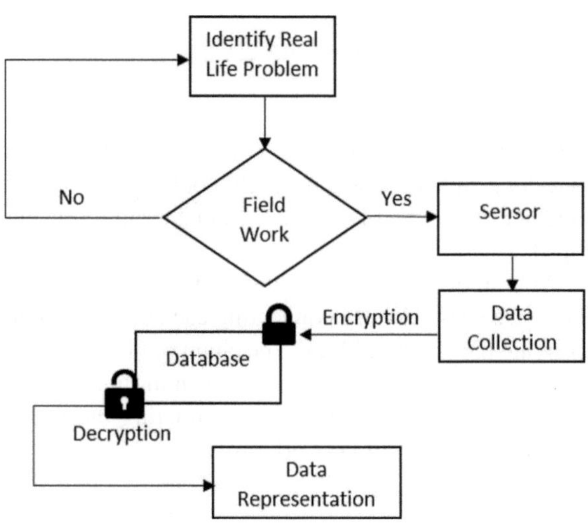

Fig. 1 Proposed workflow for IoT solutions

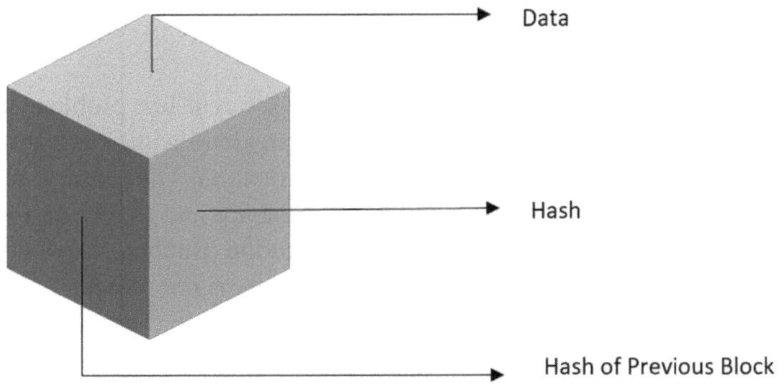

Fig. 2 Single block of blockchain

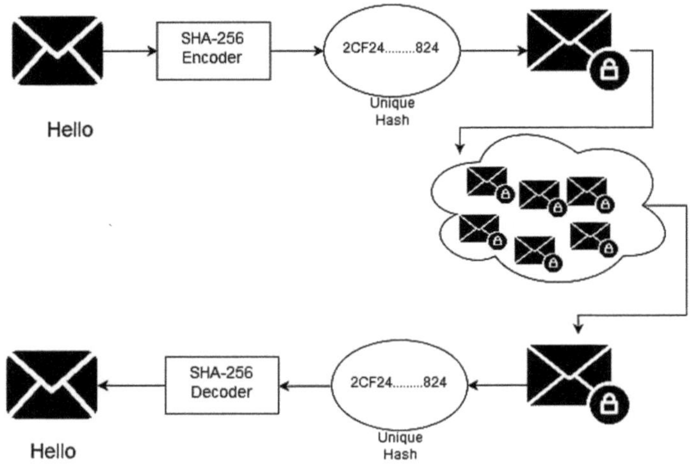

Fig. 3 Data encryption using SHA-256

256-bit (32-byte) signature text for a normal text. In this research, it generates 32-byte-long cipher text against each block by encrypting the sensor data, and the blocks together create the blockchain. Data encryption using SHA-256 has been shown in Fig. 3.

After applying SHA-256 algorithm and having the unique hash, it has been possible to create blocks. By connecting each block with one another, it has been possible to create the chain of block. The chain of blocks having hash, hash of previous block and encapsulated data has been shown in Fig. 4.

These blocks are generated eliminating the need of any kind of keys or using numerous number of network protocols.

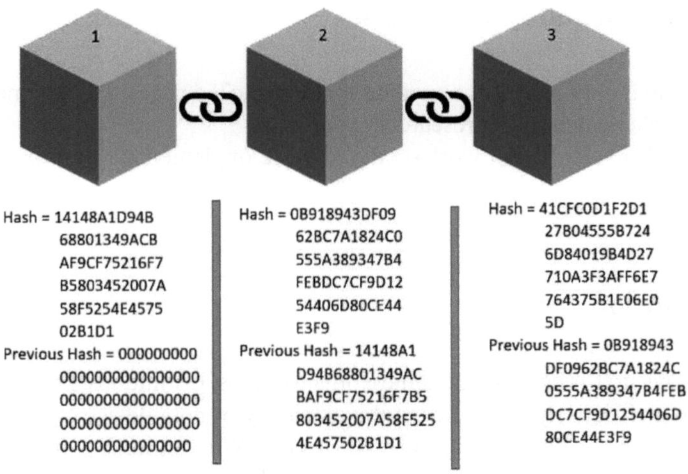

Fig. 4 Chain of data blocks

5 Case Study

Most of the IoT devices need authentication. To complete the authentication process, user needs to register with a unique username and password. Hacking a password is quite possible using password cracking. Password cracking is the process of recovering passwords from data that have been stored in or transmitted by a computer system. The multiple techniques of password cracking are given below:

- Dictionary Attack
- Brute Force Attack
- Rainbow Table Attack
- Hybrid Attack
- Guessing
- Phishing
- Social Engineering
- Malware.

Password cracking is possible to avoid by using strong password like combining letters, numbers and special characters and never repeating the passwords. Even after using a strong password, it is highly possible to face attacks and lose control on the devices. In this certain position, blockchain might be the life savior as blockchain has the power to eliminate the requirement of using a password itself by using its distributed ledger mechanism.

6 Limitation and Future Work

• This research only proposes a method to secure IoT devices, but this method has not been implemented by the researcher yet.
• The extended version will work on the real-life implementation of this proposed method.

7 Conclusion

In recent days, IoT is connecting homes, organizations, institutions, banks and many more. This connectivity is increasing the efficiency of both device and human. But this connectivity is also gaining all kind of data. Without having proper security measurement, the efficiency could easily be turned to the threat. This is the high time to think about security issues for each and every IoT devices and call for action.

References

1. Zaslavsky A, Jayaraman PP (2015) Discovery in the internet of things. Ubiquity 1–10. https://doi.org/10.1145/2822529
2. HP Study Finds Alarming Vulnerabilities with Internet of Things (IoT) Home Security Systems. https://www8.hp.com/us/en/hp-news/press-release.html?id=1909050. Last accessed 11 Dec 2018
3. Khan MA, Salah K (2017)IoT security: Review, blockchain solutions, and open challenges. Fut Gener Comput Syst 395–411. https://doi.org/10.1016/j.future.2017.11.022
4. Alaba FA, Othman M, Hashem IAT, Alotaibi F (2017) Internet of things security: a survey. J Network Comput Appl 8:10–28. https://doi.org/10.1016/j.jnca.2017.04.002
5. Granjal J, Monteiro E, Sa Silva J (2015) Security for the Internet of Things: A Survey of Existing Protocols and Open Research Issues. IEEE Commun Surv Tutor 17(3):1294–1312. https://doi.org/10.1109/comst.2015.238855
6. Roman R, Alcaraz C, Lopez J, Sklavos N (2011) Key management systems for sensor networks in the context of the internet of things. Mod Trends Appl Secur Architect Implement Appl 37(2):147–159
7. Granjal J, Silva R, Monteiro E, Silva JS, Boavida F (2008) Why is IPSec a viable option for wireless sensor networks. In: 2008 5th IEEE international conference on mobile ad hoc and sensor systems, pp 802–807. https://doi.org/10.1109/MAHSS.2008.4660130
8. Cirani S, Ferrari G, Veltri L (2013) Enforcing security mechanisms in the IP-based internet of things: an algorithmic overview. Algorithms 6(2):197–226. https://doi.org/10.3390/a6020197
9. Sicari S, Rizzardi A, Grieco L, Coen-Porisini A (2015) Security, privacy and trust in internet of things: the road ahead. Comput Netw 76(Suppl. C):146–164. https://doi.org/10.1016/j.comnet.2014.11.008
10. Zhang Z-K, Cho MCY, Wang C-W, Hsu C-W, Chen C-K, Shieh S (2014) IoT security: ongoing challenges and research opportunities. In: 2014 IEEE 7th international conference on service-oriented computing and applications. https://doi.org/10.1109/soca.2014.58
11. Huh S, Cho S, Kim S (2017) Managing IoT devices using blockchain platform. In:2017 19th international conference on advanced communication technology (ICACT). https://doi.org/10.23919/icact.2017.7890132

12. Kosba A, Miller A, Shi E, Wen Z, Papamanthou C (2016) Hawk: The blockchain model of cryptography and privacy-preserving smart contracts. In: 2016 IEEE symposium on security and privacy (SP). https://doi.org/10.1109/sp.2016.55
13. Azaria A, Ekblaw A, Vieira T, Lippman A (2016) MedRec: using blockchain for medical data access and permission management. In: 2016 2nd international conference on open and big sata (OBD). https://doi.org/10.1109/obd.2016.11
14. Biswas K., Muthukkumarasamy V (2016) Securing smart cities using blockchain technology. In:2016 IEEE 18th international conference on high performance computing and communications. https://doi.org/10.1109/hpcc-smartcity-dss.2016.0198
15. Kshetri N (2017) Can blockchain strengthen the internet of things? IT Profess 19(4):68–72. https://doi.org/10.1109/mitp.2017.3051335
16. Kiayias A, Russell A David B, Oliynykov R (2017) Ouroboros: a provably secure proof-of-stake blockchain protocol. In: Lecture notes in computer science, pp 357–388. https://doi.org/10.1007/978-3-319-63688-7-12
17. Pass R, Seeman L, Shelat A (2017) Analysis of the blockchain protocol in Asyn-chronous networks. Adv Cryptol (EUROCRYPT) 2017:643–673. https://doi.org/10.1007/978-3-319-56614-6-22
18. Puthal D, Malik N, Mohanty SP, Kougianos E, Yang C (2018) The bockchain as a decentralized security framework [future directions]. IEEE Consum Electron Mag 7(2):18–21. https://doi.org/10.1109/mce.2017.2776459

A Review of Distributed Supercomputing Platforms Using Blockchain

Kiran Kumar Kondru, R. Saranya, and Annamma Chacko

Abstract Advances in blockchain technologies like the introduction of smart contracts and providing alternative solutions to payment systems have given rise to the idea of using blockchain and smart contracts as a backbone for distributed supercomputing. Just like IPFS is used for distributed storage over the Internet, the concept of distributing computing safely came into existence. Distributed supercomputing is not new but using blockchain technology as a way of registering and rewarding the participating computers and users is novel. Here, we analyse four of the many projects, namely Gridcoin, Golem, SONM and iExec and point out their strengths and failings.

Keywords Blockchain · Distributed supercomputing · Golem · Grid · iExec · SONM

1 Introduction

Our study considers four of the many projects that try to make use of blockchain technology for the purpose of distributed supercomputing. The platforms we analyse here are Golem [1], SONM [2], Gridcoin [3] and iExec [4]. Here, we would like to stress that information related to these projects is sparse, and we have to rely on white papers, documentation, blogs, etc., for literature review. These projects are more commercially oriented, and each of these platforms offers their own unique solution to the problem of distributed supercomputing. Here we discuss the projects, their architecture and use cases. We briefly introduce the history of the attempts made

K. K. Kondru (✉) · R. Saranya · A. Chacko
Central University of Tamil Nadu, Thiruvarur, India
e-mail: kirankondru.tech@gmail.com
URL: https://cutn.ac.in/

R. Saranya
e-mail: saranya@cutn.ac.in

A. Chacko
e-mail: annjc01@gmail.com

A. K. Tripathy et al. (eds.), *Advances in Distributed Computing and Machine Learning*, Lecture Notes in Networks and Systems 127, https://doi.org/10.1007/978-981-15-4218-3_13

to utilize the underutilized CPUs of the Internet-connected computers. SETI@home [5] is one of the first such experiments. We will first introduce the concept of high-throughput computing which forms the back idea for distributed supercomputing.

1.1 High-Throughput Computing

The rapid increase in the speed of microprocessors and a greater number of computer systems being connected to the Internet ushered in a new era of distributed computing. High-throughput computing (HTC) is mainly used for scientific computing which requires a computing environment that can provide large amounts of computational power over longer periods of time [6]. This is in contrast to high-performance computing (HPC) [6]. HPC is an environment that provides a tremendous amount of computational power over a short period of time. It is usually measured in floating point operations per second (FLOPS), whereas HTC is more concerned with how many operations per month or year. More specifically, HTC is concerned with how many jobs it can complete rather than how fast an individual job can be done. HTC is well suited for scientific and engineering research where problems can be divided among multiple CPUs without the need for frequent communication of intermediate results between CPUs. HTCondor [7], developed at the University of Wisconsin and BOINC [8], developed at the University of California at Berkley are a few of the more famous HTC systems.

1.2 SETI@Home

SETI@home [5, 9] is the well-known Internet-based public volunteer computing project aimed at using the idle CPU cycles of the vast number of Internet-connected computers. The project started in 1999 at Berkley SETI Research Centre. Its main aim is to search for extraterrestrial intelligence (SETI) using the untapped CPU cycles of the Internet-connected PCs. Users of these PCs are requested to voluntarily allow their PCs to compute tasks sent by SETI server and return the results after completion. This concept called "volunteer computing" [10] succeeded.

1.3 BOINC

Berkeley Open Infrastructure for Network Computing (BOINC) [8] is an open-source platform for public resource distributed computing. BONIC is a development of original SETI@Home [5]. It supports more computationally intensive projects in various disciplines. Some of the other significant projects supported by BONIC are Predictor@home [11], Folding@home [12], Climateprediction.net [13], Climate@home

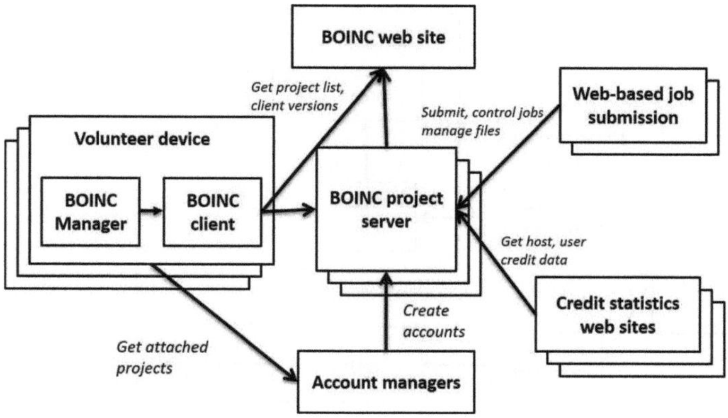

Fig. 1 Components and architecture of BOINC system

[14], CERN Projects [15], Rosetta@home [16], Einstien@home [17], etc. BOINC is mainly designed for "volunteer computing" (VC) [10]. It is the use of Internet-connected computer systems for high-throughput scientific computations. It is an open-source middleware system for VC projects. HTC encompasses many forms of computing including grid, cloud, volunteer computing, etc. VC differs from other forms of HTC in that the computers' systems used are anonymous and are not trusted, and the computers are heterogeneous both in the use of hardware and software. As such the job turnaround times are not guaranteed as the computers can go offline anytime. This is more specifically known as "device churn" [10] that is the arrival and disappearance of computers.

From Fig. 1, we can infer that BOINC consists of a project server [18] and client software [19]. Anyone with resources to share can participate in a project after registering through the project Web site. After registration, the client software is downloaded into the volunteer device taking part in the particular project. The server system is concentrated around a relational database that stores and gives out data to volunteer devices through the client software. When the volunteer device completes a task called workunit, it returns the results to the project server. Each project in BOINC has its own independent project servers. BOINC distributes "credits" to the participants on completion of the project. The credit statistics Web sites use this credit data to analyse the credits and create leader boards. The client initiates communication through HTTP. The request sent to the server is an XML document which contains details about the hardware of the volunteer device. The response from server usually contains a list of new jobs and their details (Fig. 2).

The client software consists of core client [19] which communicates with the server system and executes coordinates and monitors applications. It also contains BOINC manager. The BOINC manager provides the users with a view of the projects they are taking part in and the computation status of the respective projects. Communication between the BOINC core client and BOINC manager is via remote procedure

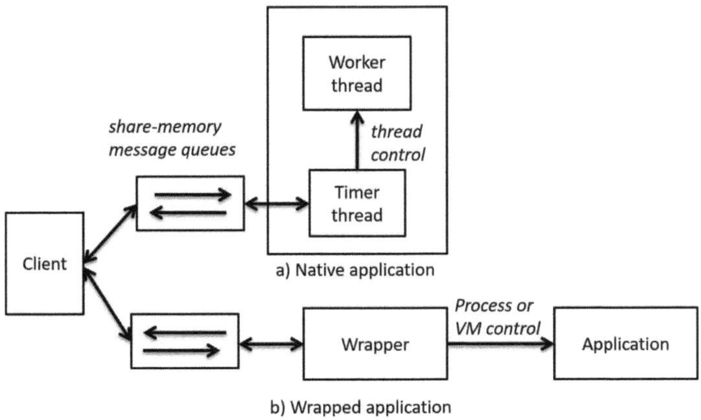

Fig. 2 Process structure of BOINC

calls (RPC) over TCP connection. The client software also contains a screensaver component which displays screensaver graphics when the system goes idle.

1.4 Drawbacks of Volunteer Computing

BOINC uses volunteer computing, but there is a limit on how many users are willingly donating their computers' idle time for the cause. The user has to pay for the power and bandwidth usage when keeping their computers on. For users to share their CPUs, there needs to be some incentive structure, a reward system. That is where blockchain and smart contracts come in. Blockchain acts as a bridge to offer a viable solution for ordinary computer users to allow their computers to be used in return for a fee. This created a new marketplace, and these newer blockchain projects are aiming at creating a marketplace for distributed supercomputing.

2 Blockchain-Based Distributed Supercomputing Platforms

The blockchain projects discussed here try to leverage the distributed nature of the blockchain network to register new members, record their contribution, rank their honesty based on their results and reward them accordingly. Smart contract execution on a blockchain is extremely costly in terms of computation as almost every node on a blockchain network has to execute the contract. Hence, the computation part is taken off-chain to a side chain. Mostly, the computation part is done in a sandbox environment like Docker containers or VMs due to safety and security issues associated with executing code from untrusted third parties. The four platforms chosen here vary widely in their approach to distributed computing. We discuss the novelty in their approach by each platform.

2.1 Gridcoin

Gridcoin is a decentralized, open-source, blockchain-based digital asset that rewards participants for participating in volunteer computing performed on the BOINC platform. The main aim of the project is to shift the computational power primarily to BOINC projects keeping the mining process secondary. As the network scales up, so too does the BOINC usage scale up in tandem. Bitcoin mining has a huge energy footprint [20]. By using alternative currencies like Gridcoin, there can be a substantial decrease in energy consumption all the while being useful for the scientific community [21].

The participant receives cryptocurrency based on the computing power contributed towards the project on BOINC. The genesis block of Gridcoin was mined in 2013 using proof-of-work [22] protocol, and later in 2014, they changed to the proof-of-research [23] a variant of the proof-of-stake protocol [22]. The BOINC work is used to measure the subsidy to be given to the participant and is a good alternative to the classic proof-of-work measuring schemes. It also complements the security of the network as any other proof-of-stake system.

The coin miners will be rewarded with small token subsidy if they fail to mine BOINC shares, but much larger subsidy while mining with BOINC. Gridcoin defines Idle Processing Potential (IPP) as the processing power of a computing device multiplied by the proportion of unused processing over time [3]. The participant receives credits when they contribute their computing power to BOINC projects. The GRC reward is distributed based on the credits [10] earned by the participant. Gridcoin consists of two types of participants. Crunchers who provide their Idle Processing Potential to the projects and stakers who secure the blockchain through the proof-of-stake protocol.

The recent average credit of a participant in BOINC can be accessed with the help of BOINC superblocks. The superblock data is converted into a Gridcoin variable called magnitude. Magnitude determines the amount of GRCResearch-Mint a participant receives. GRCResearch-Mint is distributed among participants based on their contributions towards the projects.

2.2 Golem Network

Golem [1] project began in 2014. It is an open-source project. It is aimed at providing an alternative to the existing cloud service providers which are centralized by nature, constrained by closed networks and propriety payment systems. Golem proposes a

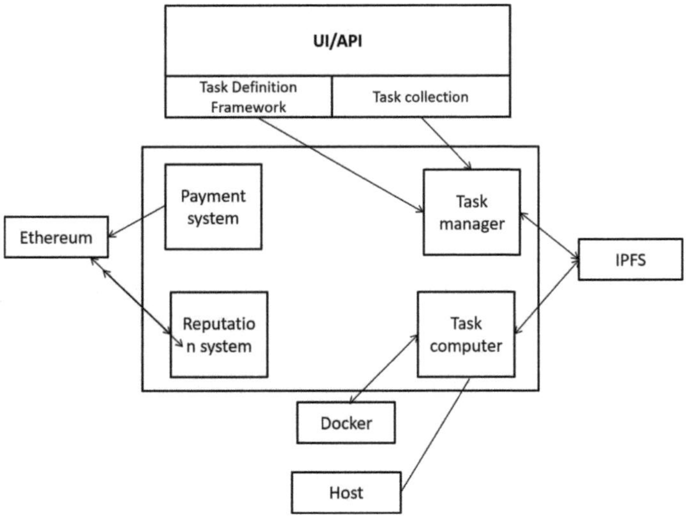

Fig. 3 Golem's high-level architecture

decentralized and user-controlled alternative. Golem tries to create a global market for distributed computing power.

The Golem network consists of three groups of people:

- "Providers" who contribute their computing resources to the Golem network
- "Requestors" who need the computing resources for computing their tasks
- Software developers.

Golem enables the requestors to become providers because there is no constant need for the resources so they can rent their resources and earn extra money. The application registry, which is an Ethereum [24] smart contract, provides a decentralized method for the developers to publish the applications. The providers have complete control over code. Payments are handled by the transaction network, and it also manages information about contributions to tasks and the results. Batch and nano-payments are accepted by Golem.

Figure 3 depicts the architecture of Golem network. The task collection contains the user's tasks. If a task is not available in the task collection, the user has to code it using task definition framework. The task manager contains all the tasks created, and it also keeps track of them. The transaction system is responsible for checking the reputation of every node in the reputation system. It rejects offers from nodes with a poor reputation. The IPFS [25] network is used for communication between the task manager and the task computer.

2.3 SONM

SONM is a powerful, global decentralized fog computing [26] platform for computing purposes. It provides a platform for users to use a large number of computing resources. Individuals with computing resources can rent their resources and can earn profit from it. It also provides an alternative to cloud computing by introducing fog computing. SONM has customers and suppliers. The customers are people and companies that create various projects and set the standards required for their projects. Customers set the capabilities and resource requirements needed for the project. Suppliers are those people and companies who own the computational resources. They earn virtual currency, SNM token in return for the computational resources they provide. SNM token is an Ethereum-based ERC-20 standard token [27]. The main feature of SONM is that it is decentralized. There are no middlemen between the customers and the suppliers. They find each other and provide resources for projects which suit them best. Machine learning, video rendering, Web hosting and scientific calculations are the main use cases of SONM.

For projects that require consensus, smart contracts [28] are implemented. Smart contracts also provide security; therefore, a major part of the market is implemented on smart contracts. The account is the Ethereum blockchain address. These accounts transfer cryptocurrencies between them. Account address helps to uniquely identify a worker in SONM. A worker is an executable module that manages the supplier's computing resources in SONM, and each worker has a unique Ethereum address. When you create an account in SONM, it generates a private key and account's address. A private key is used as a digital signature for any operation in SONM.

Every customer in SONM has one account. They can also have multiple accounts. But SONM will treat the multiple accounts created by the same user as different customers. Every supplier in SONM has an Ethereum master account which is his unique identifier. The master account receives all incomes and holds the supplier's profile and rating. It is mandatory for the account to be pointed in each worker's configuration. The computing resources are sold and bought by placing an order in the SONM market. The supplier creates ASK order to rent resources while the customer creates BID order to buy resources. On the basis of the ASK and BID orders, a deal is made between customer and supplier through a smart contract.

2.4 IExec

iExec blockchain platform aims at providing virtual distributed cloud infrastructure to high-performance computing. iExec platform uses a set of research technologies developed at research institutes like INRIA [29] and CNRS [30]. iExec heavily relies on XtremeWeb-HEP [31]. XtremeWeb-HEP is an open-source desktop-grade software which is battle-tested and has all the necessary features like data management, fault tolerance, security, accountability, multi-application, hybrid public/private infrastructure and more.

The aim of iExec is to develop a scalable high-performance manageable and secure infrastructure sidechain that will provide the new form of distributed cloud computing with Ethereum being used for payment services. Though some blockchains have support for Turing-complete language like solidity, they are severely constrained when it comes to the cost of computing, as the code in the smart contract has to be executed in all the nodes in the network. As such the (distributed) computation part is shifted off the chain to a side chain. iExec is no different from the rest of the platforms discussed herein adopting this solution.

iExec uses Ethereum for awarding resource providers with cryptocurrency thereby encouraging more and more individual computer users to participate in this marketplace and to get a monetary reward for the services provided. For big data applications, often cloud and HPC infrastructures are chosen. However, both cloud and HPC infrastructures are expensive, and the data centres which have clusters of processors tightly placed together generate an enormous amount of heat and the cooling requirements for these data centres are huge and contribute to not only huge electricity bills but also environmentally harmful as their dependence on coal-powered plants for electricity.

In the iExec platform after signing of the smart contracts [32], the consumer is required to divide the application into smaller tasks and provide them to the scheduler. Here the scheduler does not send the tasks to the providers, but the providers instead ask for new tasks where they are unoccupied. The contract that is signed in the Ethereum must contain its execution code and a description explaining its aim. This structure is very similar to Request for Comments (RFCs[33])of the Internet Engineering Task Force (IETF) [34]. These signed contracts are submitted to peers for review and approval.

In order to encourage honest producers and dissuade malicious behaviour, this platform too has a reputation system to monitor and rank workers. iExec has a novel solution based on workers' reputation, the replication of services and the incentive mechanism. Here, the consensus defines the confidence required for the outcome or the result, which, in turn, is directly related to the costs. Providers have to put upfront a security deposit (stake).

At the end of the execution, the providers, after the verification of the result, split the payments and stakes while the reputation of nodes with malicious behaviour decreases. This protocol is resilient against Sybil-like attacks [35] as a stake from the nodes is a requirement, thus attacking the system is counterintuitive. The project also provides an application store, where providers advertise ad consumers choose among different payment schemes. A market management framework is used to register bids and to provide templates. iExec supports OpenCL framework and offers API integration to run applications on a wide variety of heterogeneous platforms.

3 Challenges

The above platforms use either Ethereum for payments or have their own cryptocurrency system in place. The advantage of having a separate existing payment system like Ethereum is that it provides guaranteed distributed consensus, defining smart contracts, managing payments and guaranteeing identity. It also takes the load off from specifically developing their own solution while also having scope for integration of these projects with similar projects in the future scenario.

The biggest drawback of using existing cryptocurrencies like Ethereum is the transaction cost. To reduce this, each platform offers its own unique solution. SONM uses Ethereum only to transfer funds to their side network. Golem uses probabilistic scheme, though it has some unique requirements that have to be met in order for the scheme to work. Here only with the participation of a large number of consumers must adopt this scheme. This, then actually guarantees fair distribution of incomes but only in the long term.

Services are compulsorily executed off-chain in a separate peer-to-peer network due to the known issue of the cost of computation on Ethereum. Another big challenge for distributed computing is the verification of the completed results. This verification is integral to every platform mentioned above as there could both honest and malicious users in the network. Golem adopted three methods to verify the completed results: (1) log verification, (2) correctness checking of the results and (3) redundant computation.

There are still issues with each of the mentioned methods. Logs can be altered or deleted with ease once access is gained on that node. Correctness checking only works on some class of problems and that too for stateless applications. Malicious users can detect and act accordingly to deceive the consumer. When a large number of nodes are provided with a single duplicated task, the reward would significantly decrease making it unprofitable, and hence, dissuading participation.

The important component of verification is the reputation or economic penalties. There should be a balance between how severe the penalties are and the entry cost of the market. Golem and iExec rely on oracle which uses the above-mentioned verification system to check whether the computation is successful or not. If successful, the oracle will trigger the smart contract to execute when terms are met. The catch is that the node acting as oracle (an intermediary) itself must be trusted, as it can also provide arbitrary information irrespective of the truth. As such, the oracle directly influences the decision-making process. There are proposals for decentralized oracles to mitigate the dependence on centralized authorities.

In a decentralized environment, where a user can be a provider or resources, such network ecosystem would be composed of highly heterogeneous software and hardware platforms along with varying network speeds. Also, privacy laws vary from one country to another, whereas the nodes can be anywhere in the world. There is also the question of what happens after privileged information is leaked and who would be held accountable.

4 Conclusion

The projects discussed here approached differently to a similar concept. They are more targeted towards commercial adoption rather than as for proving the concept for academic purpose. The success and adoption of these projects are mixed but have room to grow. However, there are key issues still to be addressed by them. Issues like users overextending their rented processor time, causing damage to the hardware, or to scope for vulnerabilities return later for an attack. Human–computer interface too is important. It must be generalized enough to adopt easily but should have the flexibility to accommodate different kinds of processing tasks. Though there are challenges and issues to be addressed and not just mitigated, there is a resurgence of distributed computing at a time where more power is concentrated around big tech companies like Facebook, Amazon, Google, Microsoft, etc. And blockchain is helping these projects to provide a secure layer of trust in an untrusted environment, thereby making centralized mediators unnecessary. Though the future cannot be predicted, looking at the increasing number of blockchain platforms trying to address the problem of reliable distributed computing, the future of these technologies looks promising.

References

1. Zawistowski J (2016) The GOLEM Project. Golem Whitepaper 28
2. sonm-whitepaper @ whitepaper.io. https://whitepaper.io/document/326/sonm-whitepaper
3. Gridcoin: the computation power of a blockchain driving science and data analysis, pp 1–12
4. Fedak G, He H, Moca M, Bendella W (2018) Eduardo Alves: iExec—blockchain-based decentralized cloud computing, p 40
5. Anderson D, Cobb J, Korpela E, Lebofsky M (2002) SETI @ home. Commun ACH 45:56–61
6. Eze K (2018) High-throughput computing, pp 2–4
7. Thain D, Tannenbaum T, Livny M (2005) Distributed computing in practice the condor experience. Concurrency comput Practice Exp
8. Anderson DP (2004) BOINC: a system for public-resource computing and storage. In: IEEE/ACM International Workshop on Grid Computing, pp 4–10 (2004). https://doi.org/10.1109/GRID.2004.14
9. SETI@home. https://setiathome.berkeley.edu/
10. Anderson DP (2018) BOINC: a platform for volunteer computing, pp 1– 37
11. Taufer M, An C, Kerstens A, Brooks CL (2006) PredictorHome: a protein structure prediction supercomputer based on global computing. IEEE Trans Parallel Distrib Syst 17:786–796. https://doi.org/10.1109/TPDS.2006.110
12. Folding@Home. https://foldingathome.org/, last accessed 30 Dec 2019
13. Climateprediction.net. Last accessed 30 Dec 2019
14. Xu C, Yang C, Li J, Sun M, Bambacus M (2011) Climate@Home: crowdsourcing climate change research
15. LHC@Home. http://lhcathome.web.cern.ch/
16. Rosetta@Home. http://boinc.bakerlab.org/rosetta/. Last accessed 30 Dec 2019
17. Einstein@Home. https://einsteinathome.org/
18. BOINC. https://boinc.berkeley.edu/trac/wiki/ServerIntro
19. BOINC Client. https://boinc.berkeley.edu/wiki/BOINC_Client

20. O'Dwyert KJ, Malone D (2014) Bitcoin mining and its energy footprint. IET Conf Publ 2014:280–285. https://doi.org/10.1049/cp.2014.0699
21. Chohan U (2018) Environmentalism in cryptoanarchism: Gridcoin case study. SSRN Electron J. https://doi.org/10.2139/ssrn.3131232
22. Alahmad MA, Al-Saleh A (2018) AlMasoud FA (2018) Comparison between PoW and PoS systems of cryptocurrency. Indones J Electr Eng Comput Sci 10:1251–1256 https://doi.org/10.11591/ijeecs.v10.i3.pp1251-1256
23. Proof of Research—Gridcoin. https://en.wikipedia.org/wiki/Gridcoin
24. Wood G (2017) Ethereum: a secure decentralised generalised transaction ledger. EIP-150 REVISION. 2017, p 33
25. Benet J (2014) IPFS—content addressed, versioned, P2P file system
26. Fog Computing. https://en.wikipedia.org/wiki/Fog_computing
27. ERC-20 Token. https://en.wikipedia.org/wiki/ERC-20#Technical_standard
28. SONM Smart Contracts. https://docs.sonm.com/concepts/underlying-technologies/smart-contracts
29. INRIA. https://www.inria.fr/en/. Last accessed 11 Nov 2019
30. CNRS. http://www.cnrs.fr/. Last accessed 11 Nov 2019
31. XtremWeb-HEP: middleware for distributed data processing. https://www.projet-plume.org/en/relier/xtremweb-hep
32. Smart Contract. https://en.wikipedia.org/wiki/Smart_contract
33. Request for Comments. https://en.wikipedia.org/wiki/Request_for_Comments
34. Internet Engineering Task Force. https://en.wikipedia.org/wiki/Internet_Engineering_Task_Force
35. Newsome J, Shi E, Song D, Perrig A (2004) The sybil attack. 259: https://doi.org/10.1145/984622.984660

An E-Voting Framework with Enterprise Blockchain

Mohammed Khaled Mustafa and Sajjad Waheed

Abstract Electronic voting or E-voting system is being widely used in both corporate and government voting system for its ease of use and efficiency. It has proved some significant advantages over the paper-based one, but vital key elements of voting remain within the later system like privacy, anonymity, etc. The evolution of Blockchain technology with advanced cryptographic techniques paved the way to solve the existing drawbacks of the E-voting system with additional new features. In this paper, we proposed a scheme of the E-voting system based on permissioned or Enterprise Blockchain. The solution proposed here is suitable for online voting of small to medium-scale enterprise, student representative election in universities, or conducting an election for professional bodies. The Hyperledger Fabric (HLF) platform was used in the core of the framework along with blind signature and zero-knowledge proof (ZKP) through Idemix to establish security, un-reusability, transparency, and privacy or receipt-freeness of the voting system.

Keywords E-voting · Security · Blockchain · Privacy · Hyperledger fabric · Zero-knowledge proof

1 Introduction

Election or voting is the cornerstone of modern democracy. The paper ballots are still widely used across the world for its reliability over the years. This voting process is quite cumbersome in the modern context because voters must come to the polling station physically by spending his/her own time and money. Besides, the people who are out of the designated area cannot participate. These issues hamper the enthusiasm

M. K. Mustafa (✉)
Bangladesh University of Professionals, Dhaka, Bangladesh
e-mail: khaled95@gmail.com

S. Waheed
Mawlana Bhashani Science and Technology University, Tangail, Bangladesh
e-mail: sajjad@mbstu.ac.bd

A. K. Tripathy et al. (eds.), *Advances in Distributed Computing and Machine Learning*, Lecture Notes in Networks and Systems 127, https://doi.org/10.1007/978-981-15-4218-3_14

of youngsters and senior citizens and lead to very low turnout in the voting process. So, it may not reflect genuine public opinions all the time.

On the other hand, E-voting is a growing rapid popularity among the voters for its easiness and flexibility. It can easily attract from tech-savvy youngsters to senior citizens for the comfort to participate. Moreover, expatriates or voters staying outside the designated area can also cast their votes in favor of their candidates from any corner of the world.

In high level, an ideal E-voting system should have the following attributes:

1. Security: The system should have CIA triad (confidentiality, integrity, and availability) of security and should have the latest security features and patches installed.
2. Un-reusability: The voters can cast their votes once and only once.
3. Transparency: The over process should be transparent and fair to the stakeholders, so that they can acknowledge and accept the result of the voting system.
4. Immutability: The system should be tempering-free and protected from any undue influence. The essence is, no one will be able to manipulate the result.
5. Privacy or receipt-freeness of the voting system: One of the most fundamental attributes of a voting system is ballot or voter's privacy. It means that an adversary is not able to reveal the way a specific voter has voted. The voter opinion has to remain anonymous to the stakeholders and un-traceable against a casted vote.

The recent evolution of Blockchain technology has attracted researches for its features in distributed ledger technology (DLT) and smart contract. A Blockchain-based platform has the following criteria [1]:

(I) Immutability: Each "new block" to the ledger must come through the reference to the previous block of the ledger. Thus, an immutable chain of blocks is created; this is how the name of Blockchain comes from. In Blockchain, it is practically impossible to change the blocks once created.
(II) Verifiability: Blockchain is a decentralized and distributed ledger over multiple locations. This eliminates the single point of failure (SPoF) and ensures high availability. The participants in the Blockchain system are updated on each change about the transaction in the ledger.
(III) Distributed Consensus: Consensus protocol controls who can affix the new transaction to the ledger. A majority of the nodes must reach a common consensus while appending a new block of entries becomes part of the ledger.

Blockchain platforms are mainly divided into two types

1. Public Blockchain and
2. Private Blockchain.

The public Blockchains are permissionless where anyone can join without prior permission and can become a member of the network. Normally, the popular Blockchain technologies like Bitcoin and Ethereum are examples of it. The private Blockchains are also known as Enterprise Blockchains are always permissioned where only the authorized members are allowed to participate. The private

Blockchain is confined to a limited number of entities to serve specific purposes where security, access control is well maintained on top of basic Blockchain features. Hyperledger Fabric and Exonum are examples of this type of Blockchain.

This paper developed a scheme for small to medium-scale level voting which is confined to a limited or selective people rather than to general voting purpose. Moreover, the security and privacy of the system and voters are the essences of this scheme. Analyzing the recent research on Blockchain-based E-voting and to obtain the basic attributes of the voting system, we selected Hyperledger Fabric (HLF) for our scheme.

1.1 Our Contribution

In this paper, we proposed a Blockchain-based E-voting solution suitable for SME enterprises, student council election, or general election of professional bodies. Here, we employ HLF in the core of the system and advanced cryptographic features like zero-knowledge protocol (ZKP) and blind signature are amalgamated with HLF for privacy for the voters. We assume that the participants are familiar with the online voting system having the necessary access devices.

1.2 Outline

The rest of this paper is organized as follows. In Sect. 2, we will share our view on related works of E-voting. In Sect. 3, we introduce the techniques used in our solution for a better understanding of the system. In Sect. 4, we will present the structure or the framework of our scheme and its components. In Sect. 5, we will describe the transaction flow of an E-voting in our scheme. In Sect. 6, we will discuss the outcome of the proposed solution. Finally, in Sect. 7, we will conclude our analysis along with the suggestions of future work.

2 Related Work

Developing a transparent and privacy-preserving E-voting system is quite complex as there are several attributes to ensure. Therefore, different approaches along with varying technical solution are explored to achieve the desired quality of E-voting system. Here, we only review the papers based on the Blockchain-based E-voting system. Researchers had employed many cryptographic techniques like the blind signature, ring signature, zero-knowledge proof, Paillier cryptosystem, etc., with Blockchain to develop a full proof E-voting platform.

Blind signature using Certified Authority (CA) over the public key was proposed by Sheer [2]. Here, every node authenticates the signature and appends it to the ledger, and the voter gets his confirmation ID. The use of Blockchain technology ensures that the voters can verify that their votes were taken into account, and it is cast accordingly. The proposed solution achieves a certain degree of anonymity by using blind-signed public keys, but receipt-freeness is missing.

An Ethereum-based E-voting platform was implemented by Koç [3]. The system did not employ any cryptographic techniques; hence, any participant can see the contents of transactions. Moreover, for every call to smart contract, voter needs to incur some cost. The system lacks anonymity and receipt-freeness too.

In the proposed model by Liu [4], the voter generates two pairs of keys in two stages. Here, the first public key to the CA is along with the data for authentication. Once it is endorsed by the controlling peers, then other keys are generated for the final voting. This model satisfied most of the criteria except the receipt-freeness.

Yu [5] proposes the use of the Paillier cryptosystem and the short-linked ring signature (SLRS) as cryptographic techniques in his approach. The approach satisfies almost all the requirements for an E-voting system except the receipt-freeness can be bypassed a pool of encrypted zeros that are added to the encrypted voter's voice to bring some randomness. The use of both ring signature and the blind signature was proposed by Yan [1] in an Ethereum-based platform for boardroom voting. The proposed scheme achieves the privacy of the voters and block reusability but it based on the public Blockchain system. So we can conclude as below:

1. None of the approaches can fully satisfy all the attributes of an ideal E-voting system.
2. Voter's privacy or receipt-freeness seems to be the most complex part to manage.
3. The approaches can be used for the voting purpose only.
4. No such approaches for SME level voting.
5. None of them use off-the-shelf product that can be used for other purposes.

To overcome the limitations and to satisfy our objectives, we are proposing an E-voting model based on Hyperledger Fabric, which is an open and standard platform by Linux foundation [6, 7] and employed blind signature and zero-knowledge protocol (ZKP) inside HLF component using Chaincode to achieve the E-voting attributes.

3 Used Technologies

3.1 Hyperledger Fabric

The Hyperledger Fabric (HLF) has a modular and flexible architecture. It supports open and standard protocols. Figure 1 below here shows a high-level representation [6, 7] of HLF including its different functionality.

HLF has the following basic modules:

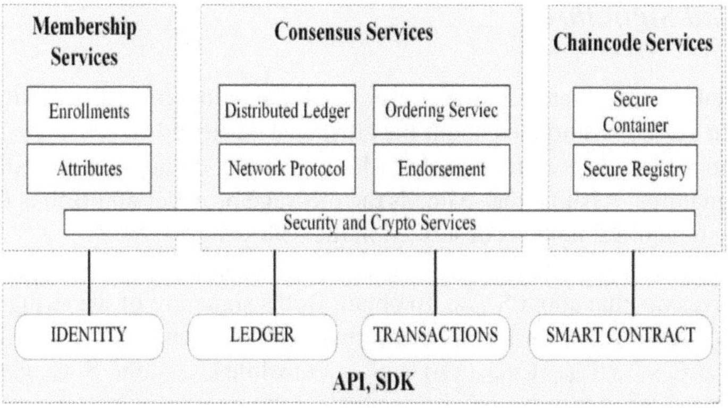

Fig. 1 Hyperledger Fabric (HLF) reference architecture

(a) Membership services: This module takes care of authentication and access control. It identifies which certification authorities (CAs) are trusted to define the members of a trusted domain and roles of peer in the network.
(b) Chaincode services: Smart contracts mainly contain the business logic of HLF. Chaincode is the superset of the smart contracts, i.e., the smart contracts manage the business logic while Chaincode manages the smart contracts defined in it. All transactions run by Chaincodes are stored on the ledger.
(c) Consensus services: It permits digitally signed transactions to be proposed and validated by network members. In HLF, the consensus protocol is pluggable and interconnected to the endorse-order-validation process. The ordering services represent the consensus system in HLF. The guideline to authorize the validity of a transaction by any peer is defined in the endorsement policy.

Inside the above module, the key elements of HLF are:

Ledger: Ledgers store all immutable historical records of every transition.

Node: Nodes are the logical entities of a Blockchain network. There are three types of nodes:

- Clients are used to submit transactions to the network through the users.
- Peers commit transactions in Blockchain and preserve the ledger state.
- Orderers create communication channel between clients and peers.

Channels are isolated communication route for transferring confidential data between multiple network members. Activities on a channel are only visible to associated and authorized entities.

Endorsers invoke smart contracts to validate transactions and send the results back to the applications.

In addition, in our scheme, we used the Idemix feature of HLF which is available with HLF version 1.2 or higher. Idemix is briefly discussed in the later subsection.

3.2 Blind Signature

Blind signature [8] scheme allows a user to obtain signatures on attributes from a signer even he/she has no idea about the attributes he signed.

In our model, it plays a crucial role by hiding voters' choices on the ballots while getting signatures. Assume that Alice is the message provider and Bob is the signer. The blind signature scheme is described in the following.

Bob has a signing function S'_{Bob} which inverse is S_{Bob}. It satisfies $S_{Bob}(S'_{Bob}(x)) = x$, but gives no clue about S'_{Bob}. To obtain Bob's signature of the string s without revealing it, Alice relies on a computing function C_{Alice} and its inverse C'_{Alice} and satisfies that $C'_{Alice}(S'_{Bob}(C_{Alice}(x))) = S'_{Bob}(x)$ while C_{Alice} and S'_{Bob} give no clue about x. The signing procedure is presented as follows:

1. Alice sends $C_{Alice}(s)$ to Bob.
2. Bob receives $C_{Alice}(s)$ and signs it using S'_{Bob} to obtain $S'_{Bob}(C_{Alice}(s))$. Then, he sends $S'_{Bob}(C_{Alice}(s))$ back to Alice.
3. Alice uses C'_{Alice} to get $S'_{Bob}(s)$ according to $C'_{Alice}(S'_{Bob}(C_{Alice}(s))) = S'_{Bob}(s)$. Thus, Alice gets $S'_{Bob}(s)$, the signature of s signed by Bob, without revealing s.

3.3 Zero-Knowledge Proof (ZKP)

A zero-knowledge proof is a unique method where a user can prove to another user that he/she knows an absolute value, without actually conveying any extra information. The main essence behind this concept is to prove possession of knowledge without revealing it. In HLF, privacy-preserving authentication and transfer of certified attributes can be done using Idemix [9, 10]. It provides the following:

Anonymity: Sending transactions without having to reveal your identity.
Unlinkability: Sending multiple transactions without revealing that all the transactions come from the same source.

3.4 Idemix

Idemix is a cryptographic protocol suite [10] of HLF, which offers privacy-preserving features such as anonymity, i.e., conducting a transaction without revealing the identity of the user. Idemix technology is built from a blind signature scheme and efficient ZKP of signature possession. All of the cryptographic building blocks for Idemix were published at the top conferences and journals and verified by the scientific community. Idemix employs the Camenisch–Lysyanskaya (CL) signature scheme [11] specifically designed to have efficient zero-knowledge proofs.

Fabric's implementation of Idemix consists of three components:

- Idemix crypto-package with cryptographic algorithms (key-gen, signing, verification, and ZKP).
- MSP for verifying the transactions using the Idemix crypto-package.
- A CA service for distributing E-cert credentials using the Idemix.

HLF has three components with Idemix those are:

- The Fabric SDK is the API for the user.
- Fabric CA is the Idemix issuers.
- Idemix MSP is the verifier.

Idemix ensures unlinkability for both the user and verifiers since even the CA is not able to link proofs to the original credential. Neither the issuer nor a verifier can tell whether two proofs were derived from the same credential or two different ones.

4 The Framework

With used technology, here we propose a framework to manage the E-voting process through Enterprise Blockchain features and in permissioned environment to achieve transparency and integrity throughout the process. Hyperledger Fabric is the main driving engine of the framework. The HLF maintains a business network connects the different stakeholders of the voting system through the client software development kit (SDK). Figure 2 shows the proposed framework along with their different component.

In the figure, voters are connected with the framework for casting votes. The traditional parties of a voting process like polling agents and polling admin (presiding officers) are directly involved to manage the voting process. The returning officers will be the ultimate authority to declare the result of the election and the quality of the overall process. The access level and authority of different parties will be defined through MSP.

All the desired voting logics are coded through smart contract, and Chaincode (CC) manages and interconnects the smart contract as required. CC utilizes cryptographic techniques like blind signature and ZKP (with the help of Idemix of HLF) to ensure the desired criteria of the voting system.

Moreover, to certify fairness and transparency, the framework allows connection from any external monitor agent. The Blockchain-based ledger in the framework remains immutable, and the authorized peers can access the transactions for auditing purposes.

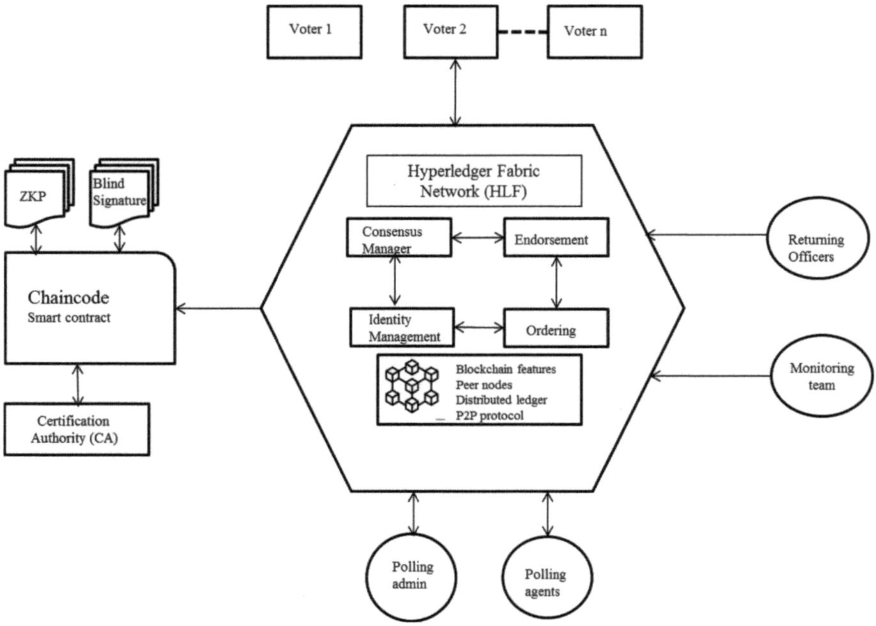

Fig. 2 E-voting framework based on Hyperledger Fabric

5 Transaction Flow

HLF has a flexible, modular, and configurable architecture and supports open and standard protocols. The framework employs the execute-order-validate pattern which decouples the execution and endorsement of a transaction from its total ordering and committing to the ledger and opens the door for parallel execution of independent transactions [7].

Every voter needs to connect with the voting platform through Fabric SDK. Before participating in the voting process, all the voters have to register in the voting system with their credentials. Here, the voters have to obtain their public key pairs from the certification authority (CA) and sign up through Membership Service Provider (MSP).

Figure 3 shows the transaction flow inside the framework for an E-voting process. The brief steps are discussed as per the following steps:

1. The pre-registered voters will collect their electronic ballot paper with their credentials through Fabric SDK. They will encrypt their ballot and identity and will submit it to the endorsing peer with blind signature.
2. The endorsing peers (will play the role of polling agents and polling officers) receive encrypted voting information from the voter and verify it using the Chaincode, MSP, and voter's signature. Here, the endorsing peer will only verify the eligibility of the voter and will have no clue about the content in the ballot.

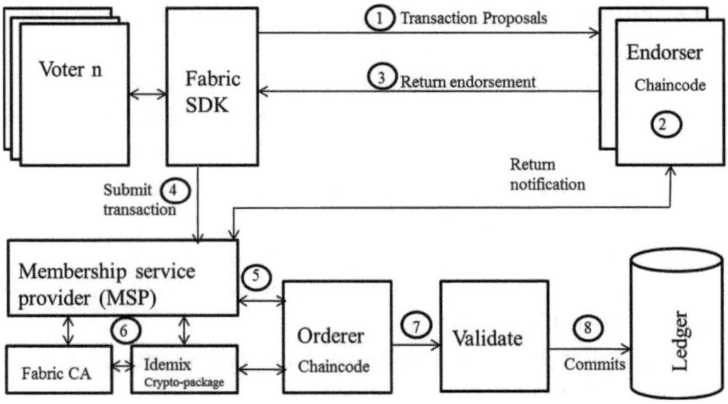

Fig. 3 E-voting transaction flow in Hyperledger Fabric framework

3. Once verified, the endorsing peers execute the transaction into the ledger and send signed read–write set response back to the voters.
4. The voters receive and verify the response from endorsing peers. They now submit the responses to Idemix crypto-packages through SDK for casting the vote.
5. The Idemix MSP then verifies the proposal and sends the transaction proposal using Fabric CA and ZKP to different orderer peers [10]. The orderer services will send the voting proposal to MSP to use the Idemix crypto-packages (with ZKP).
6. The orderer peers will verify the transaction requests from Idemix through different attributes. The orderer peers will receive multiple transactions for the same voting request thus it cannot practically trace the true identity of the voter even though it was verified by the peer. Moreover, the peers will have no ideas about the content of the ballot.
7. The ordered peers will only confirm the transaction proposal and will convey the ballot information to validating peers.
8. Once the peers validate transactions, ballot content will be appended to the blocks of the ledger. The transactions are then committed to the state database.

After a stipulated time, the process will close. The returning officer will send the voting database to smart contract for the final result.

6 Discussion

The proposed model can meet the desired attributes of an E-voting system as mentioned earlier (security, un-reusability, transparency, immutability, and privacy or receipt-freeness).

Blockchain utilities can make a framework verifiable and immutable thus the transparency is achieved. In the inclusion of HLF, it enhances the security and access control of the system. The components of HLF like MSP, Fabric CA, and SDK guarantee a high level of security and can eliminate any chance of manipulation.

The main challenge lies in the privacy or anonymity of voters against casted votes. The HLF version after 1.2 has the Idemix feature which can validate a transaction without revealing the content of the transaction. Along with ZKP, Idemix makes sure that the voter's choice will remain anonymous and un-linkable, i.e., it can only be possible to guess up to a certain level about a voting decision but can never guaranty about of the voter's real ID.

Moreover, the system has some additional basic advantages over the traditional Ethereum-based voting system.

In the framework, the smart contracts for Chaincode can be written on traditional programming languages like Java, GO, etc. So, additional programming language skill is not required for the developers.

HLF is an off-the-shelf product that is maintained by the group of tech-giants like IBM, Linux Foundation, etc. So, the continuous development and expert support will be available.

The platform can be used for a boardroom meeting where individuals want to give their opinion anonymously. Moreover, corporate offices can run an internal survey among the employee without the help of a trusted third party.

Lastly, with such platform, anyone can convey his/her message fearlessly without the risk of identity trace back.

7 Conclusion

In this paper, we have presented a framework for Blockchain-based E-voting model with Hyperledger Fabric to conduct a medium-scale voting requirement for SME, corporate offices, or universities. The purpose of the model was to develop a voting system to ensure security, un-reusability, transparency, and immutability along with the privacy of the voter. The inclusion of blind signature and ZKP through Idemix with permissioned Blockchain technology delivers the accomplishment of desired criteria. We believe that the platform has the potential to conduct large-scale voting like the national election.

7.1 Future Work

The platform requires further research to prepare for the large-scale voting and faster transactions. A detail forensic analysis is required on the implemented system to identify the real-time drawback of the implemented system. Some future use cases like an anonymous survey or board room decision, sharing compliance evidences preserving, etc., can be explored for future works.

References

1. Zhu Y, Zeng Z, Lv C (2018) Anonymous voting scheme for boardroom with blockchain. College of Information and Electrical Engineering, China Agricultural University, Beijing, 100083, China
2. Sheer H, Freya G, Apostolos A, Raja N, Markantonakis K (2018) E-Voting with blockchain: an E-voting protocol with decentralisation and voter privacy
3. Koç AK, Yavuz E, Çabuk UC, Dalkılıç G (2018) Towards secure e-voting using ethereum blockchain. In: 6th international symposium on digital forensic and security (ISDFS), Antalya, 22–25 March 2018. https://doi.org/10.1109/ISDFS.2018.8355340
4. Liu Y, Wang Q (2017) An e-voting protocol based on blockchain. IACR cryptology ePrint archive
5. Yu B, Liu JK, Sakzad A, Nepal S, Steinfeld R, Rimba P, Au MH (2018) Platform-independent secure blockchain-based voting system. In: Chen L, Manulis M, Schneider S (eds) ISC 2018. LNCS, vol 11060. Springer, Cham, pp 369–386. https://doi.org/10.1007/978-3-319-99136-820
6. https://www.hyperledger.org/projects/fabric
7. https://hyperledger-fabric.readthedocs.io
8. Chaum D (1983) Blind signatures for untraceable payments. Adv Cryptol pp 199–203
9. https://www.hyperledger.org/blog/2019/04/25/research-paper-validating-confidential-blockchain-transactions-with-zero-knowledge-proof
10. https://hyperledger-fabric.readthedocs.io/en/release-1.2/idemix.html
11. Camenisch J, Lysyanskaya A (2004) Signature schemes and anonymous credentials from bilinear maps. Ann Int Cryptol 2004:56–72

Survey of Blockchain Applications in Database Security

Vedant Singh and Vrinda Yadav

Abstract Blockchain is an implementation of distributed computing which provides decentralized solutions for systems that work in a centrally governed manner. It provides services such as smart contracts to establish consensus, immutability of data and decentralization of data. Due to these services that blockchain puts to the floor, it becomes a viable option for the data storage systems that thrive for security, privacy and confidentiality in an automated and decentralized manner. This paper investigates the current state of research in the field of blockchain as a service in providing security to databases. Five major types of data storage implementations which include blockchain as a tool to provide security solutions are discussed in this paper. Comparison studies of these implementations are done with respect to the security solutions provided. Further, some research questions are discussed which need to be addressed.

Keywords Information security · Blockchain · Distributed storage · Distributed computing

1 Introduction

Blockchain was introduced as the foundation for bitcoin [1], a decentralized peer-to-peer digital currency. The linked block structure of blockchain facilitated maintenance of a decentralized structure which was demanded by a system like bitcoin. The blockchain structure also provides solutions to problems like maintaining the order of transactions and avoiding the double-spending problem.

V. Singh (✉) · V. Yadav
Centre for Advanced Studies, Dr. A.P.J. Abdul Kalam Technical University, Lucknow, Uttar Pradesh, India
e-mail: singhvedant111@gmail.com

V. Yadav
e-mail: vrinda@cas.res.in

A. K. Tripathy et al. (eds.), *Advances in Distributed Computing and Machine Learning*, Lecture Notes in Networks and Systems 127, https://doi.org/10.1007/978-981-15-4218-3_15

Blockchain in its early days was restricted in the field of finance and banking [2]. The evolution of various cryptocurrency platforms based on blockchain frameworks came into being. Blockchain grabbed interest from the researchers and experts of various fields like the Internet of Things [3], education, manufacturing, governance [4] and security. Blockchain provides a distributed ledger technology with which a centrally governed system is made decentralized.

This paper discusses about the application of blockchain technology in the field of database security. Database security has issues related to integrity, availability and confidentiality of data. In conventional database security, the users completely rely on trusted third parties [5] that provide the database service for storing and transferring data securely. This review paper attempts to provide an insight into the various research work in sub-area of using blockchain for providing data security.

Blockchain is a *distributed chained append-only timestamped ledger* [6]. This structure of blockchain helps in designing an independent peer-to-peer network in which each individual can interact with other members of the network without the need of a central trusted authority. As the name suggests, blockchain is a chain of blocks. Each block is linked together cryptographically with the previous block in the chain. The blocks in a blockchain store hash of previous blocks hence maintaining traceability of data. If an entity tries to manipulate the transactions in a block, then it will be required to update the hash value which will need the consensus of the network and causes high computational power consumption. This makes blockchain robust against such attacks. The blocks also store an arbitrary one time nonce which is a 4 byte value which defines the level of difficulty for adding a block in the blockchain. A blockchain block also contains transactions that are log details of the interaction between different entities in the blockchain network. Figure 1 illustrates the general structure of the blockchain block along with the chaining of the block with its previous block.

The paper is organized as follows: In Sect. 2, discussion about the different blockchain- based database security solutions is done in detail. Sect. 3 deals with future scope of research which could happen in the field of database security using blockchain through some research questions. Section 4 provides a conclusion to the paper.

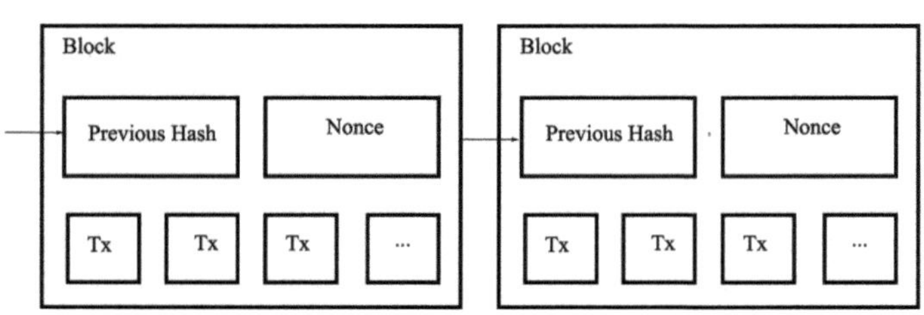

Fig. 1 Conventional blockchain block

2 Blockchain-Based Database Security Solutions

Databases require critical attention to provide a means to securely store, manage and analyze huge amount of complex data to find out trends and patterns that provides cognitive insight into purpose of data storage. There are various security challenges and issues linked with the security of databases that are broadly classified into the following:

1. Privacy and integrity
2. Consistency of data
3. Access and permission control
4. Intrusion detection and damage control.

To study the various applications of blockchain in data security, an in-depth survey was done. Research papers on the usage of blockchain to facilitate security of database or general data were included in the study. The most recent research papers were considered from various sources such as IEEE, Elsevier, Springer Nature and PLOS ONE. The analysis was documented according to the contribution and the future work possible in the paper. In our study, five main research papers were identified based on the usage of blockchain in providing solutions to all the security issues mentioned above, which is the main focus of this paper. A comparison study of these five security solutions were done on the basis of the data security concerns solved by these solutions and the type of data storage they use as a base to provide security. The comparison is presented in Table 1.

Table 1 Comparison of the security solutions provided by the implementations discussed

Implementation	Privacy	Integrity	Consistency	Access cntrol	Intrusion detection	Local server	Cloud server
Personal data management platform by Zyskind and Nathan [7]	✓	×	×	✓	×	✓	×
BlockDS by Do and Mg [8]	✓	×	✓	✓	×	✓	×
Blockchain-based cloud storage by Gaetani et al. [9]	×	✓	✓	×	×	×	✓
ChainFS by Tang et al. [10]	✓	×	×	×	×	×	✓
MedRec by Azaria et al. [11]	✓	✓	✓	×	×	✓	×

2.1 Blockchain and Off-Blockchain-Based Solution

Zyskind and Nathan [7] proposed a personal data management platform constructed by combining blockchain and off-blockchain storage. The main focus was on providing privacy to the users of the secure storage platform. The system is based on trusted computing and focuses on providing privacy to its users. It is made up of three entities, *mobile phone users* who download the data and use the application , *service providers* who provide the secure database service and *nodes* which are entities entrusted to maintain the blockchain. These entities maintain a distributed private key-value data store in a distributed hashtable(DHT) which is a off-blockchain way of storage. Figure 2 is an adaptation of the decentralized platform proposed by the authors. The blockchain accepts two types of transactions, T_{access} and T_{data}. When a user signs up to use the service of this secure decentralized database, a T_{access} transaction is sent to the blockchain which consists of a shared identity for the user and the permissions associated with it. Data like location data and sensor data generated by the user's phone are encrypted using an encryption key which is shared and are sent using the T_{data} transaction to the blockchain which routes it to the DHT storage. The hash value of the data is stored on the public ledger (blockchain), hence creating a link to the data stored in the DHT. The data can be queried by the user and the service using the T_{data} transaction. Digital signature is verified and the transaction is checked whether it belongs to the service or the user. If the digital signature belongs to the service, its data access permissions are verified. A user can then issue a modification to the permissions which are granted to the service by seding a T_{access} transaction containing a fresh set of permissions.

Do and Ng [8] proposed a similar method (**BlockDS**) in which a DHT stores references to the data which the user prefers to securely store on the system. The actual data are encrypted on the client-side. Then, the encrypted data are sent to the system. The data are broken into multiple parts and stored in a distributed manner. These data are not directly stored by the blockchain; however, it stores the access key for the users. This access key is the link to the DHT storage to access the metadata for the data. Hash function is calculated using the access key. The system uses distributed proof-of-retrievability to maintain the integrity of data stored in the system.

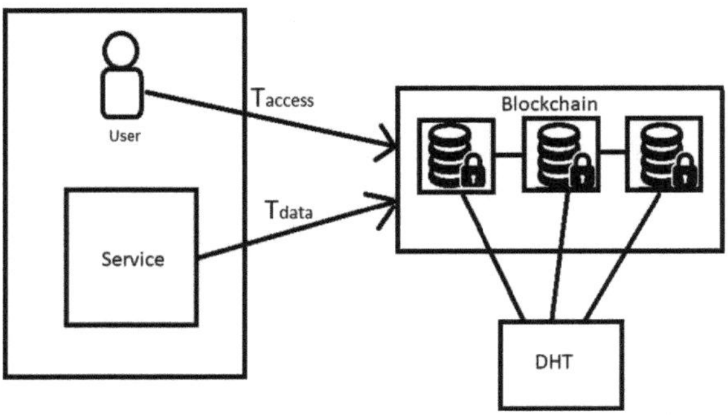

Fig. 2 Adaptation of the decentralized platform given by Zyskind et al. [7]

2.2 Cloud Security Using Blockchain

Gaetani et al. [9] proposed an effective cloud database based on blockchain to provide the service of data on cloud. Evidences of database operations are stored securely which cannot be repudiated. The system uses two layers of blockchain. The first layer, which is a *Mining Rotation-based Blockchain*, works on the cloud federation level. It logs the operations on the cloud which are then executed on the distributed *DB replicas*. Each cloud federation member contains one miner to sign the messages using a public/private key pair. A *mining rotation-* based mechanism is used to achieve consensus. The division of time is in the form of rounds. In each round, a miner node which is selected as the main node which receives new operations to sign.

The second layer, which is the *PoW-based blockchain*, links with the first layer of the blockchain using a technique named as *blockchain anchoring technique*. It links a particular part of the first layer with the second layer. A *witness transaction* is sent over to the second layer and stored as an irreversible and unchangeable transaction. The hash of the first layer uptil the latest operation on it is sent via this transaction to the second layer. These hashes can work as a proof to validate the integrity of the data in the first layer.

Another implementation of cloud security using blockchain is ChainFS by Tang et al. [10] which hardens the blockchain integrated cloud against forking attacks. When a cloud server is under forking attack, the result of some common query can be presented as different views to each client. The use of blockchain is used to store only necessary and minimal information for logging of file operations and distribution of keys. The system is implemented on S3FS and ethereum and is integrated with Amazon S3 cloud storage and clients which are on Filesystem in Userspace (FUSE).

ChainFS consists of three parties : a server, a client and a blockchain. The client manages the private files for the users and stores these data files securely through a Filesystem in Userspace (FUSE) client. The overview of the ChainFS system is illustrated in Fig. 3. The remote parties interact with the FUSE client on two planes: data storage and key management. The data plane consists of a Secure Untrusted Data Repository (SUNDR) server. The SUNDR server has two parts: a *consistency server* which logs all Remote Procedure Calls (RPCs) and *block store* that stores the files. On the key management plane, the FUSE clients interact with the public key directory. The linking between identity of the client with the public key is stored in

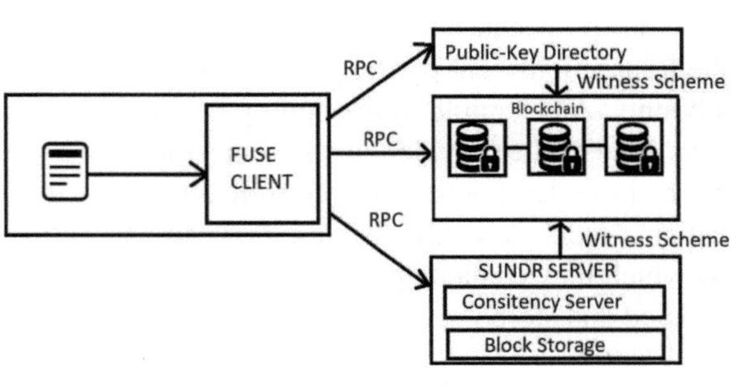

Fig. 3 Adaptation of the ChainFS system architecture [10]

this plane. Blockchain acts as a witness for the operation log which is maintained on the consistency server. Whenever a user sends an operation request, a transaction is sent by the log server to the blockchain. The transaction consists of the encoded log entry for the new operation and its index. Similarly, blockchain is also used to replace the gossiping protocol in CONIKS directory service by being a witness scheme. CONIKS is a key transparency scheme used in the key directory. Each digest produced by CONIKS is sent as a transaction and stored on the blockchain. Hence, the forking attack must have to be done on the blockchain along with the actual system to completely fork the cloud database as blockchain acts as a backup for the key directory as well as the operation logs.

2.3 Permission and Access Control Management Using Blockchain

MedRec [11] is a medical record management system designed by Azaria et al. at MIT. MedRec harnesses the power of blockchain to provide security to electronic medical records. Using the service of *smart contracts* which the blockchain architecture provides, certain state transitions like change in rights for viewership of data or addition of new data to the database are made automated. Representations of the existing medical records are created using ethereum smart contracts. These are stored in individual nodes in the blockchain network. Three types of smart contracts are used:

1. *Registrar Contract (RC):* Resolves a ethereum address for the patient using the identification strings provided.
2. *Patient–Provider Relationship Contract (PPR):* These are issued among the patient and the provider which includes a collection of data pointers and respective associated permissions that identify the patient records held by the provider.
3. *Summary Contract (SC):* It holds a list of references to PPRs which represent the engagements of all the participants with the other nodes in the system.

Figure 4 represents the architecture for the MedRec system. The Registrar Contract (RC) is used to resolve the patient's information to their ethereum address. Then, a patient–provider relationship is established using the PPR contract by the provider node. The provider node also establishes a link between PPRs and SCs of respective patients. A special miner works between the provider and the blockchain to mine the transactions sent by the provider which are for updating the PPRs in return of a bounty. An example of the bounty is to return the average levels of iron in blood tests done by the provider on different patients in a week. The bounty query is signed by the provider; hence, it is safe from alterations done by malicious entities. On the Patient Node's side, the SCs are updated and notifications are sent to the patients on which they can act upon. Based upon the response of the patients, the PPRs are updated and stored in the blockchain.

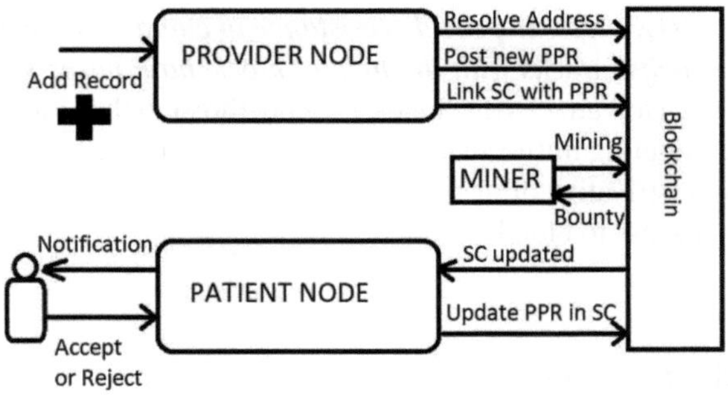

Fig. 4 Adaptation of the structure of the MedRec system [11]

3 Discussion

Although there exists various techniques which provide efficient solutions to the majority of database security problems, there still exists limitations which further need to be addressed. We identified a few research questions from our study that are listed below and can be answered in future research work.

RQ1: ***How to facilitate data intrusion detection in database using blockchain?***

The concept of hashing data, storing the hash in a block and linking the block via the hash values provided by blockchain, can be used to detect unwanted intrusions in system. Unique signatures of precious data on the database can be stored in a blockchain environment, and the blocks can be monitored for change. Any unauthorized change in the consistency of the chain will mean an intrusion has occured.

RQ2: ***How to effectively achieve damage control after database intrusion using blockchain?***

After an intrusion occurs, a rollback is required to return the database to its original state. This function can also be easily achieved by leveraging the distributed architecture of blockchain. For example, if the hash of a block is changed due to a data breach in the database, it has to be reverted back to its previous true value to maintain the consistency of the chain. Along with this, the compromised data have to be retrieved and secured in the database. To provide this, a blockchain-based solution is can be provided which can be both robust and secure.

RQ3: ***How to counter the sybil attack in DHT-based blockchain implementations?***

The blockchain and off-blockchain implementations which included the use of DHT were found to be susceptible to sybil attacks. A solution for this flaw can be found out to strenghten the security of blockchain and off-blockchain solutions.

RQ4: ***How to provide a blockchain solution which is resistant against blockchain based attacks?***

ChainFS solved the problem of forking attacks on blockchain-based clouds by hardening the blockchain against double spending. Hence, smart blockchain-based implementations can be used to protect the system against attacks on blockchains such as the sybil attack, 51% attack, timejack attack, finney attack, etc.

RQ5: ***How to provide resistance to cloud and local data storage servers against security attacks with the help of blockchain based solutions?***

DDoS attacks on clouds are a big threat to the cloud availability. Using distributed computing traffic analysis is done to prevent the DDoS attacks on cloud. Blockchain is a derivative of distributed computing and can be used as a tool to facilitate DDoS hardness to cloud.

4 Conclusion

In this paper, the focus was on blockchain as a security service for online databases. First, the blockchain technology was discussed along with its various features which can be exploited to provide peer-to-peer secure solutions to problems which are generally solved using a centrally governed architecture. Then, all the different studies in the field of implementation of blockchain in solving different database security issues were reviewed and documented. Five major implementations were identified, studied and compared. Finally, some ideas were given for further research that can be done for increasing the use case domain of blockchain in database security as well as improving the existing technology.

References

1. Nakamoto S et al (2008) Bitcoin: a peer-to-peer electronic cash system
2. Eyal I (2017) Blockchain technology: transforming libertarian cryptocurrency dreams to finance and banking realities. Computer 50(9):38–49
3. Dai H-N, Zheng Z, Zhang Y (2019) Blockchain for internet of things: a survey. arXiv preprint arXiv:1906.00245
4. De Filippi P (2018) Blockchain: a global infrastructure for distributed governance and local manufacturing
5. Hussain A, Xu C, Ali M (2018) Security of cloud storage system using various cryptographic techniques. Int J Math Trends Technol (IJMTT) 60(1):45–51
6. Zheng Z, Xie S, Dai H-N, Chen X, Wang H (2018) Blockchain challenges and opportunities: a survey. Int J Web Grid Serv 14(4):352–375
7. Zyskind G, Nathan O (2015) Decentralizing privacy: using blockchain to protect personal data. IEEE security and privacy workshops. IEEE, pp 180–184
8. Do HG, Ng WK (2017) Blockchain-based system for secure data storage with private keyword search. In: IEEE World Congress on Services (SERVICES). IEEE, pp 90–93
9. Gaetani E, Aniello L, Baldoni R, Lombardi F, Margheri A, Sassone V (2017) Blockchain-based database to ensure data integrity in cloud computing environments
10. Tang Y, Zou Q, Chen J, Li K, Kamhoua CA, Kwiat K, Njilla L (2018) Chainfs: blockchain-secured cloud storage. In: 2018 IEEE 11th international conference on cloud computing (CLOUD)
11. Azaria A, Ekblaw A, Vieira T, Lippman A (2016) Medrec: using blockchain for medical data access and permission management. In: 2016 2nd international conference on open and big data (OBD). IEEE, pp 25–30

Prevention and Detection of SQL Injection Attacks Using Generic Decryption

R. Archana Devi, D. Hari Siva Rami Reddy, T. Akshay Kumar, P. Sriraj, P. Sankar, and N. Harini

Abstract Today, Internet services have become an indispensable part of everyone's life. With Internet becoming platform to offer services, traditional core services like banking and ecommerce are now being provided in Internet platform, and making security for these services have become necessary. The openness of Internet has increased risk. In certain huge organization's cyber security, it is usual to notice major issues because of vulnerabilities in their system or software. By performing a depth analysis, it has come to know that the cause of vulnerabilities is either the developers or the design process. Research studies suggest that more than 80% of the active Web sites are vulnerable to SQL injection. The Web sites that unnecessarily expose server information have been recorded as nearly 67%. The Web sites that does not secure session cookies is approximately 50%. About 30% of the Websites secure their communications, for sending sensitive information of users. Hence, a study on mitigation of attacks is compelling. SQL injection targets interactive Internet services that employ database and aims to exploit vulnerability occurring at database layer

---

R. Archana Devi (✉) · D. Hari Siva Rami Reddy · T. Akshay Kumar · P. Sriraj · P. Sankar · N. Harini
Department of Computer Science and Engineering, Amrita School of Engineering, Coimbatore, Amrita Vishwa Vidyapeetham, Coimbatore, India
e-mail: archanadevi4688@gmail.com

D. Hari Siva Rami Reddy
e-mail: haribhamireddy@gmail.com

T. Akshay Kumar
e-mail: thalladaakshay696@gmail.com

P. Sriraj
e-mail: srirajpalitha@gmail.com

P. Sankar
e-mail: sankarpappu.17@gmail.com

N. Harini
e-mail: n_harini@cb.amrita.edu

A. K. Tripathy et al. (eds.), *Advances in Distributed Computing and Machine Learning*, Lecture Notes in Networks and Systems 127, https://doi.org/10.1007/978-981-15-4218-3_16

in the three-tier architecture. This paper proposes a solution to detect and prevent SQL injection attacks using a model which is inspired from the concept of generic decryption. Results obtained confirm the extraordinary performance of the proposed scheme against the existing prevention techniques.

Keywords SQL injection · Database security · Mitigation · Prevention · Generic decryption

1 Introduction

Many applications that do not validate user input and are susceptible to SQL injection [1]. Attackers capitalize on mere flaws to attempt a malicious activity on backend database like extraction of sensitive info, DoS attack, etc. Many Web applications are subject to rapid changes in application development phases with short turnaround requirements which make it vulnerable to SQL injection attacks. Conventional methods of detection and prevention suffer from the problem that they suffer from performance degradation due to high degree of overhead or providing better performance measures by compromising security. The paper had introduced a methodology for SQL injection mitigation, prevention, detection and for removing malicious inputs. SQL injection happens by inserting a portion of spurious SQL query through inputs not validated from user to valid Web requests. A successful SQL injection interferes with data confidentiality, data integrity and data availability of information in databases. Based on statistical research, this threat has high impact on the online business. Hence, it is necessary to find a proper solution to detect and prevent SQL attacks. Section 2 presents review of literature, Sect. 3 presents the present approach, Sect. 4 presents results and discussions, and finally, Sect. 5 presents concluding remarks.

2 Literature Review

This section talks about various works related to SQL injection and then defines the proposed work.

2.1 Related Work

Many research works [2] present a comparative study on SQL injection attack types and mitigation strategies. Radhika et al. [2] presents an analysis of SQL attacks using many dimensions in Web application. Buehrer et al. [3] proposes a method by performing a validation on parse tree to prevent such attacks. Kosuga et al. [4] proposes a general analysis against SQL injection using an automation tool by checking syntax and semantics. Jovanovic et al. [5, 6] present a detailed static analysis on the usual Web application vulnerabilities. Fonseca et al. [7] tested and compared various tools against SQL and cross-scripting attacks. McClure et al. [8] developed a technique called as SQL DOM which permits usage of dynamic queries without loss of security. Bulusu et al. [9] feature the major differences between SQL injection attacks and lightweight directory access protocol attacks (LDAP). Harini et al. [10] discusses an authentication scheme capable of mitigating different forms of attacks on Web sites. Pearson III et al. [11] proposed a methodology based on design review that permits designers to examine user friendliness of GUI and user satisfaction. Wang et al. [12] proposes a scheme for Web crawlers [13, 14] and scanners. Fonseca et al. [15, 16] proposes a prototype to demonstrate the process of injecting realistic attacks and verifying the performance of the existing security mechanisms. Wei et al. [12] propose a novel methodology by combining static application code analysis combined with dynamic validation to defend against the attacks targeted at stored procedures.

2.2 *Summary of Findings*

The existing techniques which are currently used by researchers for detecting, mitigating and preventing SQL injection attacks (all the three together) are compromising either on security or on time complexity. This paper proposes a technique for the same which aims at reducing the time and space complexity and also does not compromise on security.

2.3 *Problem Statement*

To develop a scheme based on generic decryption for effective dynamic detection and prevention of SQL injection attacks.

2.4 *Proposed Work*

The approach adopts a generic decryption [1] process based on implementation. A program is run in a protected virtual environment, and the program is halted for virus detection at regular intervals. At some point, the virus must reveal itself in order to execute the payload. Hence, it is effectively detected. Experimentation revealed the scope for further enhancement. The same was performed by modifying the scheme to include a component "view" [17] that could perform on par with virtual protected environment.

3 Proposed Model

Since views are temporary data taken from a table, it is similar to a protected virtual environment because when the attack is done on the views, it does not have any effect on the original table. Also, we can delete the view after its use is over. The architecture diagram and overall functioning of SQL detection engine are depicted in Fig. 1.

The architecture in Fig. 1 follows three-tier client–server architecture in which the client interacts with the application server through a Web page. An authorized client tries to access data from the server by entering a request in the Web page. The generic decryption method acts as the middle ware which filters the SQL query from the user request as legitimate and malicious. If the query is filtered as legitimate, the client is given access to the database by returning valid response. Otherwise, the request is dropped. The various levels of general three-tier client–server architecture are shown in Fig. 2.

The ultimate aim of this model is to check whether the input query when run on the view will give the same output as the view before running the query. To do this process, the number of the rows from the output of the input query is taken, and the hash values for the sensitive fields are calculated to check the authenticity of the inputs. Expected row count is one as there should exist only one row in the view. Hence, if the row count returned is indeed one and their respective hash values for sensitive columns of the created view are same before and after running the query, then it is proved that the input query is not a malicious one. Only such validated queries are allowed to run on the original database, otherwise the input query which is found out to be the malicious string is stored in an attack table in the database

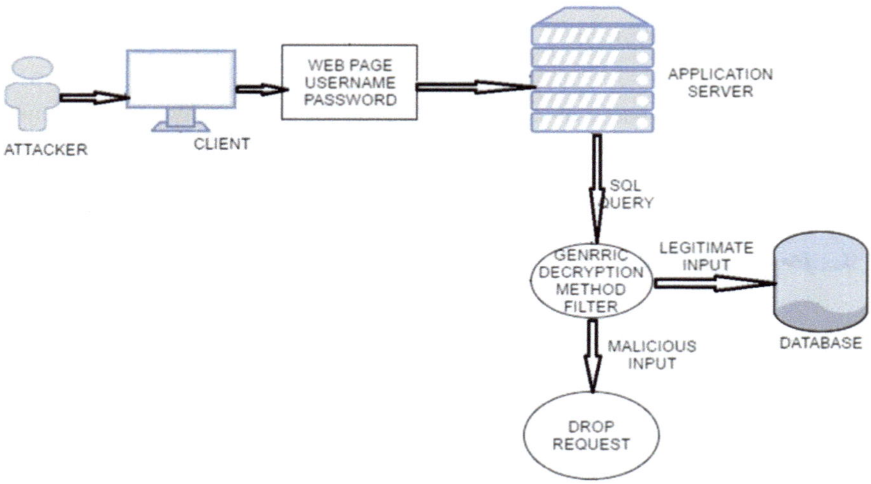

Fig. 1 Architecture diagram of detecting and preventing SQL injection attacks

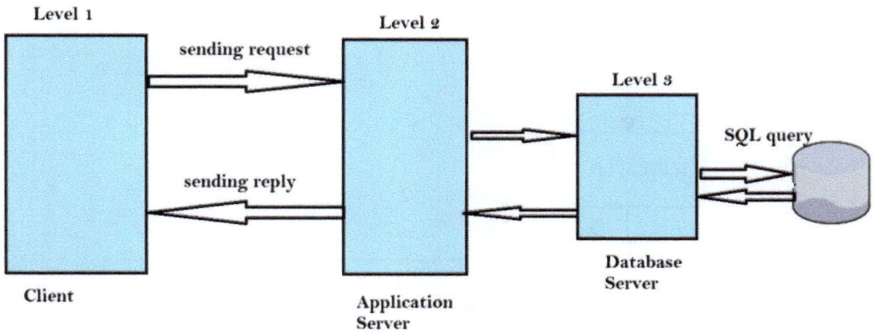

Fig. 2 Three-tier client–server architecture

along with the IP address of the attacker and current time stamp. The work flow of the model is depicted in Fig. 3.

A sample code snippet for view creation is:

```
CREATE VIEW name_of_view AS
SELECT col1,col2,…coln
FROM name_of_table
WHERE predicate;
```

The experimentation with the view clearly revealed that the performance was dependent on the size of view. For performance enhancement, a dimensionality reduction was done to ensure minimal set of sensitive parameters (cannot be altered by third parties). For security reasons, the scheme prevents original data from being altered as view alteration does not have any effect on original data. It is ensured that all sensitive attributes in the table are hashed with hashing algorithms to preserve data integrity. An extensive simulation of this proposed scheme clearly revealed its resistivity to SQL injection attacks. The results of experimentation are presented in Sect. 4.

4 Results and Discussion

The approach was successfully detecting and preventing SQL injection attacks. Figures 4, 5, 6 and Fig. 7 are some of the sample screenshots run on the test Web site which was created as a part of this project. A sample dataset of a view created from the existing table in database which contains only sensitive columns is depicted in Fig. 4. Out of the attributes from table, it can be seen that category id, phone no and path of photo are chosen as sensitive columns.

Response obtained from the Web site with valid URL as highlighted in Fig. 5.1 is presented in Fig. 5.

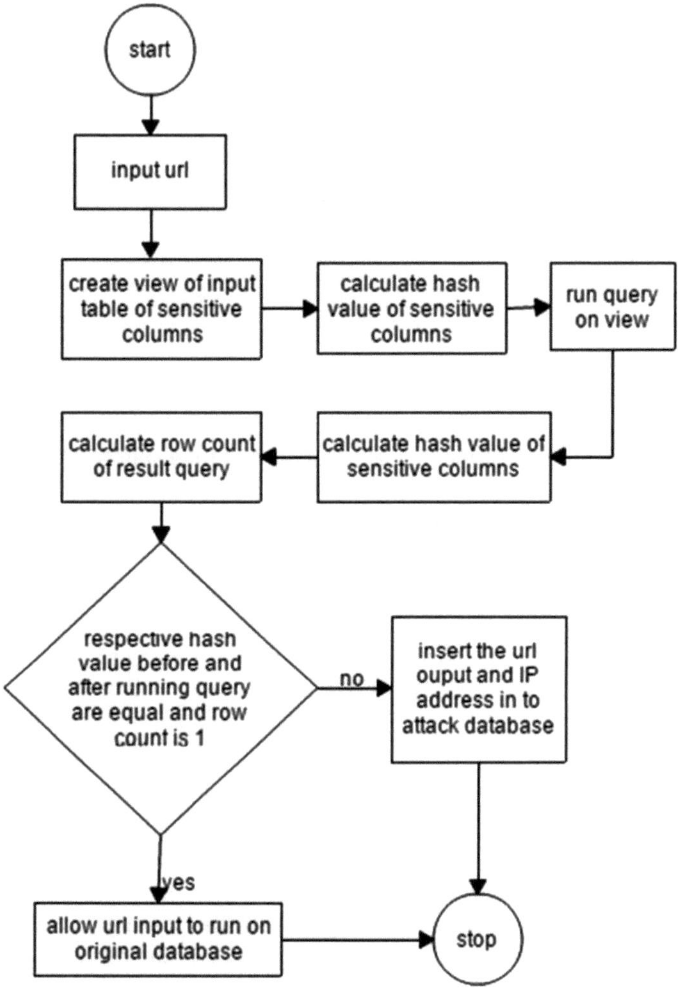

Fig. 3 Work flow diagram

Results obtained from the Web site with malicious request as shown in Fig. 6.1 are presented in Fig. 6.

All the malicious events on database server were directed to an attack database, and this dataset was used by other filters in the network to prevent further attacks on the system. Few sample records collected during experimentation process are depicted in Fig. 7.

On testing the system with all the attributes of the database, the overhead in terms of time and space complexity showed an exponential growth. With the reduced set of attributes, the system could detect malicious events with reduced time and space complexity. However, to have clear metrics on the reduction factor, a detailed investigation is required.

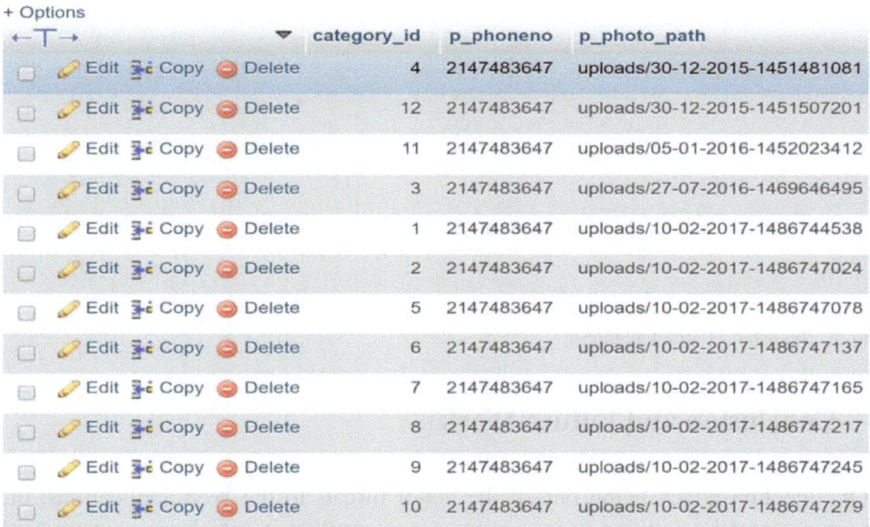

+ Options

	category_id	p_phoneno	p_photo_path
Edit Copy Delete	4	2147483647	uploads/30-12-2015-1451481081
Edit Copy Delete	12	2147483647	uploads/30-12-2015-1451507201
Edit Copy Delete	11	2147483647	uploads/05-01-2016-1452023412
Edit Copy Delete	3	2147483647	uploads/27-07-2016-1469646495
Edit Copy Delete	1	2147483647	uploads/10-02-2017-1486744538
Edit Copy Delete	2	2147483647	uploads/10-02-2017-1486747024
Edit Copy Delete	5	2147483647	uploads/10-02-2017-1486747078
Edit Copy Delete	6	2147483647	uploads/10-02-2017-1486747137
Edit Copy Delete	7	2147483647	uploads/10-02-2017-1486747165
Edit Copy Delete	8	2147483647	uploads/10-02-2017-1486747217
Edit Copy Delete	9	2147483647	uploads/10-02-2017-1486747245
Edit Copy Delete	10	2147483647	uploads/10-02-2017-1486747279

Fig. 4 View with sensitive columns

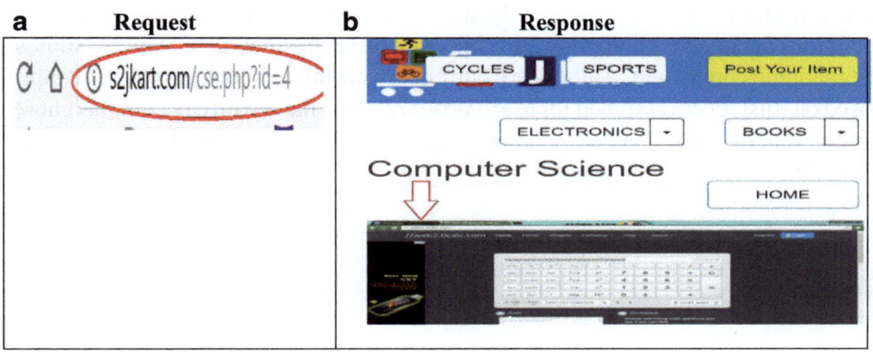

Fig. 5 Request and response from the Web page for valid input

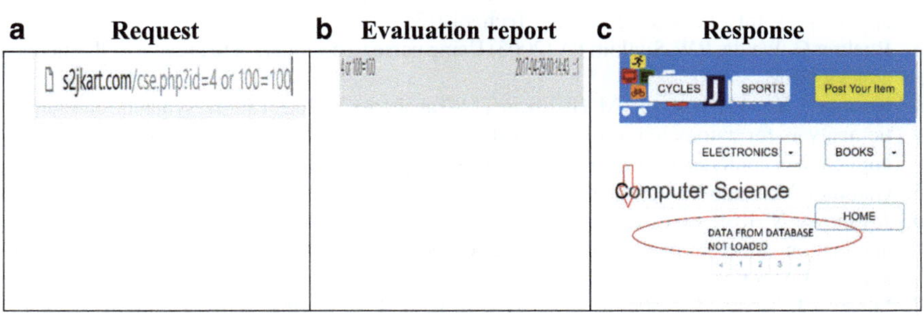

Fig. 6 Request and response from the Web page for invalid input

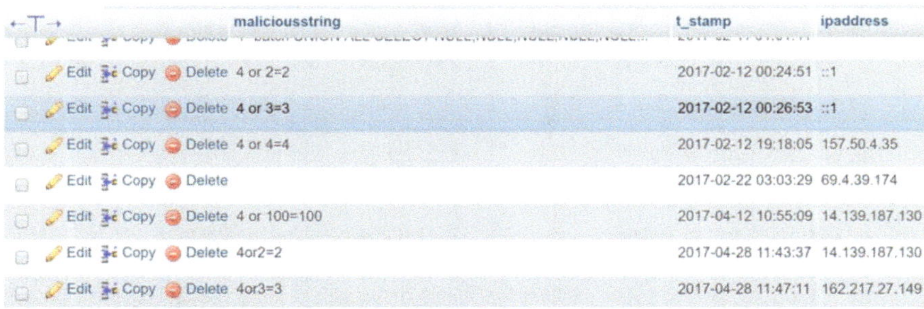

Fig. 7 Snapshot of attack database

5 Conclusion and Future Work

SQL injection attack being one of the major threats to the Web applications, the paper presented a solution based on generic decryption in which views were used to simulate virtual protected environment, and SQL injection was detected. This model was successfully implemented with reduced time and space complexity and zero compromise on security. Finally, a list of attackers was created who have used malicious queries for attempting SQL injection by saving their details like ip address, the malicious query, etc. Authors are currently working on blocking spurious inputs based on trust scores assigned for user. As future work, the sensitivity columns chosen for creating the view can be even more reduced. This method could also be used for mitigating cross-scripting attacks.

References

1. Jespersen MS, Hargreaves FP (2012) Malicious software. University of Southern Denmark, Denmark
2. Radhika N, Vanitha A (2014) Multidimensional analysis of SQL injection attacks in web applications. Int J Innovative Sci Eng Technol 1(3)
3. Buehrer G, Weide, BW, Sivilotti PA (2005) Using parse tree validation to prevent SQL injection attacks. SEM
4. Kosuga Y, Kono K, Hanaoka M, Hishiyama M, Takahama Y (2007) Sania: syntactic and semantic analysis for automated testing against SQL injection. In: 23rd annual computer security applications conference, IEEE Computer Society, pp 107–117
5. Common Vulnerability and Exposures, http://www.cve.com
6. Jovanovic N, Kruegel C, Kirda E. (2006). Precise alias analysis for static detection of web application vulnerabilities. In: Proceedings of the 2006 workshop on programming languages and analysis for security
7. Fonseca J, Vieira M, Madeira H (2007). Testing and comparing web vulnerability scanning tools for SQL injection and XSS attacks. In: 13th Pacific Rim international symposium on dependable computing (PRDC 2007)
8. McClure RA, Kruger IH (2005) SQL DOM: compile time checking of dynamic SQL statements. in software engineering, 2005. In: Proceedings. 27th International Conference on ICSE 2005

9. Bulusu P, Shahriar H, Haddad HM (2015) Classification of lightweight directory access protocol query injection attacks and mitigation techniques. In: 2015 international conference on collaboration technologies and systems (CTS), IEEE
10. Harini N, Padmanaban TR (2016) 3CAuth: a new scheme for enhancing security. Int J Netw Secure 18(1):143–150
11. Pearson III E, Bethel CL (2016) A design review: concepts for mitigating SQL injection attacks. Little Rock, AR
12. Wang X, Wang L, Wei G, Zhang D, Yang Y (2015) Hidden web crawling for SQL injection detection. Beging, China
13. Raghavan S, Garcia-Molina H (2001) Crawling the hidden web. In: Proceedings of 27th international conference on very large data bases. Morgan Kaufmann, Roma, Italy
14. Bergman MK (2000) The deep web: Surfacing hidden value. Technical report. BrightPlanet LLC, Dubai
15. Fonseca J, Vieira M, Madeira H (2009) Vulnerability & attack injection for web applications. In: IEEE/IFIP international conference on dependable systems & networks, 2009. DSN'09. IEEE
16. Fonseca J, Vieira M, Madeira H (2014) Evaluation of web security mechanisms using vulnerability & attack injection
17. Available: https://en.wikipedia.org/wiki/View_(SQL)

Prevention and Detection of SQL Injection Using Query Tokenization

R. Archana Devi, C. Amritha, K. Sai Gokul, N. Ramanuja, and L. Yaswant

Abstract One of the most serious security vulnerabilities in the current scenario is SQL injection. It stands first in the OWASP top 10 vulnerability attacks. Lack of input validation is one of the main reasons for the cause of these types of attacks. Data can be stolen from the database by the means of SQL injection. Most of the user inputs are going directly to database. An attacker can obtain the data which he does not have access to with the means of SQL injection. The paper aims in developing a method that detects and prevents SQL injection attacks.

Keywords SQL injection attack · Prevention · Detection and tokenization

1 Introduction

The latest applications that we use daily are mostly web-based. And most of these applications are accessible over the Internet. As it is exposed to the Internet, the security challenges faced by these applications also increases. Most of the user information are saved in database. And these data can be accessed by using SQL. When we

R. Archana Devi (✉) · C. Amritha · K. Sai Gokul · N. Ramanuja · L. Yaswant
Department of Computer Science and Engineering,
Amrita School of Engineering, Amrita Vishwa Vidyapeetham, Coimbatore, India
e-mail: r_archanadevi@cb.amrita.edu

C. Amritha
e-mail: cheruvatethamritha@gmail.com

K. Sai Gokul
e-mail: saigokulkatragadda@gmail.com

N. Ramanuja
e-mail: ramanuja007238@gmail.com

L. Yaswant
e-mail: lakkarajuyaswant@gmail.com

A. K. Tripathy et al. (eds.), *Advances in Distributed Computing and Machine Learning*, Lecture Notes in Networks and Systems 127, https://doi.org/10.1007/978-981-15-4218-3_17

use SQL to attack the Web application to obtain, manipulate or delete user information, it is known as SQL injection attack. Section 2 presents review of the literature, Sect. 3 presents the proposed methodology, Sect. 4 presents results and discussions, and Sect. 5 presents concluding remarks.

1.1 Background

Many companies use Web applications to provide better user experience and services. All the information generated by the organization or provided by the users are stored in databases. Web applications interact with the database to store or retrieve or update any information. The information provided in the input boxes of the HTML pages are sent in background to the databases as values. Injection occurs when a user provides a SQL query instead of providing a valid information. The applications that use dynamic query approach to communicate with the databases are prone to the SQL injection attack. Those applications have high probability that an attacker gains access to the database.

1.2 Problem Statement

This paper proposes a method to detect and prevent SQL injection attacks in Web applications keeping in mind the vulnerabilities by validating the input using query tokenization and prevent the SQL injection using SQL prepared statement.

2 Literature Survey

Sarjiyus and El-Yakub [1] have depicted a system in which a query is monitored to verify if any extra character other than the actual parameter is added by the user. The appropriateness of the parameters is verified. The advantage of this method is that it is very easy to implement, and in detecting attacks, it is highly efficient and effective. But the disadvantage is that it can only be used with PHP which is not used in modern times. Al-sahafi [2] has depicted many methods for detection and prevention of SQL injection such as AMNESIA which is analysis and monitoring for neutralizing SQL injection attacks. This approach aims to detect invalid SQL queries before they are run on the database. Mishra [3] has approached the SQL injection detection problem by analyzing the incoming traffic and applying machine learning classification as SQL injection or plain text. Naïve Bayes classifier and gradient boosting classifier are the two ML classification algorithms used. The disadvantage of this method is that accuracy is less, and in terms of usability and efficiency, it can be modified.

Shahriar et al. [4] developed an SQLI attack detection framework based on the information supplied from the server side in prior. Their approach is based on shadow queries and certain information in HTML forms. This is used for client side. Raut et al. [5] suggest a model that checks anomalous SQL query structure using string matching. It also looks into matching the outline and address managing. They make use of Aho–Corasick configuration. This method can be used for different databases. Qian et al. [6] explain all types of SQL injection attack. It also claims that it can be prevented by validating the type and SQL parameters. But it does not explain in depth any solutions to prevent SQL injection attacks.

Amutha et al. [7] proposed a scheme for detecting and preventing SQL injection attacks with an algorithm using pattern matching which takes a minimum of O(n) time. NTAGW ABIRA Lambert et al. [6] suggest an approach in which the original query, the injected query is tokenized which produces many array elements. The lengths of the resulting arrays are compared with the two queries to detect if injection is there. Voitovych et al. [7] use few inbuilt scripts of the webpage to check for an SQL attack. It is mainly developed to work with MySQL. Radhika et al. [8] present an analysis of SQL attacks using many dimensions in Web application. Appiah et al. [9] proposed an SQLIA detection framework which is based on signature. It uses a hybrid method of fingerprinting pattern matching to differentiate valid and invalid queries. Jhala et al. [12] presented a method that prevents only bypass authentication type of SQL injection also called tautology-based SQL injection.

3 Proposed Solution

This section focuses on the existing methodologies and proposed work.

3.1 Existing Methodology

A SQL injection attack occurs when user input includes SQL keywords, so that the intended function of the application is changed by the generated SQL query. Space, single quotes and double dashes are required for performing SQL injection attacks. Single quotes can be avoided if and only if the data entered by user is a number.

```
mysql> select * from students where rollno='cse16405';
```

The injection can be done like this:

```
mysql> select * from students where rollno='cse16405'or 1=1;
```

Even though there is no space between 'cse16405' and 'or', the query would work properly and all the information regarding the students would be retrieved. For user input of type integer, it is not necessary to use single quotes.

3.2 Proposed Work

The proposed method consists of tokenizing original query and a query with injection separately, the tokenization is performed by detecting a space, single quote or double dashes, and all strings before each symbol constitute a token.

After tokens are formed, they all make an array for which every token is an element of the array. Two arrays resulting from both original query and a query with injection are obtained, and their lengths are compared to detect whether there is injection or not. If it is not SQL injection infected, then the data is passed on as SQL prepared statement through parameterized query to prevent SQL injection if gone undetected. And if data is correct, the user would be able to access the program else would be prevented from getting entry. Figs. 1 and 2 show the various stages that detect and prevent the SQL injection attack.

The various stages in Figs. 1 and 2 are explained below.

STAGE 1: The input parameters are inserted by the users. The values inserted by the users may or may not be affected by SQL injection. In the case of this webpage, username and password are the two parameters to be inserted by the user.

STAGE 2: Tokenization is a way to break a string into several parts based on the delimiters given. In the case of SQL injection detection, the symbols [", /, −] are used as delimiters since SQL injected queries cannot be created without the use of these symbols. After tokenization, the query is divided into tokens and is stored in an array. In case of Java, split method can be used for tokenization.

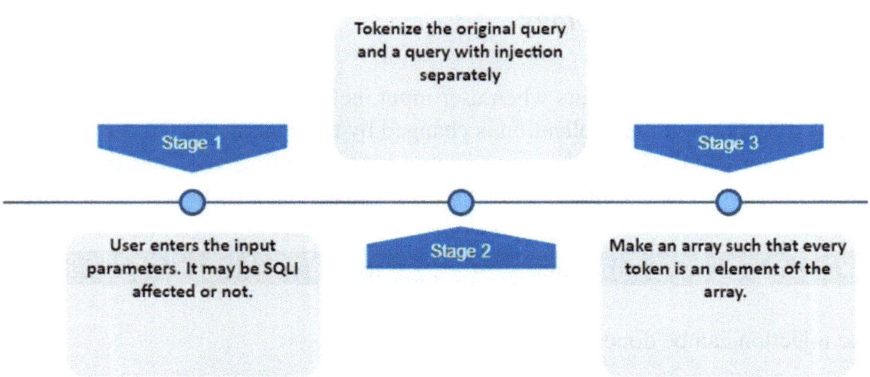

Fig. 1 System architecture

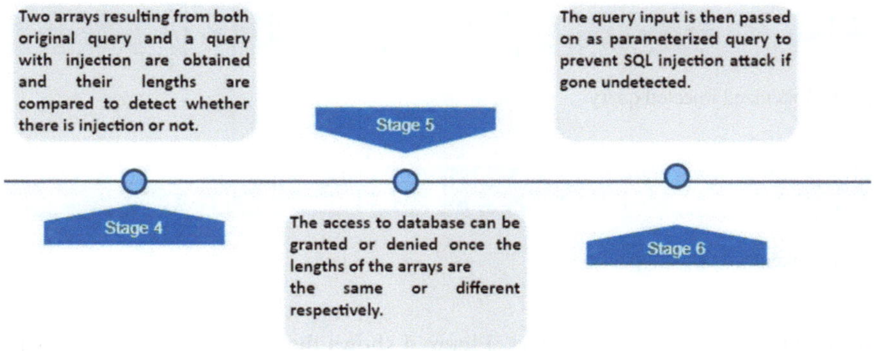

Fig. 2 System architecture

STAGE 3: First array is created from tokenizing the original unaffected query. Second array is created from tokenizing the query made from the input given by the users.

STAGE 4: Length of the two arrays is taken, and they are compared. If length of two arrays is same, then it can be concluded that there was no SQL injection effect on the input given by the user. If the length of the array generated by input query exceeds the length of array generated by original query, then it can be confirmed that the data entered by the user is SQL injection infected.

STAGE 5: If the length of the arrays is same, then the data will be granted access to create a connection to database. If they are not same, this connection is denied, and the user will be requested to insert values again.

STAGE 6: Prepared statement is used to execute the input parameters as query. Because sometimes SQL injection attack may go undetected (it never happens), then the parameterized query method will prevent the injected query from affecting the database. In parameterized query, each parameter is considered as one whole thing and no matter what exists inside the parameter it is considered as its part.

Once the input is entered, the stages mentioned above detect and prevent SQL injection as proposed.

Suppose let the original password be **usrpassword**. The original query will be:

```
mysql> select username, password from users where username = 'username' and password = 'usrpassword';
```

Let the injected input be **'usrpassword' or 1 = 1**, then the injected query will be:

```
mysql> select username, password from users where username = 'username' and password = 'usrpassword'or 1=1;
```

Tokenization is done on the query, the tokenized array is as shown in Fig. 3, and the length of the query containing user input is 13. The original query is also

Fig. 3 Tokenized injected query

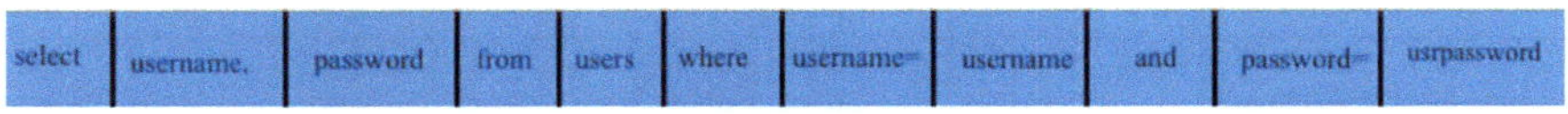

Fig. 4 Tokenized original query

tokenized with the same delimiters. Figure 4 shows the image of array containing tokenized original query. The length of the original query is 11. Since both lengths are not same injection is detected.

4 Results and Discussions

The proposed model was successfully detecting and preventing the various types of SQL injection attacks as tabulated in Figs. 5, 6 and 7. Figures 5, 6 and 7 are some of the sample screenshots run as a part of implementation.

Response obtained for a valid input is highlighted in Fig. 5.

Response obtained for an invalid input is highlighted in Fig. 6.

Response obtained for an invalid input is highlighted in Fig. 7.

Table 1 lists the various types of SQL injection [11]. This table compares the outcome of the existing model with the proposed model for different types. In the existing model [10], it is only preventing bypass authentication and injected additional query, but it does not prevent the second-order SQL injection, union injection

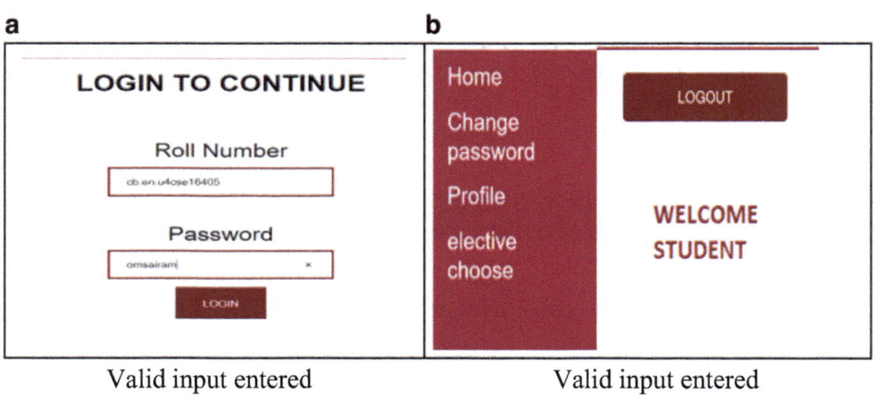

Fig. 5 Request and response for a valid input (no SQL injection)

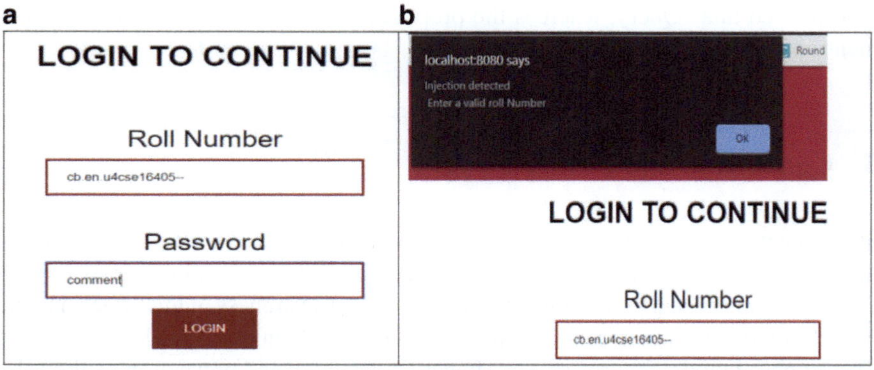

Invalid request entered-an example of second order SQL injection

Malicious input detected which prevents from successful login.

Fig. 6 Request and response for an invalid input (with SQL injection)

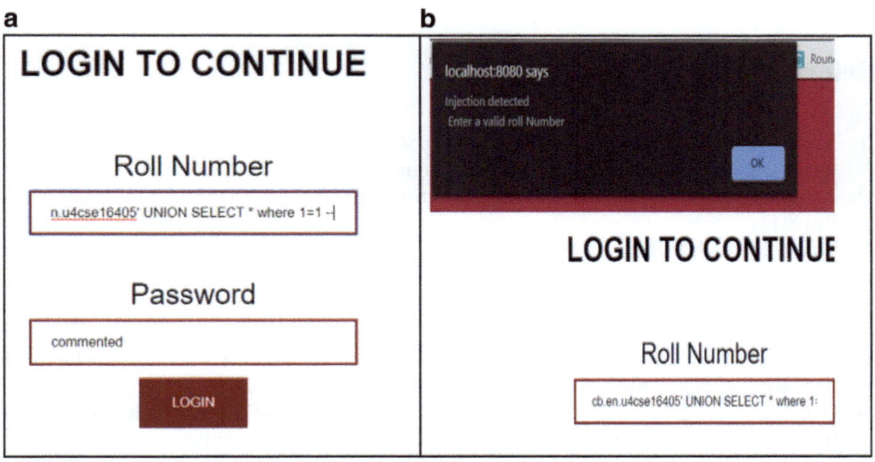

Invalid request entered-an example of UNION SQL injection.

Malicious input detected which prevents from successful login.

Fig. 7 Request and response for an invalid input (with SQL injection)

Table 1 Outcome comparison

SQL injection types	Existing model's outcome	Proposed model's outcome
Bypass authentication [12]	Prevented	Prevented
Injected additional query [13]	Prevented	Prevented
Injected alias query [13]	Not prevented	Prevented
Second-order SQL injection [13]	Not prevented	Prevented
Injected union and all union query [13]	Not prevented	Prevented

and injected alias Query, whereas the proposed system prevents all the five types mentioned in the table.

5 Conclusion and Future Work

SQL injection is the most serious and dangerous of all the vulnerability attacks in the recent times. It caused many data breaches and imposed huge losses to the companies and organizations financially. Despite the awareness of this attack and the poor development strategies, SQL injection attacks are being found in large numbers. The proposed model focuses on SQL injection detection and prevention methods by the use of prepared statements and query tokenization. In future, work can be done to combine query tokenization and prepared statement as one process instead of two separate processes.

References

1. Sarjitus O, El-Yakub MB (2019) Neutralizing SQL injection attack on web application using server side code modification. Int J Sci Res Comput Sci Eng Inf Technol 5(3)
2. Alsahafi R (2019) SQL injection attacks: detection and prevention techniques. Int J Sci Technol Res 8(1)
3. Mishra S (2019) SQL injection detection using machine learning. In: Master's projects, SJSU scholar works, May 2019
4. Shahriar H, North S, Chen WC (2013) Early detection of SQL injection attacks. Int J Netw Secur Appl IJNSA
5. Raut S, Nikhare A, Punde Y, Manerao S, Choudhary S (2019) A review on methods for prevention of SQL injection attack. Int J Sci Res Sci Technol 6(2)
6. Qian L, Zhu Z, Hu L, Liu S (2015) Research of SQL injection attack and prevention technology. In: International conference on estimation, detection and information fusion, IEEE 2015
7. Prabakar MA, Kartikeyan M, Marimuthu K (2013) An Efficient technique for preventing SQL injection attack using pattern matching algorithm. IEEE international conference on emerging trends in computing, communication and nanotechnology, IEEE 2013
8. Ntagwabira L, Kang SL (2010) Use of query tokenization to detect and prevent SQL injection attacks. In: International conference on computer science and information technology, vol 2, IEEE 2010
9. Voitovych OP, Yuvkovetskyi OS (2016) SQL injection prevention system. In: International conference "Radio electronics and infocommunications" (UkrMiCo). Kiev, Ukraine IEEE, Sept 2016
10. Radhika N, Vanitha A (2014) Multidimensional analysis of SQL injection attacks in web applications. Int J Innov Sci Eng Technol 1(3)
11. Appiah B, Opoku-Mensah E, Qin Z (2017) SQL injection attack detection using fingerprints and pattern matching technique. In: 8th IEEE international conference on software engineering and service science (ICSESS), IEEE 2017
12. Jhala K, Shukla UD (2017) Tautology based advanced SQL injection technique a peril to web application. In: National conference on latest trends in networking and cyber security, Mar 2017
13. Yasin A, Zidan NA (2016) sql injection prevention using query dictionary based mechanism. Int J Comput Sci Inf Secur 14(6)

Investigating Peers in Social Networks: Reliable or Unreilable

M. R. Neethu, N. Harini, and K. Abirami

Abstract Social networking sites have made its impact in converting a passive reader into content contributor. This has turned online social networks into a commercial sphere. The site demands creation of public profiles. The main goal of online social networks is to share content with maximum users. This enormous sharing of information can lead to malicious users leverage the privacy of an individual. An attempt has been made in this paper to improve privacy preservation by calculating the reliability of a user, can entrust on his/her peer using the content sharing, Machine Learning and tone analysis. The recorded metrics and transaction details are stored in IPFS.

Keywords Online social networks · Reliability · Machine learning · Tone analysis · IPFS

1 Introduction

Social networks have drawn attention in both academia and industry. The exponential increase in popularity of social network has generated extremely large scale user data. The usage of learning algorithms on this huge amount of user data allows disclosure of private information to hackers. As means of protecting user data privacy, protocols like P3P (Platform for Privacy Preferences Project) are used by Websites and browsers. This specification brings Web users an option to decide if and under what circumstance personal information is to be disclosed. A notification is generally sent to users when sites privacy policy changes. However, this protocol cannot

M. R. Neethu · N. Harini (✉) · K. Abirami
Department of Computer Science and Engineering, Amrita School of Engineering,
Amrita Vishwa Vidyapeetham, Coimbatore, India
e-mail: n_harini@cb.amrita.edu

M. R. Neethu
e-mail: cb.en.d.cse17014@cb.students.amrita.edu

K. Abirami
e-mail: k_abirami@cb.amrita.edu

A. K. Tripathy et al. (eds.), *Advances in Distributed Computing and Machine Learning*, Lecture Notes in Networks and Systems 127, https://doi.org/10.1007/978-981-15-4218-3_18

effectively prevent indirect disclosure of privacy in social network which can be obtained by intelligently combining unrelated user data. The recently reported incidents on security breaches compromising user privacy explicitly convey the pitfalls in the existing security models in terms of preserving privacy of personal data. Presently, social media networks accumulate user data and sell it to targeted advertisers (main source of revenue). Reports on exploitation of data to manipulate people and influence public opinion toward voting also have been reported in the recent past. It can be seen that the social media does not give user a real choice or awareness of what data about them is stored. The paper presents a method to facilitate sharing of personal information with reliable connects.

Significance of the Work: The exponential growth in the number of users associating themselves with popular social networks as shown in graph (Fig. 1) demands a security framework that enhances privacy concerns of individual participants in social networking sites. A thorough study of the existing schemes to protect privacy security of user data was undertaken. Detailed review of the current state of the art in various aspects of social networks and the research challenges related to online social networks suggests that although many research work have addressed this issue, a significant research is still required in the following areas (social graph analysis, social media search and management, predicting traffic demands, mobile social networks, and service architectures for social network platforms). Research reveals that challenges like frequent outages (fail-whale-twitter's downtime icon) at times of significant world events and severe criticism due to privacy concerns (more prominent in Facebook), etc., lead to user frustration. The fact that the borderline between public space and private space of social media space is still unclear, and this

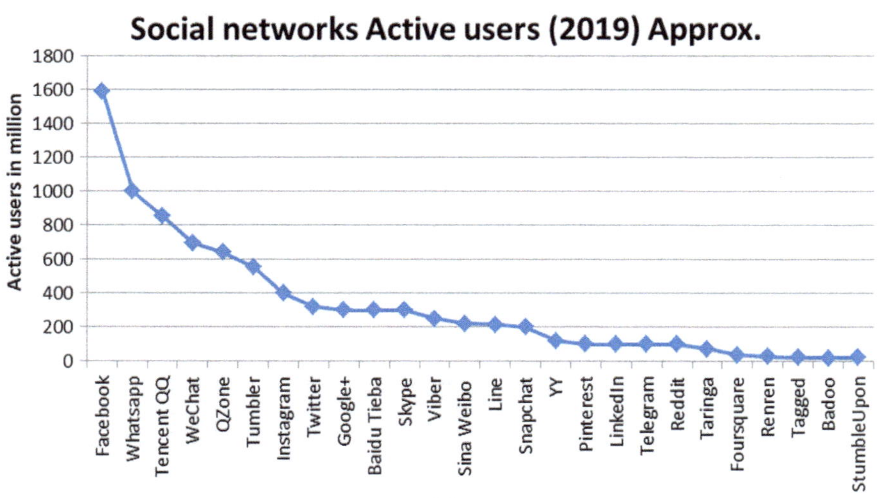

Fig. 1 Active users in social networking sites

is a major threat to securing communications in social networks. Providing security is challenging because of the following reasons:

- Legitimate users personal information could be stolen by a cyber-criminal, putting user's identity and accounts at risk.
- The personal information (phone number, email address, location, etc.) user shares in social networks could give cyber criminals enough scope to infer the legitimate user credentials. Geotagged photographs are used by hackers to understand the whereabouts of a targeted user.
- It is very common to find many social network users updating their status information on a regular basis. It is possible that social networking sites collect the data and share it with third parties.
- One has to be aware of the threats when installing and uninstalling new application in system/mobile. This could be used to extract personal information.
- Updation of status and adding unknown friends can lead to becoming a victim of scam.
- It is important to understand that uninstalling applications from the system or mobile created using social network accounts at times may not delete complete user details—that is the app owner will still have access to this preserved information [1].

2 Related Work

SNS is a kind of Internet service specifically used for connecting people with similar interests, background, and activities. The prominence of this service has led to shrink in economic and geographical borders. SNS generally contain data pertaining to an individual regarding his/her career interest contacts friendship, relationship, social skills, etc. This has made the media to become the key source for data collection. The availability of huge amount of personal data shared in this medium has made this as a desirable target for attackers. Today, most of the popular social networks are based on centralized architecture, meaning the service provider takes control over user's information by storing them in centralized infrastructure. Recent events have shown that the service provider, third-party applications in addition to malicious users (also careless legitimate users), introduce new privacy risks.

Threats in Social Networks: Today, enormous people rely on social networking services for gaining reputation by a community. The reputation of the user gain also influences the status of the user and credibility in actual life. Popular sites like Facebook, flicker allow billions of users to share their personal information and multimedia data with their very close, close, and far connects. Alternately, the malicious user can also distort data using different tools to ridicule legitimate users [2]. Threats in social network can be broadly classified as multimedia content threats, traditional threats, and social threats. SNS environment is no exception to traditional threats like cybil attacks, fake profiles, spamming, hijacking. The social relationship feature of this platform facilitates interaction between attackers and legitimate users through ways of expressing sympathy, love, and care offering online gifts, etc. The recorded activity is then used as a source for spying, blackmailing, etc. The impacts of these threats include reputation lost, cyber harassment, data ownership loss, depression, confidential information disclosure. Most of the threats happened due to the fact that users are not concerned about the importance of personal information disclosure, the policies and legislations are not capable enough to deal with types of social network threats, lack of tools and appropriate authentication mechanisms to deal with information and connection between users. So the need for the hour is thus standards and sophisticated technologies supported by social network service providers.

Machine Learning for Classification: User behavior prediction is a growing interest area in research on social networks based on data collected from social networks [3]. Data from social networks are imbalanced which make behavior prediction challenging. Machine learning algorithms construct feature set after preprocessing make classification based on text data for user behavior prediction much hassle free [10]. SVM, Naïve Bayes, decision tree, etc., are few supervised algorithms used in for this purpose.

InterPlanetary File System (IPFS): IPFS is a file system that can store and track, files, versions, respectively. It also delineates how files traffics within a network, and this is the reason for IPFS being distributed storage system [4]. IPFS enables a new and permanent Web that augments the existing Internet protocols [5]. IPFS represents the file content itself to address by using cryptographic hash and is used as address. An HTTP request would look like http://10.20.30.40/folder/file.txt. An IPFS request would look like /ipfs/QmT5NvUtoM5n/folder/file.txt.

3 Proposed Work

Social network is a platform where the increase in data is unpredictably high and more users are added to the platform daily. The tremendous hike in number of users leads to increase in number of posts in social networks. As the behavior of each user varies, the type of posts also varies and as a result the issues related to abusive or negative behavior of the users increases. Thus, there arises the need of analyzing the data. The proposed architecture (Fig. 2) depicts the flow of the system with modules: data collection, analysis manager, security manager.

Data Collection: Collection of data from social networks is involved with this module. There are various APIs associated with social networks. They provide interface for collecting data. For example, graph API explorer of Facebook API. Text data (comments and posts) is collected and stored in data collection module. Each peer connected with the users who have posted some text in users profile is taken here.

Transliteration: Transliteration is included as the users today tend to type their regional language in English, and for analysis, all the text data should be in any one language [6, 7]. In the proposed system, a transliteration engine used to convert Tamil to English is integrated (Fig. 3)

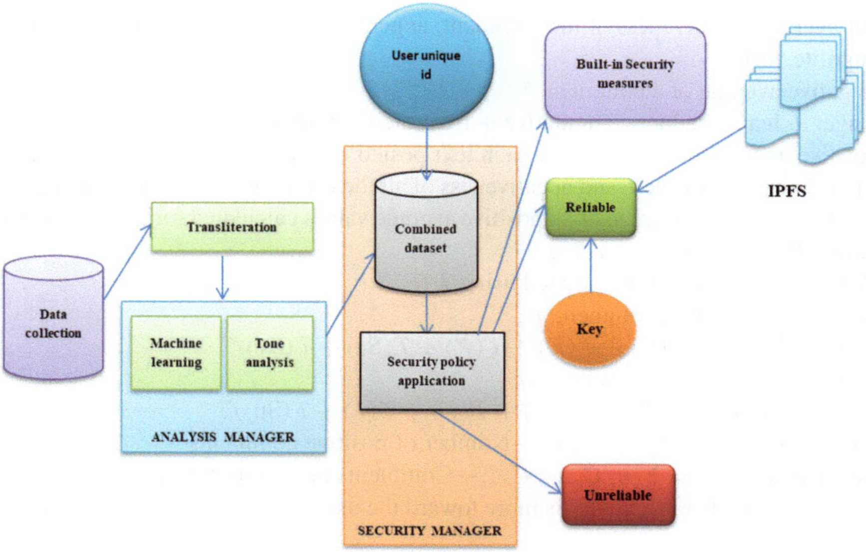

Fig. 2 Proposed architecture

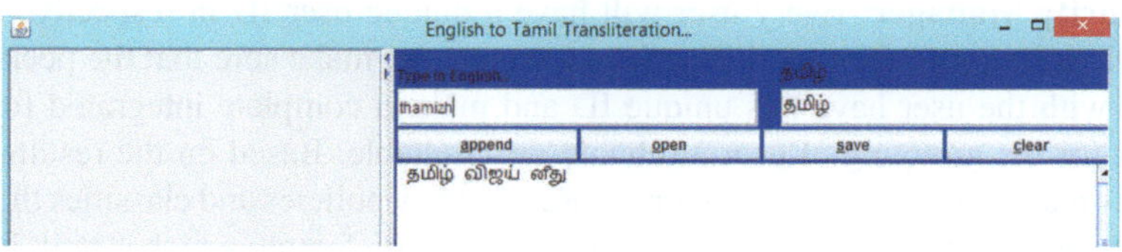

Fig. 3 Transliteration engine

Analysis Manager: The module is very important as the whole system accuracy is based on the proper analysis of the dataset. The analyzer module is where the machine learning classification, tone analysis, and transliteration take place. The text data collected from social network is classified as positive /negative using machine learning algorithm (SVM gave better results). But just machine learning classification alone will not support proper classification of user. Thus, we incorporate tone analysis for analyzing the emotion associated with the post and thus calculating the reliability of the peer.

Tone analysis is performed with the help of IBM BLUEMIX WATSON SERVICE called TONE ANALYSER [6]. The tool returns mathematical values for the tones present in the text that is passed to the analysis manager. Tones: anger, fear, joy, sadness, analytical, confident, tentative, excited, frustrated, impolite, polite, sad, satisfied, sympathetic tones that we consider to calculate average positiveness of the comment: joy, analytical, confident, excited, polite, satisfied, and sympathetic.
Positive average of a comment:
(Joy + analytical + confident + excited + polite + satisfied + sympathetic)/7.

For average negativeness of the comment: anger, fear, sadness, tentative, frustrated, impolite, sad.
Negative average of a comment:
(anger + fear + sadness + tentative + frustrated + impolite + sad)/7.
Positive and negative average of each text posted by a peer on user's wall is calculated. Then, positiveness and negativeness of all the text content posted is calculated. Positiveness is the average of all positive average values calculated for the comments. Similarly, for negativeness also.
Mathematically, tones are notated using T
Positive average of a comment: p
$T(J) + T(\text{An}) + T(C) + T(E) + T(P) + T(\text{Sa}) + T(\text{Sy})/7$
Negative average of a comment: n
$T(\text{AG}) + T(F) + T(\text{Sd}) + T(T) + T(F) + T(I) + T(\text{Sn})/7$
Positiveness : $P = \sum p/C_i$; C_i—Number of positive comments
Negativeness : $N = \sum n/C_{n-i}$; C_{n-i}—Comments that are negative.
If $N > P \parallel 0$, then negativity is more toward the user by the peer; thus, reliability is 0.
Else $\parallel$ 1, positivity of the user is more; thus, reliability is 1.
As in [8], SVM yields more accuracy in text classification for the process.

Security Manager: Every user will have a unique user ID in respective social network linked with their mail IDs. Security manager make sure that the peers connected with the user have this unique ID and make a complete integrated form of dataset for the grouping of user a reliable or unreliable. Based on the results from analysis manager, security manager proposes security policies and classifies the user. Security policies help the user in setting proper security feature which is built in with the social networks.

IPFS: User profile information is stored in IPFS which is protected by cryptographic hash method, and only peers who are reliable will be having access to key that open the user profile for viewing.

4 Results and Discussion

Online social networks witness a rise in user activity, and cyber criminals exploit this spur in user engagement levels to spread negative impacts on user. This compromises system reputation and degrades user experience [11]. In our system, we collect social network data from user profile over a period of time for primary analysis. From this data collected, we filter out post with negative content and human annotations. We use supervised learning models like naïve bayes, SVM , j48, and random forest-based models [9]. For our dataset , SVM returns the most accurate results in classification (Fig. 4). Basic supervised classification alone cannot give an appropriate result that helps us in classifying the user. Sentiment analysis measures tonality of text by identifying and assessing expressions people use to evaluate or appraise persons, entities, or events. A prominent example for analyzing the polarity of texts is media

POSTITVENESS		NEGATIVENESS	
openness	0.719038	Anger	0.018973
conscientiousness	0.279874	Disgust	0.042955
extraversion	0.613473	Fear	0.013415
joy	0.794777	Sadness	0.165241
Agreeableness	0.622743	Emotional Range	0.316637
positive averege	0.605981	negative averege	0.1114442

(a) Tone analysis results

Comment	negative average	positive averege	SVM classification
1	0.060146	0.794777	1
2	0.29849425	0.001159	0
3	0.2916865	0.018558	0
4	0.29505325	0.009106	0
5	0.28697825	0.001063	0
6	0.0047445	0.96323	1
total average	0.206183792	0.297982167	1
N<P, thus negative. thus unreliable			

(b) User classification using tone analysis results and SVM results

User unique ID	UserName	Classification
310301959122966	Praveen Muraleedharan	Unreliable
959007964197027	Sathvaiit Baburai	Reliable
822478851173781	Privanka Sunil Nair	Reliable
829327860415813	Dharani Teiaswini	Reliable
1071926086178960	Sripadh Thota	Reliable
10200220325546600	Aniu Javaprakash	Unreliable
10207719316364200	Neethu Nedungadi	Unreliable
898632173557885	Athira Sathish	Reliable
10152891420014100	Prahlad Kakkad	Unreliable
923347444422852	Drissia Thottungal	Unreliable
838670146223956	Sruthi Thottungal	Reliable
1149252745135140	Subhash Kuloor	Reliable
1108888409192460	Pradip Krishnan	Reliable
988657834504188	Sundar Dassan	Unreliable
975631215802683	Rumaniva Shahad	Unreliable

(c) Classyfying peers Final results

Instances:	2000
Attributes:	1157
Test mode	10-fold cross-validation Kernal: K(x,y) = <x,y>^2.0
Correctly Classified Instances	1656 82.8%
Incorrectly Classified Instances	344 17.2%
Kappa statistic	0.656
Mean absolute error	0.172
Root mean squared error	0.4147
Relative absolute error	34.4 %
Root relative squared error	82.9458 %
Total Number of Instances	2000
TP Rate	0.828
FP Rate	0.172
Precision	0.828
Recall	0.828
F-Measure	0.828
MCC	0.656
ROC Area	0.828
PRC Area	0.772
Confusion matrix	

a	b	classified
833	167	a = neg
177	823	b = pos

(d) SVM results

Fig. 4 Results consolidated

negativity, a concept that capture the over selection of negative over positive news, the tonality of media stories, and the degree of conflict. We try to take up the same tonality measurement in the posts in social network so that the degree of conflict in the post can be revealed and can be used in measuring trust score. The IBM tone analyzer analyzes shorter Web data, such as email messages, tweets, or even longer documents such as blog posts. The results include the tone or emotion shown in the text like anger, fear, joy, sadness, analytical, confident, tentative. We classified these emotions to calculate the negativeness and positiveness of the text post and classifying user as reliable or unreliable (Fig. 4).

5 Conclusion

Social networking Websites have witnessed thriving popularity with their ability to facilitate social interaction and communication. Popular sites like Facebook, Google plus, twitter facilitate online sharing of information and methods to develop new friendships. The work presented in the paper aims to improve security and privacy concerns using computed reliability factor as a moderator for sharing information. The purpose of the study is to investigate the effect of privacy concern based on factors like behavior, tone with reference to dataset collected from social networks. Participants of social networking sites generally have different perception toward privacy concerns. The idea proposed in this paper could be integrated with deep learning models to predict the perception with more accuracy for better results. This implementation integrated with blockchain technology is now currently being worked upon to evaluate the final impact on user privacy concerns.

References

1. Neethu MR, Harini N (2018) A model to ensure business ethics in social networks. Int J Sci Technol Res 2(5):79–91
2. Rathore S et al (2017) Social network security: issues, challenges, threats, and solutions. Inform Sci 421(5):43–69
3. Sapountzi A, Psannis KE (2018) Social networking data analysis tools & challenges. Fut Gener Comput Syst 86:893–913
4. Chen Y, Li H, Li K, Zhang J (2017) An improved P2P file system scheme based on IPFS and Blockchain. In: 2017 IEEE international conference on big data (big data). IEEE, pp 2652–2657
5. Ali MS, Dolui K, Antonelli F (2017) IoT data privacy via blockchains and IPFS. In: Proceedings of the seventh international conference on the internet of things. ACM
6. Sowmya Lakshmi BS, Shambhavi BR (2019) Automatic English to Kannada back-transliteration using combination-based approach. In: Emerging research in electronics, computer science and technology. Springer, Singapore, pp 159–170
7. Marouf Al Ahmed (2019) Recognizing language and emotional tone from music lyrics using IBM Watson Tone Analyzer. In: IEEE international conference on electrical, computer and communication technologies (ICECCT). IEEE
8. Neethu MR, Harini N (2018) Securing image posts in social networking sites. Comput Vis Bio Inspired Comput 2(5):79–91
9. Feng X (2017) An improved text classification model for mobile data security testing. In: 2017 IEEE 2nd advanced information technology, electronic and automation control conference (IAEAC). IEEE
10. Neethu MR, Harini N (2020) Secure request response (SRR): a framework to classify trust/distrust relationships in social networking. Comput Eng Technol 2(5):307–317
11. Kayes I, Iamnitchi A (2017) Privacy and security in online social networks: a survey. Online Soc Networks Media 3–4(5):1–21

A Novel Study of Different Privacy Frameworks Metrics and Patterns

Sukumar Rajendran and J. Prabhu

Abstract Learnability has impacted data privacy and security exposing two sides of a coin. A breach in security eventually leads to loss of privacy and vice versa. Evolution of technologies has put forth new platforms simplifying data derivation and assimilation providing information on the go. Even though different policies and metrics are in place, the objective varies along with the factors determined by technological advancement. This paper describes existing privacy metrics and patterns while providing an overall view of different mathematical framework privacy preserving. Furthermore, maintaining trust and utility becomes a challenge in preserving privacy and security as different techniques and technologies for assimilation of information are readily available without any restraints.

Keywords Privacy · Trust · Utility · Differential privacy

1 Introduction

Data security and privacy has and will gain the utmost attention in the forthcoming years. There is an impact of privacy in the physical world due to an exponential explosion of the IoT devices. These devices have integrated themselves with the day-to-day activities of human life and assimilate a large amount of data. Combined with technological enhancements like big data, cloud and high- performance computing, the impact on quality and acquisition of data, security, and digital privacy increases the probability of risks. New challenges arise in both securing and protecting data while accommodating confidentiality and concealment. Moreover, advances in data analytics blended with artificial intelligence allow extracting, predicting patterns, and

S. Rajendran
School of Information Technology and Engineering, VIT, Vellore, Tamil Nadu, India

J. Prabhu (✉)
Department of Software System and Engineering, School of Information Technology and Engineering, VIT, Vellore, Tamil Nadu, India
e-mail: j.prabhu@vit.ac.in

A. K. Tripathy et al. (eds.), *Advances in Distributed Computing and Machine Learning*, Lecture Notes in Networks and Systems 127, https://doi.org/10.1007/978-981-15-4218-3_19

useful knowledge from data. A massive initiative for the data-intensive applications is required while threats to security and privacy of the data are increasing. Damage or change of integrity affects not only the individuals but also organizations. Even with big technological responsibilities, big data and IoT are credited with providing the solution for complex problems in areas like medicine, finance, education, and transportation. The storage, sharing, and analysis of data have magnified personal privacy threats. The data breaches have exposed personal information of millions and revealed vulnerabilities. For effective decision-making, data mining may seem harmless from the perspective of private data exposure. Massive data collected are expected to be interconnected manually or autonomously through IoT. Although there are common laws and privacy-preserving technologies GDPR [18] to provide the protection, they become blurred due to different privacy policies.

1.1 Privacy

Privacy has different definitions, whereas it depends and changes on the needs of the different real-world applications. Techniques like anonymity, unlinkability, disclosure, and transparency play a major role in preserving privacy. Shalev-Shwartz determines privacy preservation based on different actors determined by the learning rule [20]. Shalev-Shwartz defines a learning problem if there exists a learning rule A and a monotonically decreasing sequence $\epsilon_{\text{cons}(m)}$such that $\epsilon_{\text{cons}(m)} \overset{\lim x \to \infty}{\to}$

$$\forall D, E_{s\tilde{D}^m}\left[F(A(S)) - F^*\right] \leq \epsilon_{\text{cons}}(m) \qquad (1)$$

A learning rule A for which this holds is denoted as a universally consistent learning rule. The basics of privacy identify four basic groups of activities that affect privacy based on three primary actors:

1. Collecting knowledge,
2. Processing of knowledge,
3. Distribution of knowledge,
4. Intrusions of knowledge.

The three primary actors are as follows:

1. Data subject—the privacy of the entity described in the data needs protection.
2. Data curator—a trusted collector of private information,
3. Data recipient—the trust of an adversary not guaranteed under any circumstances.

Each of the actors plays a significant role either in protecting or leaking privacy individually or combined as shown in Fig. 1. Different identifiers (quasi-identifier, microdata), generalization, suppression, anatomization, perturbation techniques as k-anonymity [19], syntactic privacy definition [14], l-diversity [15], t-closeness [14], personalized privacy, ϵ—differential privacy have their impact on three major actors within their specific domains. The confidence level of inference is directly associated

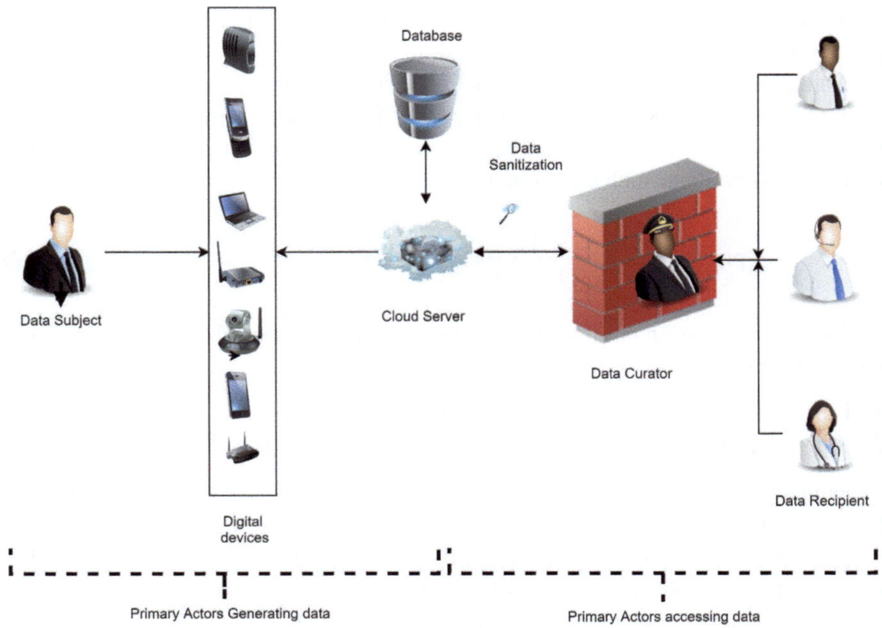

Fig. 1 Primary actors

with the damage of a privacy breach not whether the adversary obtains the sensitive value right. Principles to be followed in privacy design strategy [11]

1. Purpose limitation
2. Data minimization
3. Data quality
4. Transparency
5. Data subject rights
6. The right to be forgotten
7. Adequate protection
8. Data portability
9. Data breach notifications
10. Accountability and compliance.

Padakandla et al. [16] proposed the problem of protecting privacy while providing an accurate response to the queries from the sanitized database. The author key contribution is to define the amount of information preserved and privacy trade-off in the differentially private database sanitization mechanism. By considering the popular L1 distortion metric modelling, a generic distribution on the data and a measure of fidelity, this optimization problem prevents complexity with the number of constraints growing exponentially in the parameters of the information-theoretic framework. Padakandla et al. [16] fully characterized deep connection between the minimum expected distortion and a fundamental construct. Ehrhart theory states a simple closed-form computable expression for the asymptotic growth of the optimal privacy-fidelity trade-off to infinite precision.

Ahmad and Mukkamala [1] denotes First Privacy Axiom, such that it is not possible to build a computer with (*R*, *W*) = (0,0) for all data. A system that does not require data to be read or written can be considered to have a value of 0. It is not necessary to evaluate such a system as it does not function in terms of usefulness

$$\text{privacy} = \log \frac{1}{\epsilon(1-\pi\tau)} = -\log(\epsilon) - \log(1-\pi\tau)\,\text{bits} \tag{2}$$

The above equation is suggested in a two-part information sharing exchange as a measure of total privacy,

1. Errors of the channel described in (*R*, *W*),
2. Inherent privacy because of an unequal relationship of trust.

Ahmad and Mukkamala [1] proposed an output model by defining an information privacy agreement. The definition is rendered by assigning (*R*, *W*)-tuple at the EPA level to each identifier at an attribute stage. They also propose to describe the number of bits of privacy, and split it into three parts,

1. At the level of attributes, it is like the efficiency of the channel.
2. At the EPA level, it is like a service-level agreement (SLA), and a structured negotiating process should be used to take care of it.
3. It should rely on the trust with the immediate users of the data at the level of trust.

Ahmad and Mukkamala [1] stated the reason for not allowing a cascade of trust is that the Internet is made up of autonomous systems and there is no exhaustive way of ensuring trust beyond the limits of two neighbouring autonomous systems. Ahmad and Mukkamala [1] investigated trust and privacy in terms of bits that can be used to evaluate the optimal privacy controls to be implemented when sharing data. Kifer and Machanavajjhala [13] defined that a mathematical privacy framework called Pufferfish helps experts with no privacy experience to develop robust data sharing concepts for their data sharing needs. The system for Pufferfish can be used to test concepts of privacy. A new definition of privacy is with the objective of excluding and preserving attackers whose prior views are incompatible with unbounded continuous attributes and aggregate knowledge. The system is used to analyse and provide conditions under which the notion of balance between concepts of privacy is kept by certain forms of composition. The Pufferfish system assumes that each person is connected to a maximum of one record

$$P\left(M(\text{Data}) = \omega | s_i, \theta\right) \le e^{\epsilon} P\left(M(\text{Data}) = \omega | s_j, \theta\right. \tag{3}$$

Kifer and Machanavajjhala [13] arose a rigorous question in accordance with data adversary based on privacy definitions:

1. What comprises an inability to preserve privacy?
2. What is the intensity of the adversary?

Table 1 Different design strategies

Types	Patterns	Tactics
Data-oriented	Minimize—confine as much as possible for the evaluation of personal data	Select, exclude, strip, destroy
	Separate—personal data to be evaluate apart as much as possible	Isolate, distribute
	Abstract—limit as much as possible the detail in which personal data is processed	Summarise, group, and perturb
	Hide—conceal personal data, by without making it publicly known as well as unlinkable or unobservable	Restrict, obfuscate, dissociate, mix
Process oriented	Inform—notification in a timely and adequate manner to be given to subjects about evaluating their personal data	Supply, explain, notify
	Control—sufficient control over the processing of their personal data is given to data subjects	Consent, choose, update, retract
	Enforce—entrusting in a privacy-friendly way to authenticate personal data and adequately enforce this	Create, maintain, uphold
	Demonstrate—exhibit processing personal data in a privacy-friendly way	Record, audit, and report

3. Whose objective is it to trade-off privacy?
4. What adjunct data are accessible to the adversary even without access to the database being referred to (Table 1)?

2 Privacy Metrics and Patterns

The different privacy metrics are defined and determined by privacy patterns.

1. Uncertainty
2. Information gain or loss
3. Data similarity
4. Indistinguishability
5. Adversary's success probability

6. Error
7. Time
8. Accuracy or precision.

2.1 Privacy Patterns

The different privacy patterns provide us with a conflicting context of problems that arise due to the patterns. The context of patterns being disparate, they are still incidental on each other (Table 2).

3 Different Kinds of Privacy

3.1 Differential Privacy

Dwork et al. [8] defined DP as two datasets $D1$ and $D2$ are neighbours, denoted by $(D1, D2) \in N$, if they dissent in the value of one tuple. A distributed mechanism M satisfies differential privacy if $\forall$ set of outputs $S \subseteq \text{range}(M)$, and every pair of neighbouring datasets $(D1, D2) \in N$,

$$\Pr[M(D1) \in S] \leq e^{\epsilon} \Pr[M(D_2) \in S] \tag{4}$$

3.2 Hedging Privacy

Kifer and Machanavajjhala [10] applies single prior privacy is a diluted definition of privacy and strengthening it protects data publishers allowing partial deterministic output even though the beliefs about the data model are poor (i.e. the rule can certify the output as a true answer even though it is the true answer) the model is good.

$$S = \sigma_{(i,t)}) : h_i \epsilon H, t \epsilon \tau \cup \sigma_i : h_i \epsilon H \tag{5}$$

$$S = \left\{\sigma_{(i,t)} : h_i \in H, t \in \tau\right\} \cup \{\sigma_i : h_i \in H\} \tag{6}$$

Table 2 Different privacy patterns

Privacy pattern	Context
Anonymity set	The problem of this pattern infringes privacy that users can be distinguished by tracking and analysing the behaviour. The need is to eliminate the distinguishability by aggregating the different entities into a single set. In case the entities are probably too small to be stripped away of their key features which also leads to loss of functionality
Strip invisible metadata	There is additional data attached to the user content during the sharing of services, and publishing this data could lead to a potential threat with a loss of privacy. Identification, possible erasure, and inaccessibility of the metadata after publication irreversible and lead to loss of functionality
Pseudonymous identity	The relation of original identity and pseudonymous identity will reveal sensitive information exposed during interaction in the online communication and forums. Pseudo-anonymity identity mapping and its full usage can reveal or weaken the content
Protection against tracking	The tracking of personally identifiable information and visitors of Websites through software tools protocols or mechanisms such as cookies affect the privacy and anonymity. Other mechanisms by Web service providers for user monitoring may still work providing user profiling
Pseudonymous messaging	Incorporating all types of correspondence through messages, articles, message sheets, newsgroups, and this data could be used to manufacture client profiles. The exchange of information through pseudonyms, the trusted servers, can be misused by sending an offensive message
Onion routing	Encrypting and encapsulating the data in different layers limits the linkability and knowledge of the path between the sender and receiver. If there are a few communicating, the anonymity set is not significant and multiple layers of encryption will put the sender and receiver at risk
Privacy-aware network client	The familiar security strategies help the client to make more educated choices that lead intermediary to distinguishing changes in security.
Privacy icons	Visual clues that a user can understand with a common meaning shared by the community while the policies are lengthy and cannot be accessed at the first glance
Privacy colour coding	The user can have visual clues and prevent possible non-optimal privacy setting while misrepresentation and social and cultural aspects render the pattern completely useless
Attribute-based credentials	Authentication of attributes confirms a trait that all data are discovered and not viewed as a privacy-preserving solution and require ample processing power
User data confinement pattern	This pattern might be utilized at whatever point the accumulation of individual information with one precise and real reason still represent a significant dimension of danger to the clients' security. This may include the use of Trusted Platform Modules or cryptographic algorithms

(continued)

Table 2 (continued)

Privacy pattern	Context
Obligation management	The aim is to make multiple parties are obligatory of and adhered with organizational policies as personal and sensitive data are shared between a series of parties who store or process that data. privacy preferences and policies are communicated and adhered to among organizations sharing data
Sticky policies	Sticky policy to be taken into account when sharing sensitive data across multiple recipients with strong enforcement to enable trackability. The size of the data is altered as well as the policy cannot be updated after sharing the data
Personal data store	The need for a personal data store arises when the data subjects lose control over the data stored and operated by the third party
Identity federation does not track	Correlation of the identity of the end-user with service provider data is possible through deriving the information from the identity provider. This pattern avoids the learning of the relationship or acquiring additional information from different adversaries
Use of dummies	The use of dummies is used when there is a need in altering the content of action and generate camouflage from the real actions. This pattern requires excess resources to perform dummy actions which may degrade the quality of service
Data breach notification pattern	The pattern is used to control access and monitor whether there the data breach has occurred and a notification has been sent to the user. This pattern needs to be followed to ensure the incident manager takes adequate measures to control by other means
Privacy dashboard	A dashboard with summaries of the user personal data where deletion and correction of stored data is with the user
Federated privacy impact assessment	Identity management scenarios (that is, when the roles of the identity provider and the service provider are separated). The consequences depend on the results of the privacy impact analysis
Trustworthy privacy plugin	The service provider does not need any more to access detailed consumption data in order to issue reliable bills
Selective access control	To share content with precaution to different kinds of the user with varying social proximities. Granular configurations are time-consuming or tedious while configuration based on kinds of users or the content's context
Ambient notice	A user may not have the idea that an application given explicit permissions is distributing personal trait over a long period of time with the user being provided notice only at the time of consent, unnecessary notice to be avoided to balance against the concerns of an attacker's opting the user in without their knowledge

(continued)

Table 2 (continued)

Privacy pattern	Context
Reciprocity	Users may leave the group due to unfairly spread workload, inequalities. Lawful consent needs to be given which may be overlooked by the controller to coerce participation while content generation is continuous
Obligation management	The data subject may not approve of the way data are being shared and accessed by multiple parties even though privacy preferences and policies are communicated and adhered among organizations sharing data
Aggregation gateway	A detailed measurement of service attribute according to the demand load reveals information when repeated. To avoid brute force attack, large range of measured and homomorphic encryption schemes are included which demand computational power
Active broadcast of presence	Aims at acquiring real-time data without regretting the user has provided too much data or constant request. To provide users with relevant information without infringing the user's privacy or providing irrelevant data
Layered policy design	By layering the privacy policies with highlights on summary of notices to users while controller should ensure the policies are updated simultaneously
Policy matching display	Users be able to simultaneously configure policies with different providers new and old
Added noise measurement obfuscation	To alter the measured value by adding noise so no inference can be made by the adversary
Anonymous reputation-based blacklisting	Being a desirable property with privacy, it may increase the lack of fear in users and allow them to misbehave as they cannot be identified
Informed consent for Web-based transactions	Controllers store potentially identifiable or sensitive information about a user with informed consent user resistance in disclosing personal information because of uncertainty and the fear of their privacy being undermined. The controller may not wish to disclose the Website ability to track users without their knowledge
Masquerade	Content generation is impacted negatively due to user responding differently under active supervision. Users may reveal some subset of information and filter out the rest with regard to identifiability
Location granularity	The collection and distribution of location data to third parties increases greater chances of users being re-identified. To lessen this by implementing certain levels of granularity
Incentivized participation	User increasing their participation in the network leading to bad experience and privacy concerns in turn limiting the users alienate them. Avoiding this through social encouragement and positive reinforcement

(continued)

Table 2 (continued)

Privacy pattern	Context
Outsourcing [with consent]	Utilizing external sources to process data does not provide the means to oversee as external sources are not allowed to directly communicate with the client
Discouraging blanket strategies	Not to simplify privacy settings while sharing diverse content which could over expose or provide content which is at least valuable
Selective disclosure	To tailor services to share information upto the level the users are comfortable with
Buddy list	To connect to or add a user to user-maintained directory based on intuition and social context
Data breach notification pattern	To have a control mechanism to determine and notify whether personally identifiable information is exposed
Reasonable level of control	Users have their own conception to be comfortable of what data are collected or shared and not distributed while not trusting the service
Enable/disable functions	User to directly determine whether to allow or exclude certain functions from collecting data
Asynchronous notice	Tracking storing or distribution of the user information with or without prior user consent
Negation of privacy policy	Users privacy settings may be omitted by the controller as universal setting cannot cater to individual requirements
Payback	User motivation is necessary either through monetary benefits or through other means keeping a smooth flow of information
Encryption with user-managed keys	Excluding the service provider from decrypting the user information by allowing the user to manage the keys
Obtaining explicit consent	Users can overlook consequences as they are presented with selective disclosure as data are collected the same way for every user while spending lesser time in making decisions
Policy matching display	To reduce the burden of deciding the privacy aspects with integrating new services increasing the chances of error in their privacy-preserving decisions
Decoupling and location information visibility	Users can define and share generating content in one place, or where contextually relevant, the granular privacy settings for the fine-grained location configuration of their content
Single point of contact	Securely distributed data need specialized privacy management with reliable credentials and lawful consent-based authentication system supervised and provably sound

3.3 *Blow-Fish Privacy*

He et al. [10] improvised Pufferfish framework with a rich interface with a better interexchange between privacy and utility. Data publishers are allowed to extend the differential privacy by policy that specifies what information are to be kept secret and what constraints that are to be known about the data. The constraints also give added protection against adversary who are aware of correlating data.

$$\Pr\left[M\left(D_1\right) \epsilon S\right] \leq e^{\epsilon} Pr\left[M\left(D_2\right) \epsilon S\right] \tag{7}$$

$$\Pr\left[M\left(D_1\right) \epsilon S\right] \leq e^{\epsilon \cdot d} G^{(x,y)} Pr\left[M\left(D_2\right) \epsilon S\right] \tag{8}$$

3.4 Adversarial Privacy

Huang et al. [12] defined adversarial privacy algorithm allowing domain experts to plug in generating various data

$$P(t|O) = e^{\epsilon} P\left(t'\right) + y \tag{9}$$

Choosing the class of adversaries P is the primary difficulty in applying adversarial privacy [12]. If P is chosen to be the set of all possible adversaries, then no algorithm can be both useful and adversarially private [3].

3.5 Localized Information Privacy

Jiang (2018) proposed three types of LDP as follows: ϵ—Localized differential privacy (LDP)

$$\frac{\Pr(M(x) = y)}{\Pr\left(M\left(x'\right)\right) = y} \leq e^{\epsilon} \tag{10}$$

ϵ—Mutual information privacy (MIP)

$$\sum_{x,y \in D} \frac{Pr(X = x, Y = y)}{Pr(X = x) Pr(Y = y)} \tag{11}$$

ϵ—Localized information privacy (LIP)

$$e^{-\epsilon} \leq \frac{Pr(X = x)}{Pr(X = x | Y = y)} \leq e^{\epsilon} \tag{12}$$

3.6 Local Differential Privacy

Chen et al. [5] proposed privacy guarentees for utilizing the locally private protocols

$$P[A(l) \in O] \leq \exp(\epsilon) \cdot P\left[A\left(l'\right) \in O\right] \tag{13}$$

The proposed blended approach combines the common models as one local and the trusted curated with improvements in utility while bridging the gap between physical and theoretical engineering challenges [2].

3.7 Distributional Privacy

Blum et al. [4] motivated by learning theory proposed an alternative privacy definition which can guarantee that two databases differ in all their elements and are queried from some distribution and allow exponentially small probability for privacy violations. Although the two databases are neighbours, they are drawn at random without replacement

$$\Pr\left[A\left(D_1\right) \epsilon E\right] \leq e^{\epsilon} \Pr\left[A\left(D_2 \epsilon E\right)\right] \tag{14}$$

Blum et al. [4] showed that over continuous domain it is not possible to release simple class of queries. The algorithm is also efficient for releasing synthetic data.

3.8 Prediction Privacy

Dwork and Feldman [7] showed how to derive relatively strong generalization guarantees from uniform prediction stability. These guarantees are stronger than classical notions of stability while not being as strong as those of differential privacy.

$$E_{S,S' \sim P^n}\left[(\epsilon_{S'}[l(M(S))])^k\right] \leq e^{k^2 \epsilon} E_{S \sim P^n}\left[(\epsilon_S[l(M(S))])^k\right] \tag{15}$$

3.9 Group Privacy

Dwork et al. [8] suggests, the control and analysis of privacy loss obtained by the groups as allowed by differential privacy. For (ϵ,0)-differentially private mechanisms with the size of the group, the strength of the privacy guarantee drops linearly. The author defined $M\colon (X \times Y)n \times X \rightarrow Y$ be an (ϵ, γ)-differentially private prediction algorithm and. For every pair of dataset S,S$\epsilon(XxY)n$ differing in at most k elements and all xϵX:

$$D_{\infty}^{ke^{k-1}\delta}(M(S,x)\|M(S,x)) \leq k\epsilon \quad (16)$$

4 Protecting Privacy of Things, Cloud, and Fog

4.1 Location Privacy

Ullah et al. [21] proposed an enhanced semantic obfuscation technique (ESOT) to preserve privacy and provide security in IoT due to the issues in location privacy. ESOT selects the obfuscation level based on the user privacy preference. A random location is chosen by effacing or adding noise to the user's actual location to deviate from privacy attacks based on the degree of obfuscation level. ESOT preserves the balance between privacy and service quality by offering a reasonable gap between the initial and obfuscated locations [6].

Fawaz et al. [9] picturized as existing apps, A&A libraries and OS controls are ineffective and inefficient considering the profiling threats. The author proposes a LPDoctor a location preserving tool to allow the android user to effectively access the controls while maintaining the apps functionality. Things make human life more productable while they are resource constrained and pose new challenges to privacy and security [17]. The challenges are at the point of collection storage and hoe to react to the real-time changes in the physical world. They further introduce privacy and security issues due to network delays and jitters. Things generate tremendous data with volume and velocity that need to be assimilated due to time-constrained applications.

The different privacy issues are

1. Linkable transactions,
2. Disclosure of user personal information,
3. User/device movement analysis, and
4. User/device transaction histories.

Things are not capable of storing or processing massive data and transfer it to remote cloud for further analysis [17]. The urgency for processing the data for time-constrained applications is immediate leading to analysis of data at the network edge. This introduces a new concept of fog which is just an extension of the cloud (Table 3).

Table 3 IoT patterns for privacy [17]

Privacy patterns	Protection objective
User-defined privacy policies	Intervenability
Consent manager	Transparency, intervenability
Deactivator/activator	Unlinkability, intervenability
Privacy dashboard	Transparency, intervenability, availability
Lightweight pseudonym system	Unlinkability
Geo-Location	Unlinkability
Sticky policies	Intervenability
Pliable group signatures	Integrity, authentication, data minimization
Flexibility of security and privacy mechanisms	Intervenability
Bootstrapping identity mechanism	Integrity, authentication

5 Conclusion

Balancing the need for accessing and harnessing the data without exposing or eliminating individuals while maintaining the notion of common good pertains as the primary challenge in this era of things. Privacy-preserving increases the challenge in maintaining a delicate balance between the utility of data and the privacy of individuals. Defining and proposing a joint approach under what circumstances and definitions are privacy and fairness are simultaneously achievable and need not be traded for each other. Algorithmic decision-making systems may be biased with discriminatory outcomes such that they fail to protect the assailable members in the target population which results in decreasing the effectiveness of privacy protections. Although different techniques, approaches, and patterns to preserve privacy have been proposed, there is no standard measure to quantify privacy metrics. They depend on various factors like combining data from different sources where privacy breach can occur. Usually, when privacy mechanisms are in place data are obfuscated before being shared with third parties. The problem arises when data from different sources are shared and the obfuscation mechanisms return conflicting results. IoT allows sensitized objects to interact with other user/devices; these interactions among objects allows for the creation of relationship graphs as such in OSN. It presents an unparalleled challenge to privacy as an ardent opponent may not only learn to look at the contact pattern between apps and users but providing information about family members, acquaintances, user desires, personal preferences, and online activity and choices. IoT environment allows BYOD (bring your own devices) that may be faulty, insecure, or illegitimate. As such the data collected from such IoT devices may not be trustworthy and cannot be used in critical decision making. A typical approach to protecting privacy is to inject fake traffic into the network, particularly from global adversaries. Nevertheless, this strategy becomes more disruptive as the network density increases in IoT scenarios. Interference's are growing, the signal-to-noise ratio is decreasing, packet collisions and re-transmissions are becoming more frequent, and the channel's timeliness, efficiency, and throughput is falling precipitously. Consequently, to avoid these problems, the provision of location privacy based on fake traffic injection needs to be carefully revamped. The adversary will have the ability to eavesdrop on devices to cover wider and geographically dispersed areas without being physically present in those regions.

References

1. Ahmad, A., Mukkamala, R.: A novel information privacy metric. In: Information Technology–New Generations, pp. 221–226. Springer, Berlin (2018)
2. Avent, B., Korolova, A., Zeber, D., Hovden, T., Livshits, B.: {BLENDER}: Enabling local search with a hybrid differential privacy model. In: 26th {USENIX} Security Symposium ({USENIX} Security 17), pp. 747–764 (2017)
3. Bebensee, B.: Local differential privacy: a tutorial. arXiv preprint arXiv:1907.11908 (2019)
4. Blum A, Ligett K, Roth A (2013) A learning theory approach to noninteractive database privacy. Journal of the ACM (JACM) 60(2):12
5. Chen, R., Li, H., Qin, A.K., Kasiviswanathan, S.P., Jin, H.: Private spatial data aggregation in the local setting. In: 2016 IEEE 32nd International Conference on Data Engineering (ICDE), pp. 289–300. IEEE (2016)
6. Colesky, M., Hoepman, J.H.: Privacy patterns (2017). https://privacypatterns.org
7. Dwork, C., Feldman, V.: Privacy-preserving prediction. arXiv preprint arXiv:1803.10266 (2018)
8. Dwork, C., Roth, A., et al.: The algorithmic foundations of differential privacy. Found. Trends® Theor. Comput. Sci. **9**(3–4), 211–407 (2014)
9. Fawaz, K., Feng, H., Shin, K.G.: Anatomization and protection of mobile apps' location privacy threats. In: 24th {USENIX} Security Symposium ({USENIX} Security 15), pp. 753–768 (2015)
10. He, X., Machanavajjhala, A., Ding, B.: Blowfish privacy: tuning privacy-utility trade-offs using policies. In: Proceedings of the 2014 ACM SIGMOD international Conference on Management of Data, pp. 1447–1458. ACM (2014)
11. Hoepman, J.H.: Privacy design strategies (the little blue book) (2018)
12. Huang, C., Kairouz, P., Chen, X., Sankar, L., Rajagopal, R.: Generative adversarial privacy. arXiv preprint arXiv:1807.05306 (2018)
13. Kifer D, Machanavajjhala A (2014) Pufferfish: a framework for mathematical privacy definitions. ACM Trans. Database Syst. (TODS) 39(1):3
14. Lin, B.R., Kifer, D.: Towards a systematic analysis of privacy definitions. J. Privacy Confidenti. **5**(2) (2014)
15. Machanavajjhala, A., Kifer, D., Gehrke, J., Venkitasubramaniam, M.: l-diversity: Privacy beyond k-anonymity. ACM Trans. Knowl. Discov. Data (TKDD) **1**(1), 3–es (2007)
16. Padakandla, A., Kumar, P., Szpankowski, W.: The trade-off between privacy and fidelity via ehrhart theory. IEEE Trans. Inform. Theory (2019)
17. Rauf, A., Shaikh, R.A., Shah, A.: Security and privacy for iot and fog computing paradigm. In: 2018 15th Learning and Technology Conference (L&T), pp. 96–101. IEEE (2018)
18. Regulation GDP (2016) Regulation (eu) 2016/679 of the european parliament and of the council of 27 April 2016 on the protection of natural persons with regard to the processing of personal data and on the free movement of such data, and repealing directive 95/46. Off. J. Eur. Union (OJ) 59(1–88):294

19. Samarati P (2001) Protecting respondents identities in microdata release. IEEE Tran. Knowl. Data Eng. 13(6):1010–1027
20. Shalev-Shwartz, S., Shamir, O., Srebro, N., Sridharan, K.: Learnability, stability and uniform convergence. J. Mach. Learn. Res. **11**, 2635–2670 (2010)
21. Ullah I, Shah MA, Wahid A, Mehmood A, Song H (2018) Esot: a new privacy model for preserving location privacy in internet of things. Telecommun. Syst. 67(4):553–575

A Comparative Analysis of Accessibility and Stability of Different Networks

Subasish Mohapatra, Harkishen Singh, Pratik Srichandan, Muskan Khedia, and Subhadarshini Mohanty

Abstract These days, there is a wide assortment of system available, it is growing exponentially, and to analyze the network performance, different type benchmarking tools are used, which allows scientists and system managers numerous choices to work with. To investigate the system execution scarcely, many components are there which influences the exhibition of PC systems. Those elements will decrease the quality of service of the network. The central point is latency, packet misfortune, re-transmission, throughput and queuing delay. In any case, this assortment keeps an eye on troublesome choice procedure of the proper device. Besides, in some cases, clients are compelled to attempt a few devices so as to locate an ideal outcome giving tools which will ascertain a given measure, so they need to figure out how to control various instruments and how to decipher acquired outcomes. This paper offers a benchmarking tool which will monitor network and help in load handling of application on the individual route, test various possibilities of data in params (permutate params) with response behaves by the network, analyze network performance of the hosted application irrespective of containerization and representation of performance with the help of graph, which will guide to select this tool over others.

Keywords Component · Ping · Monitoring · Report changes · Request–response delay · Network congestion · Bottlenecking · Time-series database (TSDB) · Routes

---

S. Mohapatra (✉) · H. Singh · P. Srichandan · M. Khedia · S. Mohanty
Department of Computer Science and Engineering, College of Engineering and Technology, Bhubaneswar, India
e-mail: subasish.mohapatra@gmail.com

H. Singh
e-mail: harkishensingh@hotmail.com

P. Srichandan
e-mail: pratik.srichandan185@gmail.com

M. Khedia
e-mail: muskan.khedia2000@gmail.com

S. Mohanty
e-mail: sdmohantycse@cet.edu.in

A. K. Tripathy et al. (eds.), *Advances in Distributed Computing and Machine Learning*, Lecture Notes in Networks and Systems 127, https://doi.org/10.1007/978-981-15-4218-3_20

1 Introduction

Usually, analysts and additionally organize heads and/or network service provider need to measure different network indicators to optimize the performance. A few devices are accessible for this reason; however, not every one of them has similar qualities and assesses similar exhibition parameters. Now and then, the proposed client is obliged to utilize some of these instruments so as to finish trials and find huge outcomes. Our testing tool performs these tests with all the required parameters and to the network or device under test, and this benchmarking tool is more effective and appropriate than any other. This tool is less complicated and reduces the time in the evaluation process, for which researchers and/or network administrators and/or network service provider does not require learning to use different tools. For example, Gamess and Velásquez [1] had to use two different performance tools and write their own benchmarks in order to evaluate the forwarding performance on different operating systems. This paper presents a network performance tool which can be used on the currently used network benchmarking tools, as a way to help researchers and network administrators to make a decision about the more appropriate tool for their needs. Additionally, this paper outlines desirable features, accuracy, and performance of the network.

2 Backgrounds

Botta et al. [2] present a careful audit of available information move limit estimation instruments as a bleeding edge. They have separated the instruments into three unique classes: (1) start to finish limit estimation devices, (2) accessible transmission capacity estimation devices, and (3) TCP throughput and mass exchange limit estimation devices. They additionally present an instrument called bandwidth estimation tool (BET) and contrast it and other execution apparatuses, to the extent precision and absolute time used to complete the estimation technique. Ubik and Král condense their association in data transfer capacity estimation devices; they revolve around finding the size and area of bottlenecks.

They present an arrangement of from beginning to end transfer speed estimation apparatuses dependent on different properties, including choosing transmission capacity (introduced data transfer capacity) versus throughput (accessible transfer speed), in the computer systems of the sender and receiver, etc. They additionally portray properties of a few picked devices and present results of estimation with one apparatus (way load) on a quick situation and delayed consequences of combined estimations with a few instruments over a genuine quick long-separation organize.

Gupta et al. [3] play out an exploratory relative examination of both inactive and dynamic data transfer capacity estimation instruments for an 802.11-based remote system and derive that for the remote system an uninvolved procedure gives more noteworthy exactness. Strauss et al. depict Spruce, a fundamental instrument for estimating accessible transmission capacity, and afterward, contrast it and other existing apparatuses over numerous unmistakable Internet ways. Their examination depends on precision, disappointment designs, test overhead, and usage issues.

Li et al. [4] present WBest, a remote data transmission estimation instrument proposed for exact estimation of the accessible transfer speed in IEEE 802.11 frameworks, clarifying that an enormous segment of the present apparatuses was not expected for remote systems. They describe the calculation used by results. They additionally contrast their device and others like pathChirp [5] and pathload [6].

3 Objective and Motivation

Our aim is to show a clear depiction of this tool that can help to comprehend the network accessibility and reliability and analyze the essential work with the help of a GUI-powered highly scalable testing tool that monitors the performance of the routes in any Web/routing—application. This tool would perform a series of networking algorithms and calculations which are required to find the real-time state of routes in an application. It monitors the routes at regular intervals and analyzes the response and the time involved in the same. The moment, the delta in response rises above a threshold limit, an alert would be sent to the admin of the respective manager. Additionally, we figure out desirable features for a throughput and bulk transfer capacity measurement tool since our goal is to develop a new network benchmarking tool from scratch. During the analysis of each step, features that are helpful and efforts involved in answering the questions related to the behaviour in each step are recorded carefully.

This tool will help the developers to monitor their virtual machine instances, Web applications and their routes on a regular basis. During this monitoring, secondary calculations like jitter, response length, are done. These calculations are periodic and based on time-series and hence are stored in a time-series database. Graphs related to all the calculations along with their historic values are plotted in an interactive user interface, so that the developer can understand and study the behavior over time and draw the necessary conclusions. This is particularly useful for critical applications which are of high value to the organization.

Approach

1. Linux ping commands
    a. min/avg/max/mdev ping values
    b. Flood ping
    c. Jitter
2. Monitoring
3. Response monitoring
    a. Report response changes
    b. Req–res delay

4. Network congestion.

Detailed Explanation

1. **Linux ping command**

All Unix-based system supports basic ping command right from their shell. This can be made into use by running it through a sub-process and pipe out the necessary details of our needs and scrap them accordingly. However, it is important to note that ping works on the raw IP as a pong from the hosted service which runs as default in any network card. Hence, this method is just to check the server response time, requests performance, and could be helpful in checking bottlenecking states. In a nutshell, this helps to understand the accessibility of any remote computing instance. Following ways can make this process more helpful: **min/avg/max/mdev ping values**: These are the general ping values after the ping sub-process would end its successful execution. This can be made to run in count mode so that only a particular number of pings are made to the mentioned IP (Fig. 1).

Flood ping These would send a high number of requests (often in thousands or lakhs) to the IP route specified. This leads to more accurate results but requires sudo if the minimum interval is set to <0.2 (default). Consider the example below, the count is set to 10,000 requests with the interval set as 0 (since not mentioned) and hence requires sudo (Fig. 2).

Jitter Jitter is the amount of variation in latency/response time, in milliseconds. Reliable connections consistently report back the same latency over and over again. Jitter is the measure of variety in idleness/reaction time, in milliseconds. Reliable associations consistently report back a similar latency again and again. Lots of variation (or 'jitter') is an indication of problems (Fig. 3).

Example: Consider 5 ping samples: 136, 184, 115, 148, 125. Then, jitter found in the sample can be calculated by averaging consecutive differences in the ping response time as follows:

1. 136–184, difference = 48

```
$ ping -c 4 google.com
PING google.com (74.125.135.100) 56(84) bytes of data.
64 bytes from plus.google.com (74.125.135.100): icmp_req=1 ttl=128 time=251 ms
64 bytes from plus.google.com (74.125.135.100): icmp_req=2 ttl=128 time=180 ms
64 bytes from plus.google.com (74.125.135.100): icmp_req=3 ttl=128 time=179 ms
64 bytes from plus.google.com (74.125.135.100): icmp_req=4 ttl=128 time=179 ms

--- google.com ping statistics ---
4 packets transmitted, 4 received, 0% packet loss, time 3005ms
rtt min/avg/max/mdev = 179.569/197.734/251.433/31.005 ms
```

Fig. 1 Output of executing ping command with a count of ten in Linux terminal

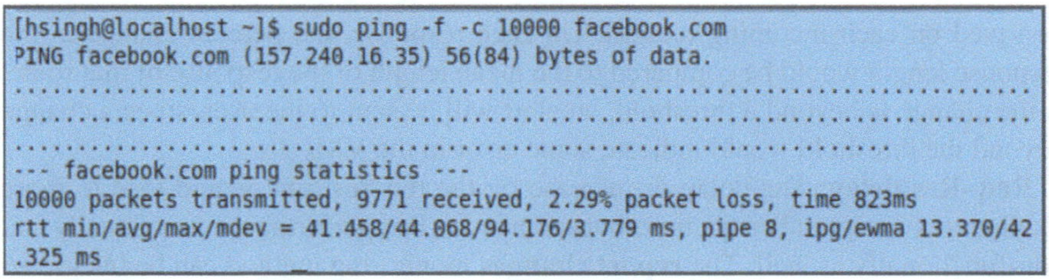

```
[hsingh@localhost ~]$ sudo ping -f -c 10000 facebook.com
PING facebook.com (157.240.16.35) 56(84) bytes of data.
.........................................................................
.........................................................................
...............................................................
--- facebook.com ping statistics ---
10000 packets transmitted, 9771 received, 2.29% packet loss, time 823ms
rtt min/avg/max/mdev = 41.458/44.068/94.176/3.779 ms, pipe 8, ipg/ewma 13.370/42
.325 ms
```

Fig. 2 Output of executing flood ping command in Linux terminal

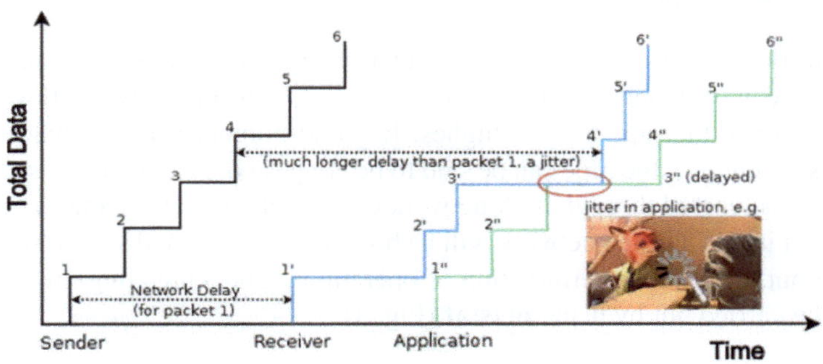

Fig. 3 Jitter graph for the assumed ping value

2. 184–115, difference = 69
3. 115–148, difference = 33
4. 148–125, difference = 23.

Summing up all these differences gives us 173. So, the average difference in the ideal network point, i.e., the jitter is 173/4, or 43.25.

Jitter tests hold great importance in helping to analyze the stability of a network connection connected to any remote computing resources. The more jitter, the more is the instability or the network connection.

2. **Monitoring**

In a world of complex and sophisticated servers with several concurrent routes, monitoring remains an issue especially with the multi-deployment instance, handled by Kubernetes. This part aims to monitor the specific route and to alert the user in case of a downstate.

3. **Response Monitoring**

a. **Report changes**: Extension of the above point in which the response of the particular route would be saved as a time-series and a mean of the response length would

be saved on each incoming request. On each response, the percent change in the response length would be compared to the mean length of the response of that route. If the change is beyond a threshold, an alert will be sent to the user, since a change beyond the threshold would indicate some error in the VM.
b. **Req–Res delay**: The delay or increase on the req–res time would lead to more likeliness of a request being a time-out and hence a bad performance functionally and non-functionally as well. The **report changes** monitoring method can be integrated into this module since we are dealing with routes and requests, so there is no sense to create another module for the cause. **Note**: We have to create a separate database to store the values of response length to reuse it next time the user starts bench-routes.

4. **Network Congestion**

Network congestion is the situation that occurs when the entire bandwidth of the network is occupied. For better understanding, consider a water flow pipe. When the water in the pipe has the highest Reynold number (in the situation of highest pressure output), the pipe can be said to be congested, i.e., any other stream of water cannot be passed through it. A network can be related in the same way. Hence, any request in a congested network will be bound to get slow and sometimes may lead to time-out, resulting in termination of operations. Checking congestion in a network can be carried out by using **netstat** (Fig. 4).

Data flow
The flow of data right from the requesting the bench-routes service by the front UI until the final response to the front UI is as follows. (All lines indicate bi-directional propagation, and/indicates a directory as related with the high-level model.) (Fig. 5).

```
[hsingh@localhost ~]$ netstat -p -a
(Not all processes could be identified, non-owned process info
 will not be shown, you would have to be root to see it all.)
Active Internet connections (servers and established)
Proto Recv-Q Send-Q Local Address           Foreign Address         State       PID/Program name
tcp        0      0 localhost:ipp           0.0.0.0:*               LISTEN      -
tcp        0      0 localhost:27017         0.0.0.0:*               LISTEN      -
tcp        0      0 localhost.localdo:56182 server-99-86-55-2:https ESTABLISHED 4167/chrome --type=
tcp        0      0 localhost.localdo:51766 ec2-52-3-109-96.c:https ESTABLISHED 4167/chrome --type=
tcp        0      0 localhost.localdo:43774 104.20.50.118:https     ESTABLISHED 4167/chrome --type=
tcp        0      0 localhost.localdo:47136 bom05s15-in-f19.1:https ESTABLISHED 4167/chrome --type=
tcp        0      0 localhost.localdo:56864 sb-in-f188.1e10:hpvroom ESTABLISHED 4167/chrome --type=
tcp        0      0 localhost.localdo:48976 104.244.42.8:https      ESTABLISHED 4167/chrome --type=
tcp6       0      0 localhost:ipp           [::]:*                  LISTEN      -
udp        0      0 224.0.0.251:mdns        0.0.0.0:*                           4167/chrome --type=
udp        0      0 224.0.0.251:mdns        0.0.0.0:*                           4167/chrome --type=
udp        0      0 224.0.0.251:mdns        0.0.0.0:*                           4044/chrome
udp        0      0 0.0.0.0:mdns            0.0.0.0:*                           -
udp        0      0 localhost.localdo:56247 kul01s09-in-f78.1:https ESTABLISHED 4167/chrome --type=
udp        0      0 0.0.0.0:bootpc          0.0.0.0:*                           -
udp        0      0 localhost:323           0.0.0.0:*                           -
udp        0      0 0.0.0.0:42112           0.0.0.0:*                           -
udp        0      0 localhost.localdo:60362 bom07s20-in-f14.1:https ESTABLISHED 4167/chrome --type=
udp6       0      0 [::]:mdns               [::]:*                              -
udp6       0      0 [::]:38405              [::]:*                              -
udp6       0      0 localhost:323           [::]:*                              -
raw6       0      0 [::]:ipv6-icmp          [::]:*                  7           -
```

Fig. 4 Processes running in a Linux system

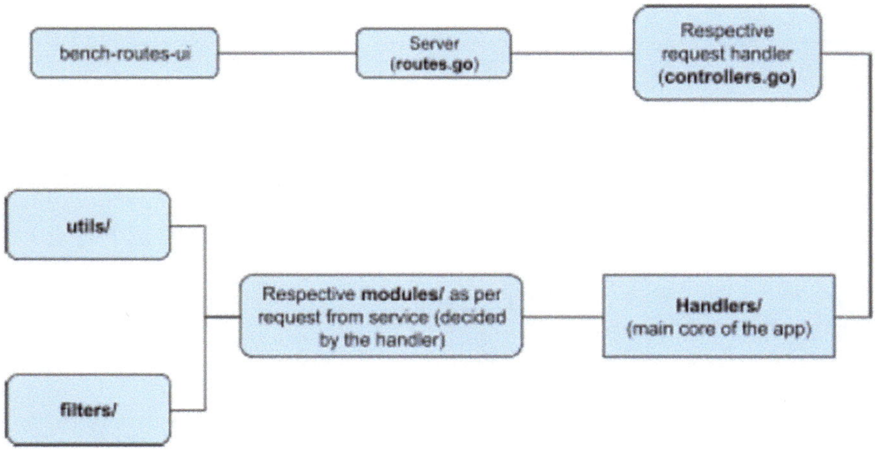

Fig. 5 Data flow in the software model

4 Implementation

The implementation of the above-proposed idea would be carried out in the Go programming language. This project will launch itself on boot and work as a daemon service. The bench-routes UI will be responsible for the graphical aspect of the project.

Facebook
See Figs. 6 and 7.

Gitlab
See Figs. 8, 9 and 10.

5 Conclusion

In this paper, we exhibited the consequences of an analysis where we also assessed and thought about a few current systems benchmarking instruments. But none of them satisfied the requirements of benchmarking and monitoring the routes at specific intervals for the user. From the graphs generated using the automated tool, we can analyze the stability of the remote Internet connection and performance of the routes based on the request. The tool also alerts the user in case of any failure of any of the routes specified and also during any critical circumstances of the routes. This allows the user to fix the problem as soon as the route or the software crashes. The database used to develop the tool is a time-series database. This has great significance in the project. Because of the database, we have tried to reduce the time complexity to store

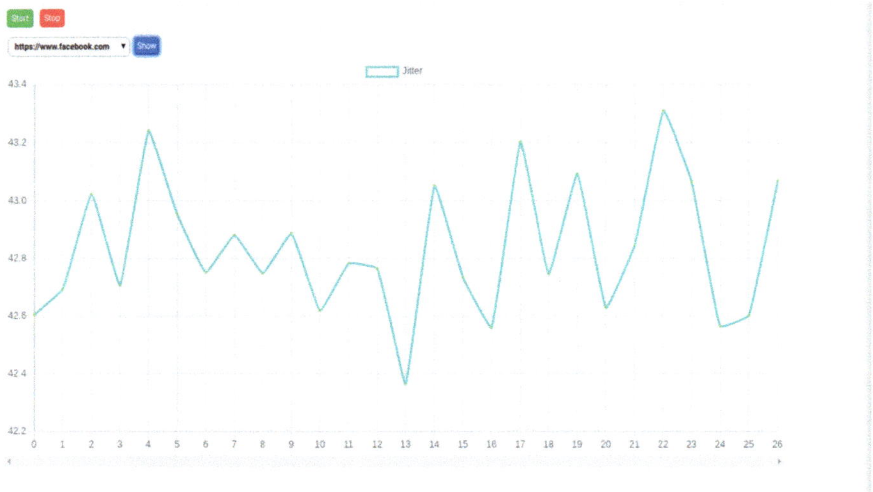

Fig. 6 Jitter of facebook.com from a local server versus normalized time

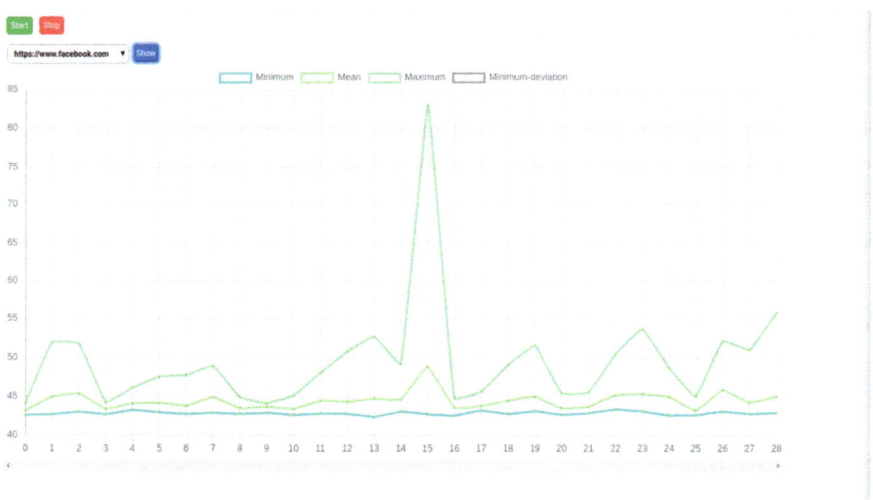

Fig. 7 Ping (min/avg/max/mdev) value of the facebook.com from a local server versus normalized time

the data or to draw the time-series values graph from O(n^n) to O(n). The minified form of tsdb also helps in reducing the size of the file. The concept of tsdb has not been adopted by any of the software mentioned above in the paper. As shown in our paper, most of the tools do not have specific support for ping, jitter, monitoring and benchmarking facilities separately, as this enables the user to analyze the data in the easiest way possible. So, we additionally plan to adapt the tool that we will produce

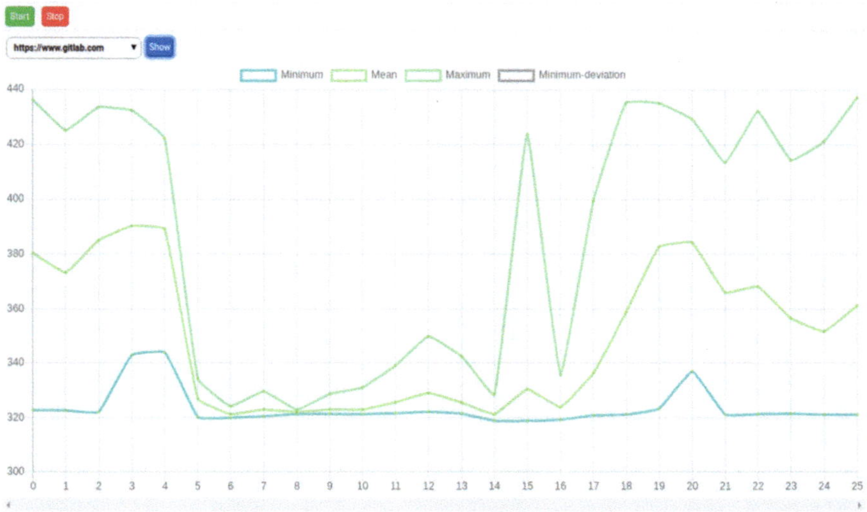

Fig. 8 Ping (min/avg/max/mdev) value of the gitlab.com from a local server versus normalized time

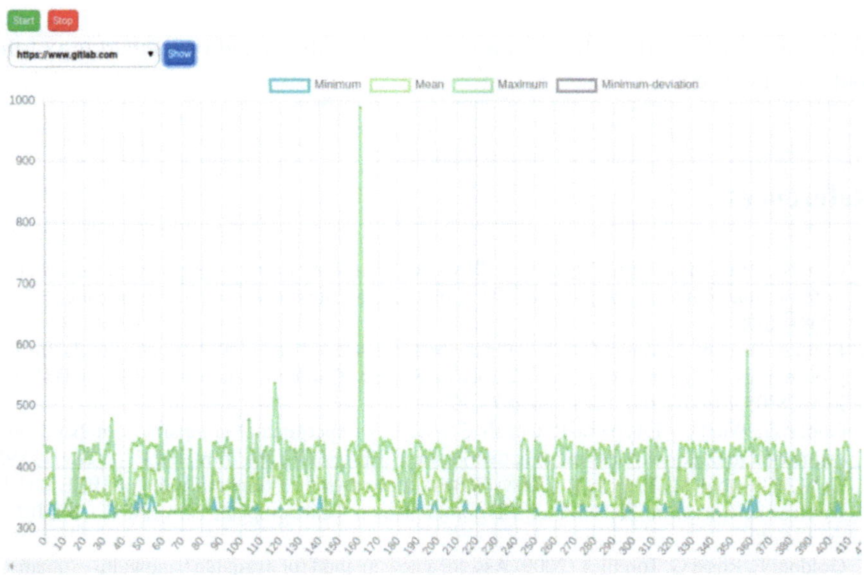

Fig. 9 Ping (min/avg/max/mdev) value of the gitlab.com over the long duration from a local server versus normalized time

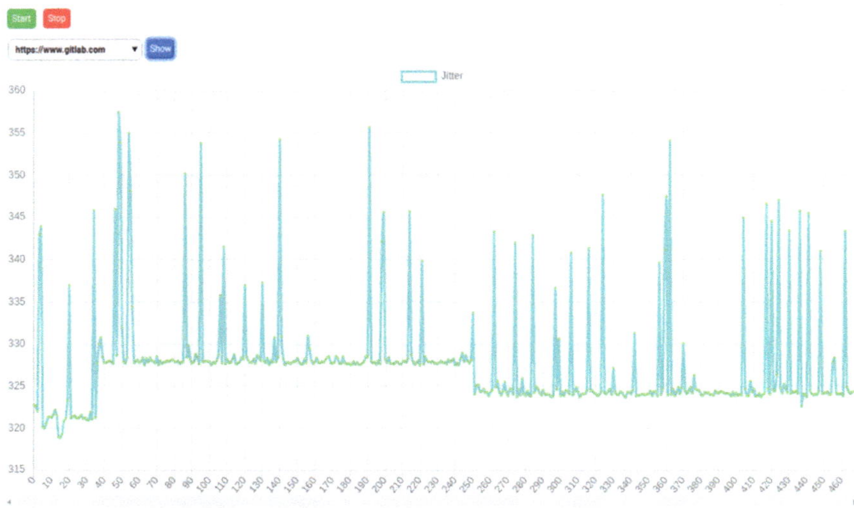

Fig. 10 Jitter value of the gitlab.com over the long duration from a local server versus normalized time

for use in fulfilling all the features for the better analyses and benchmarking of the routes by the user.

References

1. Gamess, Eric, and Karima Velásquez. "IPv4 and IPv6 forwarding performance evaluation on different operating systems." In XXXIV Conferencia Latinoamericana de Informática (CLEI 2008). 2008
2. Botta A, Pescape A, Ventre G (2005) On the performance of bandwidth estimation tools. In: The proceedings of the international conference on Systems Communications (ICW'05, ICHSN'05, ICMCS'05, SENET'05), IEEE, pp 287–292
3. Gupta D, Wu D, Mohapatra P, Chuah CN (2009) Experimental comparison of bandwidth estimation tools for wireless mesh networks. In: IEEE INFOCOM 2009, IEEE, pp 2891–2895
4. Li M, Claypool M, Kinicki R (2008) WBest: a bandwidth estimation tool for IEEE 802.11 wireless networks. In: 2008 33rd IEEE conference on local computer networks (LCN), IEEE, pp 374–381
5. Goldoni E, Rossi G, Torelli A (2009) Assolo, a new method for available bandwidth estimation. In: The proceedings of the international conference on internet monitoring and protection, IEEE, pp 130–136
6. Jain M, Dovrolis C (2002) Pathload: a measurement tool for end-to-end available bandwidth. In: Proceedings of passive and active measurements (PAM) workshop

Performance Analysis of Network Anomaly Detection Systems in Consumer Networks

P. Darsh and R. Rahul

Abstract There has been a significant increase in the amount of smart home appliances such as intelligent thermostats, Internet-connected baby monitors, lights being used; these devices have been a target for a number of cyber-attacks till date. Using only fundamental network information such as source and destination ports, packet sizes, TCP flags, time between subsequent packets, necessary features can be extracted to detect the aforementioned anomalies. This paper aims to analyze some critical operating system performance metrics on detecting such anomalies. It also consists of the taxonomy of various approaches to classify anomalies and detailed description on capturing and cleaning network packets.

Keywords Anomaly detection · IoT · Machine learning

1 Introduction

It has been projected that there would be a huge increase in the number of IoT devices by 2020; many of these IoT devices are fundamentally insecure [1]. With this increase in number of IoT devices, there has also been an uptick in the number of attacks targeting such IoT devices.

Sweeping attacks on key Internet services around the world have been launched with botnets powered by IoT devices such as security cameras and wireless routers, topping 1.1 terabits per second [2]. These IoT devices basically act as network probes from where attackers can hear and see what is going on inside the network in fact resources of these devices can also be used to mine cryptocurrencies, etc.

Network anomaly detection is a wide topic which boasts numerous research, articles, surveys as well as books [3–5]. Major part of it aims at thwarting attacks on commercial networks [6]. There is very little research done on how the hardware performance for various anomaly detection methods.

P. Darsh (✉) · R. Rahul
School of Computing Science and Engineering, Vellore Institute of Technology, Vellore, India
e-mail: darshkpatel@gmail.com

A. K. Tripathy et al. (eds.), *Advances in Distributed Computing and Machine Learning*, Lecture Notes in Networks and Systems 127, https://doi.org/10.1007/978-981-15-4218-3_21

Outlier detection has been a huge part of research by the statistical community [3, 7–16], but with the numerous recent advancement in machine learning, it has been playing a notable role in anomaly detection too.

While it seems attractive theoretically, this technique has its own set of disadvantages, such as the intrinsic complexity of the system to the high false positive rates, determining which event triggered the alarm, etc. These problems need to be addressed before wide adaptation of anomaly-based detection systems.

The primary aim for this paper is to provide the readers with a proof of concept system for detection of anomalies in consumer networks and also evaluating the performance metrics for the most commonly available hardware boards.

1.1 Distinction from Existing Research

Existing research on anomaly detection algorithms for network traffic has been targeted toward commercial scenarios. Out of the multitude of attacks consumer network face, only a few are a real concern for consumer networks.

For example, distributed denial-of-service (DDoS) attacks are generally aimed at high profile websites and essential infrastructure for fortune 500 companies, and many researchers have taken up the challenge to mitigate such attacks, also the fact that a very large portion of the traffic for such DDoS attacks come from "Zombie" devices taken hostage by nefarious actors are found in consumer networks.

Detecting anomalies on a consumer network can have the following challenges:

- Lack of network infrastructure: Most consumer networks do not have managed switches with a dedicated monitoring port; thus, packet capturing tools need to be installed on users computers or a low power device connected the network to monitor network.
- Public datasets are not available: There exist publicly available, labeled datasets with millions of records, and attack types have been made available for research purposes [17]. In fact, research has been done on the quality of such datasets. But, no consumer network data has been collected or published for obvious privacy reasons.
- Less data points: Consumer networks do not have multiple subnets containing thousands of devices generating a multitude of network traffic over various protocols which can be analyzed for patterns. For example, in a typical commercial environment, one can analyze traffic on the DC and then analyzing user login data or the average overall bandwidth use which correlates work hours. This cannot be done on a typical consumer network.

1.2 Anomaly-Based Intrusion Detection

It refers to finding uncommon patterns or irregularities in network traffic that do not adhere to the expected behavior. These patterns are often referred as anomalies or outliers [3].

1.3 Tools Used for This Research

The well known network sniffer `wireshark` [18] was utilized to intercept network data which was then stored as a PCAP file. This PCAP file which consists a lot of unnecessary data was then processed, compressed, and converted to CSV files using `netcap` [19]. `matplotlib` [20], which is a python library which was then used to plot the graphs identifying useful features in the data. Python library `pandas` [21] was then utilized to extract these features from the data. Machine learning library `scikit learn` [22] was utilized for training ML models.

2 Processing Data and Packet Capturing

As mentioned earlier in Sect. 1.1 packet capturing and processing has been one of the most daunting tasks in this area of research. More so, for networks without the dedicated hardware to do so. Typical consumer routers do not have a "monitor port" like most commercial switches do. Thus, a low power dedicated device such as a Raspberry Pi can be used for capturing and storing network traffic. A similar network monitor device has been mentioned in [23]. A capture device can be setup in two ways inside a local network

- As a server which listens to packet capture data sent via dedicated programs installed on individual devices in the network. This configuration is not suitable for the purpose of this paper as this paper focuses on monitoring traffic from IoT devices, programs which capture network traffic and send that to a server cannot be installed on these devices. Refer Fig. 1.
- As a network bridge which sits between the router and devices. This allows the capture device to capture all the traffic across the network. Refer Fig. 2.

For the purpose of this paper, a total of 1.5 GB of network data were captured from eight different consumer networks each containing various IoT devices such as smart lights, voice assistants, and IP cameras. The captured data was then manually cleaned for training the models.

Also, synthetic network traffic for generating anomalies such as Internet sweeps, performing DDoS attacks using devices on the network was also generated. Network traffic from well-known malware such as Mirai whose source code readily available was also simulated.

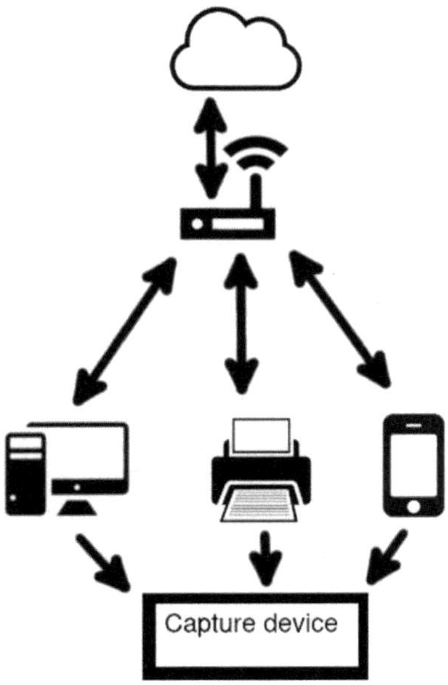

Fig. 1 Capture device listening for capture data sent via programs installed on network devices

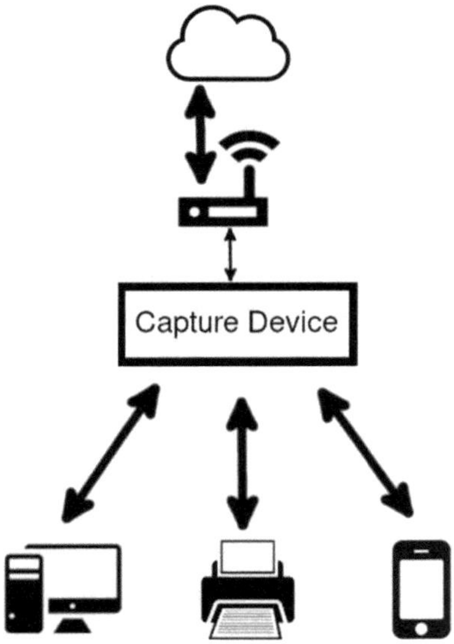

Fig. 2 Capture device acting as a network bridge

3 Anomaly Detection Methods

Detailed introduction to various approaches is out of scope of this paper; existing literature can be reffered to [24].

3.1 Statistical Anomaly Detection

"An anomaly is an observation which is suspected of being partially or wholly irrelevant because it is not generated by the stochastic model assumed" [25].
Thus, the occurrences with a low probability of being generated are anomalies. Prime advantage of this technique over the others is that it does not require "Prior Knowledge" of the network's normal activity [4]; thus, it can provide accurate results about anomalous activity [25]. One promising approach to detect network anomalies is an entropy-based approach. Entropy is a measure of the uncertainty or randomness associated with a random variable. If it was more random, it contains more entropy [26].

The primary drawback of this approach is that attackers can "train" the detection model until the traffic is considered as normal according to the statistical model. This approach also requires a lot of training data which is difficult to obtain in a consumer setting.

3.2 Supervised Classification-Based Anomaly Detection

This approach is a supervised learning approach where model is trained using a labeled dataset, which then tries to classify new data into categories based on the training data. "Linear classification tries to find a line between the classes" [25], but the "classification boundary may be nonlinear" too [25] as seen in Fig. 3. These techniques have a low false positive ratio subject to suitable thresholds [25]. The prime drawback for these is that they are highly dependent and biased on the training data and thus generally cannot identify anomalies it has not seen before, which defeats the purpose of this research (Fig. 4).

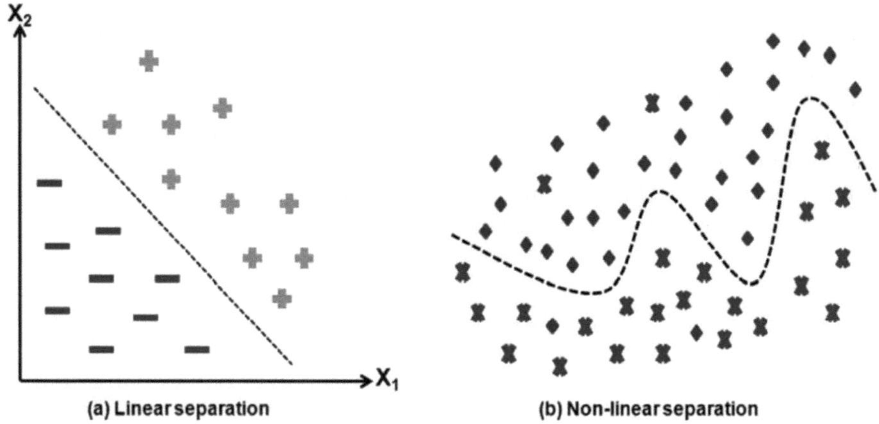

Fig. 3 Classification-based anomaly detection from

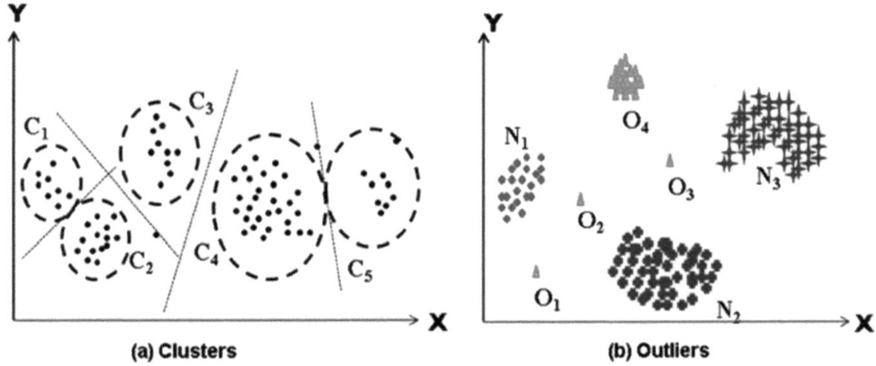

Fig. 4 Clustering and outlier-based anomaly detection

3.3 Clustering and Outlier-Based Novelty Detection

Unsupervised Approach Fundamentally, grouping data into various sets of similar objects is called clustering. This has been represented in Fig. 3a. In Anomaly detection, the primary assumption made is "larger clusters are normal" [25]. The rest of the clusters can be considered as anomalies.
In Fig. 3b, the points which do not fit into any of the clusters are considered outliers (anomalous data points).
It is also worth nothing that unsupervised learning approach works with unlabeled data. Due to most unsupervised learning models using both outlier detection and clustering, the computation complexity can be quiet high [25].

Semi-Supervised Approach Semi-supervised learning is prevalent in scenarios where there is very little anomalous data available to train the model, but non-anomalous data is readily available; thus, it is trained on a "single class" of data and then detects "novelties."
Thus, this is also commonly referred to as novelty detection [27]. One-class support vector machine classifiers (OCSVM) are favorable in case of anomaly detection, in this approach as they do not require pre-labeled data sets which are expensive or difficult to obtain [28].

The performance and accuracy of the models one-class support vector machine and isolation forest has been compared in this paper. The models were trained on the features mentioned in Table 2 using the captured data.
The results have been mentioned in Tables 1 and 2 and have been discussed in Sect. 4

Table 1 Data points collected

S. no.	Feature name	Feature description
1	Protocol	Layer 3 protocol: IP, UDP, TCP
2	Source IP	Packet source IP address
3	Destination IP	Packet destination IP Address
4	Source port	Packet source port
5	Destination port	Packet destination port
6	Frame length	Length of the captured frame
7	Time	Time of said packet
8	TTL	Packet time to live
9	TCP flags	Flags: RST, SYN, ACK, FIN

Table 2 Features extracted from I

Feature	Feature description
1	Ratio between TTL and payload size
2	Average payload size in N seconds
3	Average TTL in N seconds
4	Number of DNS queries in T seconds
5	Average number of ACK packets in T seconds
6	Average number of RST packets in T seconds
7	Average number of SYN packets in T seconds
8	Average number of FIN packets in T seconds
9	Average time between two frames over T seconds
10	Average number of IPV4 frames in T seconds

4 Anomaly Detection Results

All the models were optimized to use the best possible hyperparameters. The dataset is randomly split into training and test sets with a 80:20 ratio. Anomalous data points are tested directly on the trained model.

Anomalous detection rate is defined as the fraction of anomalous data points classified as anomalies from the total number of anomalous data points the classifier was tested upon, denoted by `DR(a)`

$$\text{DR(a)} = \frac{\text{Number of anomalous data points marked positives}}{\text{Total number of anomalous data points}}$$

Non-anomalous detection rate is defined as the fraction of non-anomalous data points not classified as anomalies from the test dataset split denoted by `DR(na)`

$$\text{DR(na)} = \frac{\text{Number of non-anomalous data points marked negatives}}{\text{Total number of non-anomalous data points}}$$

Anomalous false positive rate is defined as the fraction of anomalous data points marked as negatives from the total number of anomalous data points the classifier was tested upon. `FPR(a)`

$$\text{FPR(a)} = \frac{\text{Number of anomalous data points marked negatives}}{\text{Total number of anomalous data points}}$$

Non-anomalous detection rate is defined as the fraction of non-anomalous data points classified as anomalies from the test dataset split. `FPR(na)`

$$\text{FPR(na)} = \frac{\text{Number of non-anomalous data points marked as positives}}{\text{Total number of non-anomalous data points}}$$

One-Class Support Vector Machine

One-class SVM outperforms unsupervised models such as isolation forest. Also, similar to isolation forest, optimizing the hyperparameters helps overcome the problems due to imbalanced classes.
One-class SVM requires a small sample size to train the model and proves to be very accurate in most of the cases [29].
It is also worth noting that a semi-supervised learning model with high accuracy requiring a small sample size to train deems to be an ideal model for anomaly detection.

Unix Versus Windows

It is clearly evident from the results of the tests that detecting network anomalies for both the operating systems are the same , given a 5% margin for errors. This can be due to the fact that the underlying implementation of common protocols such as SSH, FTP is the same, regardless of the OS it runs on (Table 3).

Table 3 Embedded device specifications

Board	CPU	Cores	Frequency (GHz)	Memory	OS
Raspberry Pi	ARM Cortex-A53	4	1.2	1 GB LPDDR2	Debian Kernel 4.4.34
MacBook Pro	Intel i5	4	2.7	8 GB LPDDR3	macOS 10.13.2
Intel UP	Intel atom z8350	5	1.4	4 GB DDR3L	Ubuntu kernel 4.15.0
Intel NUC	Intel i7	4	3.5	8 GB DDR4	Ubuntu kernel 4.15.0

5 Hardware Analysis

As discussed previously, a packet analysis device must be placed inside the network which would capture, clean, and analyze the network data. It becomes essential that such a device is low powered but also capable to perform the tasks discussed previously.

It is essential for the analysis device to have low power consumption, noise and enough performance to run basic machine learning models. For the scope of this survey, the following devices were tested.

5.1 Experiment Setup

The time metrics were measured using the system clock on individual systems. TensorFlow and Caffe used were compiled on each system with the compile time flags "-funsafe-math-optimizations -ftree-vectorize" for benchmarking performance The memory metrics were measured using the "ps" command. Power consumption for the devices was measured using Texas Instruments INA219 power monitor IC by attaching it to the powerline. The power consumption was measured over the period of 10 min for idle consumption and for 5 min over multiple inferences using the Caffe model for under load metric (Table 4).

Table 4 Power consumption

Metric	Raspberry Pi	MacBook pro	Intel UP	Intel NUC
Idle (w)	1.9	13.1	12.7	14.3
Under load (w)	10.8	33.4	31.2	32.5

Table 5 Timing metrics on Caffe2 model

Metric	Raspberry Pi	MacBook Pro	Intel UP	Intel NUC
Import package(s)	1	0.3	0.8	0.7
Load model(s)	23.3	6.2	14.7	4.1
Inference time(s)	3.51	0.9	1.01	0.5
Total time(s)	27.81	7.40	16.51	5.30

5.2 Memory Footprint

SqueezeNet on Caffe2 Model consumes avg. 4000 MB on the Intel NUC, avg. 2000 MB on both MacBook Pro and Intel UP while it consumed only 532 MB on Raspberry Pi.

5.3 Power Consumption

The Raspberry Pi Board powered by a standard 5 V 2 A power USB power supply consumes the least power when it is under load, while both the MacBook Pro, Intel NUC and Intel UP have similar power consumption.

5.4 Hardware Summary

The Raspberry Pi being a low powered board provides subpar inference times on the standard Caffe model, but consumes drastically less power in general than the other devices. Intel UP provides an optimal mix of power consumption and performance if there is adequate utilization of its resources. Thus, for a dedicated network traffic analysis device, cheap low powered boards such as Raspberry Pi or Intel's UP can be utilized according to the amount of data to be analyzed (Table 5).

6 Conclusion

As this research demonstrates, network anomalies can be detected with high accuracy without requiring large labeled datasets. The models prove to accurately detect common network anomalies such as DDoS attacks and Internet sweeps being performed by network devices. Other "Novelties" in network traffic were also flagged which demonstrates the ability to detect Zero-day attacks on or using consumer devices. Semi-supervised models such as one-class SVM are clearly the forerunners for practical purposes due to their high detection rates and the fact that they do not need extensive labeled datasets which are unavailable for consumer networks. A deep learning approach using deep auto- encoder networks [30] might also prove effective as semi-supervised ML models have shown promising results in this research. Handling more categorical variables and unbalanced classes in semi-supervised learning models offers a promising direction for future research, especially for securing the next generation of IoT devices.

References

1. Doshi R, Apthorpe N, Feamster N (2018) Machine learning ddos detection for consumer internet of things devices. CoRR abs/1804.04159. http://arxiv.org/abs/1804.04159
2. Kolias C, Kambourakis G, Stavrou A, Voas J (2017) DDoS in the IoT: Mirai and other botnets. Computer. https://doi.org/10.1109/MC.2017.201
3. Chandola V, Banerjee A, Kumar V (2009) Anomaly detection: a survey. ACM Compu Surv. https://doi.org/10.1145/1541880.1541882
4. García-Teodoro P, Díaz-Verdejo J, Maciá-Fernández G, Vázquez E (2009) Anomaly-based network intrusion detection: techniques, systems and challenges. Comput Secur. https://doi.org/10.1016/j.cose.2008.08.003
5. Patcha A, Park JM (2007) An overview of anomaly detection techniques: existing solutions and latest technological trends. Comput Networks. https://doi.org/10.1016/j.comnet.2007.02.001
6. Lo CH, Ansari N (2013) Consumer: a novel hybrid intrusion detection system for distribution networks in smart grid. IEEE Trans Emerg Topics in Comput 1(1):33–44. https://doi.org/10.1109/tetc.2013.2274043
7. Yu M (2012) A nonparametric adaptive cusum method and its application in network anomaly detection. Int J Adv Comput Technol. https://doi.org/10.4156/ijact.vol4.issue1.32
8. Chan PK, Mahoney MV Arshad M (2003) A machine learning approach to anomaly detection. Department of Computer Sciences
9. Chhabra P, Scott C, Kolaczyk ED, Crovella M (2008) Distributed spatial anomaly detection. In: Proceedings of IEEE INFOCOM. https://doi.org/10.1109/INFOCOM.2007.232
10. Lu W, Ghorbani AA (2009) Network anomaly detection based on wavelet analysis. Eurasip J Adv Signal Process. https://doi.org/10.1155/2009/837601
11. Mahoney M, Chan P (2004) Learning rules for anomaly detection of hostile network traffic. https://doi.org/10.1109/icdm.2003.1250987
12. Manikopoulos C, Papavassiliou S (2002) Network intrusion and fault detection: a statistical anomaly approach. IEEE Commun Mag. https://doi.org/10.1109/MCOM.2002.1039860
13. Simmross-Wattenberg F, Asensio-Pérez JI, Casaseca-De-La-Higuera P, Martín-Fernandez M, Dimitriadis IA, Alberola-López C (2011) Anomaly detection in network traffic based on statistical inference and α-stable modeling. IEEE Trans Depend Secur Comput. https://doi.org/10.1109/TDSC.2011.14
14. Wang K, Stolfo SJ (2010) Anomalous payload-based network intrusion detection. https://doi.org/10.1007/978-3-540-30143-1_11
15. Xiuyao S, Mingxi W, Jermaine C, Ranka S (2007) Conditional anomaly detection. IEEE Trans Knowl Data Eng. https://doi.org/10.1109/TKDE.2007.1009
16. Zhang Z, Li J, Manikopoulos CN, Jorgenson J, Ucles J (2001) HIDE: a hierarchical network intrusion detection system using statistical preprocessing and neural network classification. In: Proceedings of IEEE workshop on information assurance and security

17. Creech G, Hu J (2013) Generation of a new IDS test dataset: time to retire the KDD collection. In: IEEE wireless communications and networking conference (WCNC 2013). https://doi.org/10.1109/WCNC.2013.6555301
18. wireshark. https://www.wireshark.org/
19. Netcap (https://githubcom/dreadl0ck/netcap)
20. matplotlib (https://matplotliborg)
21. pandas (https://pandaspydataorg)
22. scikit-learn (https://scikit-learnorg)
23. Mukerji A, Rothstein J (2012) Detecting anomalous network application be havor
24. Agrawal S, Agrawal J (2015) Survey on anomaly detection using data mining techniques. Procedia Comput Sci 60:708–713. https://doi.org/10.1016/j.procs.2015.08.220
25. Bhuyan MH, Bhattacharyya DK, Kalita JK (2014) Network anomaly detection: methods, systems and tools. IEEE Commun Surv Tutor. https://doi.org/10.1109/SURV.2013.052213.00046
26. Navaz ASS., Sangeetha V, Prabhadevi C (2013) Entropy based anomaly detection system to prevent ddos attacks in cloud. CoRR abs/1308.6745. http://arxiv.org/abs/1308.6745
27. Arunraj, NS, Hable R, Fernandes M, Leidl K, Heigl M (2018) Comparison of supervised , semi-supervised and unsupervised learning methods in network intrusion detection system (NIDS) application. Anwendungen Und Konzepte Der Wirtschaftsinformatik (AKWI)
28. Rassam MA, Maarof MA, Zainal A (2014) Adaptive and online data anomaly detection for wireless sensor systems. Knowl Based Syst 60:44–57. https://doi.org/10.1016/j.knosys.2014.01.003
29. He X, Mourot G, Maquin D, Ragot J, Beauseroy P, Smolarz A, Grall-Maës E (2014) Multitask learning with one-class SVM. Neurocomputing. https://doi.org/10.1016/j.neucom.2013.12.022
30. Schreyer M, Sattarov T, Borth D, Dengel A, Reimer B (2017) Detection of anomalies in large scale accounting data using deep autoencoder networks. CoRR abs/1709.05254. http://arxiv.org/abs/1709.05254

Remote Automated Vulnerability Assessment and Mitigation in an Organization LAN

Nishant Sharma, H. Parveen Sultana, Asif Sayyad, Rahul Singh, and Shriniwas Patil

Abstract The process of making the network of an organization (an enterprise or an institution) fool-proof and secure holds great importance in today's interconnected world. Any possible backdoor entry into any of the workstations in a large inter-connected organizational network has the potential of a severe security breach that could prove fatal to the integrity of the organization. Strong and automated solutions are required that provide as a complete package, a sound vulnerability assessment, and satisfactory mitigation should a compromise in any workstation is detected. This paper explores the need and creation of such a complete security package.

Keywords Network vulnerability · Vulnerability assessment · Cyber attacks · Vulnerability mitigation

1 Introduction

The assessment of vulnerabilities in a computer network in an organization is aimed at finding and subsequently remedying the security holes that exist in any of the systems in the network which may be exploited by a malicious user. The central idea is to hold off an attack from a malignant user for as long as possible. The attacker may carry out a wide variety of attacks on a system in the network. An up to date assessment and monitoring tool may assist us in anticipating and remedying a possible security risk. There are various types of vulnerability assessment scans such as: (1) Host-based analysis that finds vulnerabilities in a particular workstation or a firewall etc., (2) application-based analysis that finds vulnerability in the web infrastructure, (3) network-based analysis for vulnerabilities like a default login page

N. Sharma (✉) · H. Parveen Sultana (✉) · A. Sayyad · R. Singh · S. Patil
School of Computer Science and Engineering, Vellore Institute of Technology, Vellore 632014, India
e-mail: nishant.sharma2018@vitstudent.ac.in

H. Parveen Sultana
e-mail: hparveensultana@vit.ac.in

A. K. Tripathy et al. (eds.), *Advances in Distributed Computing and Machine Learning*, Lecture Notes in Networks and Systems 127, https://doi.org/10.1007/978-981-15-4218-3_22

for all routing devices in a network etc., (4) database-based analysis for vulnerabilities in the database such as the possibility of an SQL injection, and so on.

The process of vulnerability assessment of various assets in an organization usually follows the steps outlined as: (1) The identification of the critical assets in an organization, (2) the security vulnerabilities that affect these systems, (3) the quantification of such threats and ranked in order: Highly critical to Low, and (4) emphasis of the necessity of a remedial action based on anticipated threat [1].

The entire process of automated vulnerability assessment and mitigation as used in the paper is summarized as follows. An attacker aims at finding the vulnerabilities of a workstation in a large organizational network and aims at compromising the entire network by gaining access to the workstation. To counter such a threat on a workstation in the LAN, an organization may deploy a system that does both the automated vulnerability assessment and a corresponding remedial action if required. This paper explores various types of attacks on a vulnerable system including the ransomware attack, the SQL injection attack, MD5 hash-based reverse password determination, and brute force password identification. Beginning from a manual vulnerability assessment policy, an automated vulnerability assessment scanner OpenVAS 9 is introduced and studied for usage as a reliable vulnerability assessment tool. A mitigating script written in Python programming language that solves one of the detected vulnerabilities is presented to make a case for a complete remote and automated vulnerability assessment and mitigation package as a host-based solution.

2 Related Work

To understand cybersecurity, it is important to learn the basic concepts of ethical hacking and the defense mechanisms in place to counter known cyber-threats. This is the initiating step in order to create stronger defense mechanisms for the threats that may be posed by competent attackers in the future. State of the art, open source and current vulnerability scanning tools are discussed in [2]. In [3], the authors claim that, of the many available vulnerability scanners in the marketplace, none of them formalize satisfactorily that the assessment process for vulnerabilities in the system never compromises the system on its own. The authors attempt to provide a solution to this issue. The case made for automated vulnerability assessment in [4] is one of the earliest voices in support of automated testing. The authors of [5] extend an already existing open source tool "Nikto" (a web application vulnerability scanner [6]) to enhance the functionality for vulnerabilities in session management in network communication. The authors of [7] compare and analyze performance of two popular network scanning tools: Nmap and Nessus. The authors of [8] did a sample review on SQL injection technique. The authors of [9] aimed at providing the behavioral analysis of a crypto ransomware, and at creating an intrusion detection system for it.

3 Network Vulnerabilities and Attacks: Simulation of a Ransomware Attack

Network vulnerabilities such as a compromised workstation leave the organizational network in a vulnerable state such that an attacker may succeed in exploiting these vulnerabilities and gaining illegal access to the organizational network. One such attack that exploits the vulnerabilities of a host in a network is the ransomware attack. In the ransomware attack, the attacker(s) lock the system of an end user until a ransom is paid. In this paper, a basic recreation of such an attack is performed.

For the simulation of ransomware attack, an Ubuntu 18.04 LTS virtual machine is set up on Oracle VirtualBox on an Ubuntu 18.04 LTS host machine. A deliberately weak password is set up for the virtual machine. A study presented in [10] states that the password 'password' is used by more than 3 million accounts. This demonstrates the lack of motivation among people to secure their workstation and such a problem gets aggravated in the context of a large organizational LAN.

The paramiko Python library [11] allows us to do a remote ssh login into a target machine. A list of popularly used passwords is maintained in the Python script that attacks the remote target. The attacking host machine initiates an attack by trying to log into the remote virtual machine with an IP and username pre-known to the host and using iteratively, a password from the list of commonly used passwords. After a successful match of the password is done, the first step completes and the script has logged onto the target machine. Using the sftp_ client() of the SSHClient() of the paramiko library, the target directory is traversed and files (i.e., their content) in that directory will now be encrypted using a suitable encryption method.

The encryption of the files (in one by one manner) in the simulation is done by using the AES algorithm [12]. Using the AES algorithm as demonstrated in [13], the contents of a file are encrypted using an input password as a key. The file is divided into chunks of size 64KB and an AES encrypter is created with Cipher Block Chaining mode (such that each plaintext chunk gets XOR-ed with the previous ciphertext chunk) and a random initialization vector.

The ransomware program in the simulation shows the following flow. The program initiates and an attempt at logging in is begun. The program displays a 'Still Searching' animation while iterating through the list of popularly used passwords and reattempting the login process. Once a successful match is found, it displays a 'Cracked!' animation and displays the list of files from a pre-coded target folder. For example, the 'Documents' folder is always found at /home/$user/Documents in an Ubuntu OS. Hence, all the files in the Documents directory in the Ubuntu 18.04 guest operating system are listed as a Python list. The user is then asked to Encrypt or Decrypt a file. A suitable option may be selected as per the current program flow, and so here, the Encrypt option is chosen. The user is then required to enter the password that is used as the AES key. Using this key, the chunks of the file are encrypted. The encryption process is thus completed. After the successful encryption of the file (or files), some showmanship is performed. The original file is deleted from the remote machine and an encrypted file with a changed name to signify an attack is placed in that same location. The file contents can be retrieved back using the same steps in reverse that are used for encryption using AES and the password key.

4 Vulnerability Assessment: Exploiting the Vulnerabilities of a System Manually

A VulnOS:2 (Vulnerable Operating System) package is available at [14]. VulnOS:2 has been explicitly introduced with vulnerabilities that can be exploited in order to learn penetration testing skills. The central idea in this ethical hacking experiment is to understand how the vulnerabilities can be exploited in a system that has not been kept up to date.

For this experiment, a virtual machine is set up for the VulnOS:2 on the Ubuntu 18.04 LTS host. A host only IP address is assigned to the VulnOS:2 virtual machine to facilitate communication between the VulnOS:2 guest machine and the Ubuntu 18.04 host. Using the similar approach as demonstrated in [15], the following process follows. The nmap tool [16] is first used to find the IP address of the VulnOS:2 virtual guest machine from the Ubuntu host. The IP address of the guest machine is opened on the browser and a set of hyperlinks are followed to arrive at a database page that is intentionally severely compromised. Using sqlmap a successful SQL injection attack is performed on the target database and a successful database takeover happens. As a result of the database takeover, the entries stored into the database are revealed. The passwords are stored 'securely' as an MD5 hash. MD5 hash-based secured password of a user is successfully cracked using an online MD5 cracking tool, the likes of which are easily and commonly available on the internet. Knowing the username from the database, SQL injection attack and the cracking of the MD5 hash to reveal the password can then be used to do a remote SSH login into VulnOS:2 from the Ubuntu host.

4.1 Disadvantages of Manual Vulnerability Assessment

Manual vulnerability assessment as conducted in this section serves as a good understanding to know the basics and know-hows of the vulnerability assessment as a concept: its need and importance to maintain security and the integrity of the organization. However, such a technique is lengthy, tiresome, and not explicit. It leaves unattended use cases as areas that can be exploited. It is also not possible to standardize manual vulnerability assessment. This creates a need to adopt automated vulnerability assessment techniques.

5 Automated Vulnerability Assessment: A Case Study on 'OpenVAS 9'

5.1 Creating a Vulnerable Windows Based Virtual Machine

Although Microsoft claims to end its support for their Windows 7 OS in January 2020 [17], the market share forecast for Windows 7 usage in the world in the year 2020 is just over 25% [18]. This is a significant market share for an OS that will not be receiving security upgrades from January 2020. Following the central theme of creating a weak password 'password,' we set up a virtual machine with Windows 7 Operating System to conduct automated vulnerability assessment and remote mitigation (in later sections) in the context of a workstation in an organizational LAN. With the understanding of the popularity of windows operating system in India, such a choice of OS to conduct the later experiments seems a rational one.

Using the Host Network Manager of the Virtual Box, a host only IP is assigned to the Windows 7 virtual machine. In the advanced settings of the Windows Firewall, a new custom inbound rule for a request from the host machine to the guest machine's IP address is created to facilitate the two-way communication of the Windows OS virtual machine with the Ubuntu 18.04 LTS Linux host. Following this step, an ICMP echo ping between host and the guest in both directions is successfully performed. The Ubuntu 18.04 host is going to act as the Management and Mitigation System that runs the automated vulnerability assessment tool to find the vulnerabilities of the Windows 7 OS virtual machine and after the assessment and based on the quantified risk, either offer a warning or take remedial action to prevent an anticipated attack.

5.2 Installation of OpenVAS 9

To simulate the automated vulnerability assessment style, Open Vulnerability Assessment Scanner (OpenVAS 9) [19] is installed on the host Ubuntu 18.04 LTS machine. After the OpenVAS tool is installed and made up to date with the latest vulnerability database, the feeds synced, the relevant libraries and packages installed, it is ready to perform the vulnerability assessment tasks. OpenVAS 9 performs its vulnerability assessment tasks using a web-based interface called the Greenbone Security Assistant. Greenbone Security Assistant for the OpenVAS 9 is made available at the host machine at the web socket 'localhost:4000.' The feed status of the Greenbone Security Assistant is as shown in Fig. 1.

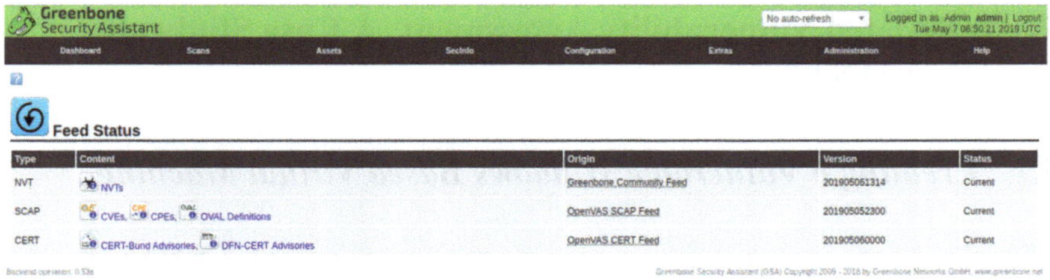

Fig. 1 The feed status of the NVT, SCAP, and CERT contents for the OpenVAS 9 Greenbone Security Assistant is current

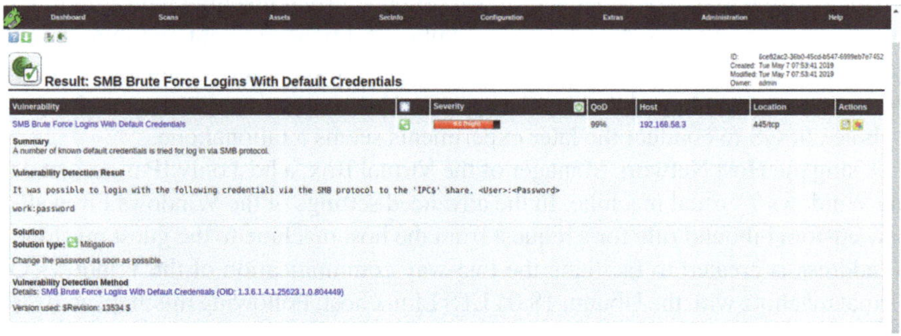

Fig. 2 Brute Force identification of the password by OpenVAS 9 with Severity 9.0 and QoD 99%

5.3 *Vulnerability Assessment in Windows 7 OS Using OpenVAS 9*

Using the OpenVAS Greenbone Security Assistant, a remote and automated vulnerability assessment scan task is initiated on the Windows 7 guest OS using its IP address. This scan task is successfully able to brute force into the weak login password 'password' set on the Windows 7 guest OS as shown in Fig. 2. Some of the characteristic features of the detection are the 'Severity' metric which is a 9.0 on a 10.0-point scale and classified as a 'High' for this vulnerability. The second metric is the 'Quality of Detection' (QoD) metric which is 99%. QoD provides a metric for the reliability of the detection of a vulnerability.

6 Vulnerability Mitigation: Remote Automated Script for Mitigation of Vulnerability in Target Workstation

A vulnerability as reliably detected and as severe as the one shown in Fig. 2 compromises not only the workstation but also the organization's LAN as a whole. In such a compromised situation, a remote action may be desired. The central idea is that if a severely compromised workstation be reliably detected by the remote vulnerability assessment tool, its authority be overridden by the remote server authorized for the assessment and mitigation of vulnerabilities in the LAN.

In context of the weak password 'password' that is set on the Windows OS virtual guest machine, a remote patch to mitigate this security vulnerability is created in Python which is then remotely executed by the Ubuntu 18.04 host server on the remote and compromised Windows 7 guest OS. The script updates the weaker password to a strong password. It uses the NetUserChangePassword() [20] to update the weaker password to a newer and a stronger password created by the server. The script as an addition also verifies the logon process with the newly created password to return success if the new password successfully logs onto the system.

When the Python script is run, the weak password 'password' on the Windows 7 guest OS is successfully updated to the stronger newer password. When the OpenVAS assessment scan task is run again on the Windows 7 OS guest machine, the updated report shows not the previously detected 'SMB Brute Force Logins with Default Credentials' vulnerability as shown in Fig. 2 and that the vulnerability has been satisfactorily mitigated.

7 Integration of the Automated Vulnerability Assessment Scanner and the Mitigation Script

The first half of the integration between the OpenVAS 9 (aimed at automated assessment of the vulnerabilities) and the Python script (representing an automated mitigation procedure) is the generation of assessment report by the OpenVAS 9 scan task wizard and the availability of the complete report to the Python script that decides on a remedial action. The second part of the integration is the actions (or inactions) performed by the Python script based on the QoD parameter and the Severity of the detection. In case of the vulnerability detected in Fig. 2, the Python script should intervene and override the authority of the workstation user. The updated password and the details of the compromised system can then be forwarded through a secure channel to the administrative authorities for further action. This two-fold integration can serve as a complete package against any attack that may follow from such a vulnerability.

8 Conclusion

In the era when network security is always under investigation and when vulnerabilities in any one of the systems in a large organization's LAN can lead to severe security breach, this paper attempts at exploring the problems of network vulnerabilities and the ways to provide automated solutions using an integration of the automated vulnerability assessment scanner and an automated remedial system as a package. When such a competent package is used in the organization and is deployed periodically, the possibility of compromised workstations is significantly reduced. However, which vulnerabilities require immediate action, and which shall be considered benign will be a continuous research endeavor. Future work may involve delving into this question and also testing such complete automated packages in the real-world organizations (enterprises or institutions) while maintaining the trust (and reliability) of the organization that implements such a system.

References

1. The Benefits of a Vulnerability Assessment. https://www.hitachi-systems-security.com/blog/the-benefits-of-a-vulnerability-assessment/. Last accessed 6 Dec 2019
2. Wang Y, Yang J (2017) Ethical hacking and network defense: choose your best network vulnerability scanning tool. In: International conference on advanced information networking and application workshops (WAINA), vol 31. IEEE, pp 110–113
3. Guo F, Yu Y, Chiueh TC (2005) Automated and safe vulnerability assessment. In: Annual computer security applications conference (ACSAS05), vol 21. IEEE, p 10
4. Liu C, Richardson DJ (2000) Ethical hacking and network defense: automated security checking and patching using Testtalk. In: International conference on automated software engineering, Proceedings ASE, vol 15. IEEE, pp 261–264
5. Lukanta R, Asnar Y, Kistijantoro AI (2014) A vulnerability tool for session management vulnerabilities. In: International conference on data and software engineering (ICODSE). IEEE, pp 1–6
6. Introduction to the Nikto Web Application Vulnerability Scanner. https://resources.infosecinstitute.com/introduction-nikto-web-application-vulnerability-scanner/#gref. Last accessed 6 Dec 2019
7. Im SY, Shin SH, Roh BH (2016) Performance evaluation of network scanning tool with operation of firewall. In: International conference on ubiquitous and future networks (ICUFN), vol 8. IEEE, pp 876–881
8. Mukherjee S, Sen P, Bora S, Pradhan C (2015) SQL injection: a sample review. In: International conference on computing communication and networking technologies (ICCCNT), vol 6. IEEE, pp 1–7
9. Almashhadani AO, Kaiiali M, Sezer S, OKane P (2019) A multi-classifier network-based crypto ransomware detection system: a case study of Locky ransomware. In: IEEE Access, vol 7. IEEE, pp 47053–47067
10. How Hackable is your password? https://edition.cnn.com/2019/04/22/uk/most-common-passwords-scli-gbr-intl/index.html. Last accessed 6 Dec 2019
11. Welcome to Paramiko. http://www.paramiko.org/. Last accessed 6 Dec 2019
12. Advanced Encryption Standard. https://en.wikipedia.org/wiki/Advanced_Encryption_Standard. Last accessed 6 Dec 2019
13. File Encryption And Decryption Using Python. https://blog.eduonix.com/software-development/file-encryption-decryption-using-python/. Last accessed 6 Dec 2019
14. VulnOS:2. https://www.vulnhub.com/entry/vulnos-2,147/. Last accessed 6 Dec 2019
15. VulnOS:2 Walkthrough. https://medium.com/@Kan1shka9/vulnos-2-walkthrough-16b70b9fbe17. Last accessed 6 Dec 2019
16. nmap(1)–Linux man page. https://linux.die.net/man/1/nmap. Last accessed 6 Dec 2019
17. Support for Windows 7 ends in January 2020. https://www.microsoft.com/en-us/microsoft-365/windows/end-of-windows-7-support. Last accessed 6 Dec 2019

18. Windows by the numbers: Windows 10 hits late wall, stalls. https://www.computerworld.com/article/3199373/windows-by-the-numbers-windows-10-hits-late-wall-stalls.html. Last accessed 6 Dec 2019
19. OpenVAS—Open Vulnerability Assessment Scanner. http://www.openvas.org/. Last accessed 6 Dec 2019
20. NetUserChangePassword function. https://docs.microsoft.com/en-us/windows/win32/api/lmaccess/nf-lmaccess-netuserchangepassword. Last accessed 6 Dec 2019

Introspective Journal: A Digital Diary for Self-realization

Krishtna J. Murali, Kumar V. Prasanna, Nirmala S. Preethi, M. Raajesh, C. Sirish, and K. Abirami

Abstract A majority of the people nowadays have difficulties understanding their own selves. We have developed an application that allows the user to identify patterns in their behaviors, emotions, and traumas. The application allows the user to tag their experiences by emotions, people, morals, mental illnesses, etc. These tags can later be used to provide analytical insight that helps the user to understand themselves better.

1 Introduction

We as individuals end up having a multitude of unresolved issues, mental breakdowns leading up to unhealthy habits, and self-destructive coping mechanisms, as we do not understand ourselves and forget who we are over time. Henceforth, getting to know about ourselves can be done by proper documentation of our thoughts tagged

---

K. J. Murali · K. V. Prasanna · N. S. Preethi · M. Raajesh · C. Sirish · K. Abirami (✉)
Department of Computer Science and Engineering, Amrita School Of Engineering, Coimbatore
Amrita Vishwa Vidyapeetham, Coimbatore, India
e-mail: k_abirami@cb.amrita.edu
URL: https://www.amrita.edu

K. J. Murali
e-mail: cb.en.u4cse16620@cb.students.amrita.edu

K. V. Prasanna
e-mail: cb.en.u4cse16628@cb.students.amrita.edu

N. S. Preethi
e-mail: cb.en.u4cse16630@cb.students.amrita.edu

M. Raajesh
e-mail: cb.en.u4cse16634@cb.students.amrita.edu

C. Sirish
e-mail: cb.en.u4cse16646@cb.students.amrita.edu

A. K. Tripathy et al. (eds.), *Advances in Distributed Computing and Machine Learning*, Lecture Notes in Networks and Systems 127, https://doi.org/10.1007/978-981-15-4218-3_23

along with the embedded feelings, which helps us first understand our issues and later solve them. This application is a way to make life easy and convenient. It will help to keep memories of your important moments, and it is a perfect place for people of all ages to keep personal reflections. This application is used to examine one's feelings, thoughts, and ideas. It also helps in introspecting yourself, your relationship with others, and how they affect your behavior and emotion.

2 Related Work

Diaries and journals have been around for centuries. They are a great way to express your most intimate feelings. It helps people deal with stuff that people are not good at dealing with anymore. Play store has many applications, many of them even have extra features like mood tracking that your old journals and diaries could not do. A few apps would be as follows.

Evernote is an application that helps to keep journals/diaries by tagging each entry so that it eventually helps the users to organize his work for the maximum effort. This application also provides users to add documents, pictures, and audio files. The main benefit of this application is that all the data are stored in the cloud and can also be easily downloaded locally.

Daylio is a mobile application that helps the users to keep track of the user's mood. The user should tag their mood along with the entry, and these entries are stored in Google Drive in CSV document format so that the users can keep track of their own moods. This application also provides an analytical view of the users' mood through a graph that helps the users to improve their personality.

3 Literature Survey

The following section provides a review of the literature related to the benefits of diary writing. It also talks about the privacy issues and problems that were kept in mind during the design of this application.

The diary is a tool that has been used by human beings ever since they learnt how to write. While writing is generally considered as a good writing habit, even a significantly small amount of diary writing brought about a positive effect on the user's well-being [1].

The most crucial element here is writing of emotional experiences in the form of a narrative, which makes an individual feel happier within months [2, 3].

Another main feature of this application would be the ability for the user to write

The first and foremost concern was to secure the user's trust as users who trust an application more are more likely to worry less about their privacy while using the application, Joinson et al. [4] which was submitted during the year 2010. The authors provide detailed research on the problem of the correlation between a user's trust in an application and how their attitude toward privacy varies with trust. Hence, we decided that it was important for us to let the user know that we as an organization will not use their data for any business purpose or such. Later, this was considered in the system design revamp. Now, the users will be informed that their data will never reside outside their device in a decrypted format.

Keeping user privacy as the topmost priority, our application ensures that any data that move onto our databases are completely encrypted by a key of the user's choice [5, 6]. This way, we can guarantee the user that even if there is to be a data leak, their data will still be safe.

4 Proposed Work

The application allows the user to maintain a journal of their everyday experiences. The users can tag parts of the journal entries with emotions, people, mental illness, etc. These tags are the most crucial part of the application. Using the tags, the user can filter their experiences over a period of time and can understand more about how these tags play a role in their life.

For example, someone having anger issues will be able to see all the journal entries when they were angry assuming the user has properly tagged the anger as an emotion. The user can then use the filtered data to self-analyze and introspect as to when he becomes angry or as to who/what makes him angry.

The app also has a feature where the user can write the same entry in two perspectives. The perspectives would enable the user to express the same event in an emotional ("I" perspective) and a rational("He/She" perspective) manner [7]. This way the user can have an accurate record of how the events actually took place and how they perceived the event as per their emotions.

The app also provides analytic about each emotion/person/tag the user has tagged. For example, it shows the user the number of times they have been feeling sad with a particular person.

Since the user is documenting their own life in the app, it can contain the most intimate and personal details about their life. Hence, the user's data have to be as secure as possible. This is achieved by encrypting every piece of data entered by the user using a key of the user's choice. This way, data if stolen will be useless without the key of the user and hence of no value.

The architecture diagram of the "Introspective Journal" is shown below. The main parts are the client and the server containing a database. The functioning of each part is explained below. The application, once opened, authenticates the user and further prompts the user for a key passcode. The key passcode is a cryptographic key [8] which will be used to encrypt and decrypt the outgoing and the incoming data, respectively, using the AES algorithm [9]. The key passcode will dictate the working of the whole application. After every entry in the application, the entry and all its associated attributes are encrypted using the passcode and sent to the server. Similarly, all queries that are responded by the server are encrypted and only the

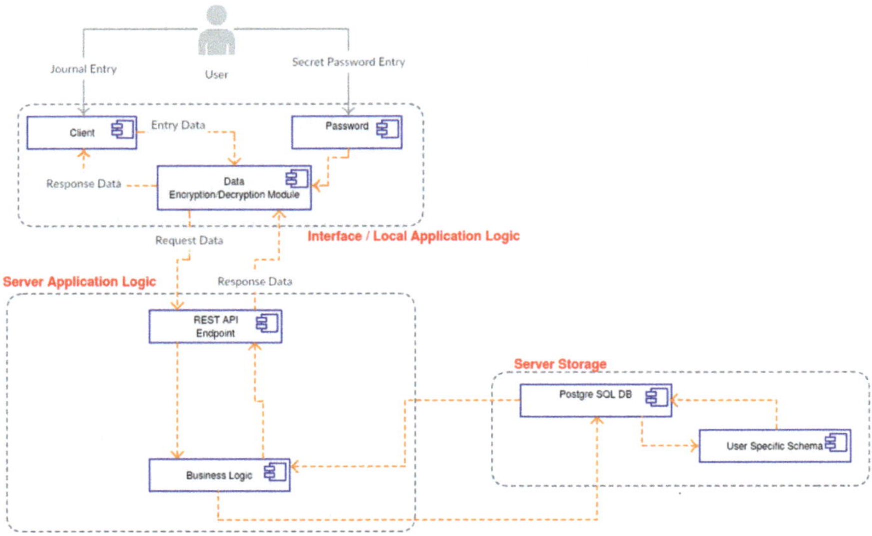

Fig. 1 Front end can be accessed as a multi-platform app or as a Web application. The working of the architecture is as follows

original key can decrypt that data. The data stored in the server will be encrypted and can only be decrypted using the user's key which is only known to the user. If the client is offline, then any data entered are stored in the local storage. When the application connects to the server again, it sends the locally stored data to the server in the usual fashion and destroys the local copy.

5 Result

Figure 1 contains the step-by-step workflow of the developed application. The user is presented with a dashboard that allows navigation to multiple screens such as analytics and the diary screen. The diary screen differs for every date. The date can be selected on the calendar, and the entries of that date will be shown. Along with each entry, the tags attached to it will also be displayed so that the user can understand the context behind each entry. The user can also add new entries if they wish to (Fig. 2).

Entries can be tagged with emotions. The android app currently provides an emotion wheel using which the user can precisely identify the particular emotion that they are feeling from a more general one like "frustrated" and "angry," respectively.

The analytics page provides the user with various analytics about their emotions and other tags. As an example, a pie chart can be used to show the percentage of emotions experienced over the past month. Various other visualizations such as bar charts can be used to indicate the count of all emotions experienced with a particular person or at a particular place.

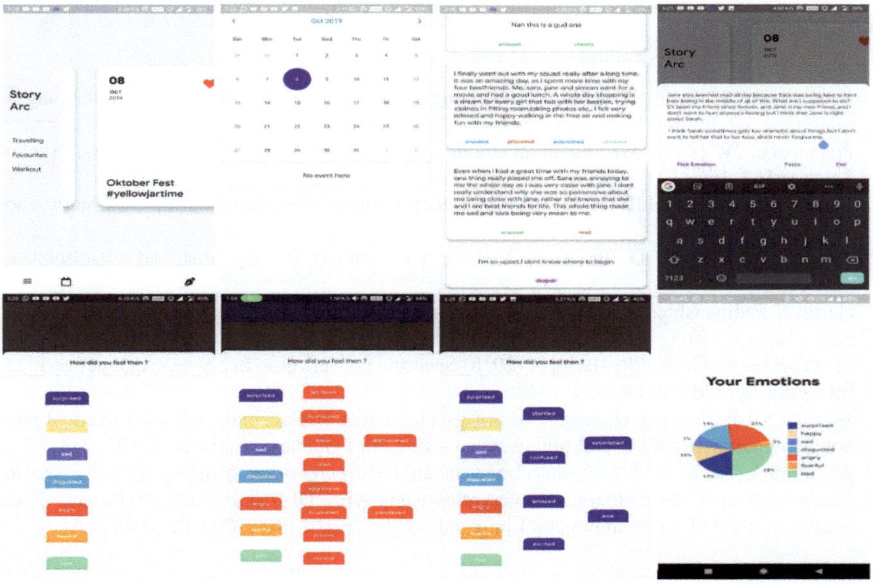

Fig. 2 Here are some snapshots of the application

The application has been successfully developed as per the requirements and the methods as mentioned in the proposed work section and the above passage.

6 Conclusion and Future Work

We have, therefore, implemented a journal application that helps the user to analyze their past emotions in terms of frequency, correlation, and underlying causes. We visualize the user's different emotions and tags that occurred over time using suitable visualizations such as pie chart and bar chart. Henceforth, they can take steps to improve themselves using the newfound insight which we have helped them to achieve. Using this application, users can document their life on daily bases and tag the documented parts with tags and such.

Our future work would be adding features like to-do lists and Kanban charts so the user can schedule their life. And also to let them tag these activities with diary entries so that the user can know what works and what does not when it comes to that particular activity. Our scope is to expand the application to take care of every introspective need of an individual user.

References

1. Pennebaker JW, Beall SK (1986) Confronting a traumatic event: toward an understanding of inhibition and disease. J Abnorm Psychol 95(3):274–281
2. Pennebaker JW (1997) Writing about emotional experiences as a therapeutic process. Psychol Sci 8(3):162–166
3. Pennebaker JW, Seagal JD (1999) Forming a story: the health benefits of narrative. J Clin Psychol 55:1243–1254
4. Joinson AN , Reips U-D , Buchanan T, Paine Schofield CB. Privacy, trust, and self-disclosure online
5. Harini N, Padmanabhan T (2016) 3C-Auth: a Nnew scheme for enhancing security. Int J Network Secur 18:143–150
6. SivaYogitha M, Gopal PR, Harini N (2018) Securing conversation in online social networking. Int J Pure Appl Math 119:1639–1648
7. Jin SR (2010) Structure characteristics of psychological displacement and its dialectical phenomenon:narratives of the multidimensional self. Chin J Guidance Counsel 28:191–233
8. Abdullah A (2017) Advanced encryption standard (AES) algorithm to encrypt and decrypt data
9. Panda M, Nag A (2015) Plain text encryption using AES, DES and SALSA20 by java based bouncy castle API on Windows and Linux. https://doi.org/10.1109/ICACCE.2015.130

Machine Learning Algorithms, Applications and Analysis

FPGA Implementation of Bio-inspired Computing Based Deep Learning Model

B. U. V. Prashanth and Mohammed Riyaz Ahmed

Abstract Computer architecture classes are subjected to power, thermal, and area limitations and suggest facilitating optimization of chip performance for a specific device technology through an optimal performance, memory subsystems, and interconnection model for various technology architectures. The goal of Bio-Inspired engineering is to use Very Large Scale Integration (VLSI) as well as analog electronic circuits to reverse engineer the human brain. This research manuscript presents brain-inspired computing architecture based on the Field Programmable Gate Array (FPGA) neuron model Leaky-Integrate-and-File (LIF) for the massively parallel bio-inspired computing system. In designing a field-programmable neuromorphic computing system, the reconfigurable and event-driven parameters are considered. In this paper an event driven FPGA system is presented which is the combination of memory arbiter for logical memory access for kernel processing. The Register Transfer Logic (RTL) results of implementation and hardware synthesis are presented as a proof of concept. The neuron model is explored in Xilinx-ISE software with utilizing VERILOG code, considering digital implementation, aiming at high-speed low-cost large-scale systems. The deep learning algorithms interact with the FPGA model developed and we carry out the simulations of neuron for the membrane potential in the corresponding receptive field, along with obtaining the characteristics plot between membrane potential and the receptive field.

Keywords Neuromorphic computing · Processor architecture · Synapse · FPGA

B. U. V. Prashanth (✉) · M. R. Ahmed
School of Electronics and Communication Engineering, REVA University,
Bengaluru 560064, India
e-mail: buv.prashanth@gmail.com; prashanthbuv@reva.edu.in

M. R. Ahmed
e-mail: riyaz@reva.edu.in

A. K. Tripathy et al. (eds.), *Advances in Distributed Computing and Machine Learning*,
Lecture Notes in Networks and Systems 127, https://doi.org/10.1007/978-981-15-4218-3_24

1 Introduction

In the year 1970 Von-Neumann architecture based hardware for digital circuit design in particular memory design was expensive. There was a need of in-expensive efficient hardware which is easy to program using standard programming interfaces [1]. According to Moore's law, maximum number of transistors densely packed on single silicon chip is doubled at regular pace. Continuing Moore's law for transistor scaling is enabled by computing with spintronics is emerging as major prospect for memory and logic circuits [2]. The trends and the progress of research in beyond complementary metal-oxide semiconductor is aimed towards to identify new techniques to use two dimensional materials and devices, neuro-inspired computing and quantum computing [3]. As seen in Fig. 1 the Bio-inspired computing is an intelligent and interactive technology, possessing an encouraging scope of research, to reverse engineer a computational architecture of a brain [4]. Convolution Neural Network Algorithms create input-output mappings using rules or weights stored in memory, are created by humans. Biological memory corresponds to a container holding data and algorithms. The complete timeline of building a intelligent computational system based on biologically inspired Artificial Neural Network (ANN) starts designing a boolean logic or functions, passes through theory of computation phase, next enters into electronics technology phase, and further into computational complexity phase and ends at a practical computation phase [5].

While dealing with intelligence the first stage is to probe into evolution, complexity, and also thermodynamics which is not an equivalent of a boolean logic or functions, during the second stage a new and novel electronics technology is required which is not an equivalent of electronics technology, which was defined during the computation phase, in the further stage the implementation complexity is not an

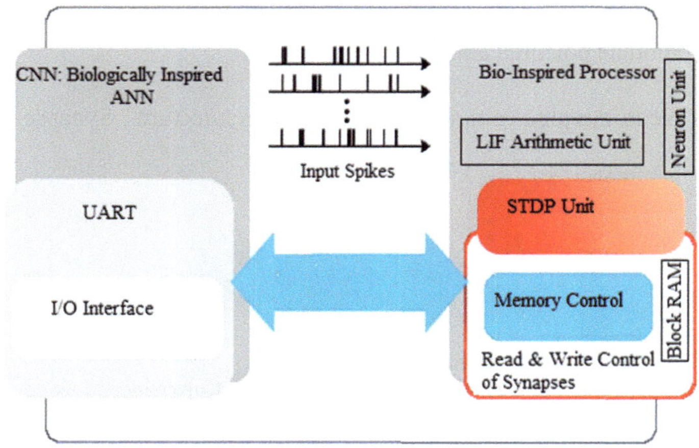

Fig. 1 Bio-inspired computing architecture

equivalent to computational complexity, and the final stage is the practical intelligence stage which is also not an equivalent practical computation stage, defined during process of computation [6]. A discrepancy between latent computing models such as Von-Neumann architectures and Boolean logic is affecting efficiency of neuromorphic computing in software and custom hardware implementation [7, 8]. This radical limitation can be overcome by designing highly efficient and coarse-grained realization of building blocks in neuromorphic computing systems with emerging spintronics devices [9, 10].

2 Problem Statement

The machine learning algorithm techniques are tightly linked with processor architecture techniques, with memory bottleneck and processor bottleneck discussed to increase efficiency and enable the inference of deep neural networks in hardware devices along with an initiative to develop a hardware-algorithm co-optimization method. To develop the fault-tolerant algorithms to which the approximate computing techniques can be applied. Further one of the lesson learnt is the dynamic range of highly accurate computations of necessary precision must be supported in any modern approximate computing technique. We further discuss about paradox of programming a neuromorphic computer. It is learnt that we need to figure out the new class of algorithms, we are also missing out some very important basic concepts. To make this problem statement more specific we further specify as several frame-based chip solutions have been developed in recent years to deploy real-time neural networks. Despite the good efficiency and accuracy of these technologies, there are still a large number of operations due to the complexity of the current network models.

3 Proposed System

Considering all of these points in the problem statement, an experimental setup is developed as depicted in Fig. 2 which illustrates the practical setup of bio-inspired computing architecture based on simple neuron model implemented on FPGA. This setup is considered as multi-core neuromorphic processor. The next generation algorithmic development should also continue in the direction of addressing integrated solutions of the spatio-temporal complexity of challenging issues that are capable of handling wide variants of datasets. In order to improve the functionality of hardware implementations, new devices with properties that more directly comport to neurobiological operations, must be directed towards enhancing the functionality of hardware implementations. The features of biological computations along with concerns of biological fidelity need to be balanced for future work in novel devices. Programmable circuits and Field Programmable Gate Arrays (FPGA's) share various necessary specifications as digital computing systems, except for tradeoff of long

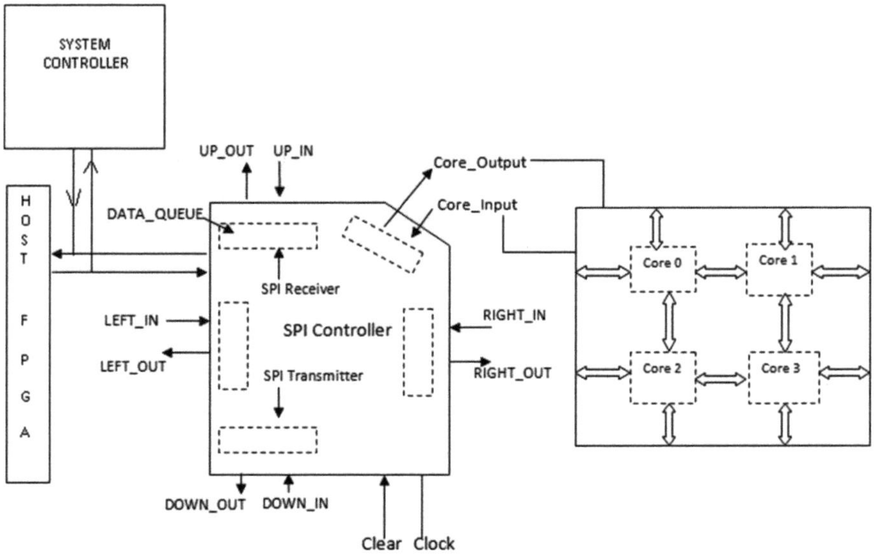

Fig. 2 Bio-inspired computing model

retention with high endurance. The challenges lie in developing low-power devices based on CMOS that goes into the third dimensions on improvements in advanced materials technology. The emerging directions of the above-stated research gaps are to develop a virtual paradigm for self re-configurable computing architectures that reach beyond Moore's law.

The next section presents the core results of bio-inspired computing architecture based simple neuron model.

4 Results Obtained

The complete architecture of the multi-core neuromorphic computing system consists of system controller based on an Advanced RISC Machine (ARM) processor, core-array of two dimensions, and a Serial Peripheral Interface (SPI) Controller. In this set up the Serial Peripheral Interface (SPI) controller emulates a driving device configured as a router. The router signal communication protocol is hardware controlled comprises of four channels for communicating the signals. The router follows two approaches which are the routing network based on static and dynamic schemes in which the latter is a switch allocator based cross bar switch, is a resource and power intensive and it requires the allocation of buffers in the form of data queues. The implementation of multi-core neuromorphic computing system is done on Altera Cyclone IV FPGA contained as a part of ALTERA DE2 Board. The VERILOG language was used to program the neuromorphic framework and compiled

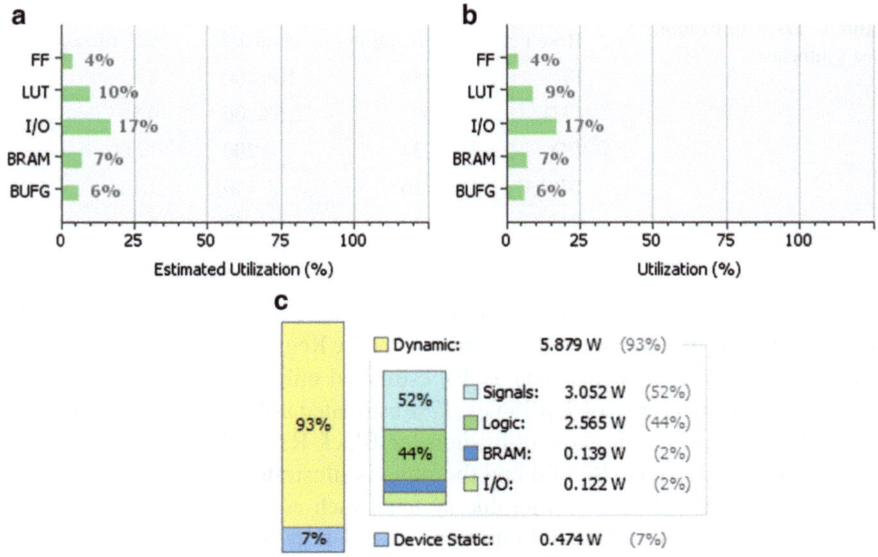

Fig. 3 **a** Estimation utilization after RTL synthesis, **b** Utilization after RTL implementation, **c** Power analysis diagram

in QUARTUS-II. The clock region block of FPGA is described in Fig. 3b–f depict internal logic regions of clock region block. The execution is based on vector less activity propagation where peak memory is 1130.086 MB with gain of 12.105 dB. The execution is carried out in Vivado-v-2014.2 integrated simulation environment and this is divided into initial four phases of execution.

1. **Phase 1**: The phase 1 is based on logic optimization task.
2. **Phase 2**: Phase 2 is based on constant propagation and in both of the phases memory utilized is found to be 897.250 dB.
3. **Phase 3**: The phase 3 is based on sweep with the same memory utilization as in phases 1 and 2.

In all of these three phases the gain is observed to be 0 dB. The logic optimization task ends at completion of phase 3.

4. **Phase 4**: The phase 4 starts with power optimization task with skipping clock gating for clocks with period <3.13 ns and applying IDT and ODT optimizations. The WRITE_MODE attribute of 0 BRAM(s) out of a total of 12 has been updated to save power. Here the peak memory is observed to be 1048.023 MB with the gain of 532.957 dB.

The Multithreading enabled for route design using a maximum of 2 CPUs in the routing task. The router is initialized and will operate in resource-optimization mode. Here the peak memory is observed to be 1113.621 MB and gain is observed to be 65.598 dB. We have to specify clocks using VERILOG syntax for sequential elements. For pure combinatorial circuits, a virtual clock is specified. By utilizing

Table 1 Logic utilization post synthesis

Resource	Utilization	Available	% Utilization
FF	4258	106,400	4
LUT	5097	53,200	9.58
I/O	34	200	17.00
BRAM	10	140	7.14
BUFG	2	32	6.25

the on-board memory the synaptic memory was implemented with in the FPGA. Figure 3a depicts the % estimated utilization after the Register Transfer Logic (RTL) synthesis for logical elements such as 4% estimated utilization for flip-flops, 10% estimated utilization for Look Up Tables, 17% estimated utilization for Input-Output (IO), followed by 7% estimated utilization for Block RAM (BRAM) and 6% utilization for Gate Buffers (BUFG) and the same is illustrated in Table 1. In Fig. 3c the power is analyzed as neuromorphic systems such as the one presented in this work takes inspiration in human brain to process data in real time with low power consumption. FPGA has significant static power consumption, so efficient power is usually calculated as the system's difference in idle state and real-time data processing. The system does not process events in Idle mode, and therefore there is no computation. On the other hand, the worst case is represented by a high input performance because the system dynamically adjusts the incoming events.

Figure 3b illustrates the % utilization after Register Transfer Logic (RTL) implementation process, here as seen from the figure the logical element Look Up Table (LUT) has a % utilization is 9% and % utilization is same for rest of the logical elements in Fig. 3a and also same is as depicted in Table 2. Figure 3c illustrates the power analysis diagram depicting the device static power and device dynamic power represented in 7% indicated in blue color and 93% indicated in yellow color respectively. The logic utilization for post synthesis and post implementation is as described in Tables 1 and 2 respectively for various logical elements such as Flip Flops (FF), Look Up Tables (LUT's), Input/output (I/O), Block RAM (RAM), and Gate Buffers.

In Tables 1 and 2, except LUT all other parameters have the same value for logic utilization post synthesis and post implementation, the reason behind this is the designed system is more of device specific rather than board specific.

Table 2 Logic utilization post implementation

Resource	Utilization	Available	% Utilization
FF	4258	106,400	4
LUT	5000	53,200	9.40
I/O	34	200	17.00
BRAM	10	140	7.14
BUFG	2	32	6.25

The simulations of neuron are for each input neuron for the membrane potential in the corresponding receptive field is calculated and is as depicted in Figs. 4 and 5 depicts the characteristics plot between membrane potential and the receptive field. In Fig. 4 a biological neuron reduces its membrane potential when it receives no excitation. This biological neuron is emulated due to the property of LIF neuron that

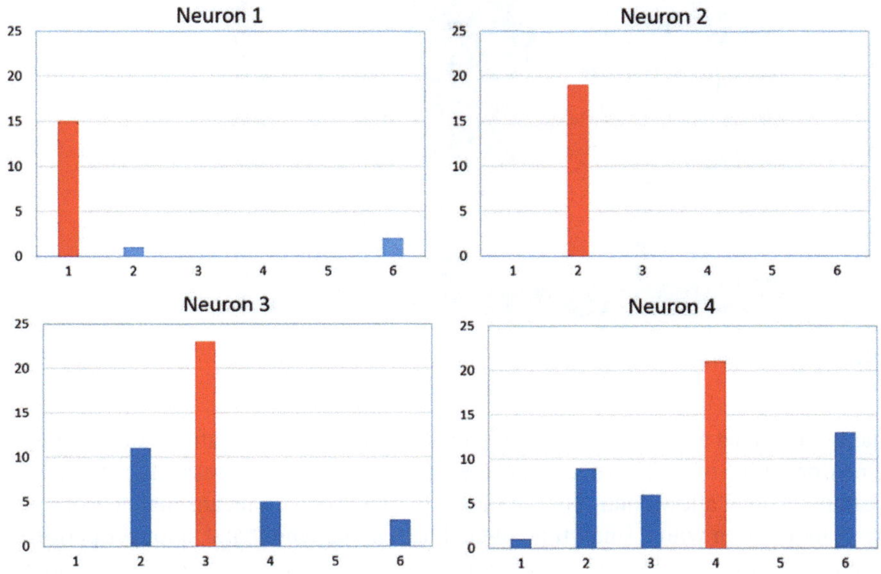

Fig. 4 The simulations each input neuron

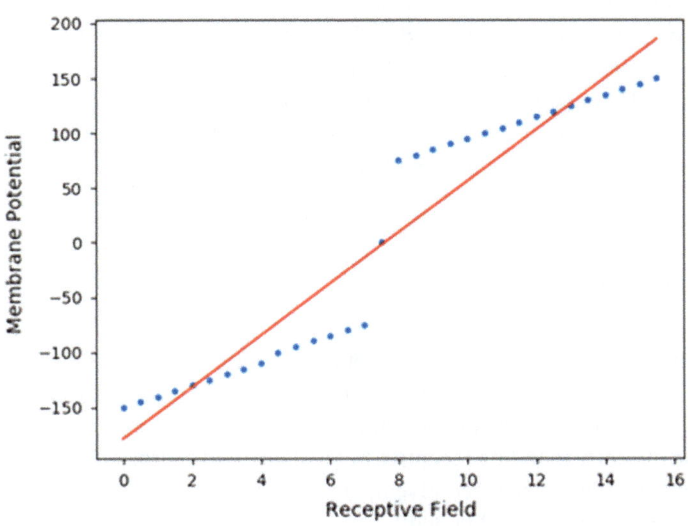

Fig. 5 Characteristics plot between membrane potential and the receptive field

Fig. 6 Experimental setup

has not received any stimulation and it does not give the scene information. In Fig. 5 a neuron's time of leakage decay and the value the neuron decreases its potential are parameters that allow control of the rate of output events. It means that neurons will reset more often with a small time of decline. On the other hand, a longer period of decline makes neurons rarely reset their potential, increasing the number of spikes in output.

In the proposed system we have developed a practical setup as illuminated in Fig. 6, which elucidates the connection of FPGA board to the work station using a JTAG adapter. The FPGA is utilized with SYSGEN as a modeling tool, is used for fixed point modeling of applications. Using SYSGEN, netlist can be generated for a given FPGA. This netlist is given as input to Xilinx ISE Tools, which generates the bit file; the generated bit file is used by the SYSGEN for hardware co-simulation. The design was described as RTL with Verilog language.

5 Conclusions

We present neuromorphic computing architecture based on the FPGA neuron model as Leaky-Integrate-and-File (LIF) neuron for the massively parallel bio-inspired computing system. The machine learning simulations of neuron for the membrane potential in the corresponding receptive field, along with characteristics plot between membrane potential and the receptive field are obtained. The theoretical calculations match with practical result obtained.

References

1. Manipatruni S et al (2018) Beyond CMOS-spin and polarization. Nat Phys 14(4)
2. Pan C (2017) Beyond-CMOS benchmarking. In: DATE 2017. IEEE, New York, pp 133–138
3. Jungwirth et al (2018) multiple anti ferromagnetic spintronics. Nat Phys 14(3):200
4. Wang Y et al (2015) Thermal modeling of STT MTJ. Microelectron Rel 55:1649–1653
5. Gale E (2014) TiO_2-memristors Re-RAM. Semicond Sci Tech 29(10):104004
6. Shalf JM (2015) Computing beyond Moore's law. Computer 48(12):14–23
7. Soman S et al (2016) Recent trends in neuromorphic engineering. Big Data Anal 1(15)
8. Allibert F et al (2001) From SOI to innovative devices. Solid-State Electron 45(4):559–566
9. Brazil R (2015) Putting a spin on it: spintronics super-fast computing. Eng Tech 10(10)
10. Schuman CD et al (2017) Framework for neuromorphic systems with Emerging Techologies. In: NanoCom '17

Road Accident Detection and Severity Determination from CCTV Surveillance

S. Veni, R. Anand, and B. Santosh

Abstract Wide variety of road types like intersections, highways poses a real challenge to the computer vision algorithms. Hence, there is a need of efficient algorithm to detect the accident on road and also evaluate the severity of the incident. This can be used to improve the emergency services response time. The work demonstrated in this paper aims to develop such an algorithm by modifying existing CCTV surveillance system. In this work, the accident is detected by the dispersion in the motion field of the vehicles during collision. Motion field of the road is obtained from the optical flow of the video frames. The moving objects in the frames are segmented and tracked. The dispersion in the angle vector of the optical flow is derived for each of the moving object. The dispersion of angle vector for each object is monitored, and deviation of the same from the threshold is determined as an accident. The harshness of the accident can be found by the range of dispersion of the motion field. The algorithm developed here is capable of detecting accidents between any types of moving objects.

Keywords Motion field · Optical flow · Farneback optical flow · Angle vector

S. Veni · R. Anand
Amrita School of Engineering, Amrita Vishwa Vidhayapeetham, Amrita University, Coimbatore 641112, India

R. Anand (✉)
Sona Signal and Image Processing Research Center, Sona College of Technology, Salem, India
e-mail: anand.r@sonatech.ac.in
URL: http://www.springer.com/gp/computer-science/lncs

B. Santosh
Robert Bosch, Department of Mechanical Engineering, Amrita School of Engineering, Amrita Vishwa Vidhayapeetham, Amrita University, Coimbatore 641112, India

A. K. Tripathy et al. (eds.), *Advances in Distributed Computing and Machine Learning*, Lecture Notes in Networks and Systems 127, https://doi.org/10.1007/978-981-15-4218-3_25

1 Introduction

Ever increasing number of automobiles on roads increases the risks of road accidents. As per the Global Status Report of WHO, road traffic incident becomes more than 1.3 million survives every year and it is the low- and middle-income countries that are hard hit [1]. It is estimated around 1214 road crashes take place every day causing one death in every 4 min in India. Though such unfortunate incidents have to be avoided, it is a fact that they cannot be completely eliminated. The response time for emergency services in such cases helps in reducing the severity of the incident. Studies suggest a 10 min reduction in response time reduces the mortality rate by one-third [2]. All the current emergency service systems are human operated and required a human presence to assess the situation and take the response action. The increasing number of accidents can be a real challenge to current system of emergency response. If a computer system can monitor the situation and determine the severity, the responsive action for an incident can be more quick and swift. Traffic surveillance systems with CCTV are in place in many countries to monitor and regulate the traffic flow. Existing systems require a human observer monitoring multiple screens from different surveillance cameras with continuous traffic. There is always a possibility to miss certain events which can result in delayed response. Moreover, today's emergency response systems like 911, 108 are not linked to surveillance systems and act based only if a call is received. With computer vision, several monitoring objectives such as congestion, traffic rule violation, and vehicle interaction can be targeted using cameras from the existing traffic surveillance systems. In particular, we are concentrating on novel technique to detect and assess the situation of risk such as accidents in multiple scenarios. For the previous few years, many researchers developed many methods to decrease the recognition rate of road accidents [3–5]. In this paper, we present a state-of-the-art computer vision technique for traffic video analysis and an outlook to future research directions. Object tracking is at the heart of this application. Object tracker algorithm assigns consistent labels to the tracked objects in different frames of a video. Vehicular flow can be determined from these tracked objects as well as the disturbances in the optical flow. Different thresholds are calculated using statistical techniques, and severity of accidents is divided into low, high, and severe. This process can be divided into three steps as shown in Fig. 1. (1) Tracking, (2) Evaluation, and (3) Classification.

The steps shown in Sect. 1 are explained in Sect. 2, and the results are discussed in Sect. 3.

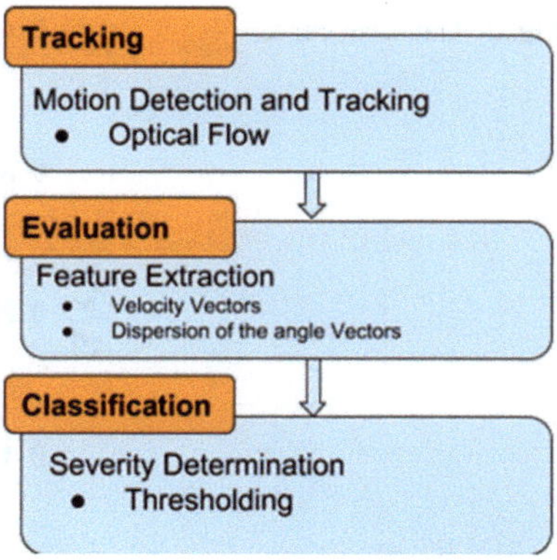

Fig. 1 Project flow for road accident detection and severity determination

2 The Material and Method

2.1 Moving Object Detection and Tracking

The moving objects in the video frame are detected by determining the optical flow of the scene. The optical flow algorithms can be broadly classified into two types (a) Sparse optical flow, which computes optical flow for a sparse feature set of the frame, for example, corners in the frame detected using Shi-Tomasi algorithm. (b) Dense optical flow computes the optical flow for all the pixels in the frame. Dense optical flow algorithm is chosen in the proposed method where the optical flow is built on three-dimensional and time-based differences of the image illumination at all pixels. The choice of algorithm is Farneback optical flow.

Farneback Algorithm for Optical Flow Method This algorithm employs "Two-Frame Motion Estimation Based on Polynomial Expansion" [6]. Each pixel neighbourhood is approximated by a polynomial.

$$f_1(x) = x^T A_1 x + b_1^T x + c_1 \tag{1}$$

The next frame from the CCTV surveillance camera will have a displacement of pixels, d, then

$$f_2(x) = f_1(x-d) = (x-d)^T A_1 (x-d) + b_1^T (x-d) + c_1 = x^T A_2 x + b_2^T x + c_2 \tag{2}$$

Assuming the pixel intensity change between the two frames is negligible, then equating (1) and (2) gives

$$b_2 = b_1 - 2A_1 d \tag{3}$$

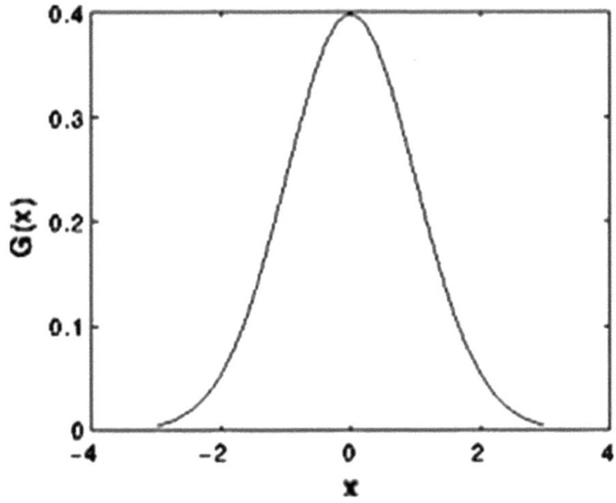

Fig. 2 Gaussian curve

From (3), the displacement vector d is calculated

$$d = 1/2(A_1 - 1)(b_2 - b_1) \tag{4}$$

From (4) optical flow vectors, (u, v) are computed.

The obtained flow field is passed through a Gaussian low pass filter to smoothen the discontinuities in the flow field.

Filtering In this work, a low pass filter is used to smooth the variations in the obtained flow field; hence, the operation is also called as smoothing. One such smoothing filter used here is the Gaussian filter. The pixel situated in the mid would take the major heaviness. This heaviness of its neighbouring pixels reductions as the three-dimensional distance amongst them and the midpoint pixel increases. A plot of 2D Gaussian kernel is shown in Fig. 2.

2D Gaussian can be represented as:

$$G_0(x, y) = Ae\left((x -_x)^2/(2_x^2) + (y -_y)^2/(2_y^2)\right) \tag{5}$$

here μ is the mean (the peak), and σ represents the standard deviation (per each of the variables x and y).

Moving Object Detection Once the flow field for the frame is generated, the moving objects are determined by examining the area of pixels where the flow values are greater. Ideally, the flow of the background road and other non-moving objects is zero, but in practice, it is observed that there is a small amount of displacement due to environmental disturbances. This disturbance is filtered out by thresholding. The flow vectors (u, v) obtained from the optical flow algorithm is divided into velocity and orientation of the flow by using the following formulate [7].

$$|v| = \sqrt{(x(i+1) - x_i)^2 + (y(i+1) - y_i)^2} \quad (6)$$

$$\theta = \tan^{-1}\left(\frac{(y(i+1) - y_i)}{(x(i+1) - x_i)}\right) \quad (7)$$

A connected component analysis of the orientation vectors of the frame gives the moving objects. The objects are labelled using this method.

Tracking The labelled moving objects found in the detection phase are compared with the moving objects found in the previous frame. Distances between the centroid's are calculated for all the objects in the current frame with that of the previous frame. The objects with the least distance with the current and previous frame are identified as the same vehicle and re-labelled accordingly. Occlusion of two moving objects is detected by comparing the areas of the previous object and current object for overlapping. If two or more of the previous objects overlap with the current frame object, then an occlusion is detected. The occluded objects are split into individual objects based on the object area's from previous frame.

2.2 *Evaluation*

The moving objects, be they, vehicles, walking pedestrians, etc., are tracked. The features that are tracked are the orientation values (θ) of the pixels for the object detected. If there is a smooth flow of traffic, every pixel of the moving object should be having a θ value close to the mean orientation of that object. When an accident occurs, it is expect to see a dispersion in the values of θ from the mean for the object pixels. The dispersion of the object θ values are calculated by determining the standard deviation of θ for the object.

$$\text{SD} = \frac{\sum(\theta - \mu)^2}{N} \quad (8)$$

where N = number of pixels, θ = Pixel orientation μ = mean Orientation of object pixels. If dispersion of pixel orientation crosses a certain threshold, then it can be concluded as an accident has occurred.

2.3 *Classification*

Using the dispersion of pixel orientation for every tracking object, accident is detected [8, 9]. By measuring how much the pixel orientations dispersed above the threshold, the severity of the incident is classified as low, medium, and high.

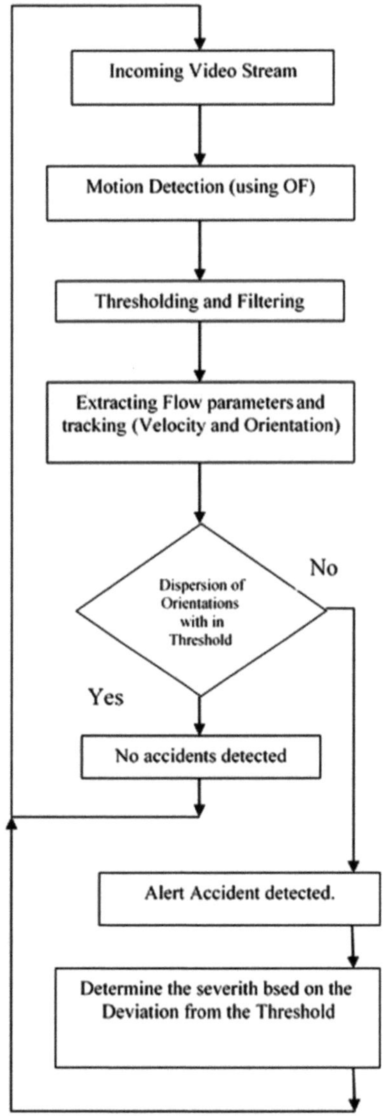

Fig. 3 Algorithm for accident detection and severity evaluation

2.4 Flow Chart

Figure 3 shows the flow chart of the work done. Here, the incoming video steam is passed through optical flow algorithm. The output of the algorithm is the flow vectors of the moving objects in the frame.

The output of the optical flow is filtered for smothering the variations. Small movements in the background are eliminated by thresholding. The velocity vectors obtained after the above step is split into its individual components to obtain the orientation and magnitude parameters. Individual objects are identified based on the orientation values obtained and tracked frame by frame. Also, the dispersion of the orientation values of the detected object is measured. This dispersion is compared with the predetermined thresholds to check if there is an accident. The extent of dispersion gives the severity of the incident.

3 Results and Discussions

The footage from the CCTV surveillance is given as an input to the algorithm. The colour frame is first converter in to a greyscale image a shown in Fig. 4a.

The greyscale image generated is passed through the Farneback optical flow algorithm. The output is a two-channel array with optical flow vectors, (u, v). The colour code representation of the flow vectors is shown in Fig. 5. Direction corresponds to Hue value of the image, and magnitude corresponds to value plane. The image is shown in Fig. 4b.

The above image consists of both the magnitude and the orientation for each pixel's flow vector. The magnitude and direction of the flow vectors are derived, and the resulting image with orientation values as the pixel data is shown in Fig. 5. For further analysis, only orientation information of the flow is considered. This is because the orientation information is invariant to the fame rate. This gives the algorithm a novel advantage of being independent of the camera capture rate. It can be observed from Fig. 6 that the moving objects have a value where as the background, which is not moving, has zero value. This image can be called as angle image. On binarization and doing a connected component analysis, the moving objects can be segmented from the image. The output of binarization of angle image is shown in Fig. 5a.

On superimposing the above data on the original image in the form of arrows, the flow of each moving object can be visualized. This is shown in Fig. 5b. Whenever there is an accident in the frame, the flow vectors tend to disperse as shown in Fig. 6.a.

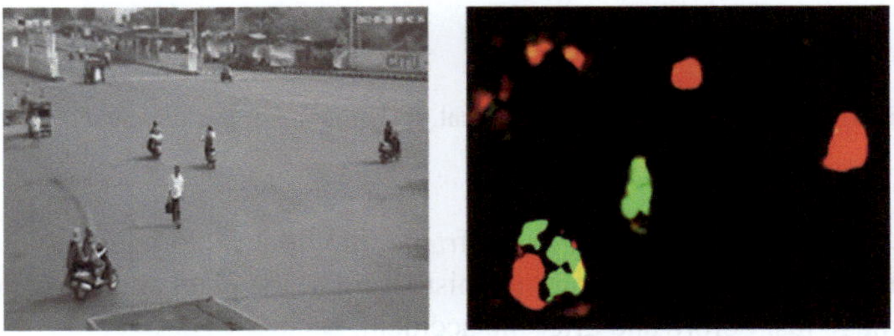

(a) (b)

Fig. 4 **a** Traffic scene at an intersection. **b** Output from Farneback optical flow algorithm

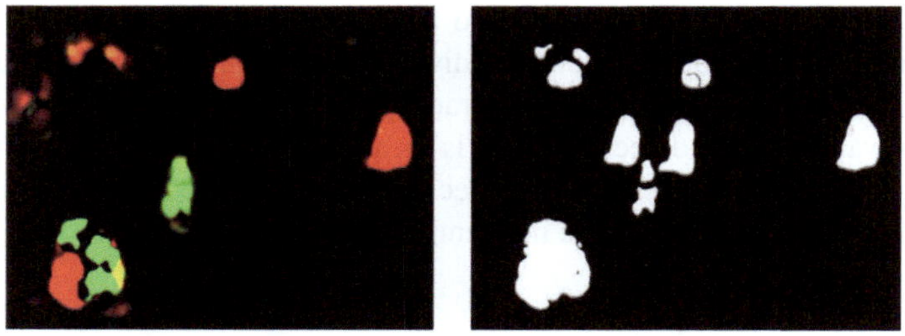

(a) (b)

Fig. 5 **a** Pixel values as angle of the optical flow at that point. **b** Frame image after binarization

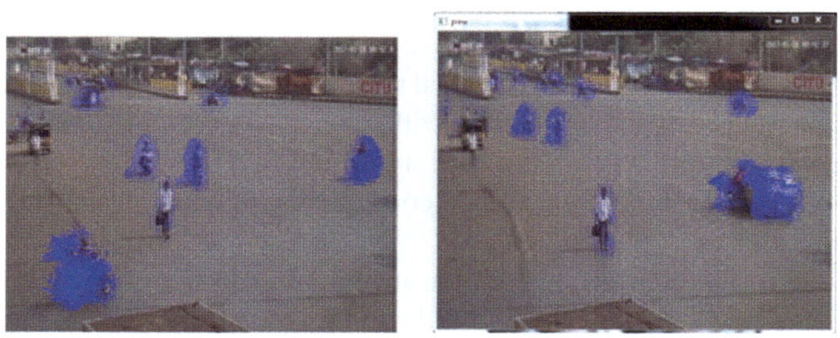

(a) (b)

Fig. 6 **a** Frame with flow vectors superimposed. **b** Accident frame with flow vectors superimposed

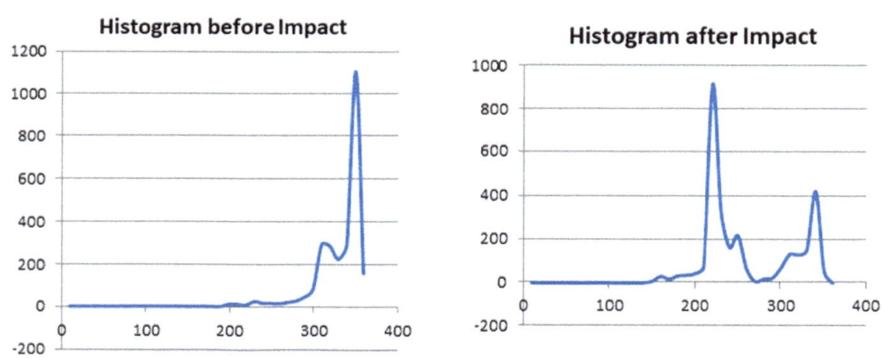

(a) (b)

Fig. 7 **a** Histogram of the vehicle before accident. **b** Histogram of the vehicle after impact

Whenever there is an accident in the frame, the flow vectors tend to disperse as shown in Fig. 6b. Figure 7a, b shows the histogram of the pixel angle component for accident image just before and after the accident.

On observing these histograms, it is clear that there is a change in the dispersion in the angle component of the object's optical flow. This dispersion is measured as

Fig. 8 Accident detected

a b

Fig. 9 **a** Accident on a freeway. **b** Flow vectors superimposed

the standard deviation of the flow vector angles. A variation in the standard deviation values to that of the normal flow is identified as an accident.

Figure 8 shows the identified accident. The severity is determined by considering how much the standard deviation deviated from the norm values. In Fig. 8, this is shown as the text.

The algorithm is run on different videos with different backgrounds and illumination conditions. The algorithm was able to detect the accident and arrive at a severity level. Figure 8 shows the accident at a junction, where Figure 9a, b, shows the accident at a freeway. Figure 10a, b shows an accident during nights.

a b

Fig. 10 **a** Accident in a intersection during night. **b** Flow vectors superimposed

4 Conclusion

The algorithm is tested on multiple accident videos taken online sites. The algorithm was able to detect the accidents successfully. The project is implemented in Visual Studio 2017, on C++. Computer vision library OpenCV 3.0 was used. The proposed algorithm not only successfully detected the accident but also was able to come up with a measure to determine the severity of the accident occurred.

References

1. World Health Organization (2015) Global status report on road safety 2015. World Health Organization
2. Sánchez-Mangas R, Garcia-Ferrrer A, De Juan A, Arroyo AM (2010) The probability of death in road traffic accidents. How important is a quick medical response? Accident Anal Prevent 42(4):1048–1056
3. Yun K, Jeong H, Yi KM, Kim SW, Choi JY (2014) Motion interaction field for accident detection in traffic surveillance video. In: 2014 22nd international conference on pattern recognition (ICPR). IEEE, pp 3062–3067
4. Manocha M, Kaur P (2014) Object tracking techniques for video tracking: a survey. Int J Eng Sci (IJES) 3(6)
5. Rout RK (2013) A survey on object detection and tracking algorithms. Ph.D. dissertation
6. Gunnar F (2002) Polynomial expansion for orientation and motion estimation. , Ph. D. dissertation. Linköping University Electronic Press
7. Maaloul B, Taleb-Ahmed A, Niar S, Harb N, Valderrama C (2017) Adaptive video-based algorithm for accident detection on highways. In: 2017 12th IEEE international symposium on Industrial embedded systems (SIES). IEEE, pp 1–6
8. Anand R, Veni S, Aravinth J (2016) An application of image processing techniques for detection of diseases on brinjal leaves using k-means clustering method. In: 2016 international conference on recent trends in information technology (ICRTIT). IEEE
9. Sabeenian RS et al (2019) Palm-leaf manuscript character recognition and classification using convolutional neural networks. In: Computing and network sustainability. Springer, Singapore, pp 397–404
10. Harsha SS, Anne KR (2016) A highly robust vehicle detection, tracking and speed measurement model for intelligent transport systems. Int J Appl Eng Res 11(5):3733–3742
11. Lee IJ (2012) An accident detection system on highway through CCTV with calogero-moser system. In: 18th Asia-Pacific conference on communications (APCC). IEEE, pp 522–525
12. Sadeky S, Al-Hamadiy A, Michaelisy B, Sayed U (2010) Real-time automatic traffic accident recognition using hfg. In: 2010 20th international conference on pattern recognition (ICPR). IEEE, pp 3348-3351
13. Ki Y-K (2007) Accident detection system using image processing and MDR. Int J Comput Sci Network Secur (IJCSNS) 7(3):35–39
14. McCahill M, Norris C (2002) CCTV in London. Report deliverable of UrbanEye project
15. Kamijo S, Matsushita Y, Ikeuchi K, Sakauchi M (2000) Traffic monitoring and accident detection at intersections. IEEE Trans Intell Transport Syst 1(2):108–118
16. Akshay S, Thomas S, Ram Prashanth A (2016) Improved multiple object detection and tracking using KF-OF method'. Int J Eng Technol 8:1162-1168
17. Megalingam RK, Nair RN, Prakhya SM (2010) Wireless vehicular accident detection and reporting system. In: 2010 international conference on mechanical and electrical technology (ICMET 2010), Singapore, pp 636–640

A Fuzzy Graph Recurrent Neural Network Approach for the Prediction of Radial Overcut in Electro Discharge Machining

Amrut Ranjan Jena, D. P. Acharjya, and Raja Das

Abstract Manufacturing of goods rely on its design methodology and the process parameters. The parameters used in manufacturing process play an important role to build a quality product. Initially heuristic techniques are used for parameter selection. Many researchers conducted research to predict the radial overcut using neural networks. Besides, fuzzy neural network gains more popularity due to presence of fuzzy system and neural network. In this paper fuzzy graph recurrent neural network architecture is used for modelling and predicting the radial over cut for an electro discharge machining information system.

Keywords Recurrent neural network · Fuzzy graph · Mean square error · Radial overcut · Electro discharge machining

1 Introduction

Manufacturing plays a major role to develop the economy of a country. Different manufacturing industries like automobiles, aeronautics, electronics, food and beverages, chemicals, textiles, and others have important role in growth of economy for a country. The production of different goods come through manufacturing processes. The manufacturing process faces numerous challenges due to randomness in the system for which the performance of the system is affected, and it is difficult to predict the performance of the manufacturing system [1]. The performance evaluation is the process of enumerating the efficiency and usefulness of the action [2]. The performance evaluation of a manufacturing system is also depending on

A. R. Jena (✉) · D. P. Acharjya
School of Computer Science and Engineering, VIT, Vellore, India
e-mail: amrut.ranjan7@gmail.com

D. P. Acharjya
e-mail: dpacharjya@gmail.com

R. Das
School of Advanced Science, VIT, Vellore, India
e-mail: rajadasrkl@gmail.com

A. K. Tripathy et al. (eds.), *Advances in Distributed Computing and Machine Learning*, Lecture Notes in Networks and Systems 127, https://doi.org/10.1007/978-981-15-4218-3_26

five performance objectives like quality, cost, speed, flexibility, and dependability [3]. In manufacturing process, machining is a key process and has immense role in the production. The performance and production of manufacturing process depend on machining processes. The outcome of the machining process is a product. The product popularity in the global market depends on its cost and quality. The manufacturing of the product depends on the machining process. However, the traditional machining processes are used to make the products [4]. When the material used for product making is very hard in nature, traditional machining fails on it. Therefore, to overawed this problem a non- traditional machining process called electro discharge machining (EDM) is used [5]. EDM does the machining work on a work-piece, if the work-piece is electrically conducted irrespective of its hardness [6]. In EDM the machining is carried out without any physical contact between the material and machining tool. Besides, electrical sparks erosion is imposed on work-piece for making the desired product [7]. Hence, the accuracy of the product is more in EDM as compared to conventional machining process. As manufacturing process faces several challenges and complexity due to randomness in the system, it is difficult to predict the response in conventional process. Therefore, soft computing techniques are applied to model and predict the response of the system [8].

Predicting product quality rely on manufacturing process and time. Therefore, manufactures uses predictive analytics in manufacturing to forecast a solution. The benefits of using predictive analytics in manufacturing are to upgrade the product quality, minimizing production time, and design cost effective product. It also helps the manufactures to maintain their assembly line machines by collecting previous machining data for analysing earlier machining process, and finding suitable pattern for machining parameters. Primarily, different linear regression techniques are used for predictive analytics [9]. Further, dealing with uncertainties in an information system various intelligent method such as fuzzy set [10], rough set [11], support vector machine [12], artificial neural network [13], and genetic algorithm [14], are presented. However, many soft computing methods like neural network [15], general regression neural network [16], neuro-fuzzy system [17], neural network-genetic algorithm [18], and non-dominated sorting genetic algorithm [19] are applied in EDM process parameters to predict response of the system.

Many researchers have used artificial neural networks for modelling and predicting the performance of EDM [20]. But data modelling faces four problems such as overfitting, overtraining, chance effects, and interpretation by applying artificial neural network on it [21]. General regression neural network (GRNN) is also used for predicting radial overcut in EDM [22]. It takes values from spared parameters in a trial and error method. But GRNN is not used any intuitive method to find optimal smoothing factor in spared parameters. Besides, many researchers used adaptive neuro-fuzzy inference system (ANFIS) in EDM to optimize the single output [23]. A system with multiple output is not handled by ANFIS. Though, multiple outputs exist in EDM, ANFIS fails to predict the multiple outputs in EDM. Similarly, genetic algorithm and neural network are combinedly applied in EDM to predict radial overcut. But genetic algorithm faces problem in parameters selection process, and lower convergence rate [24]. Likewise, non-dominated sorting genetic algorithm (NSGA) is also applied in EDM to predict the radial overcut [25]. But non-dominated sorting genetic algorithm (NSGA) faces problem to deal with continuous values. All the above methods are used to predict the response for EDM. However, factor analysis is not shown in these prediction process. Therefore, in this work we are proposing a prediction model for EDM machining process to predict the radial overcut using fuzzy graph recurrent neural network. The aim of this research work is to do the prediction by analyzing the factors through fuzzy graph recurrent neural network. We have analyzed the model in EDM information system and observed the effectiveness of the model in the form of accuracy.

The organization of remaining part of the paper is as follows. Following introduction, Sect. 2 presents the foundations of fuzzy graph recurrent neural network. Development of proposed research design is presented in Sect. 3. Section 4 presents the experimental study on EDM followed by result analysis in Sect. 5. Finally, the conclusion is presented in Sect. 6.

2 Fundamentals of Fuzzy Graph

Fuzzy graph concept comes into the picture when we consider fuzzy relations on a fuzzy set [26]. It is defined as $G : (\sigma^*, \mu^*)$ through $\sigma^* = \{u \in V : \sigma(u) > 0\}$ and $\mu^* = \{(u, v) \in V \times V : \mu(u, v) > 0\}$. Here σ represents a fuzzy subset of a finite nonempty set V and μ represents a fuzzy relation on σ. Besides, fuzzy relation μ is fuzzy reflexive and fuzzy symmetric in nature [27]. The description of the definition is presented in Fig. 1.

In the given figure $\sigma^*=\{v_1, v_2, v_3, v_4, v_5\}$, and $\mu^*=\{(v_1, v_2), (v_1, v_3), (v_1, v_4), (v_2, v_5), (v_3, v_5), (v_4, v_5)\}$ with $\mu(v_1, v_2)=0.3$, $\mu(v_1, v_3)=0.2$, $\mu(v_1, v_4)=0.3$, $\mu(v_3, v_5)=0.1$, $\mu(v_2, v_5)=0.5$, and $\mu(v_4, v_5)=0.3$.

In recurrent neural network model, fuzzy OR neuron technique is applied. Because the output of a fuzzy OR neuron facilitates the strongest path between input and output layer in the network. The operation performed by a fuzzy OR neuron is as follows. Initially the fuzzy OR neuron evaluate an AND operation between the input x_i and the weight w_{ji}. Besides, an OR operation is evaluated to know the aggregation of the dendritic inputs. Let us assume the output of the fuzzy neuron is y_j. The output y_j

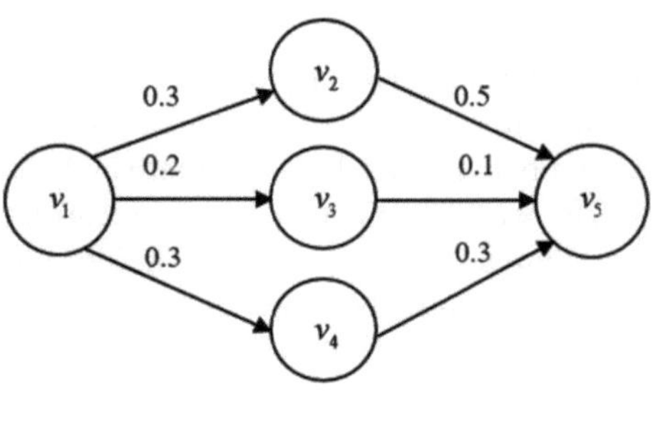

Fig. 1 An illustration of fuzzy graph

generated from a fuzzy neuron by an AND operation is presented in Eq. 1 whereas the output y_j generated from a fuzzy neuron by an OR operation is presented in Eq. 2.

$$\bigwedge_{i=1}^{n} (x_i \vee w_{ji}) = y_j \tag{1}$$

$$\bigvee_{i=1}^{n} (x_i \wedge w_{ji}) = y_j \tag{2}$$

Therefore, Eqs. (1) and (2) state that AND operation and OR are complemented to each other.

2.1 Recurrent Neural Network

Recurrent neural network (RNN) is a dynamic neural network where the signals flow in forward direction and backward direction. Therefore, the outputs generated by some neurons are feeding back to the same neurons or forwarded to the preceding layers neurons. By structure, recurrent neural network consists of a dynamic memory. The output of the memory at a certain instant reproduce the current input as well as earlier input and output those are gradually reduced. This feature makes recurrent neural network popular in dynamic systems applications. In this work, we are applying Elman's recurrent neural network [28]. This recurrent neural network consists of four layers such as input layer, hidden layer, context layer, and output layer. In this network every two adjacent layers are connected with adjustable weights. The functions of input later, hidden layer, and output layer of recurrent neural network are same like a back-propagation neural network, but exception occurs in context layer. The context layer is constructed with context nodes and feed forward mechanism. The context nodes are used to memorise some past state of the hidden nodes. The hidden nodes outputs are used as inputs for context nodes through feedback connections. Therefore, the output of the recurrent neural network depends on the aggregate of earlier states and present inputs.

Let us assume X as the input pattern, H as the hidden pattern, C be the context pattern, and O as the output pattern for a recurrent neural network. Assume O_i be the output for ith input pattern, O'_j be the output for jth hidden pattern, O'_c be the output of the context pattern, and O''_k be the output for kth output pattern. Assume $w1_{ji}$ is the weight in between jth hidden neuron and ith input neuron. Likewise, $w2_{kj}$ is the weight between kth output pattern and jth hidden pattern. Similarly, $w3_{jc}$ is the weight between jth hidden pattern and cth context pattern.

The working methodology of recurrent neural network architecture consists of four phases such as input phase, hidden phase, context phase, and output phase. The steps included in these phases for rth iteration is described below.

1. For input nodes, compute $O_i = x_i^{(r)}$ for $i, j = 1, 2, \ldots, k$
2. Compute the following for context nodes.

   In rth iteration, input to the context node is given by $c_i^{(r)} = O'^{(r-1)}_j$, where $c_i^{(r)}$ is the context node and $O'^{(r-1)}_j$ is the past state/past output of jth hidden node. Initially, for $r = 1$, $O'^{(r-1)}_j$ is 0 and $c_i^{(r)}$ is 1.
3. An activation function $f(\bullet)$ is applied in hidden layer for jth hidden node at rth iteration to compute the output of the hidden node. $f(\bullet)$ is defined by: $O'^{(r)}_j = f(g_j^r)$; where g_j^r is a linear output of jth hidden node at rth.
4. By using weight matrix, the output of the jth hidden node at rth iteration is computed as:

$$O'^{(r)}_j = f\left(\sum_{i=1}^{n} w1_{ij} O_i^{(r)} + \sum_{j=1}^{m} w3_{ij} O'^{(r-1)}_j\right). \tag{3}$$

5. By using weight matrix, the output of the output pattern is computed as:

$$O''^{(r)}_k = f\left(\sum_{j=1}^{m} w2_{jk} O'^{(r)}_j\right) \tag{4}$$

   In the learning process, the network modified the weights to minimize the error E. The modified new weight is presented by: $w_{\text{new}} = w_{\text{old}} + \eta \Delta w$. Here η and Δw represent learning rate and change of weight respectively. Change in weight at time t is computed as: $\Delta w(t) = -\eta\, \nabla E(t) + \alpha\, \Delta w(t-1)$, where η is the learning rate and α is a constant.
6. The error E is computed as:

$$E = \frac{1}{2} \sum_{r=1}^{q} \sum_{k=1}^{l} \left[o_k^{(r)} - O''^{(r)}_k \right]. \tag{5}$$

   Where, $o_k^{(r)}$ is the target output, $O''^{(r)}_k$ is the actual output, and q is length of the training sequence. For better understanding, a 3-input recurrent neural network design is presented in Fig. 2.

3 Proposed Research Design

In this section we explore the research design in detail. The research design has modelled with three segments such as data assemble and process unit, training unit, and testing unit. In data assemble and process unit, data are captured from electro discharge machining and processed through data cleaning method. Due to this

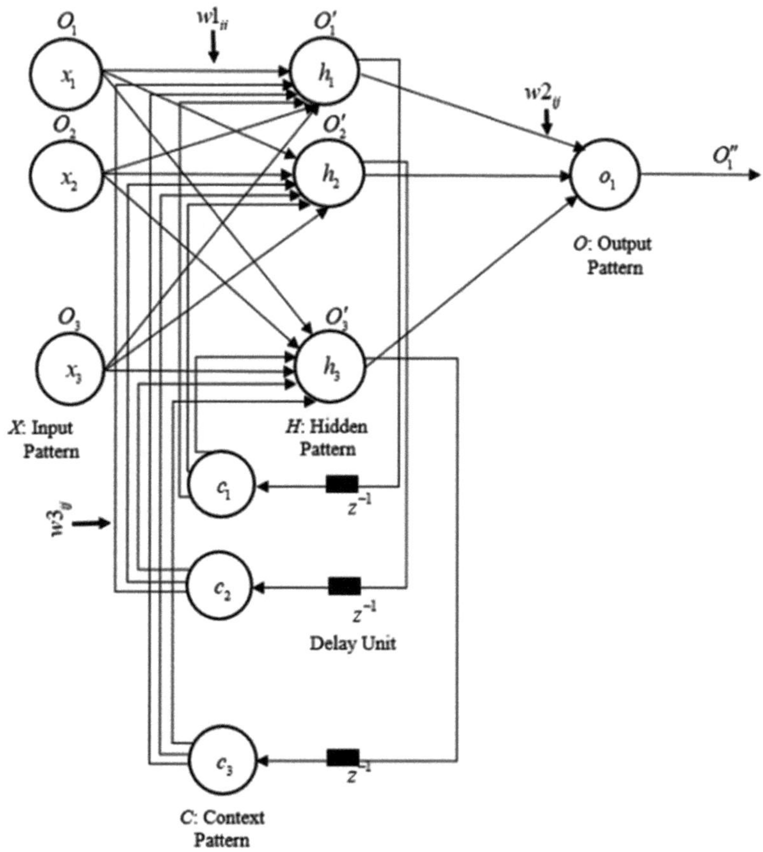

Fig. 2 A sample design of recurrent neural network

method the missing value objects, and identical objects are removed from the data set. Once data processing is finished, further, the processed data set is divided into two categories such as training data set, and testing data set. We consider 70% of processed data as training data, and 30% of processed data as testing data. The training unit consists of a 4-4-1 (4 input neurons in the input pattern, 4 hidden neurons in the hidden pattern, and 1 output neuron in the output pattern) fuzzy recurrent neural network with fuzzy graph approach to get the target output. Simultaneously, the error is reduced between target output and actual output by internal weights modification. Once the training process of the model is completed successfully, then it passed through the testing unit to find out accuracy of the model. The testing unit is analyzed with rest 30% of the processed data. Figure 3 represents the proposed research design.

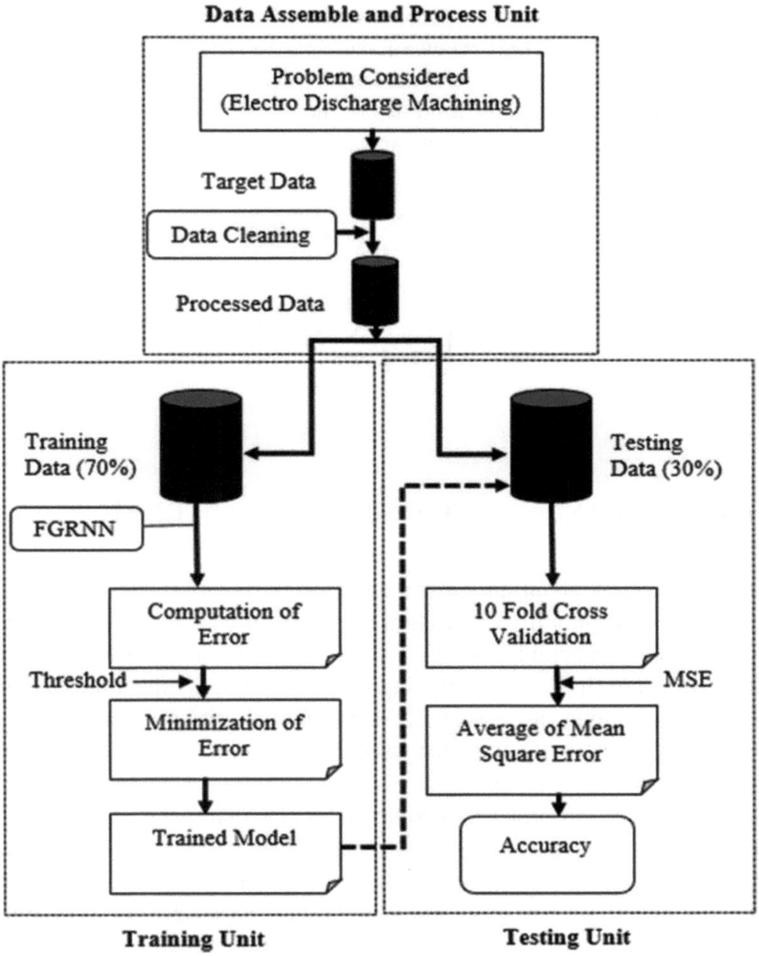

Fig. 3 Proposed research design

3.1 Research Methodology

The training unit uses fuzzy recurrent neural network to train the proposed model. The benefit of hybridized fuzzy graph with recurrent neural network is to find the strongest path of the network, and discard the weak path in the network for passing the inputs from input pattern to output pattern through hidden pattern and context pattern. Therefore, high impact paths are often used to train the model which enhance the efficiency of the model. In testing unit, it uses tenfold cross validation, and mean square error to find the accuracy of the proposed model. In tenfold cross validation, mean square error for each folder is calculated by using Eq. (6), where R_o denotes experimental radial overcut, $\hat{R}_o$ denotes predicted radial overcut, and n represents the total sample count. Further, the average of mean square errors for k folders is computed. We use this average mean square error to compute the accuracy of the proposed fuzzy graph recurrent neural network model for predicting radial over cut in EDM. Equation (7) computes the accuracy of the model, where k stands for number of folders, and MSE_j represents the mean square error for jth folder.

$$\mathrm{MSE} = \frac{1}{n}\sum_{i=1}^{n}\left(R_o - \hat{R}_o\right)^2 \quad (6)$$

$$\mathrm{Accuracy} = \left(1 - \frac{1}{k}\sum_{j=1}^{k}\mathrm{MSE}_j\right) \times 100 \quad (7)$$

4 Experimental Study on Radial Over Cut in EDM

This section explains about an experimental study on radial overcut executed in EDM. EDM is a non-traditional machining process used for machining very hard metal materials. For machining work in EDM, the work-piece has to be electrically conducted in nature. In EDM, machining of a work-piece is done through erosion process. The machining process of EDM depends on different parameters. Basically, there are two types of parameters used in machining process such as electrical parameters and nonelectrical parameters. Input current (I_p), Voltage (V), pulse on time (T_{on}), and pulse off time (T_{off}) come under electrical parameters. Flushing of dielectric fluid, and electrode rotational speed comes under nonelectrical parameters. However, it is difficult to measure the performance of the system by manual experiments. Hence, we use fuzzy graph recurrent neural network to optimize the process parameters, and measure the efficiency of the system.

Additionally, to analyse the model we have taken 500 experimental data from VIT, Vellore. For exemplary a sample data set with 15 objects is shown in Table 1. Furthermore, we divide the data randomly into two packets such as training data with 350 (70%) objects, and testing data with 150 (30%) objects. Table 1 represents the input parameters such as Input current (I_p), Voltage (V), pulse on time (T_{on}), and pulse off time (T_{off}). The measuring unit for the input parameters such as Input current (I_p), Voltage (V), pulse on time (T_{on}), and pulse off time (T_{off}) are ampere, volt, and seconds respectively. Radial overcut is the decision parameter for the EDM information system and measured in micrometre.

Table 1 Sample experimental data of radial overcut of the EDM

Objects	Input current (I_p)	Machine on time (T_{on})	Machine off time (T_{off})	Gap voltage (V)	Radial overcut (R_o)
o_1	15	280	6	60	0.21
o_2	16	280	6	60	0.21
o_3	7	320	4	40	0.25
o_4	13	340	6	60	0.24
o_5	14	340	6	60	0.23
o_6	15	340	6	60	0.22
o_7	16	100	6	50	0.22
o_8	12	150	4	40	0.27
o_9	14	100	6	60	0.25
o_{10}	13	400	4	40	0.27
o_{11}	13	100	6	40	0.24
o_{12}	13	200	6	40	0.25
o_{13}	16	100	4	60	0.22
o_{14}	13	300	6	40	0.21
o_{15}	13	500	4	60	0.23

5 Result Analysis

This section presents the result analysis of 500 experimental radial overcut data collected from VIT, Vellore, Tamilnadu, India. Prior to data analysis the sample experimental data is normalized by Eq. (8), where a_{ij} represents the attribute value of ith object o_i and jth parameter p_j. Once we get the results, this also de-normalize to actual form using Eq. (9). Table 2 represents the normalized data for the sample information system shown in Table 1.

$$a_{ij}^{\text{norm}} = \frac{a_{ij} - \text{Min}(a_j)}{\text{Max}(a_j) - \text{Min}(a_j)} \tag{8}$$

$$a_{ij} = \text{Min}(a_j) + a_{ij}^{\text{norm}} \times (\text{Max}(a_j) - \text{Min}(a_j)) \tag{9}$$

We conduct the experiments in a computer with following configurations: Windows 8 operating system, Intel core i5 processor, 8 GB RAM, and MATLAB R2018a. The model is trained by using 350 (70%) objects with four parameters. We use Levenberge-Marquardt back propagation training algorithm, and fuzzy graph recurrent neural network architecture to train the model. Once the training session is over, the testing phase is carried on 30% (150) objects, and the result is predicted for 150 objects. For illustration, the radial overcuts experimental values and predicted values are shown in Table 3 as a sample.

Table 2 Normalized sample experimental data of EDM of Table 1

Objects	Input current (I_p)	Machine on time (T_{on})	Machine off time (T_{off})	Gap voltage (V)	Radial overcut
o_1	0.9	0.38	0.43	0.6	0.45
o_2	1	0.38	0.43	0.6	0.44
o_3	0.1	0.46	0.14	0.2	0.59
o_4	0.7	0.50	0.43	0.6	0.56
o_5	0.8	0.50	0.43	0.6	0.51
o_6	0.9	0.50	0.43	0.6	0.48
o_7	1	0.04	0.43	0.4	0.48
o_8	0.6	0.13	0.14	0.2	0.66
o_9	0.8	0.04	0.43	0.6	0.62
o_{10}	0.7	0.62	0.14	0.2	0.67
o_{11}	0.7	0.04	0.43	0.2	0.55
o_{12}	0.7	0.23	0.43	0.2	0.59
o_{13}	1	0.04	0.14	0.6	0.48
o_{14}	0.7	0.42	0.43	0.2	0.44
o_{15}	0.7	0.81	0.14	0.6	0.52

Table 3 Testing analysis of radial overcut using FGRNN

Objects	Input current (I_p)	Machine on time (T_{on})	Machine off time (T_{off})	Gap voltage (V)	Experimental radial overcut	Predicted radial overcut
o_1	0.9	0.38	0.43	0.6	0.45	0.46
o_2	1	0.38	0.43	0.6	0.44	0.44
o_3	0.1	0.46	0.14	0.2	0.59	0.59
o_4	0.7	0.50	0.43	0.6	0.56	0.57
o_5	0.8	0.50	0.43	0.6	0.51	0.51
o_6	0.9	0.50	0.43	0.6	0.48	0.46
o_7	1	0.04	0.43	0.4	0.48	0.47
o_8	0.6	0.13	0.14	0.2	0.66	0.66
o_9	0.8	0.04	0.43	0.6	0.62	0.56
o_{10}	0.7	0.62	0.14	0.2	0.67	0.68
o_{11}	0.7	0.04	0.43	0.2	0.55	0.63
o_{12}	0.7	0.23	0.43	0.2	0.59	0.60
o_{13}	1	0.04	0.14	0.6	0.48	0.46
o_{14}	0.7	0.42	0.43	0.2	0.44	0.44
o_{15}	0.7	0.81	0.14	0.6	0.52	0.51

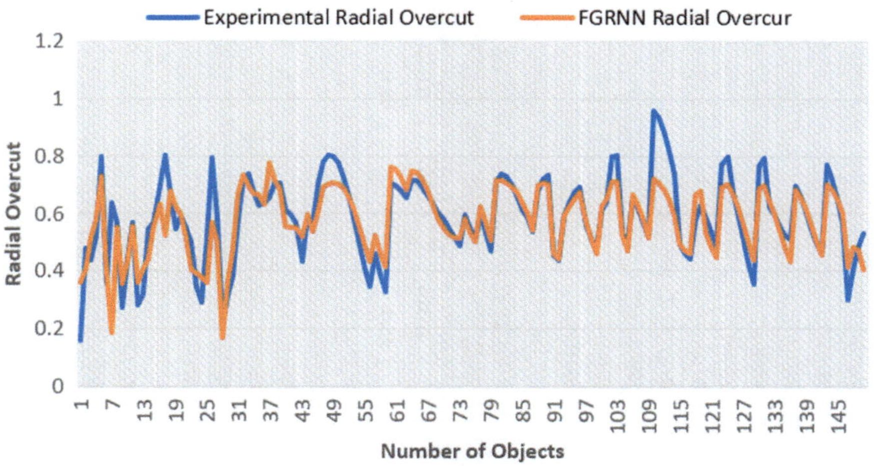

Fig. 4 Comparison between experimental and predicted radial overcut

Besides, Fig. 4 shows the experimental radial overcuts values and the corresponding predicted radial overcuts values because of FGRNN model on testing data. It reflects that, the prediction values are isomorphic to the experimental values.

5.1 K Fold Cross Validation

In this section we figure out the accuracy of the proposed model FGRNN by using K-fold cross validation. All the objects of the testing data set break into tenfolds randomly and every folder consists of 15 objects. We compute the MSE for every folder using Eq. (6). Further proceeding, we calculate the average of MSEs found for 10 folders. Finally, the accuracy of the FGRNN is calculated using Eq. (7). The MSE found for every folder and its corresponding accuracy is depicted in Table 4.

In Table 4 we have observed that, the accuracy of FGRNN model for predicting radial overcut in electro discharge machining is 96.01%.

6 Conclusion

Radial overcut plays a major role for designing a solid product. In electro discharge machining prediction of radial overcut is a challenging matter. Therefore, in this paper we have introduced fuzzy graph recurrent neural network model to predict the radial overcut in electro discharge machining. The result and analysis part of this paper shows 96.01% accuracy by FGRNN model across all the folders. The proposed FGRNN model finds the strongest path in the architecture by using fuzzy

Table 4 Accuracy of proposed model FGRNN using cross validation

Folders	1	2	3	4	5	6	7	8	9	10	Total
MSE	0.044	0.041	0.038	0.043	0.040	0.031	0.042	0.035	0.046	0.039	0.0440
Accuracy	95.6	95.9	96.2	95.7	96	96.9	95.8	96.5	95.4	96.1	96.01

AND-OR technique and updates the weights through Levenberge-Marquardt training algorithm. In this model, only the strongest paths are used to propagate the inputs from input layer to output layer through hidden layer, and the rest of the paths are ideal.

References

1. Gershwin SB (2000) Design and operation of manufacturing systems: the control-point policy. IIE Trans 32(10):891–906
2. Tangen S (2003) An overview of frequently used performance measures. Work Study 52(7):347–354
3. Slack N, Chambers S, Johnston R (2010) Operations management. Pearson education, London
4. Fenoughty KA, Jawaid A, Pashby IR (1994) Machining of advanced engineering materials using traditional and laser techniques. J Mater Process Technol 42(4):391–400
5. Davim JP (2013) Nontraditional machining processes. In: Manufacturing process selection handbook, pp 205–226
6. Abbas NM, Solomon DG, Bahari MF (2007) A review on current research trends in electrical discharge machining (EDM). Int J Mach Tools Manuf 47(7–8):1214–1228
7. Zhang JH, Lee TC, Lau WS, Ai X (1997) Spark erosion with ultrasonic frequency. J Mater Process Technol 68(1):83–88
8. Ming W, Ma J, Zhang Z, Huang H, Shen D, Zhang G, Huang Y (2016) Soft computing models and intelligent optimization system in electro-discharge machining of SiC/Al composites. Int J Adv Manuf Technol 87(1–4):201–217
9. Tso GK, Yau KK (2007) Predicting electricity energy consumption: a comparison of regression analysis, decision tree and neural networks. Energy 32(9):1761–1768
10. Roy AR, Maji PK (2007) A fuzzy soft set theoretic approach to decision making problems. J Comput Appl Math 203(2):412–418
11. Grzymala-Busse JW (1988) Knowledge acquisition under uncertainty—a rough set approach. J Intell Rob Syst 1(1):3–16
12. Wang X, Pardalos PM (2014) A survey of support vector machines with uncertainties. Ann Data Sci 1(3–4):293–309
13. Simoff SJ (1996) Handling uncertainty in neural networks: an interval approach. In: Proceedings of international conference on neural networks (ICNN'96), vol 1. IEEE, New York, pp 606–610
14. Wu J, Zheng C, Chien CC, Zheng L (2006) A comparative study of Monte Carlo simple genetic algorithm and noisy genetic algorithm for cost-effective sampling network design under uncertainty. Adv Water Resour 29(6):899–911
15. Pontes FJ, Ferreira JR, Silva MB, Paiva AP, Balestrassi PP (2010) Artificial neural networks for machining processes surface roughness modeling. Int J Adv Manuf Technol 49(9–12):879–902
16. Das R, Pradhan MK (2014) General regression neural network and back propagation neural network modeling for predicting radial overcut in EDM: A comparative study. World Acad Sci Eng Technol Int J Mech Aerosp Ind Mechatron Manuf Eng 8(4):799–805
17. Behrens A, Ginzel J (2003) Neuro-fuzzy process control system for sinking EDM. J Manuf Process 5(1):33–39
18. Rangajanardhaa G, Rao S (2009) Development of hybrid model and optimization of surface roughness in electric discharge machining using artificial neural networks and genetic algorithm. J Mater Process Technol 209(3):1512–1520
19. Deb K, Agrawal S, Pratap A, Meyarivan T (2000) A fast elitist non-dominated sorting genetic algorithm for multi-objective optimization: NSGA-II. In: International conference on parallel problem solving from nature. Springer, Berlin, pp 849–858

20. Panda DK (2010) Modelling and optimization of multiple process attributes of electro discharge machining process by using a new hybrid approach of neuro–grey modeling. Mater Manuf Process 25(6):450–461
21. Livingstone DJ, Manallack DT, Tetko IV (1997) Data modelling with neural networks: advantages and limitations. J Comput Aided Mol Des 11(2):135–142
22. Pradhan MK, Das R (2015) Application of a general regression neural network for predicting radial overcut in electrical discharge machining of AISI D2 tool steel. Int J Mach Mach Mater 17(3–4):355–369
23. Tsai KM, Wang PJ (2001) Comparisons of neural network models on material removal rate in electrical discharge machining. J Mater Process Technol 117(1–2):111–124
24. Zhang Y (2004) The research on the GA-based neuro-predictive control strategy for electric discharge machining process. In: Proceedings of 2004 international conference on machine learning and cybernetics (IEEE Cat. No. 04EX826), vol 2. IEEE, New York, pp 1065–1069
25. Kuriakose S, Shunmugam MS (2005) Multi-objective optimization of wire-electro discharge machining process by non-dominated sorting genetic algorithm. J Mater Process Technol 170(1–2):133–141
26. Rosenfeld A (1975) Fuzzy graphs. In: Fuzzy sets and their applications to cognitive and decision processes. Academic Press, Cambridge, pp 77–95
27. Mordeson JN, Chang-Shyh P (1994) Operations on fuzzy graphs. Inf Sci 79(3–4):159–170
28. Gao XZ, Ovaska SJ, Dote Y (2000) Motor fault detection using Elman neural network with genetic algorithm-aided training. In: SMC 2000 conference proceedings, vol 4. IEEE, New York, pp 2386–2392

AgentG: An Engaging Bot to Chat for E-Commerce Lovers

V. Srividya, B. K. Tripathy, Neha Akhtar, and Aditi Katiyar

Abstract Regular customer assistance chatbots are generally based on dialogues delivered by a human. It faces symbolic issues usually related to data scaling and privacy of one's information. In this paper, we present AgentG, an intelligent chatbot used for customer assistance. It is built using the deep neural network architecture. It clouts huge-scale and free publicly accessible e-commerce data. Different from the existing counterparts, AgentG takes a great data advantage from in-pages that contain product descriptions along with user-generated data content from these online e-commerce websites. It results in more efficient from a practical point of view as well as cost-effective while answering questions that are repetitive, which helps in providing a freedom to people who work as customer service in order to answer questions with highly accurate answers. We have demonstrated how AgentG acts as an additional extension to the actual stream Web browsers and how it is useful to users in having a better experience who are doing online shopping.

Keywords E-commerce · NLTK · Chatbot · Keras · Deep neural network

---

V. Srividya · N. Akhtar · A. Katiyar
School of Computer Science and Engineering (SCOPE), Vellore Institute of Technology, Vellore, Tamil Nadu 632014, India
e-mail: srividya.v2018@vitstudent.ac.in

N. Akhtar
e-mail: neha.akhtar2018@vitstudent.ac.in

A. Katiyar
e-mail: aditi.katiyar2018@vitstudent.ac.in

B. K. Tripathy (✉)
School of Information Technology and Engineering (SITE), Vellore Institute of Technology, Vellore, Tamil Nadu 632014, India
e-mail: tripathybk@vit.ac.in

A. K. Tripathy et al. (eds.), *Advances in Distributed Computing and Machine Learning*, Lecture Notes in Networks and Systems 127, https://doi.org/10.1007/978-981-15-4218-3_27

1 Introduction

The service provided to the customer by the customer service department plays a vital role in producing profits for an organization. A company with this division has to spend a lot over the resources which may be billions in order to change the customer's perception that he holds. People in customer service department have to spend most of their time in answering the questions asked by customers through telephone or applications like messaging so that the questions posed by the customers are satisfied. This outdated technique that was earlier used for customer service mainly suffers from two issues: firstly, many questions that are asked to the staff members usually are of repetitive nature, and these questions can be economically responded by using machines. Secondly, to provide the round, the clock facilities are a cumbersome task, generally for the businesses that are mostly non-global. Consequently, these virtual assistants will be of great importance and replace the customer service personnel, as they are more cost-effective and time saving. Of late, virtual assistance provided as a substitute for the service provided to customers is rapidly getting popular with businesses that are customer-oriented. The basic building block of these bots is the conversations among humans that occurred earlier in the past. These are upfront but involve the two problems, namely data scalability and the privacy of those customers' conversations. Getting in touch with a support staff who works in the customer service division to answer their queries involves a lot of waiting time. This mechanism is not very effective, and also it involves scalability issues [1–6]. Another important aspect that needs to be considered is that the privacy of the customer's conversations is at stake. These discussions are not permissible to be used as data for training. The solution to the above problem of training is finding easily available and large amounts of data related to serving the customers. These data act as a basic block to build the bot that serves as the helping agent. In this paper, we create AgentG, an influential service provided to customers which is a chatbot that manages such extensive and freely accessible data on all the e-shopping websites. There exist large e-commerce websites that showcase a great variety of product descriptions and also content that is user-generated. Some of the shopping webpages are Amazon.com, Flipkart.com and Snapdeal.com. The above existing data is extracted and provided as an input to the bot being constructed, namely AgentG. This virtual assistant helps in providing better services to the customers while shopping online along with the human staff. The extracted data is stored in a JSON file that is processed using the Natural Language Toolkit (NLTK) package available in Python. This processed data is used as training data to the deep learning model. The NLTK package is used for natural language processing (NLP).

2 Literature Survey

In [7], they demonstrate a chatbot name SuperAgent, and it is a useful chatting assistant that helps in providing service to customers. It is extremely beneficial and available widely for data on e-shopping. It can easily pull crowdsourcing styles, which contain a collection of hi-tech NLP and also merge into e-commerce websites as an additional feature extension.

In [8], they talk about a combination of artificial conversation systems along with an e-commerce site that will give unrestricted services for chatting. When user will get into the e-commerce website firstly, he can enquire about e-commerce to know the system particularly. E-commerce system sends the query of the customer to the Artificial Intelligence Markup Language (AIML) Knowledge Base System in order to get answers just by applying the algorithm for pattern matching.

In [9], it is said that Chappie is being used as a routing agent wherein it can identify the needs of user on the basis of first few chats into one of the services provided by business and then send it to an agent who has a good knowledge in that service. It examines the chats and extracts important content of the user that is similar to the likes of wit.ai website (WIT) with the help of (NLP). Then, it uses this useful information and AIML to start talking to the user.

In [10], they discuss the end result of two cases that have been studied in a large international corporation in order to test the utilization of chatbots for security in training the internal aspect of the customer data. However, it appears to be the data that is qualitative in nature which suggests that the attitude of the customers appears to be highly positive towards security while using chatbots rather than with the existing traditional e-learning courses for security training at the company.

In [11], they propose that the system helps the user to communicate with a community that consists of chatbots, having specific properties in order to go through many concepts that are mechanically produced with the help of Latent Semantic Analysis (LSA) paradigm. The knowledge of chatbots is created in order to deal with the field of cultural heritage, which is then coded into semantic space that is further created with the help of LSA, making them useful to calculate their own accuracy formulated by the customer, which is later mapped into the same semantic space.

In [12], the authors stated that with the advancement of massive open online courses (MOOC) providers, like "edX, Coursera, FutureLearn, or MOOC.ro", it is indeed problematic to find the best resources for learning. MOOCBuddy is a recommendation system for MOOC which works as a chatbot for the application named Facebook Messenger, generally based on social media profile of the user along with their interests, which can be provided as solution.

In [13], they say that the database is used to store the knowledge of the chatbot, which is accessing the core in a deeper relational database management system (RDBMS). The storage of knowledge here is the database, and interpreter here works as a function stored program and produces sets of required pattern matching. By using programing language of Pascal or Java, the interface has been built.

In [14], the authors propose the first chatbot in Bengali that is named as Golpo totally based on a feature of being language-independent and uses the natural language processing library with a process of making the machine learn. The experiment performed in this project has shown that the chatbot can give responses to the customers in real-time world. Based on the calculation of customers, Golpo can generate syntactically natural and correct Bengali responses.

In [15], the authors provide the detailed knowledge about the development and design of an intelligent and highly accurate chatbot system based on voice recognition. This paper presents a technology and a method to demonstrate and to verify a framework that has been proposed and require providing support to such a bot (a Web service).

In [16], the authors conducted a survey on various chatbots and their techniques. According to the paper, research suggested that 75% of the customers have had poor customer service experience as the chatbots have not been able to respond to all queries of the customers. It also gives a comparison among all the existing chatbots and the techniques used to build them. One of the major problems related to chatbots that result in poor performance is their inability to generate long, meaningful responses.

In [17], authors have provided an overview on the technologies that are the driving force of chatbots such as information retrieval and deep learning. They also offer some insights into difference between conversational chatbots and transactional chatbots. Conversational bots are trained on general chat logs, whereas the transactional bots are trained for a specific purpose such as any ticket booking service.

In [18], the proposed method is for chatbots to be built using deep learning technologies, which is a new area of machine learning. In deep learning, every algorithm applies a nonlinear approach on the input to learn statistical information from it. The statistical information is then used to obtain the output. For large datasets, the data is split into training and testing set. This process of applying nonlinear approach to the input is repeated until an acceptable accuracy, precision, recall or f1-measure is obtained.

In [19], the proposed system is a chatbot to interact with virtual patients so that the doctors can complete the clinical assessment of the patient with ease. This also leads to significant logistical savings. A deep learning framework is developed to improve the virtual patient's conversational skills based on the domain-specific embedding, and then a long short-term memory (LSTM) network is used to derive sentence embeddings before a convolutional neural network model selects an answer from script to a specific query. Accuracy of the system is around 81%.

3 Background Study

3.1 Json File

The input file introduced here is a JSON file which stores objects and data structures that are simple in JavaScript Object Notation (JSON) format, which is a standard data interchange format. It is mostly used for data transmission between a server and a Web application.

3.2 Natural Language Toolkit (NLTK)

NLTK is one of the most useable platforms for writing Python programs in order to work along with data that is generally human language. It provides us with easy-to-use interfaces for WordNet, along with a set of libraries for text processing like classification, stemming, tokenization, etc.

3.3 Tokenization of Words

There exist a pre-trained Punkt tokenizer in the NLTK data package generally for English in order to perform a reduction of words to their corresponding stem or root form — mostly a word's written form and lemmatization along with removal of noise, stop words.

3.4 Bag of Words

As we have gone through and completed the phase of text pre-processing, now we have changed the given text into an array or a vector consisting of meaningful and logical numbers. A bag is a collection of elements where multiple occurrences of elements are allowed [6, 20]. One of the most common ways of text-based representation is bag of words that gives the deeper insight into the appearances of the words in the document. It generally includes the two of the things: a collection of words that are known and a count of known words those are present. Here, we are only considering if the words that we know are present in the given document, and if they are there then what is their position of occurrence.

3.5 Deep Learning

The deep learning paradigm uses the concept of multiple layers of nodes in order to extract features from raw input at higher levels. This is done in a progressive manner by the neural network.

A feedforward network is also known as deep feedforward network, or multilayer perceptron (MLP). In general, the process of approximation of some function $f*$ can be said to be the aim of a feedforward network. For instance, consider the case of a classifier whose goal is to map the given input "x" to the category or class label "y" represented by the equation $y = f*(x)$. In terms of a MLP, the mapping is defined using a parameter θ as $y = f(x, \theta)$ whose value is learnt by choosing the best method of approximation.

4 Methodology

According to the statistics on the usage of chatbots, more than 67% of the clients throughout the world use a chatbot for customer support. E-commerce being one of the major customer-based businesses also has started providing chatbots for customer service. In this work, an interactive chatting machine AgentG is created that intelligently answers the queries related to e-commerce websites or products.

In this section, more information about the dataset, the proposed model, pre-processing steps and the libraries that are used is provided.

4.1 Dataset

The project needs a corpus that is fed into a model for training purpose. The dataset described below is a JSON file named intents.json that contains these terms like tags, patterns and responses. These tags act like the class or target for the deep learning module. The model predicts the input to belong to any one of these categories. Table 1 shows the data file.

4.2 Building the Bot-Proposed System

The proposed system is described as follows (Fig. 1):

Step 1 Creating a corpus file. A JSON file consisting of tags, patterns and responses is created for numerous topics like greetings, thank you, mobiles, laptops, Flipkart, Amazon, payments, return policy and goodbye.

Table 1 Description of the dataset

Tag	No. of patterns	No. of responses	Example
Greetings	5	3	Hi, hello
Good bye	3	3	Bye, goodbye
Thanks	3	3	Thank you, thanks
Flipkart	3	3	Big billion days, Flipkart fashion latest trend
Amazon	3	2	Great Indian Festival, freedom sale
Mobiles	3	3	Price of Oppo phone
Computers	4	3	Best desktop or notebook
Payments	3	2	Pay by cash only
Return policy	2	2	Exchange policy on a product

Fig. 1 Sequence of steps of the proposed system

Step 2 The next step is to extract the words from patterns in the JSON file using the word tokenizer. The tags in the JSON file are stored into labels.

Step 3 A bag of words is created for each word in the pattern based on the vocabulary.

Step 4 The deep neural network is then trained on these pairs of patterns to responses.

Step 5 Predicting the answers for any queries asked by the customer.

In this work, we are using the Keras library and NLTK. NLP is the area that commonly emphasizes on the common interactions amid computers and human language that is largely known as natural language processing. It is an intersection of computer science, linguistics computation and artificial intelligence system. It is a kind of way for computers to go through and understand, then analyse and finally derive proper meaning from language that can be useful. By using NLP, we can structure knowledge in order to perform automatic summarization, named entity recognition, relationship extraction, etc.

4.3 Corpus

In this work, we have mostly used a JSON file that consists of tags, patterns and responses. This JSON file contains topics related to queries such as "hi", "hello", Flipkart-related queries such as big billion days or Amazon's Great Indian festival days. Mobiles and laptops are also added as topics in the corpus. The JSON file is called intents.json. This file is read, and words, patterns and labels are extracted. This pre-processing is done by word tokenizer and Lancaster Stemmer.

We make use of random function in order to get more favourable responses from the available responses under a particular label or tag as in the JSON file. Since our model is trained by taking data from the NLTK package, any query that does not match the words in the JSON file is also answered. This feature makes the chatbot more reliable.

4.4 Keras Library and Proposed Architecture

Keras is one of the libraries used in this work. It is a high-level neural networks API that is used to implement deep learning models. In this work, a fully connected neural network with three hidden layers with eight units each and activation function of Relu is created. The model is compiled using the "rmsprop" optimizer with mean square error loss. Accuracy is used as a measure of metric. The final layer has an activation function of softmax (Fig. 2).

The activation functions used are Rectified Linear Unit (ReLu) and Softmax. The equation of ReLu can be defined as follows:

$$f(x) = \max(0, x)$$

The softmax is an activation function that normalizes an input vector of n real numbers into a probability distribution comprising the n probabilities that are proportional to the exponents of the input numbers. The equation is given by

$$S(y_i) = \frac{e^{y_i}}{\sum_j e^{y_j}}$$

5 Experiment

The experiment part of this paper discusses about the working of the model of chatbot. Figure 3 shows the stages in the workflow of the proposed model. The stages are as follows.

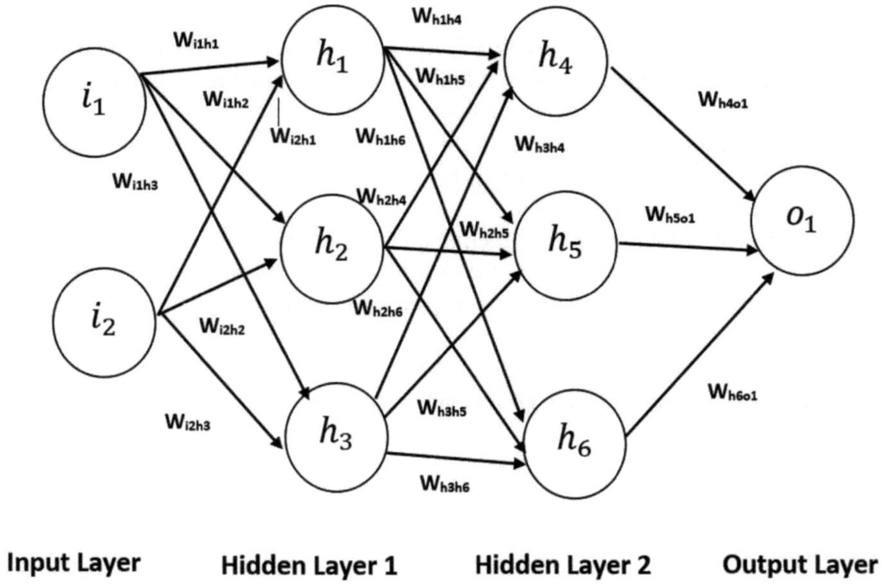

Fig. 2 Feedforward network architecture used in the work

Fig. 3 Workflow of the proposed model

Collect input from the user. In this phase, the customer or the user starts to chat with AgentG. The sentence entered by the customer is stored in an input variable. The sentence is tokenized, and root words are extracted from them. These words are then used to build a bag of words model. Once the bag of words is obtained, it is fed to the deep neural networks predict function to get a prediction for the entered input. From this prediction, the most probable tag or class of the input is found. A random response is chosen from the responses under the chosen tag. This random response is given as the reply from the chatbot AgentG.

6 Results and Discussion

The figures below show the conversation with the virtual customer service agent "AgentG". A series of questions was asked to AgentG like; which is the best camera with cost being less than Rs. 10,000? The response in this case was "Samsung galaxy

M30”. If queries related to price or exchange policy were asked, the response was “no exchange or 30 day free return time”. AgentG has a special feature that makes it capable of answering queries, which it had not faced but are based upon queries put at it in the past. The model has an accuracy of around 80%.

Figure 4 shows the conversation with AgentG where the customer greets the chatbot in different ways. The chatbot’s response was greeting the user back. These queries were existing in the corpus.

Figure 5 deals with the unseen greeting queries and the chatbot’s response to it.

Figure 6 displays the response of the chatbot when the user asks about Flipkart’s fashion, Amazon’s Great Indian festival, gaming laptops, price of a mobile phone and so on.

Figure 7 displays the response of the chatbot when the user asks about payment mechanisms, return or exchange policy and desktops.

```
Hi there I am AgentG !! How may I help you? (type quit to stop)!

You: Hi
Tag: greeting
Hi there, how can I help?

You: Hello
Tag: greeting
Good to see you again

You: How are you?
Tag: greeting
Good to see you again

You: Is anyone there?
Tag: greeting
Good to see you again

You: Good day
Tag: greeting
Good to see you again
```

Fig. 4 Greetings tag conversation with AgentG (existing queries)

```
You: hello there
Tag: greeting
Good to see you again

You: hi there
Tag: greeting
Hi there, how can I help?
```

Fig. 5 Greetings tag conversation with AgentG (unseen queries)

```
Hi there I am AgentG !! How may I help you? (type quit to stop)!

You: Hello
Hello, thanks for visiting

You: whatsup?
Have a nice day

You: New fashion in Flipkart this week?
Aliya's Dresses and Ranbir's jackets are the latest flipkart fashion trends

You: Best gaming laptops?
DELL Xps 13, HP Envy 13

You: price of oppo reno 2
Yes we have a 30 days free return and exchange policy

You: How much does an oppo reno2 cost?
It would be around Rs. 37,000
```

Fig. 6 General conversation about Flipkart and Amazon's sale

Fig. 7 Conversation about exchange policy or computers

```
You: Is there a return policy for food?
No returns or Exchanges on food products

You: When is the Great indian festival?
 Next sale is After November 14

You: Do the portal accept only cash?
We accept VISA, Mastercard and AMEX

You: Which desktop is best suited for home use?
Alienware Are 51M, ASUS TUF Gaming laptops

You: Thank you
My pleasure

You: Bye
Bye! Come back again soon.
```

7 Conclusion

We have come up with AgentG, a customer service chatbot that can be used for websites that allows online shopping. As we compare the traditional way of servicing the customer, we infer that AgentG has an advantage of being huge-scale, freely available and collaboratively collected data about the customers. Also, AgentG uses the up-to-date NLP and machine learning techniques. Based on the analysis of its usage, the outcome obtained was that AgentG was involved in helping the endwise user experience with respect to online shopping. It is an extremely convenient method to acquire information regarding the customers, specifically if the content generated by the user involves the data from the product page.

References

1. Raghuveer VR, Tripathy BK (2013) An object oriented approach to improve the precision of learning object retrieval in a self-learning environment. Int J E-learning Learning Objects 8:193–214
2. Raghuveer VR, Tripathy BK (2014) Reinforcement learning approach towards effective content. In: Proceedings of the 2nd IEEE international conference on MOOC, innovation and technology in education (MITE)
3. Raghuveer VR, Tripathy BK (2014) Multi-dimensional analysis of learning experiences over the e-learning environment for effective retrieval of Los. In: 6th IEEE international conference on technology for education (T4E), pp 68–171
4. Raghuveer VR, Tripathy BK (2015) On demand analysis of learning experiences for adaptive content retrieval in an eLearning environment. J e-Learning Knowl Soc Italy 11(1):139–156
5. Raghuveer VR, Tripathy BK (2016) Affinity-based learning object retrieval in an e-learning environment using evolutionary learner profile. Knowl Manage E-Learning 8(1):182–199
6. Jena SP, Ghosh SK, Tripathy BK (2001) On the theory of bags and lists. Inf Sci 132(1–4):241–254
7. Cui L, Huang S, Wei F, Tan C, Duan C, Zhou M (2017) Superagent: a customer service chatbot for e-commerce websites. In: Proceedings of ACL 2017, system demonstrations, pp 97–102
8. Satu MS, Akhund TMNU, Yousuf MA Online shopping management system with customer multi-language supported query handling AIML chatbot
9. Behera B (2016) Chappie-a semi-automatic intelligent chatbot. Write-Up
10. Kowalski S, Pavlovska K, Goldstein M (2009) Two case studies in using chatbots for security training. In: IFIP world conference on information security education. Springer, Berlin, Heidelberg, pp 265–272
11. Pilato G, Vassallo G, Augello A, Vasile M, Gaglio S (2005) Expert chat-bots for cultural heritage. Intelligenza Artificiale 2(2):25–31
12. Holotescu C (2016) MOOCBuddy: a chatbot for personalized learning with MOOCs. In: RoCHI, pp 91–94
13. Setiaji B, Wibowo FW (2016) Chatbot using a knowledge in database: human-to-machine conversation modeling. In: 2016 7th international conference on intelligent systems, modelling and simulation (ISMS), IEEE, pp 72–77
14. Orin TD (2017) Implementation of a Bangla chatbot. Doctoral dissertation, BRAC University; In: Li X, Liu H (2018) Greedy optimization for K-means-based consensus clustering. Tsinghua Sci Technol 23(2): 184–194
15. du Preez SJ, Lall M, Sinha S (2009) An intelligent web-based voice chat bot. In: IEEE EUROCON, IEEE, pp 386–391
16. Nuruzzaman M, Hussain OK (2018) A survey on chatbot implementation in customer service industry through deep neural networks. In: 2018 IEEE 15th international conference on e-business engineering (ICEBE), IEEE, pp 54–61
17. Hristidis V (2018) Chatbot technologies and challenges. In: 2018 first international conference on artificial intelligence for industries (AI4I), IEEE, pp 126–126
18. Kumar P, Sharma M, Rawat S, Choudhury T (2018) Designing and developing a chatbot using machine learning. In: 2018 international conference on system modeling and advancement in research trends (SMART), IEEE, pp 87–91
19. El Zini J, Rizk Y, Awad M, Antoun J (2019) Towards a deep learning question-answering specialized chatbot for objective structured clinical examinations. In: 2019 international joint conference on neural networks (IJCNN), IEEE, pp 1–9
20. Jena SP, Ghosh SK, Tripathy BK (2001) On the theory of fuzzy bags and fuzzy lists. J Fuzzy Math 9(4):1209–1220

An Improved Approach to Group Decision-Making Using Intuitionistic Fuzzy Soft Set

R. K. Mohanty and B. K. Tripathy

Abstract One of the latest models to handle uncertainty is soft set. It associates each parameter in its set of parameters to a subset of elements in the universe of discourse. In its introductory article itself, Molodtsov has shown the applications of this new tool in many different areas. But, over the years, soft sets have been used by many researchers for decision-making applications over uncertainty-based datasets. It has been observed that appropriate hybridization of models is always better than the route ones. Following this, many hybrid models of soft set have already been added to the literature by different researchers. So far, multiple attempts have been made to use intuitionistic fuzzy soft set (IFSS) for decision-making. This paper introduces the concept of interval-valued fuzzy (IVF) priority for decision-making using IFSS. It also provides an improved approach to deal with group decision-making (GDM) using IFSS and an application of IFSS in GDM.

Keywords Soft sets · Fuzzy sets · Fuzzy soft sets · IFSS · GDM

1 Introduction

Fuzzy set [1] is one of the best mathematical approaches to handle uncertainty and vagueness in data. In modern day computing, fuzzy set is indispensable for any domain of computer science. Atanassov [2] generalized fuzzy set and introduced intuitionistic fuzzy set (IFS) as a new model to handle uncertainty in 1986. Unlike fuzzy set, an IFS membership and non-membership are not necessarily one's complement of each other. So, the concept of hesitation came into the picture which is true in most real-life scenarios. If the hesitation becomes 0 (zero), the IFS reduces to Zadeh's fuzzy set. Molodtsov introduced the notion of soft sets in 1999 by bringing in topological flavour to set theory. It associates each parameter to a subset of elements

R. K. Mohanty · B. K. Tripathy (✉)
VIT, Vellore, Tamil Nadu, India
e-mail: tripathybk@vit.ac.in

R. K. Mohanty
e-mail: rknmohanty@gmail.com

A. K. Tripathy et al. (eds.), *Advances in Distributed Computing and Machine Learning*, Lecture Notes in Networks and Systems 127, https://doi.org/10.1007/978-981-15-4218-3_28

in the universe of discourse. Afterwards, many new operations have been defined by Maji et al. [3–5], Tripathy et al. [6]. It has been observed that suitable hybrid models have been found to be more efficient than their individual components. Maji et al. enhanced the boundary of soft set by introducing the notion of fuzzy soft set [3], the first hybrid model of soft set. Xu et al. [7] took forward the trend and introduced IFSS by fusing the notion of IFS and soft set. Tripathy et al. redefined soft set using characteristic function approach [6]. This approach is easier to understand and manipulate. Using the characteristic function approach, soft set operations like union, intersection and complement are redefined to make those more accurate and meaningful. Using the same approach, Tripathy et al. [6, 8–17] have defined many other hybrid models of soft set. Applications of these models are also provided in these papers. The applications are mostly based on different kinds of decision-making problems [18–23]. Tripathy et al. redefined fuzzy soft set (FSS) to systemize many operations as given in [23]. Similarly, membership function for IFSS is defined in [11]. Tripathy et al. in [23] identified many issues in [3] and handled well while introducing a new algorithm for decision-making.

The major contribution of this paper is in introducing an improved algorithm for GDM which uses IFSS, and the suitability of the algorithm is illustrated through an application. The proposed algorithm will also work fine for DM or GDM using FSS or IFSS. This paper has two more major sections in it. The following section highlights necessary definitions and notations. The next section showcases the improved approach to GDM using IFSS and an illustrative example.

2 Definitions and Notations

We denote the set of objects 'x' under consideration by C and the parameters 'e' associated with them by P. By a soft universe, we understand the pair (C, P). The power set of C and fuzzy power set of C are denoted by Pow(C) and FPow(C), respectively. Let $D \subseteq P$.

Definition 2.1 A soft set (F, D) over (C, P) is such that

$$F: D \to \text{Pow}(C) \tag{2.1}$$

Definition 2.2 A FSS (F, D) over (C, P) is such that

$$F: D \to \text{FPow}(C) \tag{2.2}$$

An IFS A over C is a pair (m_A, n_A), where m_A: $C \to [0, 1]$ and n_A: $C \to [0, 1]$, called the membership and non-membership functions of A, respectively, are such that for any $x \in C, 0 \leq m_A(x) + n_A(x) \leq 1$.

The function given by $\pi_A(x) = 1 - m_A(x) - n_A(x)$ is called the hesitation function associated with A. Let IFPow(C) be the power set associated with C.

Definition 2.3 An IFSS (F, D) over (C, P) is such that

$$F: A \rightarrow \text{IFPow}(C). \tag{2.3}$$

3 Application of IFSS in GDM

In the introductory article on soft set [24], Molodtsov mentioned that soft set can have applications in many different domains. Recently, soft set and its several hybrid models used mostly for decision-making or GDM problems. There are few articles using IFSS in group decision-making. This paper provides an improved approach for GDM using IFSS by taking priorities of parameters as IVF priority.

The application is about the ordering the candidates as per their performances before an interview panel. This is a group decision-making application as the panel consists of more than one decision-maker (three in the following case). Each of the panel members (judges) takes decision independently. The final result can be computed through the proposed algorithm using the results given by all the judges. All the judges have to use a set of predefined parameters to evaluate the candidates in different aspects like Subject Knowledge, Communication, Experience, Gaps, Reaction, etc.

As Tripathy et al. mentioned in [15] that parameters can be categorized into two types.

(a) The value of a positive parameter is directly proportional to the quality. For example, Experience is a positive parameter because more experience will enhance the preferability of the candidate.
(b) The value of a negative parameter is inversely proportional to the quality. For example, 'Breaks in study' is a negative parameter because increase of 'Breaks in study' will reduce the preferability of the candidate.

In decision-making approaches, an ignored feature is the prioritization of parameters as per the user's requirement. According to the user, if a parameter is more important than other parameters, then it should get that due importance during the decision-making also. To make the priority values more natural, this paper introduces IVF priority where priority of a parameter is expressed as an IVF value.

At first, the priorities for the parameters are to be fixed by the panel. Priority value is a real number lying in the interval $[-1, 1]$. For positive parameters, the priority value lies in $[0, 1]$, and the priority value lies in $[-1, 0]$ for negative parameters. If a parameter value does not have any effect on the decision, then the priority can be assumed as 0 (zero). The parameters which are not having any affect in decision-making can be opted out of the computation to reduce the computation cost. There are many articles in literature on parameter reduction in soft set [16].

To normalize each value from a set of values, Eq. 3.1 can be used.

$$\text{Normalized value } x_i' = x_i \Bigg/ \sum_{j=1}^{n} x_j, \quad i = 1, 2, \ldots, n \tag{3.1}$$

where x_i' is the normalized value of x_i with respect to the set of values $\{x_1, x_2, x_3, \ldots, x_n\}$.

It is very difficult to make decision or to evaluate to choose from a group of competitors from intuitionistic fuzzy values. Equation 3.2 reduces an intuitionistic fuzzy score to a fuzzy membership score.

$$\text{Score} = \begin{cases} -m(1+h), & \text{if } m < 0 \text{ and } h < -1; \\ m(1+h), & \text{Otherwise.} \end{cases} \tag{3.2}$$

where $m \to$ membership score and $h \to$ hesitation score.

Note that membership and non-membership scores can be: (i) more than 1 as the value is not exactly a fuzzy value but a summation of multiple fuzzy values; (ii) negative due to higher negative priority for negative parameter.

Most of the times, it is very difficult to give the exact priority to a parameter. So, using IVF values for priorities will provide more natural and flexible way to assign priorities to parameters. Atanassov [25] established that the IFS and IVFS are equipollent and also mentioned the way to convert one to the other model.

The following formula can be used to compute the normalized score from the ranks given by all the judges.

$$\text{Normalized Score} = \frac{\sum_{m=1}^{|J|} (N - R_m)^2}{|J| \times N^2}. \tag{3.3}$$

where $N \to$ Number of candidates, $|J| \to$ Number of judges and $R_m \to$ Rank obtained from mth judge.

The major change in the decision-making approach in this paper is the use of IVF priority instead of fuzzy priority. Now, instead of multiplying the priority value of a particular parameter to the corresponding membership, non-membership and hesitation values, the lower limit, (1-Upper limit) and (Upper limit–Lower limit) of the interval are multiplied to the corresponding membership, non-membership and hesitation values, respectively, to get the values for the priority table. In case of GDM, it is a common approach to add all scores obtained from different judges or to add the ranks obtained by all the judges to get the final result. But, in many case, the addition of ranks give same result. For example, rank sum of a candidate obtained ranks as 1, 1, 4 is same as the rank sum of 2, 2, 2. Though the rank sum is same, but, it is clear that the first candidate is the top performer in two cases, whereas second candidate is never a top performer. These type conflicts can be avoided by using normalized score formula given in Eq. 3.3.

3.1 Algorithm

Step-1 Initialize the IVF priorities of the parameters and obtain the expertise levels of the individual judges in each of the parameters as expressed by them.

Step-2 The parameter data table is constructed such that the rows in order contain the names of the parameters, their IVF priorities and priority-based ranks of the parameters followed by the level of expertise of the judges in order under each parameter.

Step-3 For $i = 1, 2, \ldots n$, repeat the following steps.

a. Input the IFSS (U, E) corresponding to judge J_i.
b. Multiply the lower limit, (1-upper limit) and (upper limit—lower limit) of the intervals for priority values with the corresponding membership, non-membership and the hesitation values, respectively, to generate the modified membership, non-membership and hesitation values and construct the priority table.
c. Three new columns are generated which contain the sum of the respective values of the parameters (membership, non-membership and hesitation) row-wise.
d. The comparison table of size $n \times n$ is constructed, where each column has three sub-columns corresponding to m, n and h values. The entries in each sub-column is computed by taking $x_i - x_j$ values for rows $i = 1, 2, \ldots n$ and columns $j = 1, 2, \ldots n$; where x represents m, n or h. Three new columns are generated which contain the sum of the respective values of the parameters (membership, non-membership and hesitation) row-wise.
e. Take the normalized values from three of the newly generated columns in the previous step using Eq. 3.1. Compute the scores for each row using Eq. 3.2 and assign rank to each row as per the obtained score.
    i. In case of ties, provide the higher rank to the one who has higher score under a higher-ranked parameter.

Step-4 Use (3.3) to build the rows of the rank table. In case of conflict, the higher score by a higher-experienced judge in a higher-ranked parameter will get higher rank.

Table 1 Parameter data table

Parameters	e_1	e_2	e_3	e_4	e_5	e_6
Priority	[0.7, 0.9]	[0.2, 0.3]	[0.3, 0.35]	[−0.5, −0.8]	[−0.4, −0.5]	0
Parameter rank	1	5	4	2	3	6
Expertise of J_1	0.9	0.8	0.7	1	0.8	0
Expertise of J_2	1	0.7	0.8	0.9	0.9	1
Expertise of J_3	0.7	0.9	0.8	0.7	1	0

3.2 Application

Let us assume that a group of six candidates $U = \{c_1, c_2, c_3, c_4, c_5, c_6\}$ have been shortlisted for an interview, and the associated parameter set is $E = \{e_1 =$ Subject Knowledge, $e_2 =$ Communication skill, $e_3 =$ Experience, $e_4 =$ Gaps, $e_5 =$ Reaction, $e_6 =$ Knowledge in Local Language$\}$. Suppose there are a total of three judges in the panel of evaluators.

The panel members provide their own expertise level in each of the parameters, and the panel as a team assigns priority (IVF) values to each of the parameters as well as rank the parameters in the order of their absolute mean value of the given priority intervals. Table 1 is the parameter data table for further computations.

The performance of all the candidates as evaluated by each judge can be expressed as an IFSS as given in Tables 2, 3 and 4. Notice that the column for e_6 is omitted as its priority is zero.

The priority tables for each judge are constructed by following the procedure mentioned in the algorithm step-3(c) from Tables 2, 3 and 4. Due to space crunch, priority table and comparison table are given only for judge J_1 in Tables 5 and 6, respectively. For other judges, the priority tables and comparison tables can be constructed similarly.

Comparison Table (6 for J_1) and decision tables for each judge are constructed by following the procedure mentioned in the algorithm (Tables 7, 8 and 9).

Construct the rank table using the ranks obtained for each candidate by all the judges as given in Table 10. Compute normalized score using the formula mentioned in Eq. 3.3. In case of conflict (same normalized score obtained by multiple candidates), this can be resolved by the approach provided in the algorithm.

Decision-making: The candidate who got best rank is the top performer. For multiple candidates, choose subsequent better performers (candidate with next higher rank) from the computed ranks.

Table 2 IFSS as evaluated by J_1

U	e_1			e_2			e_3			e_4			e_5		
c_1	0.90	0.05	0.05	0.66	0.20	0.14	0.62	0.32	0.06	0.37	0.35	0.28	0.22	0.48	0.3
c_2	0.80	0.12	0.08	0.76	0.18	0.06	0.78	0.12	0.10	0.6	0.32	0.08	0.93	0.04	0.03
c_3	0.66	0.27	0.07	0.94	0.02	0.04	0.68	0.22	0.10	0.64	0.2	0.16	0.85	0.05	0.1
c_4	0.64	0.32	0.04	0.87	0.11	0.02	0.68	0.19	0.13	0.73	0.19	0.08	0.95	0.04	0.01
c_5	0.83	0.10	0.07	0.72	0.12	0.16	0.90	0.05	0.05	0.56	0.31	0.13	0.84	0.07	0.09
c_6	0.73	0.12	0.15	0.91	0.02	0.07	0.63	0.19	0.18	0.66	0.24	0.1	0.95	0.01	0.04

Table 3 IFSS as evaluated by J_2

U	e_1			e_2			e_3			e_4			e_5		
c_1	0.90	0.05	0.05	0.62	0.17	0.21	0.69	0.21	0.10	0.30	0.09	0.61	0.36	0.45	0.19
c_2	0.85	0.10	0.05	0.69	0.23	0.08	0.83	0.08	0.09	0.63	0.25	0.12	0.74	0.10	0.16
c_3	0.64	0.23	0.13	0.90	0.04	0.06	0.73	0.21	0.06	0.35	0.14	0.51	0.27	0.12	0.61
c_4	0.69	0.26	0.05	0.87	0.06	0.07	0.65	0.19	0.16	0.47	0.51	0.02	0.41	0.39	0.20
c_5	0.87	0.10	0.03	0.77	0.10	0.13	0.94	0.03	0.03	0.39	0.60	0.01	0.35	0.55	0.10
c_6	0.78	0.16	0.06	0.84	0.04	0.12	0.57	0.25	0.18	0.41	0.37	0.22	0.49	0.36	0.15

Table 4 IFSS as evaluated by J_3

U	e_1			e_2			e_3			e_4			e_5		
c_1	0.89	0.08	0.03	0.68	0.17	0.15	0.58	0.39	0.03	0.40	0.04	0.56	0.32	0.41	0.27
c_2	0.77	0.14	0.09	0.76	0.18	0.06	0.74	0.18	0.08	0.65	0.31	0.04	0.64	0.19	0.17
c_3	0.69	0.24	0.07	0.95	0.03	0.02	0.61	0.20	0.19	0.30	0.13	0.57	0.26	0.18	0.56
c_4	0.69	0.14	0.17	0.82	0.07	0.11	0.74	0.26	0.00	0.48	0.39	0.13	0.34	0.44	0.22
c_5	0.90	0.07	0.03	0.78	0.10	0.12	0.93	0.06	0.01	0.33	0.61	0.06	0.41	0.49	0.10
c_6	0.78	0.13	0.09	0.91	0.04	0.05	0.67	0.24	0.09	0.53	0.28	0.19	0.53	0.36	0.11

Table 5 Priority Table for the judge J_1

U	μ_{e_1}	ν_{e_1}	h_{e_1}	μ_{e_2}	ν_{e_2}	h_{e_2}	μ_{e_3}	ν_{e_3}	h_{e_3}	μ_{e_4}	ν_{e_4}	h_{e_4}	μ_{e_5}	ν_{e_5}	h_{e_5}	$\sum\mu$	$\sum\nu$	$\sum h$
c_1	0.297	0.005	0.01	0.066	0.14	0.014	0.0868	0.208	0.003	−0.0888	−0.07	−0.084	−0.0418	−0.24	−0.03	0.3192	0.043	−0.087
c_2	0.264	0.012	0.016	0.076	0.126	0.006	0.1092	0.078	0.005	−0.144	−0.064	−0.024	−0.1767	−0.02	−0.003	0.1285	0.132	0
c_3	0.2178	0.027	0.014	0.094	0.014	0.004	0.0952	0.143	0.005	−0.1536	−0.04	−0.048	−0.1615	−0.025	−0.01	0.0919	0.119	−0.035
c_4	0.2112	0.032	0.008	0.087	0.077	0.002	0.0952	0.1235	0.0065	−0.1752	−0.038	−0.024	−0.1805	−0.02	−0.001	0.0377	0.1745	−0.0085
c_5	0.2739	0.01	0.014	0.072	0.084	0.016	0.126	0.0325	0.0025	−0.1344	−0.062	−0.039	−0.1596	−0.035	−0.009	0.1779	0.0295	−0.0155
c_6	0.2409	0.012	0.03	0.091	0.014	0.007	0.0882	0.1235	0.009	−0.1584	−0.048	−0.03	−0.1805	−0.005	−0.004	0.0812	0.0965	0.012

Table 6 Comparison table for judge J_1

U	c_1			c_2			c_3			c_4		
	m	n	h	m	n	h	m	n	h	m	n	h
c_1	0	0	0	0.1907	−0.089	−0.087	0.2273	−0.076	−0.052	0.2815	−0.1315	−0.0785
c_2	−0.1907	0.089	0.087	0	0	0	0.0366	0.013	0.035	0.0908	−0.0425	0.0085
c_3	−0.2273	0.076	0.052	−0.0366	−0.013	−0.035	0	0	0	0.0542	−0.0555	−0.0265
c_4	−0.2815	0.1315	0.0785	−0.0908	0.0425	−0.0085	−0.0542	0.0555	0.0265	0	0	0
c_5	−0.1413	−0.014	0.0715	0.0494	−0.103	−0.0155	0.086	−0.09	0.0195	0.1402	−0.145	−0.007
c_6	−0.238	0.0535	0.099	−0.0473	−0.036	0.012	−0.0107	−0.023	0.047	0.0435	−0.078	0.0205

U	c_5			c_6			M	N	H
	m	n	h	m	n	h			
c_1	0.1413	0.0135	−0.0715	0.238	−0.054	−0.099	1.0788	−0.3365	−0.388
c_2	−0.0494	0.1025	0.0155	0.0473	0.0355	−0.012	−0.0654	0.1975	0.134
c_3	−0.086	0.0895	−0.0195	0.0107	0.0225	−0.047	−0.285	0.1195	−0.076
c_4	−0.1402	0.145	0.007	−0.0435	0.078	−0.0205	−0.6102	0.4525	0.083
c_5	0	0	0	0.0967	−0.067	−0.0275	0.231	−0.4175	0.041
c_6	−0.0967	0.067	0.0275	0	0	0	−0.3492	−0.0155	0.206

Table 7 Decision table for judge J_1

	Membership	Non-membership	Hesitation	Score	Rank
c_1	0.411818598	−0.218648473	−0.418103448	0.239635822	1
c_2	−0.024965644	0.128330084	0.144396552	−0.028570596	3
c_3	−0.108795236	0.077647823	−0.081896552	−0.099885281	4
c_4	−0.232936326	0.294022092	0.089439655	−0.253770071	6
c_5	0.088181402	−0.271280052	0.044181034	0.092077347	2
c_6	−0.133302794	−0.010071475	0.221982759	−0.162893716	5

Table 8 Decision table for judge J_2

	Membership	Non-membership	Hesitation	Score	Rank
c_1	0.196303696	0.075889615	−0.234248788	0.150319793	2
c_2	−0.274225774	0.229121278	0.096526656	−0.300695871	6
c_3	0.039460539	0.194989107	−0.265751212	0.028973853	3
c_4	−0.137362637	−0.102759622	0.165589661	−0.16010847	5
c_5	0.264235764	−0.378721859	0.186187399	0.313433134	1
c_6	−0.088411588	−0.018518519	0.051696284	−0.092982139	4

Table 9 Decision table for judge J_3

	Membership	Non-membership	Hesitation	Score	Rank
c_1	0.10084122	0.252625153	−0.218061674	0.078851615	2
c_2	−0.309253417	0.153724054	0.15969163	−0.3586386	6
c_3	0.102733964	0.093650794	−0.281938326	0.073769322	3
c_4	−0.06508938	−0.089499389	0.126651982	−0.073333079	4
c_5	0.296424816	−0.360561661	0.140969163	0.338211574	1
c_6	−0.125657203	−0.04993895	0.072687225	−0.134790876	5

Table 10 Rank table

	J_1	J_2	J_3	Normalized score	Final rank
c_1	1	2	2	0.527777778	2
c_2	3	6	6	0.083333333	4
c_3	4	3	3	0.203703704	3
c_4	6	5	4	0.046296296	6
c_5	2	1	1	0.611111111	1
c_6	5	4	5	0.055555556	5

4 Conclusions

In this paper, a new algorithm is provided for GDM using IFSS. Earlier IFSSs were used for group decision-making by many researchers, but this paper is providing an improved approach for GDM using IFSS to provide more efficient and realistic approach for decision-making.

References

1. Zadeh LA (1965) Fuzzy sets. Inf Control 8:338–353
2. Atanassov K (1986) Intuitionistic fuzzy sets. Fuzzy Set Syst 20:87–96
3. Maji PK, Biswas R, Roy AR (2001) Fuzzy soft sets. J Fuzzy Math 9(3):589–602
4. Maji PK, Biswas R, Roy AR (2002) An application of soft sets in a decision making problem. Comput Math Appl 44:1007–1083
5. Maji PK, Biswas R, Roy AR (2003) Soft set theory. Comput Math Appl 45:555–562
6. Tripathy BK, Arun KR (2015) A new approach to soft sets, soft multisets and their properties. Int J Reasoning-Based Intell Syst 7(3/4):244–253
7. Xu YJ, Sun YK, Li DF (2010) Intuitionistic fuzzy soft set. In: Proceedings of 2nd international workshop on intelligent systems and applications, IEEE
8. Tripathy BK, Mohanty RK, Sooraj TR (2016) Application of uncertainty models in bioinformatics. In: Handbook of research on computational intelligence applications in bioinformatics, pp 169–182
9. Tripathy BK, Mohanty RK, Sooraj TR (2016) On intuitionistic fuzzy soft set and its application in group decision making. In: Proceedings of 1st international conference on emerging trends in engineering, technology and science, ICETETS
10. Tripathy BK, Mohanty RK, Sooraj TR (2016) On intuitionistic fuzzy soft sets and their application in decision-making. Lect Notes Electr Eng 396:67–73
11. Tripathy BK, Mohanty RK, Sooraj TR, Tripathy A (2016) A modified representation of IFSS and its usage in GDM. Smart Innov Syst Technol 50:365–375
12. Tripathy BK, Mohanty RK, Sooraj TR, Arun KR (2016) A new approach to intuitionistic fuzzy soft sets and its application in decision-making. Adv Intell Syst Comput 439:93–100
13. Tripathy BK, Sooraj TR, Mohanty RK (2017) A new approach to interval-valued fuzzy soft sets and its application in decision-making. Adv Intell Syst Comput 509:3–10
14. Tripathy BK, Sooraj TR, Mohanty RK (2016) Advances decision making usisng hybrid soft set models. Int J Pharm Technol 8(3):17694–17721
15. Tripathy BK, Sooraj TR, Mohanty RK (2016) A new approach to fuzzy soft set theory and its application in decision making. Adv Intell Syst Comput 411:305–313
16. Tripathy BK, Sooraj TR, Mohanty RK, Arun KR (2016) Parameter reduction in soft set models and application in decision making. In: Handbook of research on fuzzy and rough set theory in organizational decision making, pp 331–354
17. Tripathy BK, Sooraj TR, Mohanty RK, Panigrahi A (2018) Group decision making through interval valued intuitionistic fuzzy soft sets. Int J Fuzzy Syst Appl 7(3):99–117
18. Mohanty RK, Tripathy BK (2017) Intuitionistic hesitant fuzzy soft set and its application in decision making. Adv Intell Syst Comput 517:221–233
19. Mohanty RK, Sooraj TR, Tripathy BK (2017) IVIFS and decision-making. Adv Intell Syst Comput 468:319–330
20. Sooraj TR, Mohanty RK, Tripathy BK (2018) A new approach to interval-valued intuitionistic hesitant fuzzy soft sets and their application in decision making. Smart Innov Syst Technol 77:243–253

21. Sooraj TR, Mohanty RK, Tripathy BK (2018) Improved decision making through IFSS. Smart Innov Syst Technol 77:213–219
22. Sooraj TR, Mohanty RK, Tripathy BK (2017) Hesitant fuzzy soft set theory and its application in decision making. Adv Intell Syst Comput 517:315–322
23. Sooraj TR, Mohanty RK, Tripathy BK (2016) Fuzzy soft set theory and its application in group decision making. Adv Intell Syst Comput 452:171–178
24. Molodtsov D (1999) Soft set theory—first results. Comput Math Appl 37:19–31
25. Atanassov K, Gargov G (1989) Interval valued intuitionistic fuzzy sets. Fuzzy Sets Syst 31(3):343–349

Analysis and Prediction of the Survival of Titanic Passengers Using Machine Learning

Amer Tabbakh, Jitendra Kumar Rout, and Minakhi Rout

Abstract The Royal Mail Ship (RMS) Titanic is the largest liner built in 1912, of the estimated 2224 passengers and crew aboard, more than 1500 died after the ship struck an iceberg during her maiden voyage from Southampton to New York City. The dataset collected from Kaggle has information about the passengers and crew which are P-class, name, age, sex, etc., and can be used to predict if the person onboard has survived or not. In this paper, six different machine learning algorithms are used (i.e., logistic regression, k-nearest neighbors, SVM, naive Bayes, decision tree and random forest) to study this dataset and deduce useful information to know the knowledge of the reasons for the survival of some travelers and sinking the rest. Finally, the results have been analyzed and compared with and without cross-validation to evaluate the performance of each classifier.

Keywords Titanic dataset · Survival prediction · Machine learning · Cross-validation

1 Introduction

Machine learning is a subset of AI that mainly focuses on learning from past experiences and making predictions based on it. Titanic, one of the largest and most luxurious ocean liners ever built on April 15, 1912, was considered unsinkable at that time. The ship carried around 2200 passengers and crew, but it sunk and 1500 passengers and crew died. The dataset was collected from Kaggle regarding Titanic contains many information about the passengers and crew (such as ID, name, age).

A. Tabbakh · J. K. Rout (✉) · M. Rout
Kalinga Institute of Industrial Technology University, Bhubaneswar, Odisha, India
e-mail: jitu2rout@gmail.com

A. Tabbakh
e-mail: mr.amertabbakh@gmail.com

M. Rout
e-mail: minakhi.rout@gmail.com

A. K. Tripathy et al. (eds.), *Advances in Distributed Computing and Machine Learning*, Lecture Notes in Networks and Systems 127, https://doi.org/10.1007/978-981-15-4218-3_29

The information can be extracted, and machine learning can be used to predict who has survived as well as died during the disaster. In this paper, different machine learning algorithms such as logistic regression, k-NN, SVM and naive Bayes, decision tree and forest have been used for prediction of survival of passengers.

2 Related Work

Lam and Tang [2] used three algorithm, namely SVM, decision tree and naive Bayes for the Titanic problem. The authors have observed that sex was a significant feature in accurately predicting survival. The best result was 79.43% (in terms of accuracy) by the decision tree algorithm. In a similar way, Balakumar et al. [3] have used three algorithms for the Titanic prediction problem(LR, DT, and RF). It has been found that logistic regression was the best to provide an accuracy of 94.26% with selected features such as P-class, sex, age, children and SibSp. Farag and Hassan [4] focused more on feature selection and concluded that sex is highly correlated to survival and also passenger's ticket proved to be strongly correlated to survival as well. Out of the two algorithms (naive Bayes, decision trees) used for the Titanic prediction problem, naive Bayes was the best to provide an accuracy of 92.52%. On a different note, Kakde and Agrawal [5] have focused on the preprocessing of data. They have observed that the age was highly influential on the survival rate. Apart from these, $parch$ and $sibsp$ columns are combined to know the family size where if family size becomes greater than three, the survival rate decreases. Four different algorithms (logistic regression, SVM, random forest, decision trees) were used for the Titanic prediction problem, and logistic regression gave the best accuracy of 83.73%.

Singh et al. [6] have observed that the features (P-class, sex, age, children and SibSp) which are selected as more significant are highly correlated to the survival of the passengers. Out of four different algorithms (naive Bayes, logistic regression, decision tree, random forest) used for the titanic classification problem, logistic regression gave the best result in terms of accuracy as 94.26%. Ekinci and Acun [7] have tried with different feature combinations; i.e., some features are added to dataset (such as family size), and some of them were eliminated like (name, ticket and cabin). Different machine learning techniques were used, and the results were compared on the basis of F-measure on the dataset obtained from Kaggle. The best F-measure was 0.82% for voting(GB, ANN, KNN).

Nair [8] used four ML techniques(LR, NB, DT and Rf) to predict the survival of passengers. The survival rate of the female is observed to be higher than that of the male. Some features are added to dataset (mother, children, family, etc.), and two performance measures are used to compare the results (accuracy and false discovery rate) in which LR proved to be the best for this problem.

3 Methodology

3.1 Contribution

The major contribution of work is:

- Selection of appropriate features for the Titanic dataset and experimentally find out which combination will work better.
- Analysis and prediction of passenger survival using different classifiers.
- Analysis of results with and without cross-validation to avoid the effect of improper distribution of data (if any).

3.2 Classifiers Used

In this work, different machine learning algorithms (i.e., logistic regression, KNN, SVM, naive Bayes) are used to predict whether a passenger will survive or not. Most appropriate attributes selected for training and testing dataset are: P-class, sex, age, SibSp, ParCh, nickname. Here are the brief descriptions of different algorithms used:

- **Logistic Regression**: Logistic regression is used for the classification, and the target variable is categorical as well as a binary where 1 means survival and 0 means demise.
- **Naive Bayes(NB)**: NB applies Bayes theorem to build a prediction model of a classification problem. It assumes that features are independent of each other. It is called the conditional probability; by multiplying all the conditional probabilities, the probability of a class value is obtained, and the class which has the highest probability is assigned as the class of a given instance. Out of four different types of NB algorithms, the Gaussian NB algorithm is used in this work with the significant features (P-class, sex, age, SibSp, ParCh, nickname) to build the model, which are categorical (or numeric). The target value is surviving which takes two values (0 for demise and 1 for survival). Summary data included calculating the standard deviation and mean for each attribute, by class value.
- **KNN**: The KNN is a supervised learning algorithm that makes use of the class labels of training data during the learning phase as a target. It is an instance-based machine learning algorithm, where new data points are classified based on stored, labeled instances (data points). Though it can be used both for classification and regression, it is more widely used for classification purposes. The k(the number of nearest neighbors) in KNN is a crucial variable also known as a hyperparameter that helps in classifying a data point accurately. The value of k taken in our case is five.
- **SVM**: SVM is a supervised machine learning algorithm that is mostly used for classification problems. Support vector machine is a frontier that best segregates the two classes.

Table 1 Snapshot of the dataset before preprocessing

PID	Survived	P-class	Name	Sex	Age	SibSp	Parch	Ticket	Fare	Cabin	Embarked
1	0	3	Braund, Mr. Owen Harris	Male	22	1	0	A/5 21171	7.25		S
2	1	1	Cumings, Mrs. John Bradley (Florence Briggs Thayer)	Female	38	1	0	PC 17599	71.2833	C85	C
3	1	3	Heikkinen, Miss. Laina	Female	26	0	0	STON/O2. 3101282	7.925		S
4	1	1	Futrelle, Mrs. Jacques Heath (Lily May Peel)	Female	35	1	0	113803	53.1	C123	S

- **Decision Tree**: The decision tree is one of the important classification models; it classifies the problem depending on making a flowchart; it has nodes and leaves, which every node is as a feature; this feature has branches related to the outcome of the test on it; through these tests and branches, we get the leaf which is the class label, where each leaf represents one class label.
- **Random Forest**: Random forest is a type of classification model based on a decision tree model with a multitude of them and outputting the class. Random forest is used to correct the DT outcome of over-fitting from the training set.

3.3 Dataset Used

The dataset was obtained from the Kaggle Website. It has three files:

- Titanic_train.csv: it is the dataset having 891 records for training the models, and its contents are information about passengers like (ID, name, age, etc., and the target attribute is survived). Table 1 shows the snapshot of the training dataset.
- Titanic_test.csv: it is the dataset having 418 records for testing the models, and its contents are information about passengers like training dataset without the target attribute survived.
- Titanic_gender_submission.csv: it is the dataset having 418 records which are actual values for the testing dataset of survived passengers.

3.4 Preprocessing

Data Cleaning and Feature Selection: The dataset which is there on the Kaggle Website has twelve features.

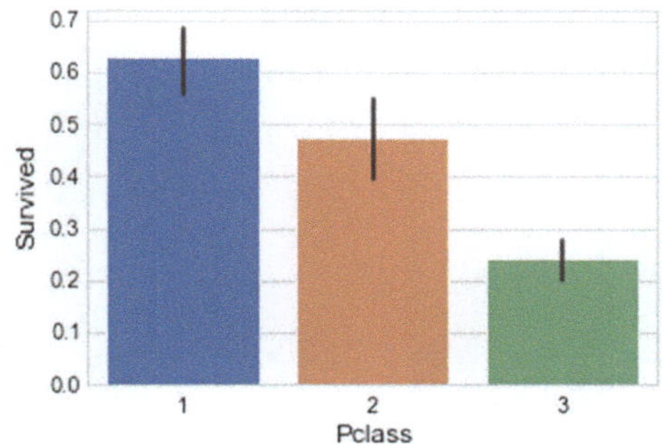

Fig. 1 Survival rate with respect to P-class

- *Passenger ID*: it represents a number of passengers; it is a counter, and it does not make a sense, so we can eliminate it.
- *Survived*: it represents the target which is needed to classify; it has two values (1,0) for survived or not survived in, respectively.
- *P-class*: it represents classes of seats and cabins for passengers; it has three numeric values (1,2,3) where the 1 for first class, 2 for second class and 3 for third class. Figure 1 shows the rate of survival of passengers for each class seats.
- *Name*: it represents names and nicknames of passengers; the nickname of the name is extracted by this feature, and those nickname (Master, Mr., Ms., Mlle, Miss, etc.) because the nickname of the name is important things in our reality.
- *Sex*: it represents the gender of passengers; the rate of survival of womens was more than that of men as at that time they used the policy of women children first. The same information is shown in Fig. 2.
- *Age*: it represents the age of passengers; in this feature, 177 instances have NA from training dataset; the missing values are handled by using the average mean of the ages.
- *SibSp and ParCh*: it represents a number of siblings, spouses, parents and children. They are relevant for the group of family units together.
- *Ticket*: it represents the ticketing system in that time, and it does not make a sense, so it can be eliminated.
- *Far*: it represents how much a passenger paid for his ticket; it does not make a sense, so it can be eliminated.
- *Cabin*: it represents seating area of the passenger; in this feature, there are lots of NA 687, and it cannot replace these by frequent value or other value because lots of cabins are not there in the dataset, and it cannot be filled by the same cabin.
- *Embarked*: it represents from which port the passenger booked his ticket; there were three ports of embarkation which are Southampton, Queenstown and Cherbourg. Two missing values are there, and it is handled by replacing them with a frequent value which is Southampton, but this feature does not make sense for our target (Survive on not Survive), so it can be eliminated and the selected features are shown in Table 2.

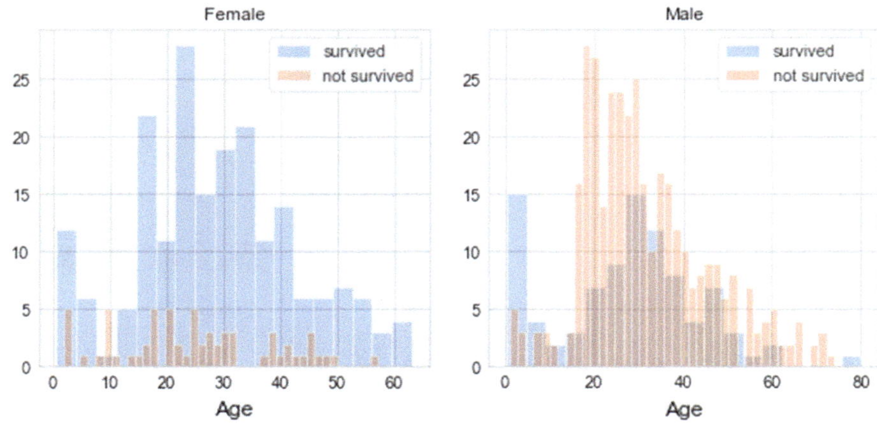

Fig. 2 Survival rate of males and females based on age

Table 2 Snapshot of the dataset after feature selection

P-class	Sex	Age	SibSp	Parch	Nickname
3	1	22	1	0	Mr.
1	0	38	1	0	Mrs.
3	0	26	0	0	Ms.

3.5 Hardware and Software Used

The experiments were carried out on Windows 10 64-bit OS, Intel(R) Core(TM)-i5-7200U@ 2.50GHz Processor, 8 GB RAM.

4 Results

In this paper, the confusion matrix is used as the metric evaluation. Accuracy is a technique to measure how much the model will predict. The model with higher accuracy is a better model. Apart from accuracy to deal improper distribution of data, we have used other parameters like precision, recall, and F1-score. Table 3 shows the results of implementation for different classification models, and Tables 4, 5 and 6 show the results of implementation for different classification models by using K-cross-validation for K equal 10, 5 and 4, respectively. From the results, it can be observed that the data in the dataset is evenly distributed as all the performance measures are giving consisitent results. We have also used k-fold cross-validation to avoid probable fluctuations in the result of classifiers for each random selection of training and testing set.

Table 3 Result for different classifiers for 891 training and 418 testing instances

Algorithm	F1_score	Accuracy	Recall	Precision
Logistic regression	95.45	95.45	94.07	93.46
KNN	86.35	86.36	80.92	81.45
SVM	95.46	95.45	95.39	92.35
Naive Bayes	98.80	98.80	97.36	99.32
DT	80.71	80.62	75.65	72.32
RF	81.49	81.33	78.94	72.28

Table 4 Results for different classifiers with $K = 10$

Algorithm	F1_score	Accuracy	Recall	Precision
Logistic regression	87.53	87.56	82.64	83.23
KNN	87.24	87.27	82.59	82.66
SVM	86.90	86.94	81.62	82.59
Naive Bayes	86.11	86.19	79.59	81.86
DT	87.01	87.06	81.35	82.80
RF	86.38	86.45	80.52	82.21

Table 5 Results for different classifiers with $K = 5$

Algorithm	F1_score	Accuracy	Recall	Precision
Logistic regression	85.66	85.79	78.58	82.76
KNN	84.49	84.57	77.84	80.58
SVM	86.12	86.25	79.42	83.26
Naive Bayes	85.51	85.71	78.48	82.88
DT	81.82	81.97	73.10	77.68
RF	83.16	83.27	75.49	79.31

Table 6 Results for different classifiers with $K = 4$

Algorithm	F1_score	Accuracy	Recall	Precision
Logistic regression	85.38	85.48	78.37	82.84
KNN	83.02	83.12	75.33	79.18
SVM	86.12	86.25	78.97	83.96
Naive Bayes	85.45	85.56	78.23	83.2
DT	80.42	80.59	71.12	76.27
RF	82.59	82.66	76.03	78.02

5 Conclusion and Future Work

In this research, we focused on preprocessing dataset and selection features and extracted main information from the name column; the final features selected as a significant feature are (P-class, sex, age, SibSp, ParCh, nickname). Naive Bayes is the best algorithm for the Titanic classification dataset where the accuracy is 98.80% as compared to other algorithms without K-cross-validation, and with using K-cross-validation to different K values equal (10,5,4), we found the accuracy is better when K equals 10 because the instances for training dataset are more than the instances for K equal 5 or 4.

In future, we will use ranking feature selection to extract some meaningful information from dataset and use deep learning algorithms to predict the survival of passengers.

References

1. Kaggle. https://www.kaggle.com/c/titanic. Last accessed 1 June 2019
2. Lam E, Tang C (2012) Titanic machine learning from disaster. In: Lam Tang
3. Balakumar B, Raviraj P, Sivaranjani K (2019) Prediction of survivors in Titanic dataset: a comparative study using machine learning algorithms. Sajrest Arch 4(4)
4. Farag N, Hassan G (2018) Predicting the survivors of the Titanic Kaggle, machine learning From disaster. In: Proceedings of the 7th international conference on software and information engineering, pp 32–37
5. Kakde Y, Agrawal S (2018) Predicting survival on Titanic by applying exploratory data analytics and machine learning techniques. Int J Comput Appl 32–38
6. Singh A, Saraswat S, Faujdar N (2017) Analyzing Titanic disaster using machine learning algorithms. In: International conference on computing, communication and automation (ICCCA). IEEE, pp 406–411
7. Ekinci EO, Acun N (2018) A comparative study on machine learning techniques using Titanic dataset. In: 7th international conference on advanced technologies, pp 411–416
8. Nair P (2017) Analyzing Titanic disaster using machine learning algorithms. Int J Trend Sci Res Develop (IJTSRD) 2(1):410–416

An AI Approach for Real-Time Driver Drowsiness Detection—A Novel Attempt with High Accuracy

Shriram K. Vasudevan, J. Anudeep, G. Kowshik, and Prashant R. Nair

Abstract Despite the sophisticated technology that could prevent accidents of vehicles on highways, many lives are claimed due to the drowsiness of drivers. According to the data reported by the NHTSA (National Highway Traffic Safety Administration) of USA, 846 succumb to death and it has become a major threat these days which is evident from 83,000 cases registered due to drowsy driving. Drowsiness is the feeling of being sleepy or being inactive towards activities and this causes a sleeping sensation which leads to closure of eyelids while driving resulting in major accidents. There are systems in action that could detect physical awaken of the driver, but detecting the drowsiness and deviation of the driver is a challenging problem in the field of transportation. Many factors could influence the drivers to fall asleep during their journeys and the chances of getting drowsy increases in night times than in the dawn and journeys did alone are even more dangerous. So, we introduce a system which is capable of monitoring the person's consciousness, acceleration pattern and the angle of vehicle's steering simultaneously to detect the deviation and drowsiness of the driver. The process of drowsiness and deviation detection of the driver has been done using image processing techniques. Our proposed system alerts the driver by comparing the acceleration pattern to the Eye Aspect Ratio (EAR) for drowsiness. It also monitors his facial movements, changes in steering angle to detect the deviation.

Keywords Drowsy detection · Image processing · Acceleration pattern · Person's consciousness · Deviation detection · Steering angle

---

S. K. Vasudevan (✉) · P. R. Nair
Department of Computer Science and Engineering, Amrita School of Engineering, Amrita Vishwa Vidyapeetham, Coimbatore, India
e-mail: kv_shriram@cb.amrita.edu

J. Anudeep
Department of Electronics and Communication Engineering, Amrita School of Engineering, Amrita Vishwa Vidyapeetham, Coimbatore, India

G. Kowshik
Department of Electronics and Instrumentation Engineering, Amrita School of Engineering, Amrita Vishwa Vidyapeetham, Coimbatore, India

A. K. Tripathy et al. (eds.), *Advances in Distributed Computing and Machine Learning*, Lecture Notes in Networks and Systems 127, https://doi.org/10.1007/978-981-15-4218-3_30

1 Introduction

Excessive tiredness, lonely travels, diseases like narcolepsy and many other factors can cause drowsiness to the driver. It affects men and women equally and is thought to affect roughly 1 in 2000 people [1]. Similarly, deviation of drivers can be caused by many factors like hallucinations, daydreaming, and speed deviation during high speed and also due to various other diseases. The facts as per records of NSF agency (National Sleep Foundation) on the drowsy driving survey was polled that one-third of the people who drive in America has fallen asleep during their drive. 37% of people (103 million in number) aren't able to control their sleep during journeys and fall asleep without any prior symptoms [2].

The Solomon curve is a graphical representation of the collision rate of automobiles as a function of deviation from average speed. The curve was based on research conducted by David Solomon in the late 1950s and published in 1964 [3, 4]. From the Solomon curve, we can see that most of the accidents occur at high speed when compared to the average speed of the vehicles. So, comparing acceleration patterns of the driver and alerting accordingly is the key aspect of our proposed system.

2 Existing Solutions

There are some of the solutions that are built in order to detect and take counter action against the driver's drowsiness.

The system proposed by Sahayadhas, A., Murugappan, M. and Sundaraj K. is a drowsy detection system that uses of facial images to detect the placement of the face and the aspect ratio of the eyelids. This system is also equipped to measure the ECG pattern of the person. All these parameters are used to generate a threshold and alert the driver [5].

Another system by Devi, M. S. and Bajaj, P. R captures video footage of the driver during his journey and makes an analysis of that footage looking for the fatigue of the driver. It finds the eyes of the driver in the video footage and calculates the level of fatigue. They measure the area of the eye to know whether the eye is open or closed [6].

All these systems try to detect the drowsiness of the driver using a single parameter and don't consider multiple aspects. But, the proposed system used facial data along with the vehicular parameters to detect the drowsiness as well as deviation from driving (Table 1).

Table 1 Comparison table for the proposed system versus existing solutions

	On-chip diagnostics	Detection of deviation	Machine learning	Image processing	Driver alerts	ECG
Devi, M. S. and Bajaj	✖	✖	✔	✔	✔	✖
Sahayadhas, A., Sundaraj	✖	✖	✔	✔	✔	✔
Proposed system	✔	✔	✔	✔	✔	✖

3 Architecture of Proposed System

The proposed system uses image processing techniques and pattern recognition algorithms to detect and recognize whether the driver is deviated or feeling drowsy while driving. It has an On-board Diagnostics (OBD) device which will fetch data about the speed and steering angle pattern of the vehicle along the journey. It is essential to monitor EAR (Eye Aspect Ratio), steering angle and acceleration pattern together because of the statistics obtained from NHTSA and Solomon curve [4]. It is even more dangerous when he starts coming to a saturation speed where the driver will be unaware of the horns, traffic signals and pedestrians. We use pattern recognition techniques to know at what speed the drivers usually feel drowsy and makes the machine more efficient at that speed. It is common that, a car might be used by n different drivers. So, we use LBPH classifiers to detect the driver's face and map it to their respective speed patterns to increase the efficiency in detecting the drowsiness of the driver (Figs. 1 and 2).

3.1 Face Recognition and Feature Extraction

Obtaining Dataset: The process of creating a dataset is done by capturing the required number of images of the driver in different angles so that the machine can be trained in an efficient way. After the dataset is created, the next step is to train the machine and store the trained data in a .yml file.

Feature Extractor: The process of feature extraction is done using LBPH (Local Binary Patterns Histogram) algorithm [7], which is basically a visual descriptor designed for computer vision. It gives adorable feature extraction outputs due to its special strategy of detection. This algorithm analyses the pixel value and also records the pattern of the binary elements present around the pixel and stores in the form of the histogram which is centralized for the training. These whole features extracted from the dataset of the person are stored in a .yml file which is later used for face detection of the driver.

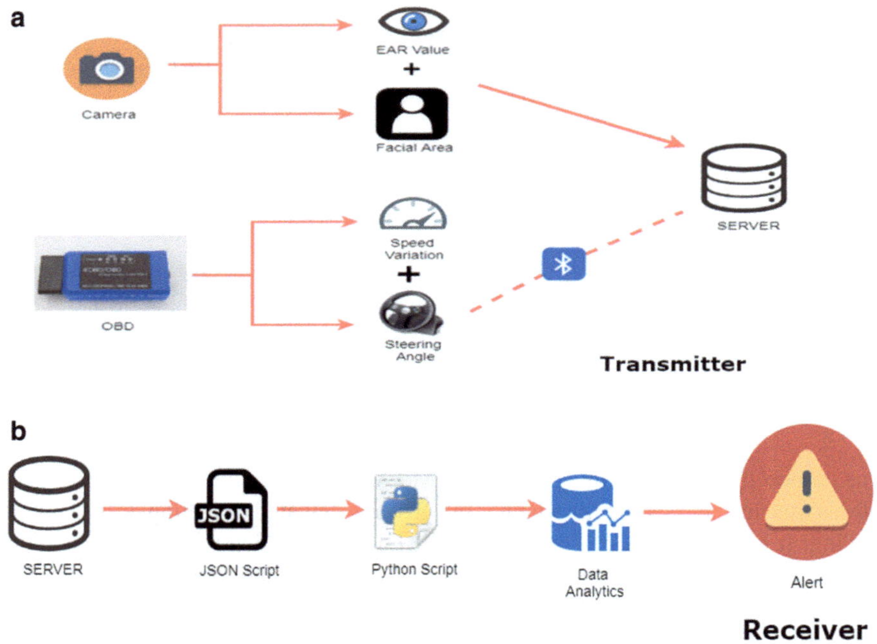

Fig. 1 **a** Architecture of transmitter part. **b** Architecture of receiver part

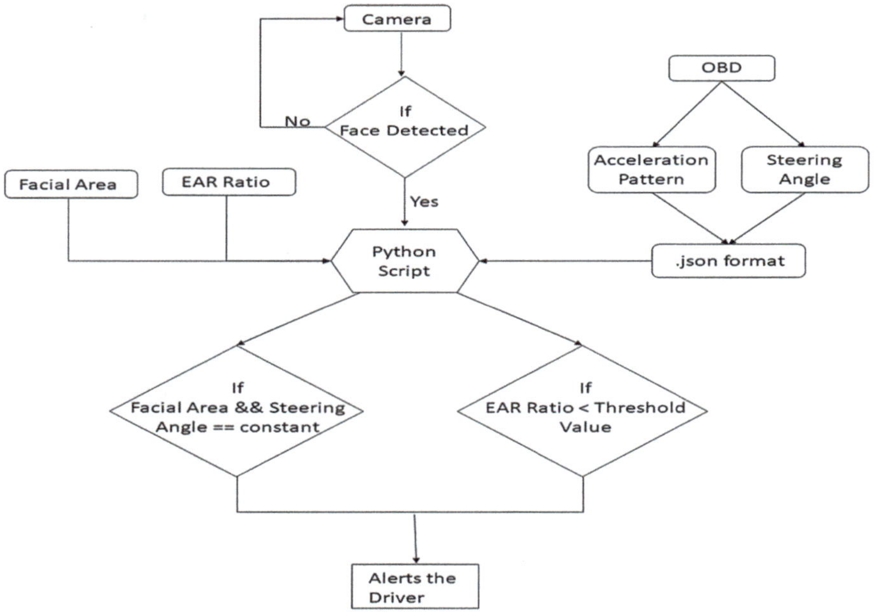

Fig. 2 Workflow of the proposed system

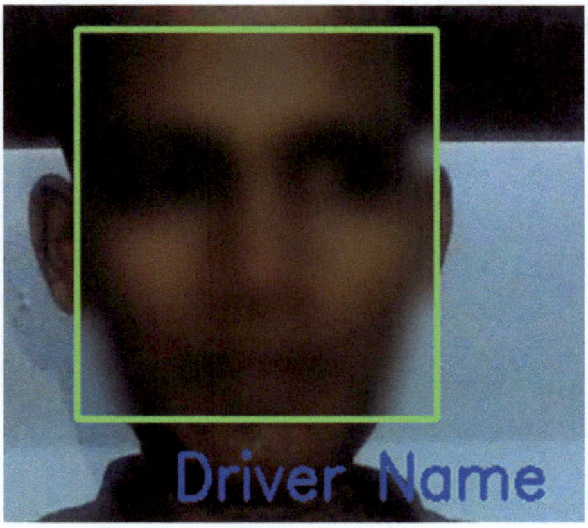

Fig. 3 Detected driver face

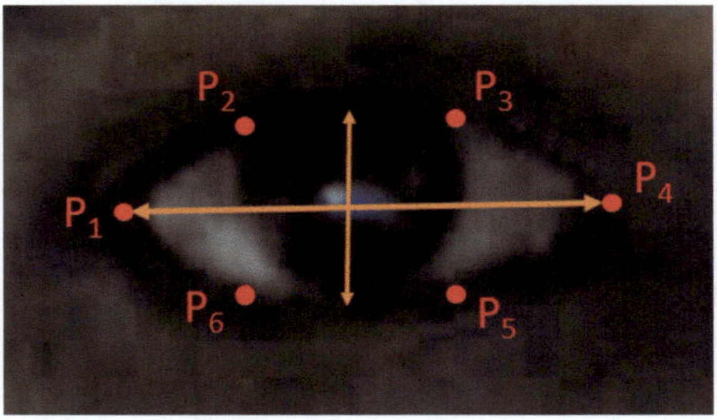

Fig. 4 Plotted coordinates

Face Detector: When the driver gets seated to start a drive it captures the face of the driver and compares with the features in the trained .yml file to determine the driver, which could trigger the database to get the pre-trained acceleration pattern of the specific driver which will hike up the accuracy in detecting the drowsiness of the driver. The detected face of the driver is shown in Fig. 3 and the driver's name is not displayed; the driver's face is blurred to maintain anonymity.

3.2 Calculation of the Eye Aspect Ratio (EAR)

The heart of the drowsiness detection process lies in the calculation of EAR ratio. When the exact portion of the eye is detected in the captured photo frame it quickly plots 6 coordinates returning to calculate the EAR to determine how wide the eye is open. The execution of the algorithm is depicted from the following figures.

In Fig. 4, the position of the eye is obtained thereby coordinates are plotted and the distances between the points that are located opposite to each other are calculated, which are substituted in the following formula in order to get the EAR value which can be threshold according to the person.

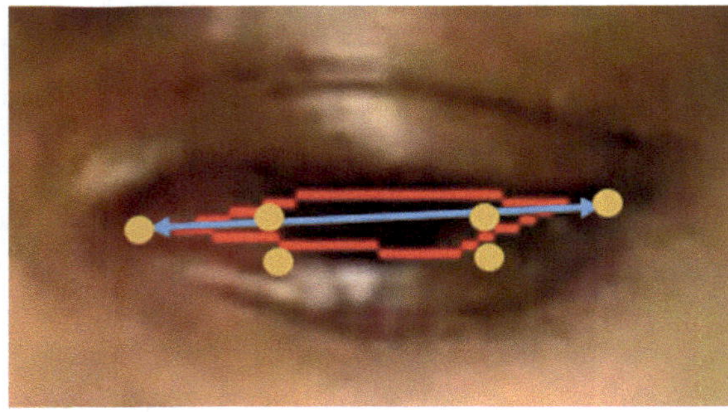

Fig. 5 Drowsy situation

$$\text{EAR} = \frac{\overline{(P2-P6)}+\overline{(P3-P5)}}{2\overline{(P1-P4)}}$$

Where,

$\overline{(P2-P6)}$ = Distance between points P2 and P6

$\overline{(P3-P5)}$ = Distance between points P3 and P5

$\overline{(P1-P4)}$ = Distance between points P1 and P4

Fig. 6 Formula to calculate the eye aspect ratio

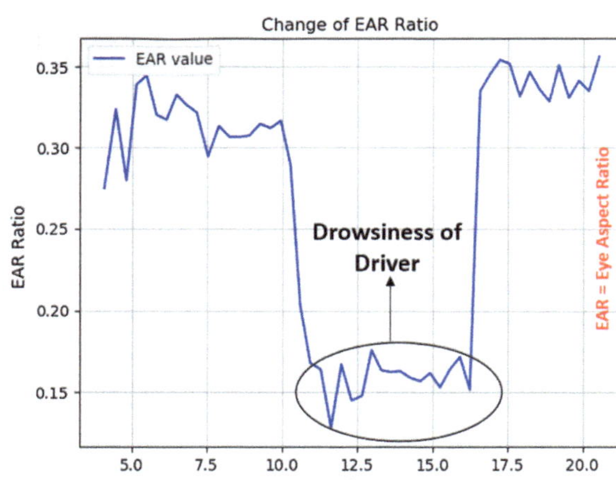

Fig. 7 EAR ratio of driver plotted with time

The formula shown in Fig. 6 is used to calculate the EAR value and Fig. 5 is the case where the driver feels drowsy and closes his eyes, the coordinates plotted on the eye comes into a position where the points P2, P6 gets collinear with P1 and P4 resulting that the distance between the points P2, P6 and P3, P5 closer to zero, so this takes the value of numerator in the formula to approximately zero, making the EAR value to fall suddenly, indicating the closure of eye and if it is prolonged then it is considered that the driver is drowsy and our system immediately alerts the driver.

From Fig. 7, we can see that the driver is feeling drowsy during the time period 11–16 s. If the EAR value is less than 0.18 (Threshold value of EAR in our case and it can be modified depending on the driver) for a threshold time period (Here we have taken it to be 4 s). Then the driver is considered to be in a drowsy state and our system alerts the driver immediately.

3.3 Deviation Detection

The above EAR calculation is for drowsy detection of the driver but for deviation detection we use the change of the facial area of driver, acceleration pattern and change in steering angle. The facial area can be found from the coordinates of the face of the driver obtained in Step 1, but the remaining two acceleration patterns and changes in steering angle are found from OBD (On-Board Diagnostics). If the facial area of the driver and steering angle is constant for a threshold period of time then it is considered as the driver is deviated and our system immediately alerts the driver. The acceleration pattern is noted to know at what speeds the driver is usually feeling drowsy and deviated.

From Figs. 8 and 9, we can see that facial area and steering angle are constant between the time period 45–49 s (here we have taken the threshold time period as 4 s) simultaneously. So we can say that the driver has deviated and our proposed system alerts the driver immediately.

From Figs. 8 and 9, we can see that the driver has deviated between the time periods 45–49 s. So, our algorithm records the speed between these time periods for improving deviation detection accuracy. (it is indicated in Fig. 10).

Here, we have taken the threshold time period as 4 s (i.e. 45–49 s). We can change the threshold time period according to the requirement of the driver (Table 2).

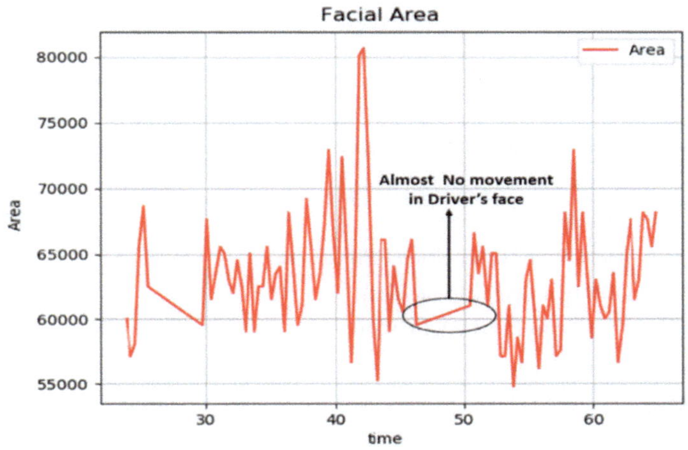

Fig. 8 Facial area plotted with time

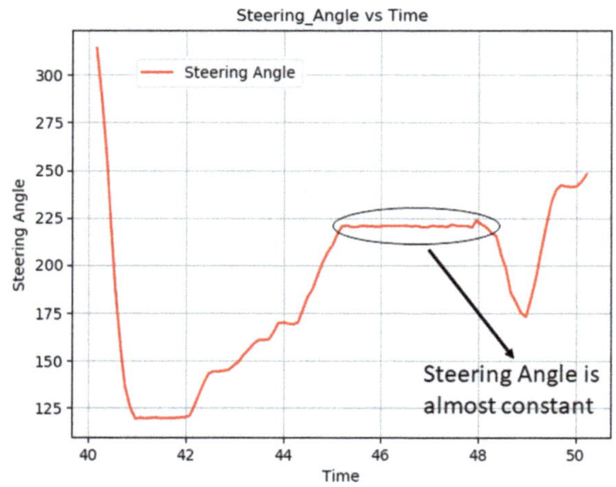

Fig. 9 Steering angle plotted with time

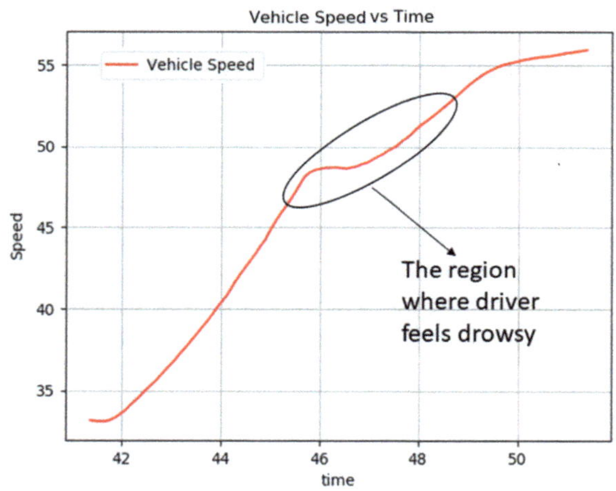

Fig. 10 Vehicle speed plotted with time

Table 2 Cost of each component and their respective quantity

S. No.	Component name	Quantity	Price
1	Intel UP^2 AI vision kit	1	$280
2	Intel UP^2 machine vision USB camera	1	$20
3	OBD (On-Board Diagnostics)	1	$12
4	USB 2.0 to TTL UART CP2101	1	$5
5	HC-05 Bluetooth	1	$7

Total cost $324

4 Hardware Used

Intel UP^2 AI vision Board: Credit card sized processing units like Raspberry pi cannot execute image processing, Pattern recognition algorithms efficiently [8]. So, for processing such algorithms we require a high end and portable computing device that works in real-time. So, we have chosen Intel UP^2 AI vision board for running our algorithm real-time (Fig. 11).

Intel UP^2 Machine Vision USB Camera: In our proposed system we have used HD webcam (Fig. 12) to acquire good frames per second and high-resolution pictures of the driver's face. The camera can be altered with respect to the driver.

On-Board Diagnostics (OBD): On-Board Diagnostics (Fig. 13) is a system that could report various statuses of the automobile i.e., for example, let the automobile be the car and OBD can retrieve the data like steering angle, speed of the car, fuel consumed since the restart of the vehicle, transmission torque levels, etc., [9] so our system grabs this data from the OBD and makes use of steering angle of the vehicle in order to monitor whether the person deviates or not and also the speed of

Fig. 11 Intel UP^2 AI vision board used in the system

Fig. 12 Camera used for capturing the photo frames of person

Fig. 13 OBD system for acquiring on-board data from vehicle

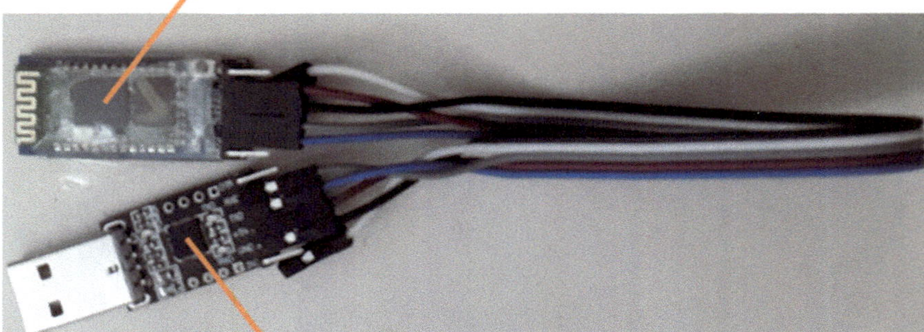

Fig. 14 Bluetooth module used for communication with server

the vehicle to improve the efficiency of the system. When the driver deviates from the normal driving state, he/she loses consciousness on road and doesn't move the steering. So, the steering angle remains likely stable without any rapid changes. So, this parameter helps the system in monitoring the deviation of the person inside the car to a far better extent.

TTL Dongle and HC-05: The module shown in Fig. 14 is used in acquiring data from OBD module and pushes it to the server using Bluetooth. The Bluetooth module HC-05 shown in Fig. 14 is programmed as the master device that automatically connects to OBD via Bluetooth. Post to the pairing action we will be able to acquire data the from OBD directly to the serial monitor in the form of JSON script which is again sent to a python script for converting into CSV format and stores each type of data(like vehicle speed, steering angle, odometer, etc.) to their respective arrays. From this data python script process accordingly and if the above-mentioned conditions are satisfied it gives an alert.

5 Conclusion and Future Scope

Many safety and security measures have been implemented in cars like airbags, over-speed buzzers, etc., but effective, rugged and frugal systems that work based on the human postures and state of consciousness are less in number [10, 11]. The proposed system is cost-effective and consumes less power to operate resulting in less energy consumption of battery power by the vehicle.

The following are some of the feature description and advancements to be done for our system:

1. Our system could effectively monitor the state of drowsiness and can report whether the person has deviated from his consciousness or not.
2. In case if the driver crosses the pre-set threshold value, our system can efficiently alert the driver and can bring back him to normal and safety state of driving.
3. The whole testing of the system is done in real-time and the prototype is working as expected for the identified test cases.
4. The system needs to be improved in the following cases;
5. Machine learning is to be introduced at higher scales (also known as Deep Learning) in order to keep track of how many times the driver had felt drowsy, at what times of day he is more likely to get deviated, etc.
6. Android application to report the feedback of the journey and precautions to be taken.
7. Proposed system as a whole cost approximately 300 to 350 USD (including marketing value) when made as a product which can be reduced.

Our system can be improved in more adverse low light conditions by using the night vision camera.

References

1. National Sleep Foundation. Narcolepsy. https://www.sleepfoundation.org/articles/narcolepsy. Last accessed 09 Sept 2019
2. Drowsy Driving. National Sleep Foundation, Facts and Stats. http://drowsydriving.org/about/facts-and-stats/. Last accessed 09 Sept 2019
3. Solomon curve. Wikipedia, Home Page. https://en.wikipedia.org/wiki/Solomon_curve
4. Figure 8-1. Deviation from average speed versus the collision rate (Solomon curve). https://i.pinimg.com/736x/fa/01/d7/fa01d7f000fdd5c9f392b4c78e5c2622–solomon-curves.jpg. Lat Accessed 09 Sept 2019
5. Sahayadhas A, Sundaraj K, Murugappan M (2012) Detecting driver drowsiness based on sensors: a review. Sensors 12(12): 16937–16953. http://www.mdpi.com/1424-8220/12/12/16937/htm
6. Devi MS, Bajaj PR (2008) Driver fatigue detection based on eye tracking. In: Emerging trends in engineering and technology. ICETET'08. First international conference on IEEE, pp 649–652. http://ieeexplore.ieee.org/document/4579980/
7. Local binary patterns. From Wikipedia, the free encyclopedia. https://en.wikipedia.org/wiki/Local_binary_patterns. Last accessed 09 Sept 2019

8. UP Squared* AI Vision X Developer Kit. Documentation and Downloads. https://software.intel.com/en-us/iot/hardware/up-squared-ai-vision-dev-kit#documentation. Last accessed 09 Sept 2019
9. OBD Solutions (2019) What data is available from OBD? http://www.obdsol.com/knowledgebase/on-board-diagnostics/what-data-is-available-from-obd/. Last accessed 09 Sept 2019
10. Janani N, Vasudevan SK, Suresh C, Srivathsan S, Shiva Jegan RD (2014) Investigation on life rescue technologies on road-airbags and anti-lock braking system (ABS). Res J Appl Sci Eng Technol 8(17):1884–1890
11. Peter M, Vasudevan SK (2017) Accident prevention and prescription by analysis of vehicle and driver behavior. Int J Embed Syst 9(4):328–336

Rising Star Evaluation Using Statistical Analysis in Cricket

Amruta Khot, Aditi Shinde, and Anmol Magdum

Abstract In cricket domain, the team players are selected on the basis of various factors like runs, strike rate, etc. Here, a concept of co-players is introduced for predicting the rising stars for better team selection. Various essential features for batsmen and bowler are evaluated. Analysis for selecting the best features is done. Using machine learning techniques, the actual prediction of top 10 rising stars is done. Also, a concept of all-rounder is introduced besides batting and bowling domain and is taken into account for the prediction. This is matched with the International Cricket Council (ICC) status or position of player.

Keywords Cricket · Statistics · Support vector machine · Data analysis

1 Introduction

A star in cricket is the one who owns a high stable profile graph, and rising star in cricket is the one who currently owns a low profile but has an ascending currier graph and may become the stars of tomorrow. So, for better team selection, it is important to predict the future stars.

A. Khot · A. Shinde (✉) · A. Magdum
Walchand College of Engineering, Sangli, India
e-mail: aditi.shinde@walchandsangli.ac.in

A. Khot
e-mail: amruta.khot@walchandsangli.ac.in

A. Magdum
e-mail: anmol.magdum@walchandsangli.ac.in

A. K. Tripathy et al. (eds.), *Advances in Distributed Computing and Machine Learning*, Lecture Notes in Networks and Systems 127, https://doi.org/10.1007/978-981-15-4218-3_31

This needs a lot of aspects to be taken into consideration. A player has to be evaluated by all means to decide whether he could be a rising star. In this work, we define a concept of co-players, i.e., a player is evaluated considering not only his own performance but also of other player he has played in a common time span. Team performances and opposite team performance are also taken into account. This is because every time the play conditions may not be same, and a player can learn playing with stars and grow his profile. Along with good batsmen and bowlers, we need all-rounders in a team. Here, we are introducing the concept of all-rounders which will have its own feature set for a player's evaluation. A different feature set for each domain (batting, bowling, all-rounder) is formulated. The scores for each player are evaluated. Analysis for selecting the best features and those that are positively co-related and negatively co-related are evaluated. A machine learning model—support vector machine—is built, and the prediction of rising stars is done for each domain.

2 Related Work

Daud et al. [1, 2] implemented machine learning algorithm as classification system for rising star concept for analysis of co-author networks in their ranking. Also, other machine learning mythologies like support vector machines (SVM), classification and regression tree, i.e., CART, were implemented for data analysis with classification in consort with other classification algorithms like Bayesian network (BN), naïve Bayesian (NB) which are generative classifiers. But these systems are not suitable for cricket domain. Also, Mukherjee proposed a gradient network strategy to implement the notion of co-players in ranking system to calculate rank of player. This strategy recommends the batsmen as co-players those face off with same bowlers, or the bowlers those played bowling with the same batsmen. Consequently, these players are graded by Google's PageRank algorithm for ranking system. The phenomenon of co-players was not implemented with consideration of other parameters like tournament played by sportsman's on his home ground or any other away grounds, types of pitches like suitable for fast bowling or slow bowling. Also, strike rate of player in tournament series, win/loss count by team in which player is playing, such metrics have not considered by author. Further, Haseeb Ahmad evaluated list of rising stars in cricket domain considering co-players performances and compared with ICC rankings. He considered various evaluation metrics for ranking and predicted the RSs using SVM. But the concept of all-rounder was never incorporated [1–3].

This work is an extension to the previous work incorporating ranking list for three domains—batting, bowling and all-rounder.

3 Proposed System

3.1 Support Vector Machine

Support vector machine (SVM) is a discriminative classifier trained for classification and analysis of data based on standard data set. And it is said to be supervised learning model, where data is plotted in space as point. Then that point based on its coordinates is classified by plotting hyperplane in two categories: (1) rising star (2) not rising star [3].

4 Feature Set Evaluation

4.1 Batting Domain [1, 4]

Name of feature	Formula	Description
Co-batsmen runs	$\mathrm{CR}(B) = \sum \mathrm{CoBR}$	Run score of co-batsman and player
Co-batsmen average	$\mathrm{CA}(B) = \frac{\sum \mathrm{CoBR}}{\sum \mathrm{CoBWF}}$	Proportion of co-batsman sum to wicket
Co-batsmen strike rate	$\mathrm{CSR}(B) = \frac{\sum \mathrm{CoBR}}{\sum \mathrm{CoBBF}}$	Proportion of co-batsman run score in balls of 100 faced
Team average	$\mathrm{TA}(B) = \frac{\sum \mathrm{TR}}{\sum \mathrm{TWF}}$	Proportion of sum of scored runs of team against total wickets of team in tournament
Team strike rate	$\mathrm{TSR}(B) = \frac{\sum \mathrm{TR}}{\sum \mathrm{BF}}$	Proportion of total score of batsman in 100 balls
Team win/loss ratio	$\frac{\mathrm{TW}}{L(B)} = \frac{\sum \mathrm{TW}}{\sum \mathrm{TL}} * \frac{\mathrm{TM}-(\mathrm{NR}+\mathrm{Tie})}{\sum \mathrm{TW}+\sum \mathrm{TL}}$	TW = Total win matches of team TL = Total loss matches of team TM = The total matches played by team NR—Out TM not resulted matches (Tie–draw)
Opposite teams average	$\mathrm{OTA}(B) = \frac{\sum_{i=1}^{n}\left(\frac{\sum \mathrm{OT}_i\ R}{\sum \mathrm{OT}_i\ \mathrm{WF}}\right)}{n}$	Average ratio of batsman score to opponent team wicket fallen

(continued)

(continued)

Name of feature	Formula	Description
Opposite teams strike rate	$\mathrm{OTSR}(B) = \frac{\sum_{i=1}^{n}\left(\frac{\sum \mathrm{OT}_i\ R}{\sum \mathrm{OT}_i\ \mathrm{BF}}\right)}{n}$	Ration of total runs of opponent team to total balls against given batsman
Opposite teams win/loss ratio	$\frac{\mathrm{OTW}}{L(B)} = \frac{\Sigma(\frac{\Sigma \mathrm{OT}_i\ W}{\Sigma \mathrm{OT}_i\ L} * \mathrm{TM\ I} - \mathrm{NR}\ I + \frac{\mathrm{Tie}_i}{\Sigma \mathrm{OT}_i\ W} + \Sigma \mathrm{OT}_i\ \mathrm{L})}{n}$	$(\mathrm{OT}_i\ W)(\mathrm{OT}_i\ L)$ – Win and loss number of matches by opponent team OT_i against a specific team TM_i, total number of matches of opposite team played against particular team

4.2 *Batting Domain [1, 4]*

Name of feature	Formula	Description
Co-bowlers average	$\mathrm{CBA} = \frac{\Sigma \mathrm{Co\ Bow\ CR}}{\Sigma \mathrm{Co\ Bow\ WT}}$	Ratio of total runs of bowlers against wickets caught by all his co-bowlers in tournaments
Co-bowlers strike rate	$\mathrm{CBSR} = \frac{\Sigma \mathrm{Co\ Bow\ TB}}{\Sigma \mathrm{Co\ Bow\ WT}}$	Ratio of balls served by bowler against wickets occupied by co-bowlers in that bowling
Co-bowlers economy	$\mathrm{CBE} = \frac{\Sigma \mathrm{Co\ Bow\ CR}}{\Sigma \mathrm{Co\ Bow\ OB}}$	Number of runs conceded per over bowled by co-bowlers
Team average	$\mathrm{TA} = \frac{\Sigma \mathrm{TCR}}{\Sigma \mathrm{TWT}}$	Proportion of overall runs to the wicket fallen of team
Team strike rate	$\mathrm{TSR} = \frac{\Sigma \mathrm{TB}}{\Sigma \mathrm{WT}}$	Ratio of total bowling over against wickets in that bowling by team
Team economy	$\mathrm{TE} = \frac{\Sigma \mathrm{TCR}}{\Sigma \mathrm{TO}}$	Ratio of sum of total run score of bowlers in team against wickets taken by bowlers
Team win/loss ratio	$\frac{\Sigma \mathrm{TW} * \mathrm{TM} - (\mathrm{NR} + \mathrm{Tie})\ \sum \mathrm{TL}}{\Sigma \mathrm{TW} + \sum \mathrm{TL}}$	TW = Total win matches of team TL = Total loss matches of team TM =The total matches played by team NR—Out TM not resulted matches (Tie–draw)
Opposite teams average	$\mathrm{OTA} = \frac{\sum \frac{\sum \mathrm{OT}_i\ \mathrm{CR}}{\sum \mathrm{OT}_i\ \mathrm{WT}}}{n}$	Ratio of sum of total run score of team including batsman, bowler in team against wickets taken by bowlers
Opposite teams strike rate	$\mathrm{OTSR} = \frac{\sum \frac{\sum \mathrm{OT}_i\ \mathrm{TB}}{\sum \mathrm{OT}_i\ \mathrm{WT}}}{n}$	Average of total overs bowled by opposite team to total wickets taken

(continued)

(continued)

Name of feature	Formula	Description
Opposite teams economy	$\mathrm{OTE} = \frac{\sum \frac{\sum \mathrm{OT}_i\ \mathrm{CR}}{\sum \mathrm{OT}_i\ \mathrm{OB}}}{n}$	Ratio of sum of runs of all bowlers against their balls in match
Opposite teams win/loss ratio	$\frac{\sum\left(\frac{\sum \mathrm{OT}_i\ W}{\sum \mathrm{OT}_i\ L} * \mathrm{TM}_i - \mathrm{NR}_i + \frac{\mathrm{Tie}_i}{\sum \mathrm{OT}_i\ L} + \sum \mathrm{OT}_i\ L\right)}{n}$	$(\mathrm{OT}_i\ W)(\mathrm{OT}_i\ L)$ – Win and loss number of matches by opponent team OT_i against a specific team TM_i, total number of matches of opposite team played against particular team

These features are calculated on the basis of formula specified, and graph analysis is performed for each of them. The results of analysis are considered for positive and negative relation of features for each domain in evaluation of rising star. Then, SVM is used to predict rising star. A rank of top players is generated on basis of prediction and RS Score.

4.3 All-Rounder Domain

For evaluating an all-rounder player, features considered are:

i. Runs (CAR)
ii. Highest score (CHS)
iii. 100s (CC)
iv. Wickets taken (CWT)
v. Stumping made (CSM)
vi. Catches taken (CCT).

But, a player has to be evaluated with respect to his co-players, i.e., more points are to be given to players who have scored well with his co-players. So, each feature is evaluated as follows:

$$\mathrm{Runs(RS)} = \mathrm{sum(Co\text{-}Player\ Runs)} - \mathrm{Runs} \tag{1}$$

Similarly, all other features are evaluated.

As, big numbers like runs, highest score cannot be directly compared with wickets taken, stampings, etc. All the features are normalized. And the final RS score for all-rounder is calculated as:

$$\mathrm{RS(AR)} = \frac{(\mathrm{CAR} + \mathrm{CHS} + \mathrm{CC} + \mathrm{SWT} + \mathrm{CSM} + \mathrm{CCT})}{5} \tag{2}$$

5 Experiments and Evaluations

5.1 Data Set

The data is taken from ESPNcricinfo for years 2006–2018 for all three domains, i.e., batsman, bowler and all-rounder. As rising star is giving chance to newly entering candidates, we are limiting age of player to 30 and also considering span of 2006–2018 to analyze data, i.e., evaluation of performance of player with respect to all other co-player.

5.2 Preprocessing

To evaluate player as rising star or not rising star, it is considered the player who had played at least 20 matches. The batsmen are graded in downward order of their total runs score in all matches. Bowlers are ranked in descending order of their wickets taken. Then, weighted average (WA) is calculated:

$$\mathrm{WA}(B) = \frac{33.33 * R + \mathrm{Avg} + 33.33 * \mathrm{SR}}{100} \tag{3}$$

$$\mathrm{WA}(\mathrm{Bow}) = \frac{25 * W + 25 + (-\mathrm{Avg}) + (-\mathrm{Eco}) + 25 * (-\mathrm{SR})}{100} \tag{4}$$

And hence, data sets for each domain are generated.

5.3 Feature Analysis

The above feature analysis can be depicted as: If the graph (red part) is inclined more toward right side, i.e., more frequency, the feature will be considered as positively correlated, and if not, then it will be negatively correlated. From above, it is noted that co-player features and team features are positively correlated, while the opposite team features are negatively correlated. Hence, in the RS score formula, all positively correlated features are added, and others are subtracted (Figs. 1 and 2).

5.4 Rising Star Prediction

The above feature analysis can be depicted as: If the graph (red part) is inclined more toward right side, i.e., more frequency, the feature will be considered as positively

Fig. 1 Co-batsman run

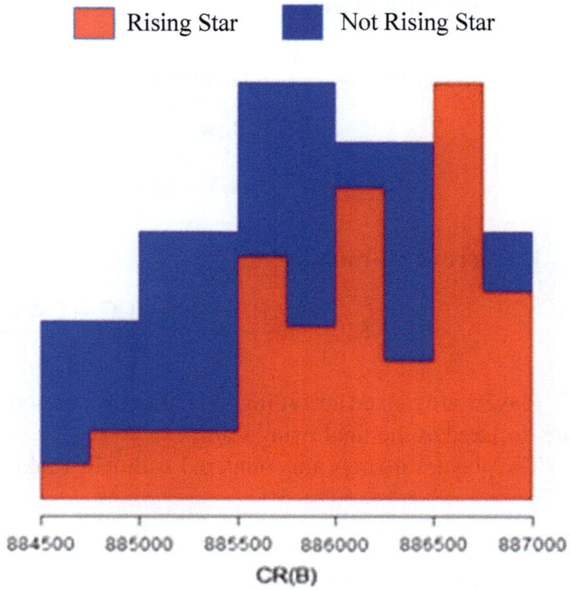

Fig. 2 Team win/loss ratio

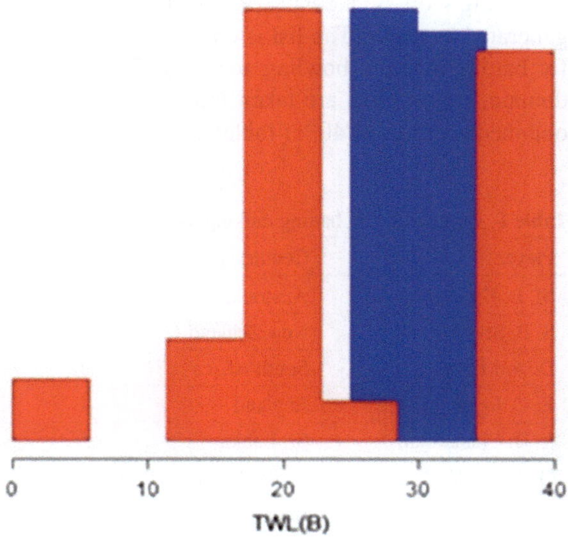

correlated, and if not, then it will be negatively correlated. From above, it is noted that co-player features and team features are positively correlated, while the opposite team features are negatively correlated. Hence, in the RS score formula, all positively correlated features are added, and others are subtracted.

RS score for bowler:

$$S(\text{Bow}) = \frac{\text{TW}}{L} + \text{OTE(Bow)} + \text{OTA(Bow)} + \text{OTSr(Bow)} - \text{CBA(Bow)} - \text{CBE(Bow)} - \text{CBSR(Bow)} - \text{TA(Bow)} - \text{TE(Bow)} - \text{TSR(Bow)} - \frac{\text{OTW}}{L} \quad (5)$$

RS score of all-rounder:

$$\text{RS(AR)} = \frac{\text{CAR} + \text{CHS} + \text{CC} + \text{CWT} + \text{CSM} + \text{CCT}}{5} \quad (6)$$

Players are ranked according to their rising star score, and top 10 for each domain are declared as the final rising stars.

The player rankings are compared with ICC rankings of the cricket sportsman.

6 Result and Analysis

The rising score is calculated for given data set, and top 10 batsman and bowlers list generated as result. The list is compared with ICC and accuracy obtained as: 60% for batting domain, bowling accuracy: 70%. On the basis of batting and bowling domain, six features are taken into consideration for all-rounder evaluation. And obtained accuracy is 40% (Tables 1, 2 and 3).

Table 1 Comparison of batting ranking list with ICC ranking

Player	Country	RS rank	Highest ICC ranking
M. L. Hayden	Australia	1	1 (2008)
S. B. Styris	New Zealand	2	16 (2008)
D. A. Miller	South Africa	3	19 (2015)
E. C. Joyce	England	4	24 (2016)
L. D. Chandimal	Sri Lanka	5	16 (2012)
Q. de Kock	South Africa	6	3 (2016)
G. Gambhir	India	7	8 (2010)
R. G. Sharma	India	8	3 (2016)
V. Kohli	India	9	1 (2013, 2018)
K. C. Sangakkara	Sri Lanka	10	2 (2015)

Table 2 Comparison of bowling ranking list with ICC ranking

Player	Country	RS ranking	Highest ICC ranking
J. J. Bumrah	India	1	1 (2018)
Dawlat Zadran	Afghanistan	2	14 (2016)
Rashid Khan	Afghanistan	3	2 (2018)
J. R. Hazlewood	Australia	4	3 (2018)
Junaid Khan	Pakistan	5	9 (2014)
K. Rabada	South Africa	6	7 (2018)
M. A. Starc	Australia	7	1 (2015)
D. W. Steyn	South Africa	8	2 (2013)
T. A. Boult	New Zealand	9	5 (2018)
M. J. McClenaghan	New Zealand	10	14 (2016)

Table 3 Comparison of all-rounder ranking list with ICC ranking

Player	Country	RS ranking	Highest ICC ranking
M. Ashraful	Bangladesh	1	14 (2018)
C. K. Kapugedera	Sri Lanka	2	16 (2013)
T. M. Dilshan	Sri Lanka	3	1 (2015)
Shakib Al Hasan	Bangladesh	4	1 (2018)
Mohammad Hafeez	Pakistan	5	2 (2018)
D. A. Miller	South Africa	6	20 (2014)
R. A. Jadeja	India	7	2 (2013)
J. C. Buttler	England	8	30 (2007)
A. D. Mathews	Sri Lanka	9	91 (2014)
B. J. Haddin	Australia	10	55 (2006)

7 Summary

Ranking list of rising stars for all the three domains is evaluated using features available. Data from 2006 to 2018 is taken into account. Considering various co-player features positively and negatively correlated are identified and RS score is formulated. SVM model is trained, and further applicable rising stars are noted. And the final list of rising stars is derived, according to their RS score. The result obtained based on computation on features states player ability depends not only on his performance but also on performance of other players and team which ensures good evaluation system.

The project can be extended for consideration of other features like team diversity, home or away ground, 50s for batsman, wickets for bowlers to incorporate better result. Methodology can also be implemented for different domains. Other methodology over SVM can also be used to improve accuracy.

References

1. Daud A, Muhammad F, Dawood H, Dawood H (2015) Ranking cricket teams. Inf Process Manage 51(2):62–73
2. Ahmad H, Daud A, Wang L, Hong H, Yang Y, Dawood H (2007) Prediction of rising stars in the game of cricket. Access 2682162
3. Li X-L, Foo CS, Tew KL, Ng S-K (2009) Searching for rising stars in bibliography networks. In: Proceedings of the International conference on database systems for advanced applications, pp 288–292
4. Daud A, Ahmad M, Malik M, Che D (2015) Using machine learning techniques for rising star prediction in co-author network. Scientometrics 102(2):16871711

A Knowledge Evocation Model in Grading Healthcare Institutions Using Rough Set and Formal Concept Analysis

Arati Mohapatro, S. K. Mahendran, and T. K. Das

Abstract A comparison of healthcare institutions by ranking involves generating their relative scores based on infrastructure, process, services and other quality dynamics. Being a top-ranking institute depends on the overall score secured against the hospital quality parameters that are being assessed for ranking. However, each of the parameters does not equally important when it comes ranking. Hence, the objective of this research is to explore the parameters which are vital one as they significantly influence the ranking score. In this paper, a hybrid model is presented for knowledge extraction, which employs techniques of rough set on intuitionistic fuzzy approximation space (RSIFAS) for classification, Learning from Examples Module 2 (LEM2) algorithm for generating decision rules and formal concept analysis (FCA) for attribute exploration. The model can be implemented using a ranking scored data for any of the specialisations (cancer, heart disease, etc.). The result would signify the connection between quality attributes and ranking.

Keywords Rough set with intuitionistic fuzzy approximation space · Formal concept analysis · Hospital ranking · Knowledge mining · Attribute exploration

1 Introduction

Healthcare sector is the prominent area which embraces the advancement of information technology indeed for their operations, modernisation and expansion. Additionally, they employ computational intelligence techniques for data analysis [1], reporting [2], medical diagnosis [3] and decision making [4, 5]. In order to compare

A. Mohapatro
Bharathiar University, Coimbatore, India

S. K. Mahendran
Department of Computer Science, Government Arts College, Udhagamandalam, India

T. K. Das (✉)
SITE, Vellore Institute of Technology, Vellore, India
e-mail: tapan.das@vit.ac.in

A. K. Tripathy et al. (eds.), *Advances in Distributed Computing and Machine Learning*, Lecture Notes in Networks and Systems 127, https://doi.org/10.1007/978-981-15-4218-3_32

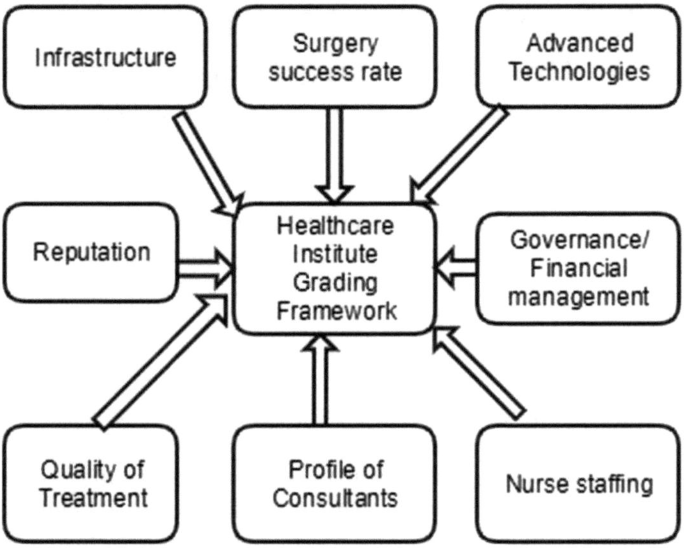

Fig. 1 Healthcare institution grading framework

service rendered by healthcare providers, research has been undergone in the direction of designing ranking framework based on various service quality [6] factors. In this study, we attempt to trace the key characteristics that make the healthcare institutions as top performers while comes ranking. With the mushrooming of hospitals, making a search of the finest hospital has been confusing in a setting where cost is not a factor as most of the citizens have adequate insurance coverage. Let us say a person is suffering from cancer; it is obvious to search for a best hospital in cancer care. It is being assessed by quality of treatment provided in the hospital as informed in public reporting, especially by government or by reputed agencies [7]. US news release the best hospitals list for sixteen specialisation, including cancer, diabetes and endocrinology, ophthalmology, etc. Few comprehensive analyses have been carried out on the subject of US news best hospital ranking. One research argues that lower 30-day mortality is the most important attribute for a hospital to come under 'best hospital' list. Another research has given importance to different parameters. It is evident that the different parameters are examined during ranking and one such frame work is represented in Fig. 1 explicitly for healthcare organisations.

2 Preliminaries

Rough set theory (RST) developed by Pawlak [8] is employable in an information system if there exists an equivalence relation between the objects, which in turn can be realised if attribute values of information system are exactly identical. However, it is rare in practice as real-life data is numerical. In order to make the prerequisite for data analysable a less strict form, fuzzy proximity relation is defined by exercising rough set with fuzzy approximation space [9], whereas intuitionistic fuzzy proximity relation applies the technique of rough with intuitionistic fuzzy approximation space.

A specific data, which needs interpretation, is developed by FCA in such a manner that the investigation is being portrayed visually depicting the association between input data and implicated attributes. Formal concept analysis (FCA) being a data representation and analysis technique developed by Wille [10] streamline the process of visual representation of data, transforms a tabular data into a lattice structure known as concept lattice. Notwithstanding the fact, data representation techniques like linked list, tree, semantic net, frames and conceptual graph prevail; distinctiveness of FCA lies in building a concept lattice for representation. RST and FCA are two competing and complimentary tools for data exploration and analysis, and hence, many researchers tried to trace the distinction in their functioning [11]. In other hands, integrating rough set and FCA can furnish a new insight into knowledge exploration [12, 13], knowledge representation and data analysis [14].

3 Proposed Research Design

This section proposes a model for knowledge mining, and the proposed framework is represented in Fig. 2. A sample data consisting of few objects is selected. Each object has a numerical score basing on which it has been ranked. The objects have few associated attributes, and each attribute is associated with a value for an object. Prior to actual analysis, attributes are being processed to confirm whether they are

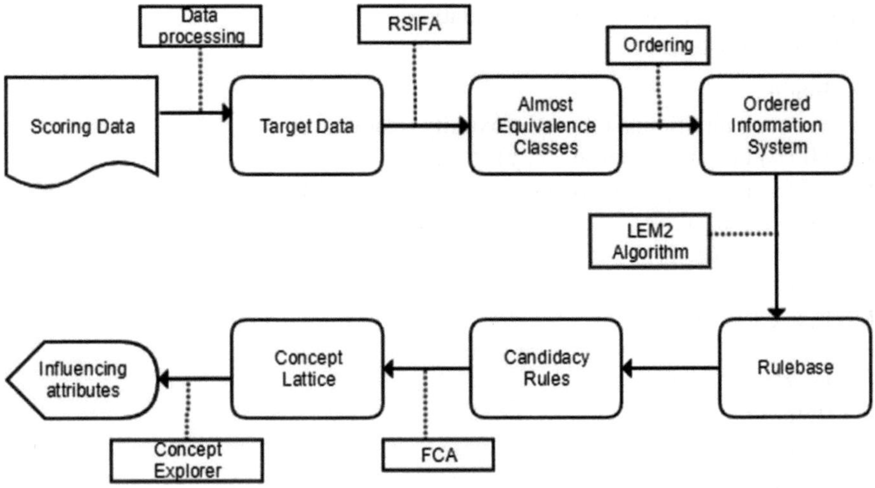

Fig. 2 Abstract view of proposed model

significant or not for classification purpose. We calculate mean and standard deviation (SD) for each set of attributes. The attributes having SD as zero are being omitted from further processing.

3.1 Phase-I (Classification)

Once the target data is prepared, each attributes are being subjected to the rough set on intuitionistic fuzzy approximation (RSIFA) space.

In this process, we compute (α, β)-equivalence class based on the intuitionistic fuzzy proximity relation as given in Eqs. (1) and (2). The membership and non-membership relations of the intuitionistic fuzzy proximity relation R to identify the almost indiscernibility among the objects x_i and x_j has been defined as below:

The degree of belongingness (μ) and degree of non-belongingness (ν) between two objects x_i and x_j,

$$\mu_R(x_i, x_j) = 1 - \frac{|V_{x_i}{}^{a_i} - V_{x_j}{}^{a_i}|}{\text{Max}(V^{a_i})} \tag{1}$$

$$\nu_R(x_i, x_j) = \frac{|V_{x_i}{}^{a_i} - V_{x_j}{}^{a_i}|}{2 \times \text{Max}(V^{a_i})} \tag{2}$$

$V_{x_i}{}^{a_i}$ and $V_{x_j}{}^{a_i}$ are the values of objects x_i and x_j for attribute a_i. Having fuzzy proximity relation, we propose an algorithm for further processing to derive the nearly equivalence classes.

Algorithm (Algorithm for classification)
Input: Intuitionistic Fuzzy proximity relation
Output: Classes

1. Begin
2. For each intuitionistic fuzzy proximity relation, do
3. {If ($\mu_R(x, y) \geq \alpha$ and $\nu_R(x, y) \leq \beta$)
4. {If ($\mu_R(x, y) \geq \alpha$ and $\mu_R(y, z) \geq \alpha$ and $\nu_R(x, y) \leq \beta$ and $\nu_R(y, z) \leq \beta$)
5. Then $C_1 = (x, y, z)$
6. else $C_2 = (x, y)$
   }
7. Else $C_3 = (x)$
   }
8. Do $C = C_1 \cup C_2 \cup C_3$
9. Return C
10. End

We derive the (α, β)-equivalence class by using the above algorithm. Further, we transform it into categorical classes by imposing order relation on the classes. Hence, an ordered information system is obtained.

Table 1 Specimen data

Objects	a_1	a_2	a_3	a_4	a_5	a_6	Decision
o_1	1	5	1	3	1	2	1
o_2	2	1	5	1	1	5	1
o_3	1	3	4	1	1	4	2
o_4	2	2	3	3	1	4	1
o_5	1	3	1	1	1	2	2
o_6	2	1	4	1	2	4	1
o_7	1	3	2	2	2	1	2
o_8	1	3	2	3	1	2	2
o_9	1	8	4	1	2	5	1
o_{10}	2	5	2	1	2	1	1

3.2 Phase-II (Rule Generation)

The outcome of phase-I; an ordered information system (OIS) serves as the input for this phase. In phase-II, LEM2 algorithm [15] is executed over OIS resulted, and we illustrate how rules are produced by an example.

Presuming an information system with ten objects $o_1, o_2, o_3, o_4, o_5, o_6, o_7, o_8, o_9, o_{10}$ and their associated attributes $a_1, a_2, a_3, a_4, a_5, a_6$ accompanying with decisions attribute as shown in Table 1, we compute the rules by applying rule generation algorithm. Each object has six conditional attributes apart from a single decision attribute. The system is processed to obtain candidate decision rules.

Employing the algorithm, a set of decision rules based on the attributes $a_1, a_2, \ldots, a_6$ are generated with their supporting objects. The candidate rules generated are represented in Table 2. For example, rule [1] is denoted as $2 \times \times \times \times \times 1$. This leads to the following decision rule: if $a_1 = 2$, then the value of the decision attribute is 1 and this pattern is satisfied by objects o_2, o_4, o_6, o_{10}. Similarly, we can derive other decision rules on considering various combinations of attributes. Hence, a rule base consisting of all the rules is composed which would be further analysed in the next phase for attribute exploration.

3.3 Phase-III (Attribute Exploration)

In continuation from last phase, a cross table of the decision rules for the entire decision classes is computed. This table has rows, which represent objects and columns, which represent attributes. The association between them is represented by a cross. We formed a cross table of the decision rules for decision class 1 as given in Table 2.

Table 2 Candidacy rule base

Rule	a_1	a_2	a_3	a_4	a_5	a_6	Decision	Supporting objects
[1]	2	×	×	×	×	×	1	o_2, o_4, o_6, o_{10}
[2]	×	1	×	×	×	×	1	o_2, o_6
[3]	×	2	×	×	×	×	1	o_4
[4]	×	×	×	×	2	×	1	o_6, o_7, o_9, o_{10}
[5]	×	×	×	×	×	5	1	o_2, o_9
[6]	×	×	×	3	×	4	1	o_4
[7]	×	3	×	×	×	×	2	o_3, o_5, o_7, o_8
[8]	×	×	1	×	×	×	2	o_1, o_5
[9]	×	×	×	×	×	2	2	o_1, o_5, o_8
[10]	1	×	×	×	1	×	2	o_1, o_3, o_5, o_8

The context table for decision class 1 which has six objects, and each object is represented as an individual row is given in Table 3. The relation between the object and the attribute is marked by cross. Afterwards, concepts, implications and association rules are generated. Finally, we generate a lattice where the nodes represent formal concepts, and the lattice diagram is shown in Fig. 3. The lattice diagram portrays the relation of subconcept and superconcept.

The subconcept–superconcept relation is transitive. It indicates that a concept is the subconcept of any concept that can be reached by travelling upwards from it. It is

Table 3 Cross table of decision class 1

	A12	A21	A22	A43	A52	A64	A65
O1	X						
O2		X					
O3			X				
O4					X		
O5							X
O6				X		X	

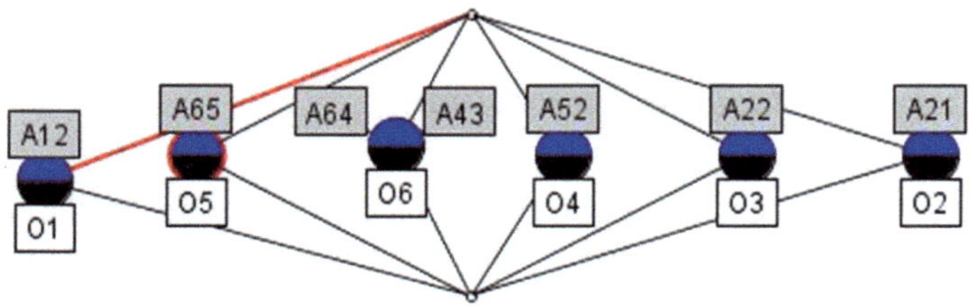

Fig. 3 Lattice diagram

evident in the figure that a concept is the pair between object and its linked attributes, and one such is represented as ({O6}, {A43, A64}). Similarly, another concept is expressed as ({O1}, {A12}).

4 Conclusion

Most of the literature discusses about ranking, which rely on the quality parameters. In those cases, an overall score is generated basing on the parameter values which are used for comparison and hence ranking. In contrary, this paper presents a framework whose working principle is quite opposite. However, in this research as in the case of reverse engineering, given the ranking data, we explore the associated attributes and their specific values, which influence ranking significantly. Thus, we have a root cause analysis statistics for various states of ranking. In the next research, the proposed hybrid model would be well explained by a contemporary data of healthcare organisations. The obtained result can be validated by the statistical approach to avoid any suspicion. Further research can be carried out by employing any of the optimisation techniques such as genetic algorithm or swarm intelligence for optimising the generated rules.

References

1. Prabhakar Kalepu RN (2014) Service quality in healthcare sector: an exploratory study on hospitals. IUP J Mark Manage 13(1):7–28
2. Lupo T (2016) A fuzzy framework to evaluate service quality in the healthcare industry: an empirical case of public hospital service evaluation in Sicily. Appl Soft Comput 40:468–478
3. Das TK, Mohapatro A (2018) A system for diagnosing hepatitis based on hybrid soft computing techniques. Ind J Public Health Res Dev 9(2):235–239
4. Das TK (2016) Intelligent techniques in decision making: a survey. Ind J Sci Technol 9(12):1–6
5. Akbar S, Akram MU, Sharif M, Tariq A, Khan SA (2018) Decision support system for detection of hypertensive retinopathy using arteriovenous ratio. Artif Intell Med 90:15–24
6. Ladhari R (2009) A review of twenty years of SERVQUAL research. Int J Qual Serv Sci 1(2):172–198
7. Gadaras I, Mikhailov L (2009) An interpretable fuzzy rule-based classification methodology for medical diagnosis. Artif Intell Med 47:25–41
8. Pawlak Z (1982) Rough sets. Int J Comput Inf Sci 11:341–356
9. De SK (1999) Some aspects of fuzzy sets, rough sets and intuitionistic fuzzy sets. PhD Thesis, IIT, Kharagpur
10. Wille R (1982) Restructuring lattice theory: an approach based on hierarchies of concepts. In: Rival I (ed) Ordered sets. Volume 83 of NATO advanced study institutes series. Springer, Berlin, pp 445–470
11. Yao Y (2004) Concept lattices in rough set theory. In: Proceedings of 2004 annual meeting of the North American fuzzy information processing society, vol 2, pp 796–801
12. Acharjya DP, Das TK (2017) A framework for attribute selection in marketing using rough computing and formal concept analysis. IIMB Manage Rev 29:122–135

13. Wang L, Liu X (2008) A new model of evaluating concept similarity. Knowl-Based Syst 21:842–846
14. Zhao Y, Halang WA, Wang X (2007) Rough ontology mapping in E-business integration. Stud Comput Intell 37:75–93
15. Grzymala-Busse JW, Stefanowski J (2001) Three discretization methods for rule induction. Int J Intell Syst 16:29–38
16. http://www.hcahpsonline.org

Image Compression Based on a Hybrid Wavelet Packet and Directional Transform (HW&DT) Method

P. Madhavee Latha and A. Annis Fathima

Abstract In this paper, a Hybrid Wavelet Packet and Directional transform (HW&DT) method is proposed to compress an image effectively. The image is pixel de-correlated using Daubechies Wavelet Packet transform and set of wavelet packet coefficients are transformed using Directional transform. The Directional transform pairs DCT-II, DCT-V, DCT-VIII and DST-VII are useful in retrieving the texture information in different directions. Then the coefficients of the hybrid transform are uniformly quantized and entropy coded using Huffman coding, generating the Bit-Stream. The performance evaluation of the proposed work is done using the compression parameters like Structural Similarity Index (SSIM), Bit-Saving (%), Peak Signal to Noise ratio (PSNR) and Mean Square Error (MSE). The experimental results confirm the improvement in Bit-Saving for the image.

Keywords Discrete cosine transform (DCT) · Wavelet packet transform (WPT) · Uniform quantization · Huffman coding

1 Introduction

With the ease of digital photography, the number of images loading on the internet and social media is increasing rapidly. According to recent statistics on the number of images uploading in social media, WhatsApp is handling billions of photos every day [1], 300 million photos per day are uploaded on Facebook [2]. This enormous increase in the number of photos on the internet needs a lot of storage space, which can be solved using compression. Compression is a technique which stores the images or data in less space by removing the redundancy. Redundancy is the irrelevant and duplicate information present in the image which can be removed without affecting

P. Madhavee Latha · A. Annis Fathima (✉)
Vellore Institute of Technology, Chennai Campus, Chennai, India
e-mail: annis.fathima@vit.ac.in

P. Madhavee Latha
e-mail: madhaveelatha.p2016@vitstudent.ac.in

A. K. Tripathy et al. (eds.), *Advances in Distributed Computing and Machine Learning*, Lecture Notes in Networks and Systems 127, https://doi.org/10.1007/978-981-15-4218-3_33

the information of that image. Inter-pixel spatial, Inter-pixel temporal, Coding and Psychovisual are the four types of redundancies.

The basic blocks of a traditional Compression technique are Pixel decorrelation, Quantization and Entropy coding. The Pixel Decorrelation in an image can be done in two ways: Prediction [3] and Transforms. Transform is the popularly used technique for the decorrelation of pixels. In literature, many transforms are used to reduce the correlation among pixels in an image like Discrete Fourier Transform (DFT), [4] Karhonen Loeve Transform, Discrete Cosine Transform (DCT) [5–12] and Discrete Wavelet Transform (DWT) [13, 14]. The second step of image compression is Quantization [15–17], which results in loss of data. The compression technique which results in loss of data due to quantization is called Lossy compression. Psychovisual redundancy is exploited using quantization. The last step in image compression is Entropy coding. This step helps in reducing the coding redundancy by encoding the information of an image efficiently into bit-stream.

Though the traditional DCT-II is often used because of its efficient energy compaction, different directional adaptive transforms are also used. In [18], the authors proposed a directional 2-Dimensional (2D) separable DCT transform. First, 1D DCT is applied along a pre-defined direction other than the horizontal and vertical directions. The second 1D transform is applied to the rearranged coefficients in a vertical direction to the first transform. Zhao et al. [19] proposed an optimized transform where the encoder can select the optimum transform from multiple transforms based on rate-distortion criteria. A secondary transform proposed in [20], is applied on the 8×8 low-frequency coefficients of DCT-II transform. Zhao et al. [21] proposed a transform which is a combination of separable multiple primary transforms and a non-separable secondary transform trained in off-line. In the literature, the traditional DCT and directional DCT transforms are used. Instead, other transforms like wavelets and wavelet packets along with the directional DCT can be used to improve the performance of the compression system.

For the compression of images, a method Hybrid Wavelet Packet and Directional Transform (HW&DT) is proposed. In this work, the image is pixel decorrelated using the hybrid transform followed by Uniform quantization and entropy coding. The proposed method is explained in Sect. 2 and the experimental results are given in Sect. 3 and this work is concluded in Sect. 4.

2 Hybrid Wavelet Packet and Directional Transform (HW&DT) Method

In this work, a Hybrid Wavelet Packet and Directional Transform (HW&DT) is proposed for the compression of images. The block diagram of the proposed method for image compression is shown in Fig. 1.

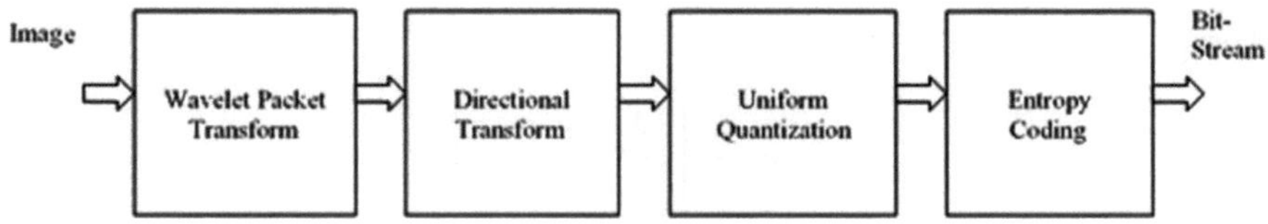

Fig. 1 The block diagram of the proposed Hybrid Wavelet packet and Directional Transform (HW&DT)

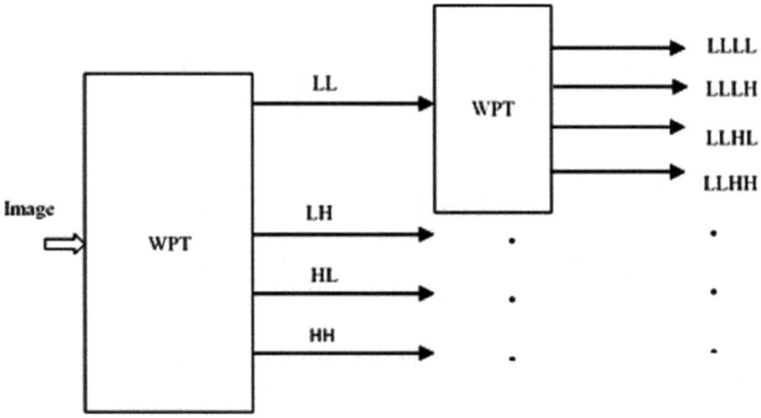

Fig. 2 The Wavelet Packet Transform (WPT) decomposition

The image is de-correlated at the pixel-level using HW&DT. Initially, the input image is decomposed up to the second level using the Daubechies Wavelet Packet transform (Fig. 2).

2.1 Directional Transform

The coefficients of the Daubechies Wavelet Packet transform are then transformed using Directional transform. The Directional transform is a set of multiple Discrete Sine and Cosine transforms taken from the 16 variants of discrete sine and cosine transforms which are generated based on different symmetric periodic extensions. These transforms are efficient in capturing the texture patterns of an image. Image can be expanded into series of Basis images, which satisfy the unitary property. The Definitions of the Basis function $T_i(j), i, j = 0, 1, 2, \ldots, N - 1$ of N-point Directional transforms used in HW&DT are given below.

$$\text{DCT-II: } T_i(j) = w_0\left(\frac{2}{N}\right)^{\frac{1}{2}} \cos\left(\frac{\pi i(2j + 1)}{2N}\right) \tag{1}$$

where

$$w_0 = \begin{cases} \frac{1}{\sqrt{2}} \; i = 0 \\ 1 \; i \neq 0 \end{cases}$$

$$\text{DCT-V: } T_i(j) = w_0 w_1 \left(\frac{2}{2N-1}\right)^{\frac{1}{2}} \cos\left(\frac{2\pi ij}{2N-1}\right) \tag{2}$$

where

$$w_0 = \begin{cases} \frac{1}{\sqrt{2}} \; i = 0 \\ 1 \; i \neq 0 \end{cases}, \quad w_1 = \begin{cases} \frac{1}{\sqrt{2}} \; i = 0 \\ 1 \; i \neq 0 \end{cases}$$

$$\text{DCT-VIII: } T_i(j) = \left(\frac{4}{2N+1}\right)^{\frac{1}{2}} \cos\left(\frac{\pi(2i+1)(2j+1)}{4N+2}\right) \tag{3}$$

$$\text{DST-VII: } T_i(j) = \left(\frac{4}{2N+1}\right)^{\frac{1}{2}} \sin\left(\frac{\pi(2i+1)(j+1)}{2N+1}\right) \tag{4}$$

The basis function of transform pairs DCT-II & DCT-II, DST-VII & DCT-V, DST-VII & DCT-VIII and DST-VII & DST-VII are as shown in Fig. 3.

The second level coefficients of the wavelet packet transform are fed to directional transform pairs DCT-II & DCT-II, DST-VII & DCT-V, DST-VII & DCT-VIII and DST-VII & DST-VII respectively. Then the resulted coefficients of the proposed hybrid transform are uniformly quantized, which helps to reduce the psychovisual redundancy. The Uniform quantizer is given in equation

$$F_Q(u,v) = \begin{cases} 0; & \text{if } |F(u,v)| < Q \\ \operatorname{sgn}(F(u,v)) X \frac{|F(u,v)|}{Q}; & \text{Otherwise} \end{cases} \tag{5}$$

where $F(u,v)$ is the transform coefficient, $F_Q(u,v)$ is the quantized transform coefficient and Q represents the quantization step size.

The quantized transform coefficients are entropy coded, which eliminate the coding redundancy. In this work, Huffman coding is used to encode the coefficients.

The Decompression process of the bit-stream is the reverse process of image compression. That is, the bit-stream is Huffman decoded followed by de-quantization. The inverse transform of these coefficients results in the decompressed image.

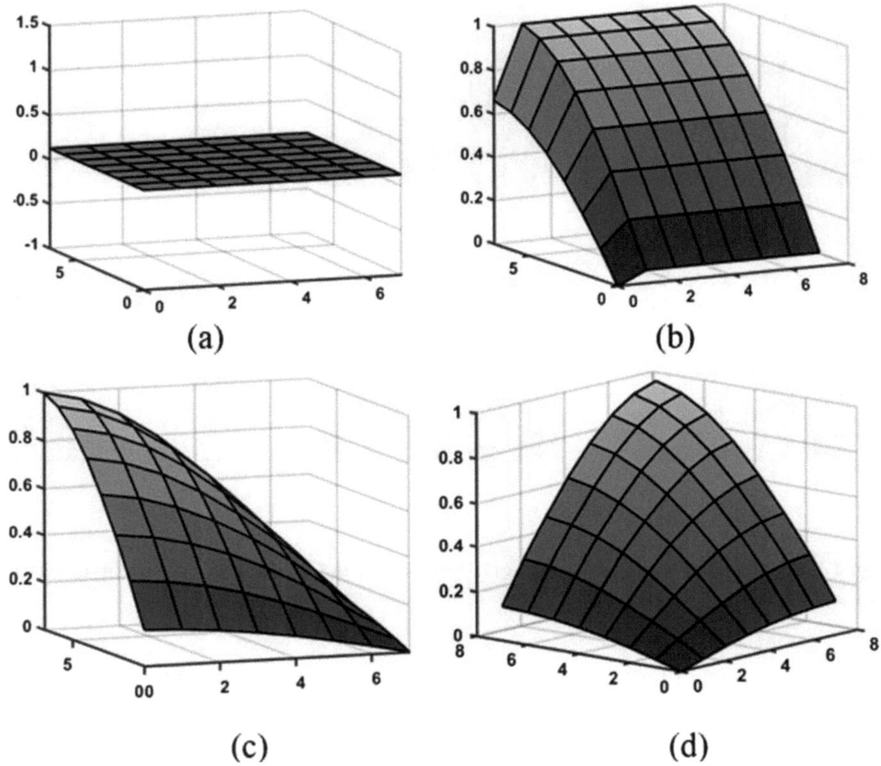

Fig. 3 The basis function plots of Directional transform **a** DCT-II & DCT-II, **b** DST-VII & DCT-V, **c** DST-VII & DCT-VIII, **d** DST-VII & DST-VII where in each transform pair, first is horizontal transform and second is vertical transform

3 Experimental Results

This section describes the performance evaluation of the proposed work. The experimentation is carried using MATLAB (version: R2019a) software platform and Intel(R) Core (TM), i5, 2.60 GHz CPU, 12 GB RAM based computer system. Standard Digital image processing images Lena, Barbara and Baboon images are considered for compression.

The decompressed images of the HW&DT method for the test images are displayed in Fig. 4.

The performance of the proposed HW&DT transform method is evaluated. The metrics like Structural Similarity index (SSIM), Mean Square Error (MSE), Bit-Saving and Peak Signal to Noise ratio (PSNR) are used for the performance evaluation of the proposed work. SSIM measures the similarity of the original and decompressed images using the covariance of the images. Mean square error gives a measure of the difference between the original and reconstructed images. Bit-Saving, which is the

Fig. 4 Decompressed images using HW&DT technique with Quantization step size (**a–c**) with $Q = 1$ and (**d–f**) with $Q = 15$

main objective, signifies the amount of memory or disc space saved by compressing the image.

3.1 *Structural Similarity Index (SSIM)*

The parameter SSIM indicates the similarity between two images. The maximum value of SSIM is '1', which indicates the two images are the same. The formula for SSIM is given by:

$$\mathrm{SSIM}(x, y) = \frac{(2\mu_x\mu_y + c_1)(2\sigma_{xy} + c_2)}{(\mu_x^2 + \mu_y^2 + c_1)(\sigma_x^2 + \sigma_y^2 + c_2)} \tag{6}$$

where x, y are the images, μ_x, μ_y are averages of x and y, σ_x^2, σ_y^2 are variances of x and y, σ_{xy} is the covariance of x and y, c_1, c_2 are constants, $c_1 = (k_1L)^2$, $c_2 = (k_2L)^2$ where L is the dynamic range of the pixel values, $k_1 = 0.01$ and $k_2 = 0.03$.

The SSIM values of the proposed HW&DT compression are lesser than DWT-DCT II and slightly higher than the DWT-DT method for the test images. The SSIM values of Lena, Barbara and Baboon images using proposed compression are as given in Table 1. In general, the SSIM values of any image depends on the texture and the direction features present in that particular image.

Table 1 SSIM value comparison of Lena, Barbara and Baboon image using different methods

Q	SSIM values of Lena image			SSIM values of Barbara image			SSIM values of Baboon image		
	DWT-DCT II	DWT-DT	HW&DT	DWT-DCT II	DWT-DT	HW&DT	DWT-DCT II	DWT-DT	HW&DT
1	0.9994	0.9991	0.9993	0.9994	0.9993	0.999	0.9997	0.9997	1
2	0.999	0.9964	0.9965	0.9995	0.9976	0.998	0.9998	0.9989	0.999
3	0.9989	0.9929	0.993	0.9991	0.9953	0.995	0.9997	0.9978	0.998
5	0.9981	0.9811	0.9813	0.9986	0.9885	0.989	0.9994	0.9945	0.995
7	0.9973	0.9669	0.9669	0.9981	0.9756	0.976	0.9992	0.9896	0.99
8	0.9969	0.9599	0.9597	0.9977	0.9716	0.971	0.9991	0.9866	0.987
10	0.9958	0.9344	0.9342	0.997	0.9671	0.967	0.9988	0.9797	0.98
11	0.9952	0.9391	0.9391	0.9966	0.9547	0.955	0.9986	0.9751	0.975
13	0.9938	0.9355	0.9363	0.9956	0.9136	0.914	0.9981	0.968	0.968
15	0.9922	0.8992	0.8993	0.9944	0.8985	0.9	0.9977	0.9544	0.955

3.2 Bit-Saving (%)

Bit-Saving is one of the main objectives of Compression, represents the amount of space saved by compression and is given by

$$\text{Bit-Saving (\%)} = \left(1 - \frac{S_o}{S_i}\right) \times 100 \tag{7}$$

where S_i is the size of the test image and S_o is the size of the compressed file.

The Bit-Saving (%) of Lena, Barbara and Baboon images using the proposed method are given in Table 2.

Table 2 depicts that the Bit-Saving (%) of the proposed method for Lena image is higher than DWT-DCT II and the DWT-DT method by 50.77% and 1.2% respectively. The Bit-Saving (%) of any test image using the HW & DT method is always higher than the DWT-DCT II and DWT-DT methods and also given in Table 2 for Barbara and Baboon images. The Bit-Saving (%) of the proposed method is slightly higher than DWT-DT for a single image but is significant if a large number of images considered for compression.

3.3 Peak Signal to Noise Ratio (PSNR)

Peak Signal to Noise Ratio (PSNR) is one of the parameters used to measure the performance of a compression system and is given by

$$\text{PSNR} = 20 \log_{10} \frac{\text{maxvalue}}{\text{RMSE}} \tag{8}$$

where maxvalue represents the maximum pixel value of the image and RMSE is Root Mean Square Error.

Table 3 depicts the PSNR values obtained for the test images using the proposed and other methods.

The PSNR of the proposed method is less than the DWT-DCT II method and is almost equal to the DWT-DT method. Though the PSNR values are lesser than DWT-DCT II, the Bit-Saving of the proposed HW&DT technique is more than that of DWT-DCT II and DWT-DT methods. In general, depending on the texture and the direction features present in an image, the PSNR values will vary as shown in Table 3 for Lena, Barbara and Baboon images.

Table 2 Comparison of Bit-Saving (%) of Lena, Barbara and Baboon images using different techniques at different quantization levels

Q	Bit-saving (%) of Lena image			Bit-saving (%) of Barbara image			Bit-saving (%) of Baboon image		
	DWT-DCT II	DWT-DT	HW&DT	DWT-DCT II	DWT-DT	HW&DT	DWT-DCT II	DWT-DT	HW&DT
1	43.93	40.36	41.78	26.46	23.35	25.29	17.51	15.95	16.34
2	45.36	52.5	53.57	28.02	36.97	38.91	19.07	28.79	29.18
3	45.71	59.29	61.07	28.79	44.36	45.91	19.84	36.58	36.58
5	46.79	67.14	67.93	29.57	54.86	55.64	20.23	45.91	46.3
7	47.14	72.64	73.68	30.35	59.53	61.09	20.62	52.14	52.14
8	47.14	75.75	76.28	29.96	62.02	63.89	20.62	54.48	54.86
10	47.5	78	78.78	30.35	67.51	68.79	21.01	59.14	59.53
11	47.5	81.21	81.32	30.35	68.83	70.23	21.01	60.31	60.7
13	47.86	83.89	84.36	30.74	68.33	69.42	21.01	63.11	63.42
15	47.86	84.64	85	30.74	72.14	73.39	21.01	65.53	65.84

Table 3 Comparison of PSNR of Lena, Barbara and Baboon images using different techniques at different quantization levels

Q	PSNR of Lena image			PSNR of Barbara image			PSNR of Baboon image		
	DWT-DCT II	DWT-DT	HW& DT	DWT-DCT II	DWT-DT	HW& DT	DWT-DCT II	DWT-DT	HW&DT
1	59.63	57.67	58.8	57.11	57.09	58.31	57.05	56.83	58.14
2	57.24	51.87	51.98	58.77	51.75	51.95	58.79	51.61	51.89
3	56.84	48.9	48.95	55.4	48.81	48.91	56.77	48.77	48.89
5	53.93	44.6	44.65	53.31	44.8	44.85	53.96	44.73	44.75
7	51.94	42	42.01	51.62	41.66	41.67	52.12	41.9	41.91
8	51.94	41.11	41.1	50.84	40.77	40.78	51.28	40.76	40.78
10	49.88	38.85	38.85	49.53	39.52	39.59	49.89	38.88	38.88
11	49.24	39.02	39.04	48.93	38.38	38.42	49.21	38.01	38.02
13	48.03	38.42	38.49	47.72	35.73	35.68	47.99	36.72	36.7
15	46.93	36.59	36.62	46.66	34.88	34.87	46.93	35.2	35.23

3.4 Mean-Square-Error (MSE)

The Mean-Square-Error (MSE) signifies the difference between the original and reconstructed images and is given by

$$e_{\mathrm{MSE}} = \frac{1}{M \times N} \sum_{x=0}^{M-1} \sum_{y=0}^{N-1} (f(x, y) - g(x, y))^2 \tag{9}$$

where $M \times N$ denotes the size of the image and f, g are the original and decompressed images.

The average MSE of the HW&DT method is 0.0005, whereas the average MSE of DWT-DT and DWT-DCT II are 0.0001 and 0 respectively. The MSE value of test image depends on the texture and the direction features present in that particular image. The MSE of proposed work is in the acceptable range and though it is slightly higher than the other methods, the HW&DT method provides better Bit-Saving (%) than DWT-DCT II and DWT-DT methods.

4 Conclusion

Image compression stores image or data in minimum space by eliminating the redundancies, which can be achieved by employing efficient transforms. In this paper, a Hybrid Wavelet Packet and Directional Transform (HW&DT) is proposed to compress an image, which employs a cascade of Daubechies Wavelet packet transform and different discrete sine and cosine transforms. The HW&DT method provides an average Bit-Saving (%) of 70.38%, whereas DWT-DCT II and DWT-DT offer an average Bit-Saving (%) of 46.68% and 69.54% respectively. The average SSIM of HW&DT is slightly higher than that of DWT-DT and lower than DWT-DCT II. The PSNR values of the proposed work are almost equal to DWT-DT and lower than DWT-DCT II. In this work, Huffman coding is used for encoding the quantized coefficients, which can be replaced with a more efficient encoding technique to improve the performance of compression.

References

1. Statista. http://www.statista.com/topics/2018/whatsapp/. Last accessed 02 Dec 2019
2. Zephoria Digital marketing. http://www.zephoria.com/top-15-valuable-facebook-statistics/. Last accessed 02 Dec 2019
3. Strutz T (2015) Context-based predictor blending for lossless color image compression. IEEE Trans Circuits Syst Video Technol 26(4):687–695

4. Leung R, Taubman D (2005) Transform and embedded coding techniques for maximum efficiency and random accessibility in 3-D scalable compression. IEEE Trans Image Process 14(10):1632–1646
5. Song HS, Cho NI (2009) DCT-based embedded image compression with a new coefficient sorting method. IEEE Signal Process Lett 16(5):410–413
6. Ponomarenko NN, Egiazarian KO, Lukin VV, Astola JT (2007) High-quality DCT-based image compression using partition schemes. IEEE Signal Process Lett 14(2):105–108
7. Wallace GK (1992) The JPEG still picture compression standard. IEEE Trans Consum Electron 38(1):xviii–xxiv
8. Ichigaya A, Nishida Y, Nakasu E (2008) Nonreference method for estimating PSNR of MPEG-2 coded video by using DCT coefficients and picture energy. IEEE Trans Circuits Syst Video Technol 18(6):817–826
9. Ngan KN, Chai D, Millin A (1996) Very low bit rate video coding using H. 263 coder. IEEE Trans Circuits Syst Video Technol 6(3):308–312
10. Kalva H (2006) The H. 264 video coding standard. IEEE Multimedia 13(4):86–90
11. Sullivan GJ, Ohm JR, Han WJ, Wiegand T (2012) Overview of the high efficiency video coding (HEVC) standard. IEEE Trans Circuits Syst Video Technol 22(12):1649–1668
12. Zhou M, Gao W, Jiang M, Yu H (2012) HEVC lossless coding and improvements. IEEE Trans Circuits Syst Video Technol 22(12):1839–1843
13. Shi C, Zhang J, Zhang Y (2015) A novel vision-based adaptive scanning for the compression of remote sensing images. IEEE Trans Geosci Remote Sens 54(3):1336–1348
14. Christopoulos C, Skodras A, Ebrahimi T (2000) The JPEG2000 still image coding system: an overview. IEEE Trans Consum Electron 46(4):1103–1127
15. Thakur VS, Gupta S, Thakur K (2017) Hybrid WPT-BDCT transform for high-quality image compression. IET Image Proc 11(10):899–909
16. Phanprasit T (2013) Compression of medical image using vector quantization. In: The 6th 2013 biomedical engineering international conference. IEEE, Amphur Muang, Thailand, pp 1–4
17. Hernández-Cabronero M, Blanes I, Pinho AJ, Marcellin MW, Serra-Sagristà J (2016) Progressive lossy-to-lossless compression of DNA microarray images. IEEE Signal Process Lett 23(5):698–702
18. Zeng B, Fu J (2008) Directional discrete cosine transforms—a new framework for image coding. IEEE Trans Circuits Syst Video Technol 18(3):305–313
19. Zhao X, Zhang L, Ma S, Gao W (2010) Rate-distortion optimized transform for intra-frame coding. In: 2010 IEEE international conference on acoustics, speech and signal processing. IEEE, Dallas, pp 1414–1417
20. Saxena A, Fernandes FC (2012) On secondary transforms for intra prediction residual. In: 2012 IEEE international conference on acoustics, speech and signal processing (ICASSP). IEEE, Kyoto, pp 1201–1204
21. Zhao X, Chen J, Karczewicz M, Said A, Seregin V (2018) Joint separable and non-separable transforms for next-generation video coding. IEEE Trans Image Process 27(5):2514–2525

On Multi-class Currency Classification Using Convolutional Neural Networks and Cloud Computing Systems for the Blind

K. K. R. Sanjay Kumar, Goutham Subramani, K. S. Rishinathh, and Ganesh Neelakanta Iyer

Abstract Identifying a currency is the easiest of task for a person. But for a blind human, it is not always the same. Although the currencies come in all shapes and sizes, it is always possible for a blind person to get confused with the currency value. We analyze the potential usage of cloud computing systems, image processing techniques along with deep learning approaches for identifying the currency based on its denomination which can be accessed by the blind via a mobile app. We examine different approaches including public cloud and private cloud systems for training and deploying the deep learning techniques. We achieve an accuracy of up to 98% in our proposed models. We propose models which can be used under a variety of situations including cost-effective approaches, efficient approaches that suite reduced local storage, etc.

Keywords Cloud computing · Deep learning · CNN · Currency classification

1 Introduction

Currency plays an important role in our life for various transactions. Even in the digital payments era, currencies still play a vital role for the economy of any country. Currencies of various countries always come with some features to help visually challenged people to detect the currency. For example, at top-right corner in each of

K. K. R. Sanjay Kumar · G. Subramani · K. S. Rishinathh · G. N. Iyer (✉)
Department of Computer Science and Engineering, Amrita School of Engineering, Amrita Vishwa Vidyapeetham, Coimbatore, India
e-mail: ni_ganesh@cb.amrita.edu

K. K. R. Sanjay Kumar
e-mail: cb.en.u4cse17149@cb.students.amrita.edu

G. Subramani
e-mail: cb.en.u4cse17117@cb.students.amrita.edu

K. S. Rishinathh
e-mail: cb.en.u4cse17145@cb.students.amrita.edu

A. K. Tripathy et al. (eds.), *Advances in Distributed Computing and Machine Learning*, Lecture Notes in Networks and Systems 127, https://doi.org/10.1007/978-981-15-4218-3_34

the Indian currency, a fixed denomination is present which the visually challenged people can feel by touching. But the marker loses its touch after repeated uses in various places.

This problem arises the need to find an efficient solution to create and implement an automated system to recognize currency in various situations such as shopping malls, metro, bus and railway station. Automated currency recognition system can be very useful for the visually challenged persons.

Implementing a currency detection model comes with many difficulties. Currency bills exhibit much more tolerance to illumination due to the lack of specular reflection. Further, they have many details that need to be identified. In addition to that, currency recognition has other sort of problems such as shape distortion due to wrinkling and/or folding. Another obstacle in existing models is the problem of currency being torn or dirty.

While a number of people train their model on local machine, we have deployed our deep learning model in the cloud. Using cloud computing for deep learning has its own advantages. The biggest benefit of cloud computing is the elimination of depending on hardware resources. Cloud offers instances of GPU processing power for the purpose of implementing the deep learning model. It also allows large datasets to be used and managed effectively. It allows deep learning models to scale efficiently at very low cost.

2 Literature Review

We provide a brief literature review in this section which also been summarized in Table 1. In [1], author proposes a solution to the problem of currency detection through RENMINGB (RMB) in which the linear gray image transform is used to reduce the noises in the image. In [2], the proposed solution involves frequency domain extraction through preprocessing, feature extraction, pattern matching, and finally, identification of the currency is done. The features mentioned in this paper are discrete wavelet transform and several general features such as color, texture, etc.This paper also suggests for fake currency identification and authentication.

In [3], the approach used is by obtaining the image and selective selection of the features to be extracted from the image then processing is done in a step-wise manner. The image passes through grayscale conversion, edge detection, image segmentation, characteristic extraction and comparison. MATLAB is the primary tool used in this research. In [4], authors propose improved scene representation using both SVM classifier which uses multiple kernel and random forest. The region of interest (ROI) selection is done using the similarities found during the training phase.

In [5], a simple Baseline PCANet approach is used. The architecture implemented consists of two stages of processing and an output layer. These stages consist of PCA filter banks. The dataset utilized is from MNIST consists of 28×28 gray images that contain handwritten digits [0–9], texture classification CUReT Dataset and object detection on CIFAR10.

Table 1 Comprehensive literature review

References	Basic objective	Approach	Learning technique	Dataset specification
[1]	Recognition of paper currency using neural networks	Linear image Gray	Transform three layer BP NN with the structure of $50 \times 30 \times 5$	100 sheets of 5 kinds of RMB
[2]	Fake currency recognition	Frequency domain feature extraction method	Support vector machines	Indian currency of at least greater than 1000×700 pixels and 150 dpi
[3]	Paper currency verification system	Image acquisition, gray scale conversion, edge detection, feature extraction, image segmentation and comparison of images	Edge detection, segmentation	All denominations of Indian currency
[4]	Image classification	PHOG and PHOW descriptors to determine ROI and then classifiers are used	Multiple kernel SVM classifier and random forest classifier	Caltech-101, Caltech-256 30 training images per category
[5]	Deep learning baseline for image classification	Random projection network and linear discriminant analysis	Two stage PCAnet	MultiPIE dataset, extended Yale B dataset, AR dataset and FERET dataset
[6]	Currency recognition system for blind people	Hamming distance matching	Oriented FAST and rotated BRIEF (ORB) algorithm	All denominations of Egyptian currencies
[7]	Currency detection and recognition	Deep learning	Convolution Neural Network (CNN) and Single Shot MultiBox Detector	7500 pictures of 1280×720 resolution pictures

In [6] the research provides an efficient method to classify currency by utilizing a sequence of steps such as preprocessing using Gaussian blurring equation to remove noise, Segmentation using Otsu Thresholding function to convert RBG to binary, region of interest extraction using ORB extraction of features and matching using Hamming Distance.

In [7], the model proposed conveys the parameters that alter and update the features for every layer of the CNN. The model is trained in two modules positioning and image classification. In [8], a survey is done to signify various classification systems. These systems tend to use pattern recognition as their major strategy to classify various objects such as currency. In [9], authors study the methodologies used to detect and identify currencies including image acquisition, preprocessing, edge detection, image Segmentation and authentication.

3 Proposed Architecture

3.1 Cloud-Based Learning Architecture

We propose three different solutions using cloud computing for the currency value prediction Fig. (1a). We propose two types of cloud models—Public and private. *Locally trained* means the training happens in private cloud. *Globally Trained* means the training happens in public cloud such as AWS and GCP. *Globally deployed* means the trained model is deployed in cloud. Finally, *Locally deployed* means the trained model is deployed alongside the mobile app right at the user end. The proposed models are:

- **Locally Trained Locally Deployed (LTLD)** Fig. (1d): This means the learning model has been trained in a private cloud and learnt model has been deployed in the client's local device storage. The main advantage is that the model does not require an internet connection for performing prediction and no restriction on dataset size.
- **Globally Trained Locally Deployed (GTLD)** Fig. (1b): A cloud-based model trained globally on the input dataset. Similar to LTLD, this trained model is also deployed locally in the device storage. This model eliminates the need for owning a cloud data center of our own. We can train in the public cloud and save money on setup and maintenance costs for private cloud infrastructure.
- **Globally Trained Globally Deployed (GTGD)** Fig. (1c): In this model, both training and deployment happen in public cloud. While this makes the client application light in weight due to the elimination of deploying, the learning model at the client device storage. However, this also incurs a constant budget requirements to maintain the deployed model in the cloud on a continuous basis.

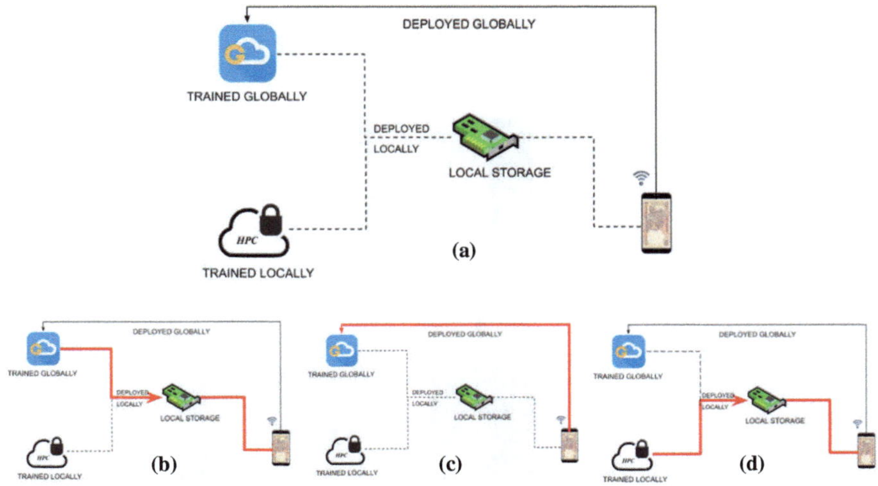

Fig. 1 **a** Overall model architecture, **b** Globally Trained Locally Deployed (GTLD), **c** Globally Trained Globally Deployed (GTGD), **d** Locally Trained Locally Deployed (LTLD)

3.2 Deep Learning Model Using CNN

Convolutional neural networks (CNN) are popularly used for image based deep learning problems which involves learning complex nonlinear features and helps in addressing multi-class classification problems, widely in literature [10].

3.3 Model Architecture

Figure 2, shows the architecture of our proposed convolutional neural network model. With the input of size $160 \times 120 \times 3$ to model, a total of around 1,732,772 parameters.

- **Conv2d_15** The CONV2d_15 layer's parameters consist of 16 learnable filters that extend through the full depth of the input volume. With an input size of $160 \times 120 \times 3$ and a filter size of 3×3, it produces 16 filtered images of size 158×118 (output: $158 \times 118 \times 16$). This layer retains almost all the information from the input.
- **Batch Normalization** A batch normalization layer is introduced to speed up the learning, stability and to reduce the amount of covariance shift. Also, batch normalization allows each layer of a network to learn by itself a little bit more independently of other layers.
- **Max_pooling2d_15** As the feature maps of convolution layers are sensitive to any modification, means a tiny movement can cause different feature map. For

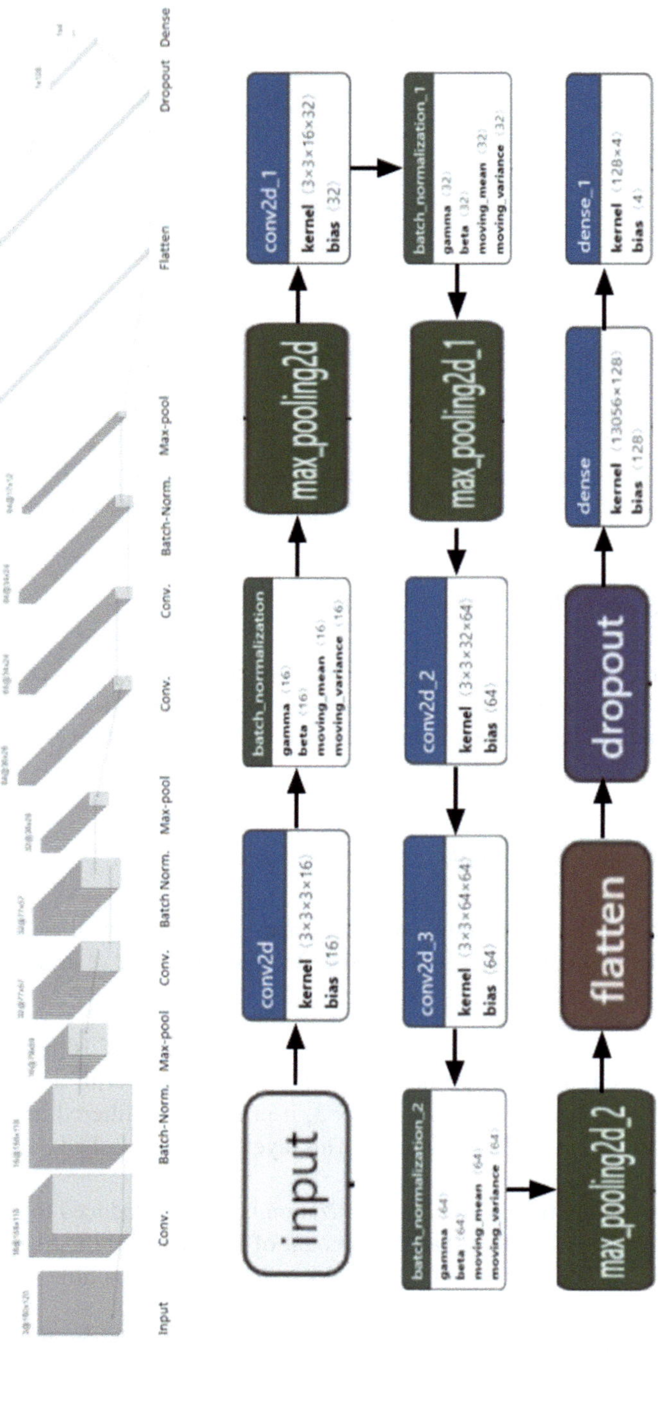

Fig. 2 Model architecture

Fig. 3 Intermediate output samples for CNN

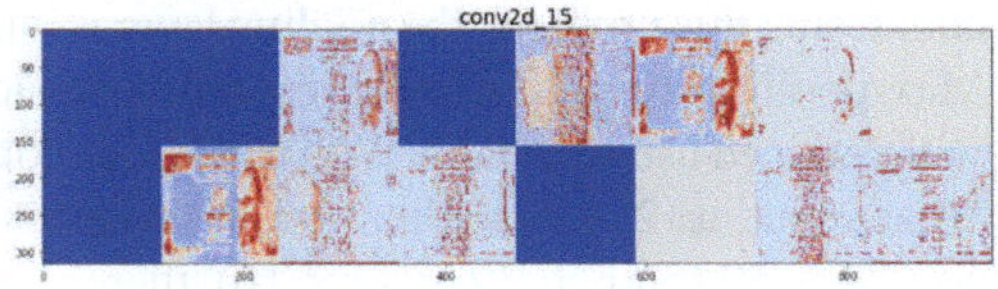

(a) Conv2d_15

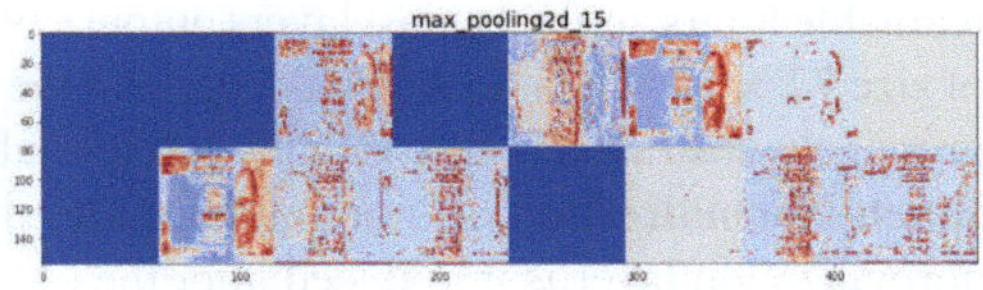

(b) Max_pooling2d_15

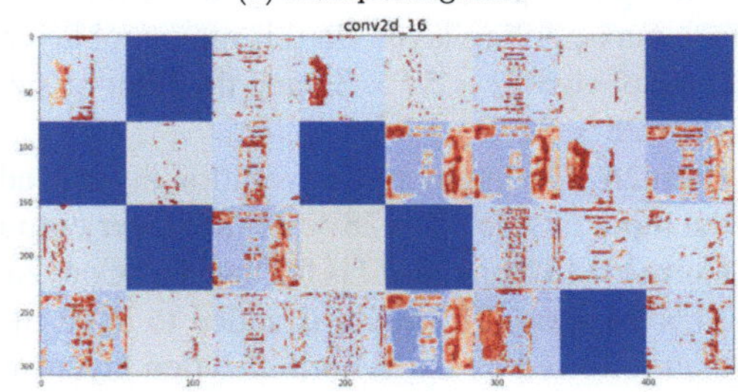

(c) Conv2d_16

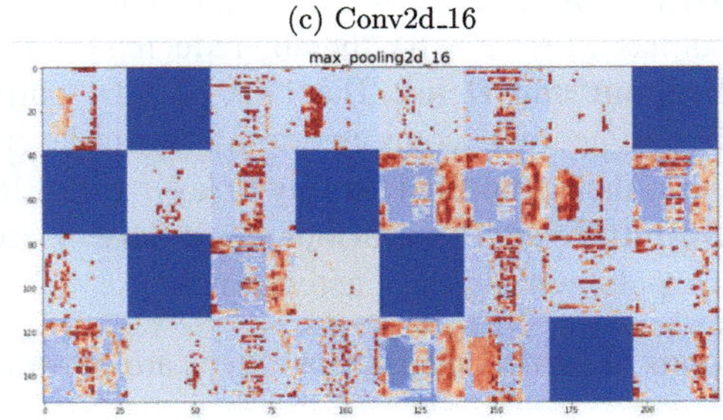

(d) Max_pooling_16

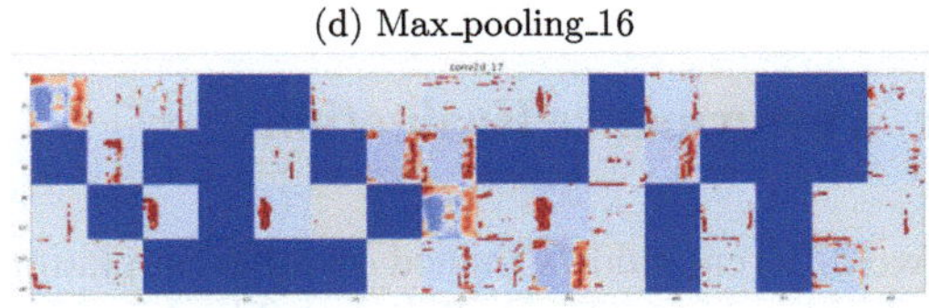

(e) Conv2d_17

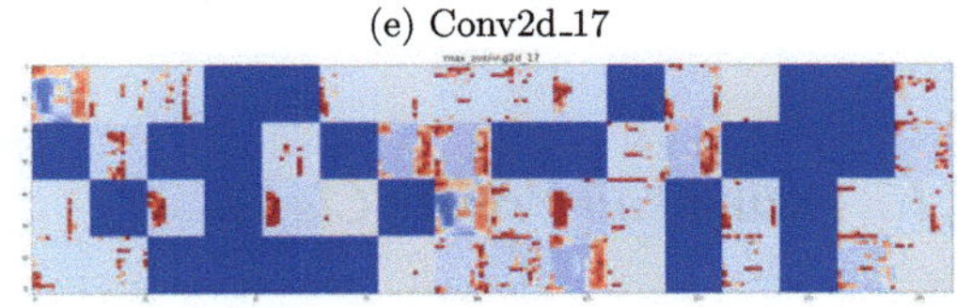

(f) Max_pooling2d_17

addressing this problem, the pooling layer is introduced after the convolution layer. Max-pooling layer is the lower resolution version of the previous convolution layer with retained useful details. With a previous convolution layer of size 158 × 118 × 16 as input, as 2 × 2 as filter size, it produces an output of 16 filtered images of size 79 × 59 (output: 79 × 59 × 16), thereby reducing the size by half.

- **Conv2d_16** Conv2d_16 layer followed by Max_pooling2d_15 consists of 32 learnable filters applied to last layers output (79 × 59 × 16) produces feature maps of size 77 × 57 × 32. As we go deeper into the layers, the activations become progressively abstract and less visually interpretable.
- **Batch Normalization** To normalize (scale) the activation of Conv2d_16, batch normalization layer is introduced after the convolutional layer.
- **Max_pooling2d_16** Generally, a convolution layer always followed by at least one max-pooling layer. So this max pool layer is introduced after the Conv2d_16 layer of size 77 × 57 × 32 and the filter of size 2 × 2 produces the output of size 38 × 28 × 32.
- **Conv2d_17** One more Convolution layer is added with 64 learnable filters. With the input of size 38 × 28 × 32, the layer outputs 36 × 26 × 64.
- **Conv2d_18** Conv2d_17 is followed by Conv2d_18 with 64 filters, the layer outputs 34 × 24 × 64. This layer is added to learn more complex features. This layer is then followed by batch normalization layer for stable training.
- **Max_pooling2d_17** With the input of previous layers size 34 × 24 × 64, this layer outputs 17 × 12 × 64, thereby reducing the number of computations by half.
- **Dropout and Dense** The 17 × 12 × 64 output of max-pooling is flattened to 13,056 neurons, connected to a dropout layer with a drop rate of 0.4. The output is then connected densely to layer with 128 neurons with Relu activation.
- **Output layer** The previous dense layer is connected densely to four neurons (outputs) with softmax activation.

Outputs of various intermediate steps are illustrated in Fig. 3.

4 Performance Evaluation

Two different models trained on different platforms, one on private cloud and other on public cloud (Google firebase) (Fig. 4).

4.1 Trained on Private Cloud

The model has been trained with a training dataset of 1500 images, each image resized to 160 × 120. The model architecture is explained in Fig. 2. The model has been trained for 270 epochs (till the desired accuracy achieved) on private cloud

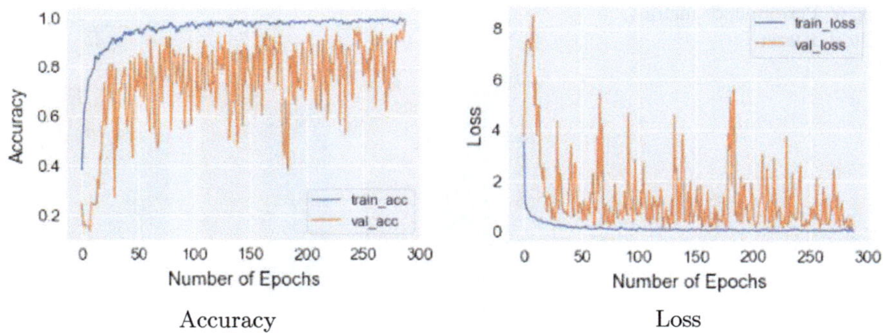

Fig. 4 Training accuracy and loss

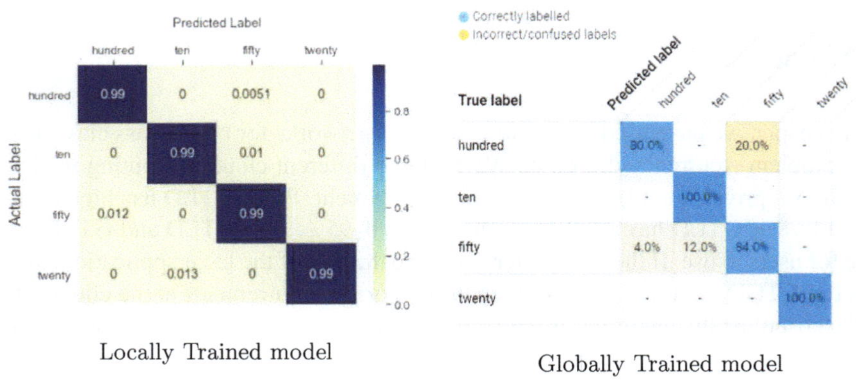

Fig. 5 Confusion matrix on training dataset

and its performance and loss values during training are shown in Fig. 4. Figure 5a confusion matrix shows the performance on train dataset.

4.2 *Trained on Google Automl*

With the limitation of the dataset size (1000 max.), Google Automl service consists of various pre-trained models that assist in achieving a certainly remarkable accuracy. The confusion matrix on the training dataset is shown in Fig. 5b.

Both the models are then tested on the test dataset (whose are outputs not known in prior). The performance metrics of both models are shown in Fig. 6. The classification report shows that locally trained model performs better than the globally trained model. One of the main reasons is the limitation of dataset size (1000) for the global model because of which it could not learn more complex features for precise prediction.

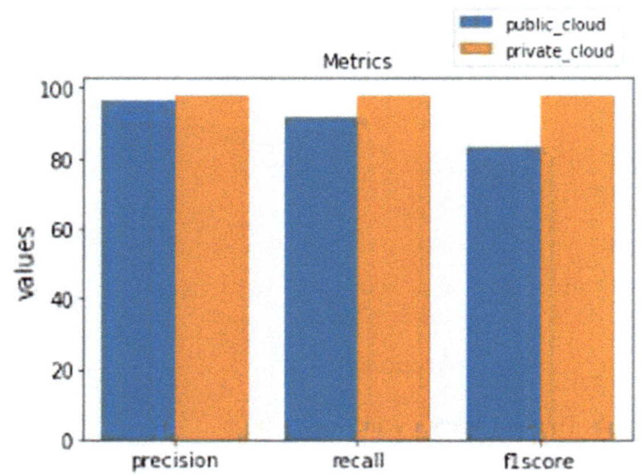

Fig. 6 Performance metrics on test dataset

5 Conclusion

In this paper, we have used convolutional neural networks for multi-class classification problem—currency detection. We explored different cloud computing models (public vs. private cloud) for training and deployment. While LTLD has high accuracy of 98%, GTLD has a similar accuracy of 90–95%. Both LTLD and GTLD are free for users to use. If the user prefer to reduce the size of the local application, they can use GTGD model which results in lower storage requirements at the client side at the cost of recurring expenses with the cloud storage service where the model will be deployed.

6 Dataset

Dataset developed for this work has been published for the benefit of public and is available in Mendeley Data [11]. We built a new dataset of 514 images of currencies belonging to four Indian denominations.

References

1. Zhang E-H, Jiang B, Duan JH, Bian Z-Z (2003) Research on paper currency recognition by neural networks. In: Proceedings of the second international conference on machine learning and cybernetics
2. Vora K, Shah A, Mehta J (2015) A review paper on currency recognition system. Int J Comput Appl 115:20
3. Mirza R, Nanda V (2012) Paper currency verification system based on characteristic extraction using image processing. Int J Eng Adv Technol (IJEAT) 1(3). ISSN: 2249-8958
4. Bosch A, Zisserman A, Munos X (2007) Image classification using random forests and ferns. IEEE

5. Chan T-H, Jia K, Gao S, Lu J, Zeng Z, Ma Y (2015) PCANet: a simple deep learning baseline for image classification. IEEE Trans Image Process 24:12
6. Yousry A et al (2018) Currency recognition system for blind people using ORB algorithm. Int Arab J e-Technol 5:1
7. Zhang Q, Yan WQ (2018) Currency detection and recognition based on deep learning. IEEE
8. Jaafar RA (2019) Survey on currency classification system. Int J Adv Res Comput Eng Technol (IJARCET) 8(8). ISSN: 2278–1323
9. Rajan GV et al (2018) An extensive study on currency recognition system using image processing. In: Proc. IEEE conference on emerging devices and smart systems, ICEDSS 2018
10. Bhavanam LT, Neelakanta Iyer G (2020) On the classification of Kathakali Hand Gestures using support vector machines and convolutional neural networks. In: International conference on artificial intelligence and signal processing (AISP) 2020, VIT, Amaravathi, India
11. Neelakanta Iyer G, Sanjay Kumar KKR, Goutham S, Rishinathh KS (2019) Indian currency dataset, mendeley data. https://doi.org/10.17632/9ccz5vsxhd.1

Predictive Crime Mapping for Smart City

Ira Kawthalkar, Siddhesh Jadhav, Damnik Jain, and Anant V. Nimkar

Abstract Crime is an ever-increasing entity. This research work aims at leveraging existing data to prevent crime in better ways than existing structures. Crime mapping technique is developed first, to provide a way of identifying and labelling crime hotspots. An algorithm for predicting crime is developed, under the domain of predictive policing. This is done with an underlying foundation of criminology theories. Finally, possible approaches to retrofit these techniques to smart cities are suggested, to provide a holistic solution to the problem of crime solving.

Keywords Smart city · Crime mapping · Predictive policing · Crime prediction · Machine learning · Clustering

1 Introduction

Crime solving has been a subject of major research interest, due to the relevance and concern of the domain in society. The advent of new technologies, along with increasing avenues for crime, opens up many possibilities for research into accurate, minimalist and cost-effective measures for crime solving. With the introduction of Smart Cities Mission in India, the deployment of these crime-solving structures will be highly useful and will add a new dimension to the concept of citizen security [5].

I. Kawthalkar (✉) · S. Jadhav · D. Jain · A. V. Nimkar
Sardar Patel Institute of Technology, Mumbai, India
e-mail: ira.kawthalkar@spit.ac.in

S. Jadhav
e-mail: siddhesh.jadhav@spit.ac.in

D. Jain
e-mail: damnik.jain@spit.ac.in

A. V. Nimkar
e-mail: anant_nimkar@spit.ac.in

A. K. Tripathy et al. (eds.), *Advances in Distributed Computing and Machine Learning*, Lecture Notes in Networks and Systems 127, https://doi.org/10.1007/978-981-15-4218-3_35

Several techniques and concepts have been researched under this domain. We look at the techniques of crime mapping and predictive policing for our research. Crime mapping is a method of identifying and denoting certain areas in a particular geographical region as crime hotspots. Predictive policing is essentially the use of predictive techniques with mathematical backgrounds in police activities to identify crimes and act accordingly.

A major gap we found in the existing literature was the gap between theoretical proposals and actual deployment of the infrastructure proposed by the theories. Implementations of the crime mapping concept need to be made such that they can be easily used by citizens and easily maintained by the police officials. Also, approaches need to be developed which are adaptive in nature to the crime of a region. Citizen perspective differs from city to city, with the status quo of the population, their demands, city ecosystem. Therefore, an implementation should be done which will suit the needs of the citizens.

Our solution aims at bridging the gaps in the existing literature by providing a very simple, holistic and pluggable component that can be easily fit into smart cities. We perform prediction of crime type and crime prediction by stacking together various classifiers. Predictions made from the data can then be utilized by police forces to deploy their resources in a more cost and time-effective manner. We also predict the location at which the crime would occur, lacking which the whole idea of predictive policing will not work.

The predictive crime mapping algorithm is proposed for crime prediction. The algorithm makes use of a crime distance metric, which we define to quantify the similarity between two crimes. The algorithm evolves based on the distance metric. Upon calculation of the distance metric, we cluster crimes and associate them to a certain type. Then, we plot the predicted crime onto our geospatial system. Finally, we consider the application of these to smart cities and suggest alternatives to fit the solutions discussed in the previous sections, to Indian smart cities.

Our work aims at understanding criminal psychology, leveraging crime data to identify crime hotspots, predicting potential crimes with their location for effective policing and application of this to smart cities, to prevent crime. Section 2 summarizes the existing research that has taken place in each of the preceding domains. Section 3 outlines the methodology for predicting crime and for mapping it. Section 4 discusses the algorithm developed for crime prediction. Results are presented in Sect. 5, followed by a concluding statement and possible further extensions to this work.

2 Related Work

In this section, we briefly discuss the existing models on crime prediction along with the mapping of crimes.

Crime Mapping

A basic research work reviewed four commonly used geographic techniques for hotspot mapping [10, 18]. The research designed a Web-based software, mapping input grids onto the geographical model to provide visualizations, whereas another research takes a qualitative approach and discusses the pros and cons of crime mapping, concerning crime management.

Taking implementation into consideration [9, 13], the researchers have created a mobile application for crime reporting and spatial mapping of crimes. The application is backed by a GIS to store the geographical entities, in a geospatial database. In another real-world implementation, we looked at considered different policing agencies and did an implementation of crime hotspots in various cities in the United States.

Some researches use a network-based model to identify criminal networks [1, 14]. It uses the label propagation algorithm on the identified network to generate crime hotspots. This facilitates network identification without the need to acquire additional data about offenders' activities.

Predictive Policing

Research on predictive policing borrows concepts from traditional policing strategies [8, 12]. It combines them with an analytical module to plan patrol. Another approach develops a model to relate crime theories with mathematical formulae. It uses the attractor–precipitator concept to define crime occurrence.

Another research takes a machine learning-oriented approach for predicting crime [2]. It proposes a machine learning model for crime prediction and focuses on time-series signals derived from data collected from target cities.

Efficient predictions with different techniques have been discussed along with comparative survey [17]. One research focuses on four data mining techniques and employs a comparative approach to test those techniques. A review-based paper conducts a comparative survey on 44 papers related to crime prediction and suggests focusing on efficient data crunching, and predictions being as fast as possible.

Prediction of Location

One research proposes an Attractor Location Predictor (ALP) model [7]. It focuses on predicting the location of the victim, rather than the location of the crime.

Another research focuses on serial crimes and aims at profiling serial crimes geographically [16]. It suggests a combination of geometric (area-based) and distance-based models.

A research paper considered explains ways for forecasting crime incidents by considering the geographical location of concern that overcomes traditional boundaries [3].

Security Implementations in Smart City Context

The concept of a smart city defines a city with connected objects (devices), engaged citizens, proper streamlining of transportation, environmentally friendly principles and optimal budget planning to be a smart city [6].

The focal research considered extensively stressed upon what makes a city smart [15]. It describes various areas of potential focus along with the measures that can be taken within each area to enable contribution towards city smartness.

One of the key challenges which will specifically be faced in the Indian context is the overhaul of the existing structures in place, to accommodate smart structures [4]. The system developed should take into consideration the various strengths and weaknesses of the city, along with the possible threats faced during development and potential scope for improvement.

Another approach conducts a comparative analysis between cities in India and abroad [11]. The author argues that combining smart principles with the existing governance of the country is a major obstacle. A system which can easily be coupled with the structures in place is, therefore, desirable.

3 Predictive Crime Mapping System

Every crime follows some kind of pattern. We therefore base our work on a modification of the routine activity theory, proposed by Marcus Felson and Lawrence Cohen. Routine activity theory requires that criminal and victim come together at someplace, thereby creating an opportunity for crime. Essentially, the criminal exploits the routine activity of the victim. The victim can be protected employing guardians, etc. We add a clause to this theory, indicating that criminals will most likely commit crimes in a search space which is familiar to them. Therefore, a combination of victim's routine activity and criminal's familiar region is utilized for further work.

3.1 Architecture of the Predictive Crime Mapping System

The architecture of the system is described in Fig. 1. The initial stage of our system starts with the utilization of criminal database. The dataset is constructed (discussed further) and mapped through various visualization techniques. The system generates interactive heatmaps, as part of its visualization. This interactive heatmap would also be beneficial for police officials to assign patrolling in that zone.

3.2 Crime Dataset Construction and Preparation

The crime data for Mumbai is a dataset of over 10,000 unique crime information from six different categories of crimes. The dataset is as shown in Table 1. The dataset was generated for the city of Mumbai by collecting various past crimes details from news articles on various websites like Mid-day, Times Of India and Mumbai Mirror and consulting local police station. The dataset is similar to Indore Police Crime Dataset which can be found online. The "dateOfCrime" stores the date and time when the crime occurred, "lat" and "lon" store the latitude and longitude of crime location,

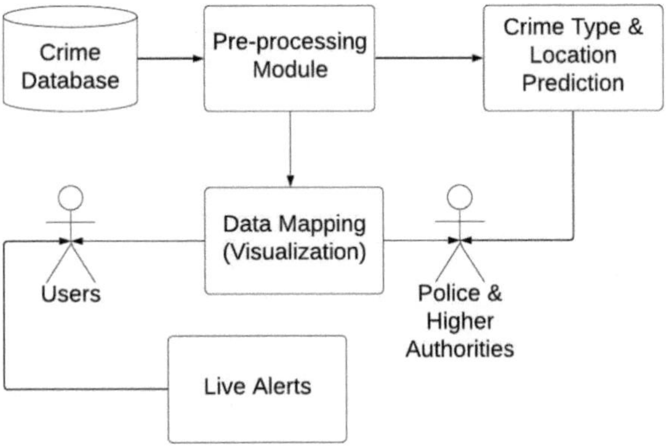

Fig. 1 Architecture of predictive crime mapper

Table 1 Snapshot of preprocessed data

dateOfCrime	lat	lon	crimeType	survey
2015-01-01 13:09	18.969839	72.999999	Violence	Train pickpocket crowded
2015-01-03 21:47	19.203346	72.958317	Accident	Train pickpocket
2015-01-07 17:14	19.114460	72.773182	Violence	Jewellery night
2015-01-07 20:42	19.145109	72.771840	Accident	Pickpocket hotel

"crimeType" stores the category of crime, and "survey" stores the features(keywords) associated with that crime records like evidence found, crime description, etc.

3.3 *Crime Prediction*

The crime prediction system is summarized in Fig. 2. It shows the whole flow starting from the raw data to prediction.

1. Preprocessing Phase: This phase focuses on the data preprocessing of the raw data. First, we clean the data, by dropping all the NaN and irrelevant features from the dataset. New features are generated from the time stamp of crime such as day, hour, dayofweek to add more relevant features to the model. Once the data is cleaned, we normalize the data using minmax normalization to get all features value on the same scale, between 0 and 1. The target column, i.e. crimeType is one hot encoded for converting it to numerical value as well as for predicting probabilities for each type.
2. Model Building and Prediction Phase: The architecture is summarized in Sect. 4.

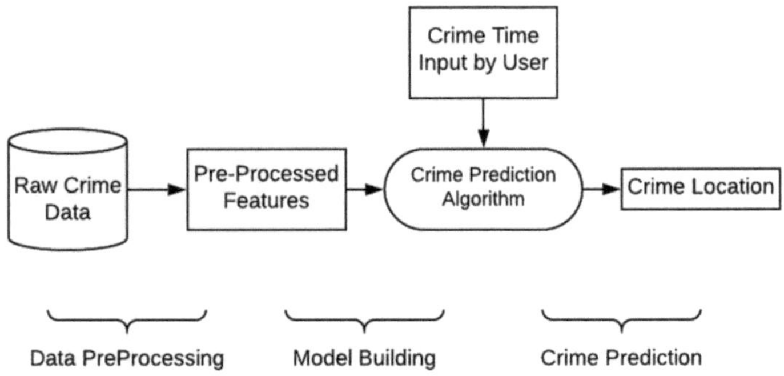

Fig. 2 Working of crime prediction component

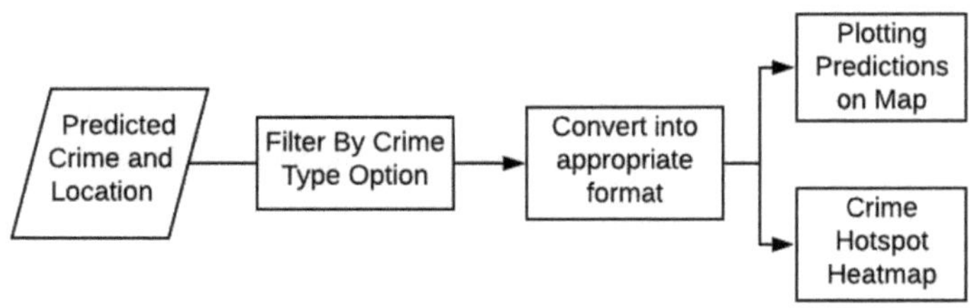

Fig. 3 Working of crime mapping component

3.4 Crime Mapping

Crime mapping can be considered as a geographical profiling of crime. The architecture of crime mapping module is explained in Fig. 3.

The crime mapping module developed is based upon a geospatial system, wherein we map the coordinates onto an actual geographical map. The module works in two parts:

1. Crime Hotspot Mapping
2. Predicted Crime Mapping.

Crime Hotspot Mapping Crime hotspots are the area on the geospatial map with high crime intensity. Once the user enters the location, we create a geofence around that point with a radius of around 2000 km. Then, we plot the heatmap of that geofence area based on the crime density of only those crime location that falls within that geofence. Areas having high crime density also known as crime hotspot are represented as red colour.

Predicted Crime Mapping Predictive crime mapping is plotting the predicted crime for better visualization and analysis for police departments. The system plots predicted crime by crime prediction model on to a geospatial map based on the location of the user. Different markers are used for different crimes along with crime details.

All possible crime predictions are mapped onto a user location within a zone of a certain radius. This is particularly useful when all possible crime opportunities need to be considered, for proper patrolling.

4 Predictive Crime Mapping Algorithm

```
Result: Crime Prediction List
clusters = empty list;
threshold = 5;
prediction list = empty list;
Crime_Matching(i, j) = Crime Distance Metric (i, j)
Take input of crime date and time
for each crimetype as crime do
    data = all crime records where Crimetype=crime
    for i in data do
        for j in data do
            if Crime_Matching(i, j) >= threshold then
                Add j to cluster i
                Mark j as visited
            end
        end
    end
    for each cluster as clusters do
        location = { } //Key-value pair
        n = 80%
        for i in range 100 do
            samples = select n random samples
            create decision tree for samples
            coord = prediction using decision tree
            location:coord += 1
        end
        result = select maximum value entry from location
        predicted location = result.key
        probability = result.value
        add predicted location, probability, crime to prediction list
    end
end
```

Algorithm 1: Predictive Crime Mapping Algorithm

The crime prediction algorithm is given in Algorithm 1. It can be divided into two parts: First part includes grouping similar crimes together, and the second includes predicting the crime location based on the similar crime groups. For grouping similar crimes together, the algorithm defines a crime distance metric, as described in Fig. 4. For similar crime observed in the dataset: $\Delta location \leq 0.1$, $\Delta datetime \leq 1$ *and* $\Delta survey \leq 1$. Hence, threshold is taken as 5.

Fig. 4 Crime distance metric

$$y = 10 * \Delta location + \Delta datetime + 3 * \Delta survey$$
$$\Delta location = \Delta latitude_{i,j} + \Delta longitude_{i,j}$$
$$\Delta datetime = \Delta hour_{i,j} + \Delta day_{i,j} + \Delta month_{i,j} + \Delta year_{i,j}$$
$$\Delta survey = \Delta survey\_token^{1}_{i,j} + .. + \Delta survey\_token^{n}_{i,j}$$
$$\Delta feature_{i,j} = |feature_i - feature_j|$$

5 Results and Discussion

Various models were evaluated using three evaluation metrics mean absolute error, mean squared error and R-squared (Table 2).

As we can observe in Fig. 5, the latitude and longitude features (crime location) are the most important feature. It is derived as the decrease in node impurity weighted by the probability of reaching that node. Higher the value, the more important the feature.

Figure 6 is the final output of the crime prediction model where nearby crimes are grouped into a crime hotspot. It represents the concentration of crimes in a certain area. The numbers indicate the frequency of crime occurrence in that specific region.

Table 2 Machine learning model performance for crime prediction

Model	MAE	MSE	R-Squared
Linear regression	0.1138	1371.96	−3411.23
Decision tree	0.1531	871.33	−7734.57
Stacked model	0.0834	288.97	−228.05

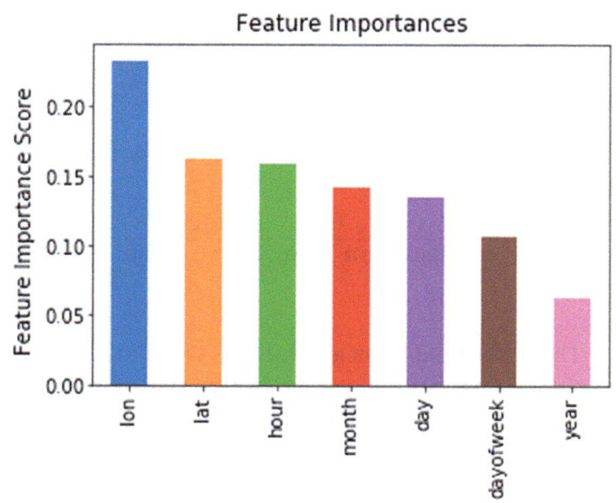

Fig. 5 Feature importance for the stacked model

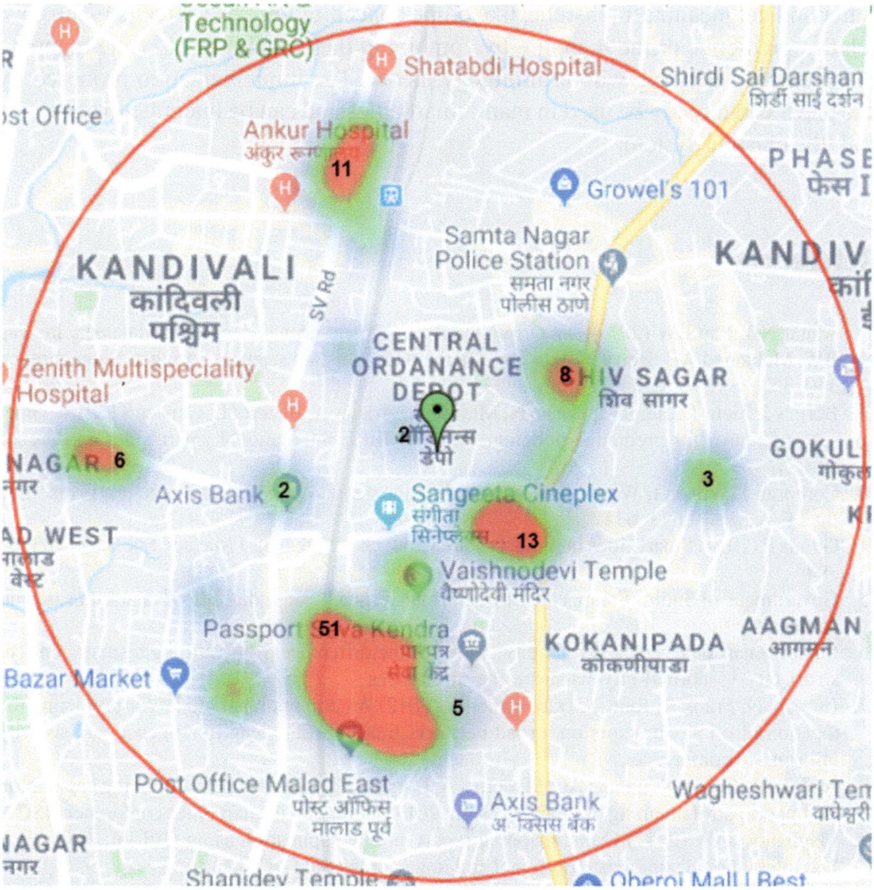

Fig. 6 Crime density for robbery

6 Conclusion and Future Scope

Different techniques such as crime mapping and predictive policing have great potential to take security to a new level, in smart cities. A perfect merger of these two methods needs to be achieved, for which this research work states an easy and reliable approach. This research not only focuses upon implementation but also derives from theoretical background to base the approach on. Founded upon the criminology theory of routine activity, the research proposes a technical implementation that will track and predict criminal activities. The research proposes a data acquisition and preprocessing approach which will then be used to predict crime according to a predictor algorithm. The performance is indicated by the results. With the development in cities, the tracked crime-related data also increases and helps the model to perform even better. Crime mapping helps to visualize the predicted crime results better and get meaningful insights like crime concentration and frequency. Thus, we achieve predictive crime mapping by combining the techniques of crime mapping and predictive policing with criminology theory. The implementation proposed by this paper can directly be used in many smart cities and can be integrated very easily with the existing system.

References

1. Anuar NB, Yap BW (2019) Data visualization of violent crime hotspots in Malaysia. In: Yap BW, Mohamed AH, Berry MW (eds) Soft computing in data science. Springer, Singapore, pp 350–363
2. Borges J, Ziehr D, Beigl M, Cacho N, Martins A, Araujo A, Bezerra L, Geisler S (2018) Time-series features for predictive policing. In: 2018 IEEE international smart cities conference (ISC2)
3. Corcoran J, Wilson I, Ware J (2003) Predicting the geo-temporal variations of crime disorder. Int J Forecast 19:623–634
4. Gunna P (2017) Smart cities in India: key areas and challenges. Int J Manage Soc Sci 4(1):386–394
5. Government of India, Smart cities mission: Government of India. http://smartcities.gov.in/content/
6. Inicio: Smart city versus normal city: what are the differences? https://blog.bismart.com/en/smart-city-vs-normal-city-what-are-the-differences
7. Iwanski N, Frank R, Reid A, Dabbaghian V (2012) A computational model for predicting the location of crime attractors on a road network. In: 2012 European intelligence and security informatics conference
8. Junior AA, Cacho N, Thome AC, Medeiros A, Borges J (2017) A predictive policing application to support patrol planning in smart cities. In: 2017 International smart cities conference (ISC2)
9. Maghanoy JAW (2017) Crime mapping report mobile application using GIS. In: 2017 IEEE 2nd international conference on signal and image processing (ICSIP)
10. Ogeto F (2018) Crime mapping as a tool in crime analysis for crime management. Int J Phytoreme
11. Sethi M (2015) Smart cities in India: challenges and possibilities to attain sustainable urbanisation. J Indian Inst Public Adm 47:20
12. Smit S (2014) Predictive mapping of anti-social behaviour. Eur J Criminal Policy Res
13. Crime mapping systems. https://www.crimemapping.com/
14. White S, Yehle T, Serrano H, Oliveira M, Menezes R (2015) The spatial structure of crime in urban environments. In: Aiello LM, McFarland D (eds) Social informatics. Springer International Publishing, Cham, pp 102–111
15. Woetzel J, Remes J, Boland B, Lv K, Sinha S, Strube G, Means J, Law J, Cadena A, von der Tann V (2018) Smart cities: digital solutions for a more livable future. Tech. rep., McKinsey Global Institute
16. Liu X, Zhang Y-B, Han J, Zhao S (2010) Finding the locations of offenders in serial crimes. In: 2010 international conference on e-health networking digital ecosystems and technologies (EDT) (2010)
17. Yu C, Ward MW, Morabito M, Ding W (2011) Crime forecasting using data mining techniques. In: 2011 IEEE 11th international conference on data mining workshops
18. Zhou G, Lin J, Ma X (2014) A web-based GIS for crime mapping and decision support. Springer, Dordrecht, pp 221–243

Dimensionality Reduction for Flow-Based Face Embeddings

S. Poliakov and I. Belykh

Abstract Flow-based neural networks are promising generative image models. One of their main drawbacks at the moment is the large number of parameters and the large size of the hidden representation of modeled data, which complicates their use on an industrial scale. This article proposes a method for isolating redundant components of vector representations generated by flow-based networks using the Glow neural network as an example to the face generation problem, which has shown effective ten times compression. The prospects of using such compression for more efficient parallelization of training and inference using model parallelism are considered.

Keywords Machine learning · Neural networks · Generative models · Embeddings · Reduction of dimensions · Image processing

1 Introduction

One of the main unresolved problems in machine learning is the generalizing ability of models: resistance to changes in input data. Such changes can lead to erroneous data classification results and other types of erroneous answers in machine learning models, which limits their applicability in conditions of insufficient data. Generative models can potentially overcome such limitations by finding dependencies in the data both with a small number of tagged data and in the absence of any markup. Such dependencies can be reused in classical machine learning models that solve prediction problems from the same domain, improving their prediction accuracy and learning efficiency.

Learning generative models can be represented as the task of extracting patterns from high-dimensional data, usually presented as a random sample from a joint distribution of dependent and independent variables. These patterns can be successfully used both for generating objects from the same distribution and for predicting certain

S. Poliakov · I. Belykh (✉)
Peter the Great St. Petersburg Polytechnic University, St. Petersburg, Russian Federation
e-mail: igor.belyh@cit.icc.spbstu.ru

A. K. Tripathy et al. (eds.), *Advances in Distributed Computing and Machine Learning*, Lecture Notes in Networks and Systems 127, https://doi.org/10.1007/978-981-15-4218-3_36

properties of given objects that are not in the original sample. Some articles also consider the probability of interpolation between hidden representations of objects from the general population, in order to obtain hidden representations of objects similar to interpolated [6, 8] or changes specific to them, such as hair color or age for objects from the distribution of facial images [6].

There are two well-known building approaches: generative models–generative adversarial networks (GANs) and likelihood-based models. Consider the subtypes of the latter approach in more detail:

1. Autoregressive models [1] for prediction sequences. The structures of such models limit the possibility of parallel computation and long-term processing of large-dimensional data, such as images and video.
2. Autocoders [1] with various optimizable functions, and, in particular, variational (Variational Autoencoder, VAE) [7] optimizing variational lower bound, evidence lower bound, ELBO is the logarithm of the likelihood of input data.
3. Flow-based generative models [3, 4, 8], which will be discussed in more detail below.

GAN architecture neural networks at the time of writing are a more popular approach for generating good quality images compared to flow-based neural networks. Comparing these two approaches, one can find the following differences between flow-based networks and adversarial ones:

- Ability to calculate latent representation and likelihood. Existing GAN models have no way to encode input into a hidden representation, nor can they calculate the likelihood of data for a model. In flow-based networks, both can be calculated directly by definition.
- An interpretable space for hidden representation of data. In GAN models, an arbitrary object from the general population cannot be represented in the hidden representation of the model, since there is no corresponding encoder for it. On the other hand, in flow-based networks, the calculation of the hidden representation for an arbitrary object occurs directly [4, 8].
- Greater computational efficiency when calculating gradients to optimize the loss function. In flow-based neural networks, this requires a constant amount of memory, regardless of the size of the input, unlike other types of neural networks.

Despite these advantages, flow-based networks also have disadvantages. These include the fact that the dimension of the vector representation (embeddings) is equal to the dimension of the simulated distribution. This limits the use of flow-based networks for the search for semantically similar objects, since it will require large resources in comparison with other methods [7, 14], and also limits their use on mobile devices and devices with a small amount of memory and computing resources. Flow-based networks can also be used to encode and generate video [9] and audio [12].

In addition, at the moment, there is no known way to semantically decompose the vector representation obtained using flow-based networks, as is done in recent studies on GAN [3].

This article discusses the proposed method to solve the problem of reducing the dimension of an embedding and some approaches for scaling training and inference of this network.

The first part of this article discusses the basic concepts of flow-based neural networks, their structure, and method of training.

The second part provides an analysis of the contents of embeddings generated by the Glow flow-based network, trained on a training set with facial images (CelebA), as well as method for their compression.

The third part presents the results of reducing the dimension of the embeddings of the Glow network using the proposed method and comparing it with known algorithms.

2 Flow-Based Networks

Flow-based neural networks were first described in [3] and finalized in [4]. We describe the basic principles of their work. Let x be a random vector of large dimension with an unknown distribution:

$$x \sim p^*(x) \tag{1}$$

Let also be given a training set D obtained by sampling from distribution (1) and a parametric model $p_\theta(x)$ with parameters θ.

In the case when the given distribution is discrete, the logarithm of the likelihood of the model is equivalent to minimizing the following functional:

$$L(D) = \frac{1}{N}\sum_{i=1}^{N} -\log(p_\theta(x_i)) \tag{2}$$

In the case where the distribution is continuous, the likelihood is optimized as follows:

$$L(D) \approx \frac{1}{N}\sum_{i=1}^{N} -\log(p_\theta(\hat{x}_i)) + c \tag{3}$$

where $\hat{x}_i = x_i + u$, $u \sim U(0, a)$ and $c = -M \cdot \log(a)$, where a is determined by the data discretization levels and M is the dimension of the objects from the sample. Both objective functions (1) and (2) measure the expected level of compression in steps or bits and are optimized by the method of stochastic gradient descent and its modifications [6].

In most flow-based generative models [3, 4, 9, 12], the distribution of data from the training set D is modeled through the distribution of hidden variable $p_\theta(z)$ transformed by a reversible transformation $g_\theta(z)$:

$$z \sim p(z) \tag{4}$$

$$x = g_\theta(z) \tag{5}$$

where z is a hidden representation and $p_\theta(z)$ has a suitable distribution density, for example, a multidimensional Gaussian distribution: $p_\theta(z) = N(z; 0, I)$. The function $g_\theta(...)$ is invertible; thus, for any object x from the distribution, obtaining a hidden representation is performed as follows:

$$z = f_\theta(x) = g_\theta^{-1}(x) \tag{6}$$

Next, we omit the θ signature from the functions f_θ and g_θ for brevity. We represent the function f in the form of a composition of functions (and similarly for the function g):

$$f = f_1 \circ f_2 \circ ... \circ f_K \tag{7}$$

For further transformations, it is necessary to use the variable substitution formula for probability density functions. In the case when the function $g : \mathbb{R} \to \mathbb{R}$ is bijective, the probability p of the corresponding area must remain unchanged after replacing the variable x by y:

$$p_Y(y)\mathrm{d}y = p_X(x)\mathrm{d}x \tag{8}$$

$$p_Y(y) = |\frac{\mathrm{d}x}{\mathrm{d}y}|p_X(x) \tag{9}$$

$$p_Y(y) = p_X(g^{-1}(y))|\frac{\mathrm{d}}{\mathrm{d}y}(g^{-1}(y))| \tag{10}$$

In the case of the vector function $g : \mathbb{R}^n \to \mathbb{R}^n$ (and the corresponding distributions):

$$p_Y(y) = p_X(g^{-1}(y))|\mathrm{det}(J(g^{-1}(y))| \tag{11}$$

$$\log(p_Y(y)) = \log(p_X(g^{-1}(y))) + \log|\det(J(g^{-1}(y))| \tag{12}$$

where J is the Jacobi matrix. The formula is logarithmic for computational stability and computational speed.

Using the variable substitution formula for equality (5), the probability density of the model according to the input data x can be written as:

$$\log(p_\theta(x)) = \log(p_\theta(z)) + \log|\det\left(\frac{\mathrm{d}z}{\mathrm{d}x}\right)| = \log(p_\theta(z)) + \sum_{i=1}^{K} \log|\det\left(\frac{\mathrm{d}h_i}{\mathrm{d}h_{i-1}}\right)| \tag{13}$$

where we define h_0 as x and h_K as z for brevity. Under the sum sign is the logarithm of the determinant of the Jacobi matrix for the functions $f_1 ... f_K$. Its value can be

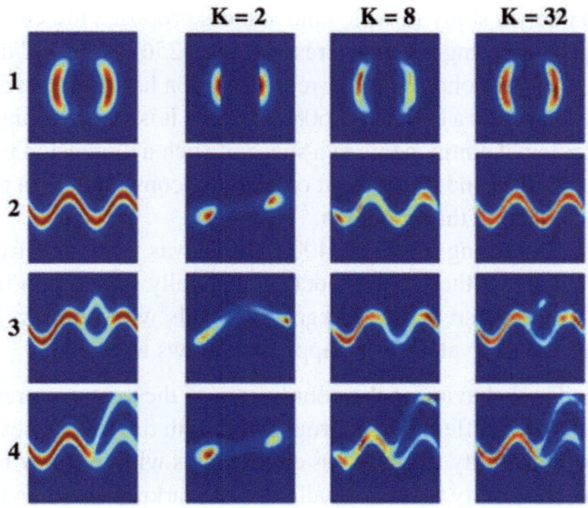

Fig. 1 Examples of restoring a given two-dimensional distribution by a flow-based network with K transformations

effectively calculated for a certain type of function, in particular for cases when the Jacobi matrix has a triangular form. For such functions, the logarithm of the determinant is calculated as:

$$\log|\det\left(\frac{\mathrm{d}z}{\mathrm{d}x}\right)| = \sum_{i=1}^{K} \log|\mathrm{diag}\left(\frac{\mathrm{d}h_i}{\mathrm{d}h_{i-1}}\right)| \tag{14}$$

Flow-based neural networks are represented as bijective layers with the log determinant defined for them. Several types of such functions were studied, which include a linear function, a rotation matrix, and others [4, 12]. An increase in the number and expressiveness of such functions in a flow-based network allows us to improve the quality of the obtained approximations of the target distributions (see Fig. 1).

3 Embeddings Dimensionality Reduction

When solving the problem of reducing the dimension, one of the generally accepted factors of compression quality is the fraction of the explained variance. When using the principal component method, it is numerically equal to the ratio of the sum of the selected eigenvalues to their total sum [11]. Without loss of generality, one can say that the fraction of the explained variance is the ratio of the sum of coordinate-wise variances of a given data set for a subset of coordinates to the total sum of variances in any selected orthogonal coordinate system. Next, we consider the case when such a system is the initial coordinates of the vectors describing the data.

To analyze the embeddings generated by flow-based networks, one of the last papers at the time of writing was chosen, in which the domain is facial images, namely a study in which the authors proposed the Glow neural network [8]. Images in the training set have a resolution of 256×256 and three color channels, therefore the corresponding vector representation has the same dimension, linearized for convenience as a list of 196,608 numbers. It is worth noting that the use of the method of principal components in a space of such a dimension requires significant computing resources and the amount of memory consumed, as a result of which it was decided to abandon this approach.

A training set D of 400 images was compiled from four randomly generated images by the Glow model by manually selecting a background in them and generating noise in the background pixels with uniform and random noise among all acceptable values. This approach allows to check:

- The behavior of the coordinates of the vector representation for images of faces with a different background and with different faces and their background.
- The ability to compress embeddings with one way to reduce dimensionality.
- The ability to use a small manual markup for the analysis of flow-based networks, instead of compiling a training set of the same orders as is required for training the model.

In addition, the choice of this approach is justified by the following observations:

- Adding noise to the whole image, depending on the amount of noise, will either worsen the quality of the face representation or create an insufficiently large variety of possible backgrounds.
- Automatic face/background segmentation is possible, but may introduce additional error.

For the compiled training dataset, vector representations were obtained using Glow, trained by the authors of the original study. Further, from all the images, the coordinate-wise standard deviation was obtained, which is presented in the form of a graph (see Fig. 2).

According to this graph, most of the coordinates have a small fraction of the total variance, namely 90% of the coordinates explains only 16% of the total variance. Further, since in Glow the hidden representation is distributed according to the normal law with zero expectation, the average of these coordinates in the entire population is also 0, and their zeroing is optimal by the Gauss-Markov theorem. This follows from the fact that, according to the least squares method, the optimal constant estimate is the average value of the target variable, which in our case is each of the replaced coordinates.

4 Flow Models Parallelism

One may find two ways to separate the model for parallel computation:

1. Layerwise—each of the three layers of the Glow model will be calculated on one of the three GPUs. This approach requires gradient synchronization, i.e., each

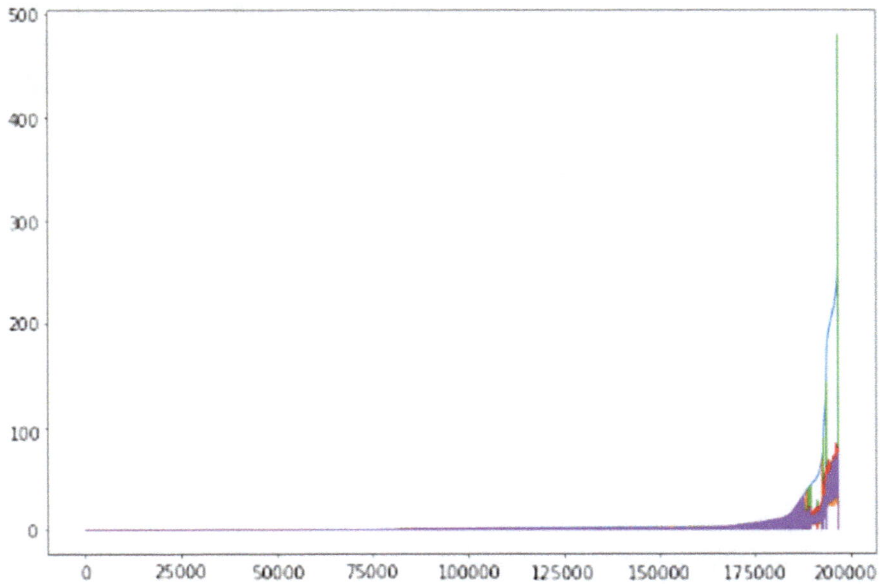

Fig. 2 Coordinate variance of vector representations sorted in order of growth. Four colors correspond to four groups of images in the training set and green color corresponds to the total variance

next layer can perform calculations only after the mini-batch has been completely processed by the previous layer (see Fig. 3).

2. Coordinate—embedding of the image is divided into N parts, for each of which a flow is made independent of the others from $K \leq L - 1$ layers, followed by combining the results of such submodels. In this approach, submodels are computed asynchronously, but their results are merged in the last layers. If the network is completely divided into N subnets with independent input and output vectors, then for each part of the resulting images the necessary coordinates from other submodels will no longer be taken into account.

Consider the first method in more detail. By dividing the model into submodels, the GPU memory requirements are reduced by several times, which implies that one can train with either a large mini-batch size or high resolution images. Moreover, despite the fact that the processing speed of one example decreases, the convergence of the loss function increases (see Fig. 4). However, the total operating time does not change much, that is, it makes sense to use this approach to generate large images. Testing this hypothesis is one of the possible developmental options.

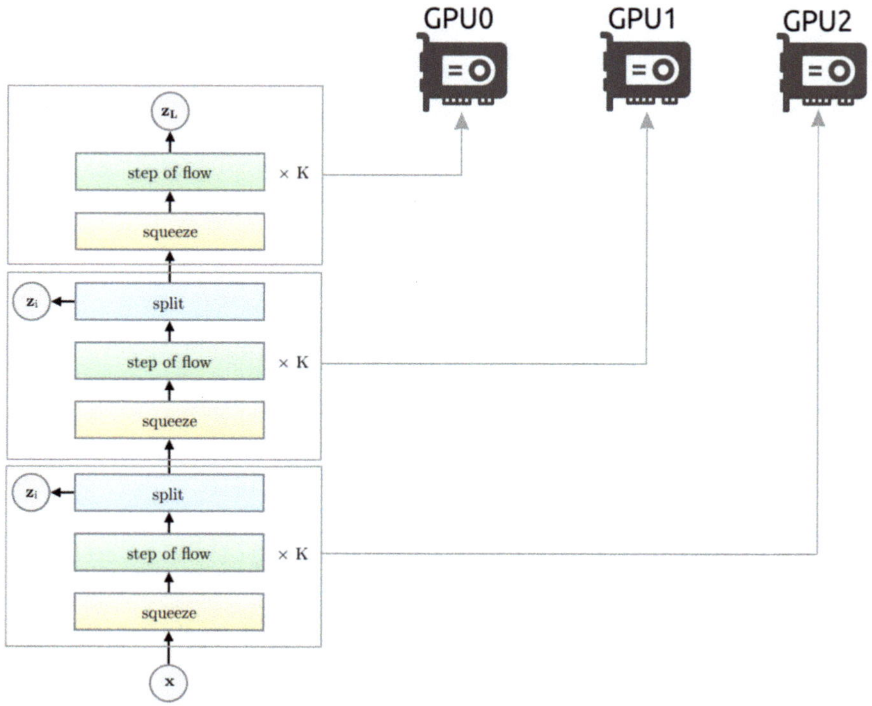

Fig. 3 Layerwise model parallelism scheme for Glow model

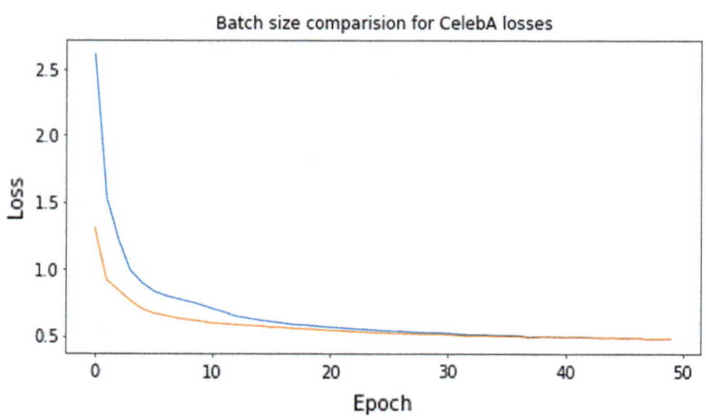

Fig. 4 The upper line stands for batch size 4, the lower one for batch size 12

Fig. 5 Example of zeroing the coordinates with the smallest dispersion: upper row is the original, lower row is the result with a modified vector representation

5 Experiments

After zeroing the least important coordinates using the described method for several randomly generated faces, the following results were obtained (Fig. 5).

As one can see, a certain part of the noise in the image has disappeared, but the overall detail has decreased. To assess the difference between compressed and uncompressed vector representations, one can apply a Glow model to them and estimate the distance between the resulting pairs of images. Often such distances are calculated on the basis of information about the visual distinguishability of colors by a person [5, 13]. To do this, they usually translate the image into the color space $L * a * b$ (CIE 1976 $L * a * b *$) or $L * u * v$, and either the Euclidean distance or the CIEDE2000 formula [13] or the more common peak signal-to-noise ratio (PSNR) [2, 10]. In this paper, the latter approach was applied for the final quality assessment.

$$\mathrm{PSNR} = 10 \cdot \log_{10}\left(\frac{R^2}{\mathrm{MSE}}\right) \tag{15}$$

where MSE is the mean squared error of coordinate-wise distance between the original image and the restored from compressed embedding. For the described zeroing method, we achieve for 100 generated images a mean PSNR = 34.51 with standard deviation 0.42.

6 Conclusions

In this article, an algorithm for compressing embeddings of flow-based neural networks was proposed. The experiment with the Glow network on the CelebA dataset shows its effectiveness. In addition, a method for model separation has been proposed for more efficient use of memory in a distributed environment for a Glow network. Both methods reduce the dimensionality curse in the use of such neural networks. In the future, one can try to reduce the size of the neural network itself and increase its inference speed, knowing about a small number of significant embedding coordinates.

Acknowledgements This research work was supported by the Academic Excellence Project 5-100 proposed by Peter the Great St. Petersburg Polytechnic University.

References

1. Ballard DH (1987) Modular learning in neural networks. In: AAAI, pp 279–284
2. Deza MM, Deza E (2009) Encyclopedia of distances. In: Encyclopedia of distances. Springer, Berlin, pp 1–583
3. Dinh L, Krueger D, Bengio Y (2014) Nice: Non-linear independent components estimation. arXiv preprint arXiv:1410.8516
4. Dinh L, Sohl-Dickstein J, Bengio S (2016) Density estimation using real NVP. arXiv preprint arXiv:1605.08803
5. Hore A, Ziou D (2010) Image quality metrics: PSNR vs. SSIM. In: 2010 20th international conference on pattern recognition. IEEE, New York, pp 2366–2369
6. Karras T, Laine S, Aila T (2019) A style-based generator architecture for generative adversarial networks. In: Proceedings of the IEEE conference on computer vision and pattern recognition, pp 4401–4410
7. Kingma DP, Welling M (2013) Auto-encoding variational Bayes. arXiv preprint arXiv:1312.6114
8. Kingma DP, Dhariwal P (2018) Glow: generative flow with invertible 1×1 convolutions. In: Advances in neural information processing systems, pp 10,215–10,224
9. Kumar M, Babaeizadeh M, Erhan D, Finn C, Levine S, Dinh L, Kingma D (2019) Videoflow: a flow-based generative model for video. arXiv preprint arXiv:1903.01434
10. Liu PY, Lam EY (2018) Image reconstruction using deep learning. arXiv preprint arXiv:1809.10410
11. Partridge M, Calvo RA (1998) Fast dimensionality reduction and simple PCA. Intell Data Anal 2(3):203–214
12. Prenger R, Valle R, Catanzaro B (2019) Waveglow: a flow-based generative network for speech synthesis. In: ICASSP 2019-2019 IEEE international conference on acoustics, speech and signal processing (ICASSP). IEEE, New York, pp 3617–3621
13. Sharma G, Wu W, Dalal EN (2005) The ciede2000 color-difference formula: Implementation notes, supplementary test data, and mathematical observations. Color Research & Application: Endorsed by Inter-Society Color Council, The Colour Group (Great Britain), Canadian Society for Color, Color Science Association of Japan, Dutch Society for the Study of Color, The Swedish Colour Centre Foundation, Colour Society of Australia, Centre Français de la Couleur **30**(1), 21–30
14. Zhang K, Zhang Z, Cheng CW, Hsu WH, Qiao Y, Liu W, Zhang T (2018) Super-identity convolutional neural network for face hallucination. In: Proceedings of the European conference on computer vision (ECCV), pp 183–198

Identifying Phished Website Using Multilayer Perceptron

Agni Dev and Vineetha Jain

Abstract Phishing is most popular in cybercrimes where a malicious individual or a group of individuals who scam users. The aim of identifying any phished website is to help the users/customers with more secure usage of online transactional websites. The research work focuses on the neural network concept which is implemented to identify phished websites. This concept is proved by multilayer perceptron (MLP)-based classification for 48 features. For result assessment, MLP is compared with other machine learning methods such as random forest, support vector machine (SVM), logistic regression and detected to have a higher accuracy of 96.80%.

Keywords Multilayer perceptron · Feature selection · Security · Phishing

1 Introduction

Phishing is basically a social engineering attack that focuses on stealing user's data and their identities.

We can identify a phished webpage through a Uniform Resource Locator (URL). A URL containing the path and file components is being taken by the attacker for phishing purposes. A URL registration is being done with a name that does not exist before and the name should not be suspicious. From the website server, the domain name is being identified. It consists of the top-level domain (TLD) which is known as a suffix and a domain name [1]. A domain name and a subdomain name put together are known as hostname. From the subdomain portions, the phisher has full control and authority to change the value. The unique part of the website domain is a free URL so only it is difficult to detect phishing domains. An attacker will choose the domain names very intelligently because it aims to convince the user and then setting free URL which makes detection difficult.

Due to the technology growth, more techniques are invented in recent times. Cybercriminals' is more often used to obtain personal information from a user. Using

A. Dev · V. Jain (✉)
Amrita University, Bangalore, India
e-mail: jain_vineetha@blr.amrita.edu

A. K. Tripathy et al. (eds.), *Advances in Distributed Computing and Machine Learning*, Lecture Notes in Networks and Systems 127, https://doi.org/10.1007/978-981-15-4218-3_37

emails, the attacker knows which individual or organization they are attacking. Sniffer helps the attackers to use web server illegally. Cybercriminals also do search engine optimization and link manipulation.

For phishing detection process based on machine learning algorithms, features are organized in different forms like URL-based features, domain-based features, webpage-based features and content-based features. Using these kinds of features helps us to identify phished and legitimate websites.

In this study, the MLP model-based classification was performed on the following 48 features [2] extracted based on features of the website in the ML Repository.

2 Related Work

Identifying a phished website is a field where many researchers have come up with a number of methodologies to overcome phished websites [3]. When a new anti-phishing technique is implemented, the attackers identify it and create a new phishing technique.

Yasin Sönmez and et al. defined 30 features of a phishing attack with 11,000 websites and proposed in a classification model. They describe all the 30 features of how phishing analysis data is classified. The input and output parameters are classified for phishing websites in the dataset based on the extreme learning machine (ELM) [4].

Rami M. Mohammad and et al. explained the effect on features for the detection of phishing websites. Using CANTINA [5], a computerized developed tool helps in feature extraction automatically without any human efforts. For feature selection, also many experimental results have been done. From those, nine features were meant to be best and the error-rate has decreased for almost all the algorithms.

Wa'el Hadia and et al. proposed a new fast way for associative classification algorithm for detecting phishing websites. They developed a new algorithm known as a fast associative classification algorithm (FACA) [6,] and it overcomes usually used associative classification (AC) algorithms as CBA, CMAR, MCAR and ECAR. The reason FACA classifier algorithm performs high classification accuracy, and outputted F1 is its way of forecasting all the test instances with all general rules, whereas in other AC algorithms they only use one rule at a time to forecast test instance.

Rami M. Mohammad and et al. proposed a model based on artificial neural network (ANN) [7] for predicting phished webpages. A webpage is identified as phished or legitimate when a feedforward neural network is trained by the back-propagation algorithm. Two traditional techniques are being explained: they are blacklist approach and heuristic approach. Website patterns and relationships are found from feature extraction with a technique named data mining. When the learning rate and momentum value are being changed, the result seems to indicate the success of using neural network's impact on phishing websites.

Mahmoud Khonji and et al. proposed different feature selection methods in the phishing domain name. For doing a classification problem [8] to achieve accuracy with respect to machine learning is a crucial part of phishing website data.

Joshua Saxe and et al. proposed a deep learning method for Fast and Format-Agnostic Detection for malicious web content phished pages [1]. Their model directly operates on a language-agnostic stream for the extraction of HTML files with a regular expression. When compared to bag-of-words, their model captures locality with hierarchical scales. They introduce two methods of inspector and master network. With traditional knowledge of domain with bias, it provides effective handling with different sizes with the help of a hierarchical spatial scale.

Phishing is not about email these days, the attackers are focusing on alternative attack vectors such as messaging, business and social media sectors. The reason phishing is difficult to stop because it is targeting the weakest link of the security chain that is the people.

Many new methodologies have been introduced in the previous works for identifying the phished website but all those methods have not proven its existence in the later future. Feature selection is one of the major challenges when choosing the optimal feature for model building. The reason for choosing MLP is basically in real-time, data will not be the same kind as our model is trained so the model has to identify its structure and act according to identify whether its phished or legitimate website. Neural networks in phishing website detection are the future.

3 Methodology

Phishing dataset is labelled as phish and legitimate since it is a classification problem [2]. These two labels are known as classes. A phished dataset will be used in the training phase to find out the best feature known as feature selection [9]. Each website is built with certain rules, so a public dataset is collected with all the rules or features with which we can identify whether the particular website is phished or not.

A wide variety of features are used for the ranking mechanism. The websites which have high-rank scores are meant to be legitimate sites [10]. One of the major used ranking services for classification for the phishing website is the correlation between features. After collecting meaningful information from the data, now we have to detect legitimate or fraudulent domain names.

In our study, we will be using multilayer perceptron (MLP) because this method is powerful and helps us to achieve a high success rate or high accuracy.

4 Implementation

The proposed work is using a multilayer perceptron algorithm. The algorithm is used on a phishing website dataset. The high-level architecture of the proposed system is mentioned in Fig. 1.

Checking the feature: There are '*n*' number of webpages and we cannot assure that all the websites are legal and trustworthy. So in order to identify the hacked website, there are some features or rules to identify those webpages. Using those rules, we can differentiate the webpage whether it is a phishing or legitimate.

Feature selection: To work on the optimal feature, the feature selection method is being done, and there are many feature selection techniques like the Gini index, entropy, information gain, chi-square method, correlation, etc. Depending upon the dataset, we have to choose the feature selection method.

K-fold cross-validation: It is also known as rotation estimation, which is a validation technique to determine the result with statistical analysis. The dataset will be split into 'K' number of times, and each time a part of the data will be kept as the

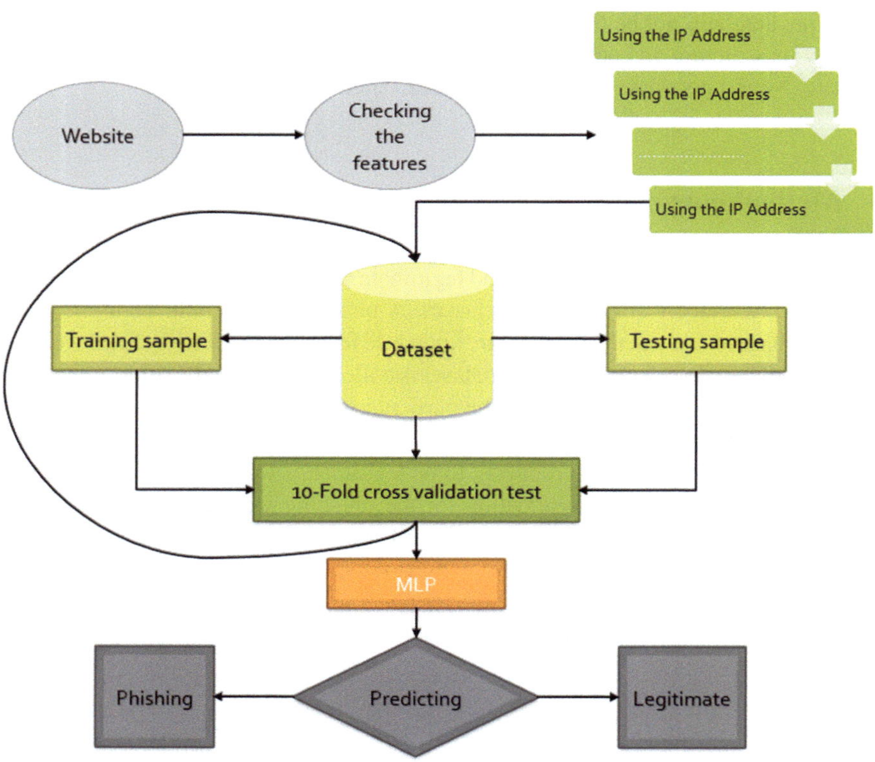

Fig. 1 Architecture diagram

validation and will verify with the rest of the data. By doing this, we can say that our model is well trained across all portions.

4.1 Dataset

The dataset used for phishing website detection is collected from PhishTank and Alexa [2]. It consists of 48 features of 10,000 webpages, and each 5000 is phishing webpage and legitimate webpages. 5000 phishing websites are being taken from PhishTank, and the other 5000 legitimate websites are taken from Alexa to verify the effectiveness of the proposed method.

This dataset is the latest updated information from January to May 2015 and from May to June 2017.

Phishing Webpage Source: PhishTank, OpenPhish
Legitimate Webpage Source: Alexa, Common Crawl.

The main process in the phishing webpage is to work on its features and how effectively it is handling the dataset. Each features or say it as rule defines the webpage, we have about 48 rules, and each individual rule contains 5000 phished webpages and 5000 legitimate webpages. From this dataset, our problem is basically as classification of two classes/labels of phished or legitimate.

4.2 Feature Selection

Feature selection is done for deciding which variable we can use in our model and which feature has a high impact on the classes. The quality of the feature which we are choosing is very important. Feature selection becomes difficult due to computational constraints and the elimination of noisy data.

In this research, we have tried three feature selection process of

Gini coefficient
Entropy
Correlation.

Gini coefficient and entropy are commonly used for impurity detection for a binary classification problem (Figs. 2, 3 and 4).

Gini index (minimize the probability of misclassification)

$$\text{Gini} = 1 - \sum_{j} p_j^2 \tag{1}$$

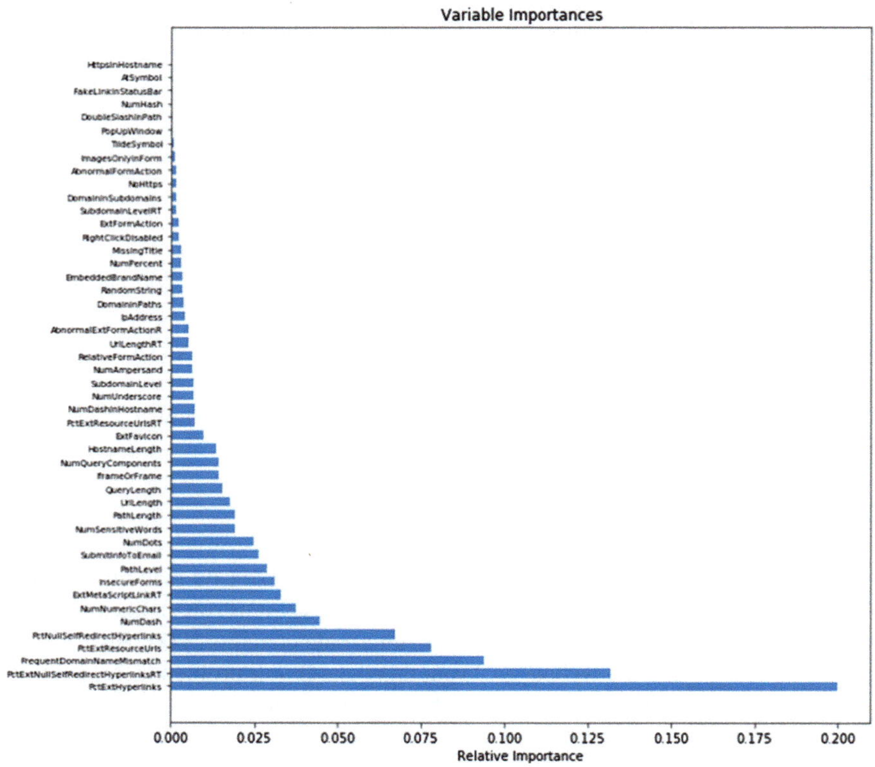

Fig. 2 Gini index feature selection

Entropy (measure impurity)

$$\text{Entropy} = -\sum_{j} p_j \log_2 p_j \tag{2}$$

Correlation (Evaluate subset of features)

$$\text{CFS} = \max_{x\in\{0,1\}^n} \left[\frac{\left(\sum_{i=1}^{n} a_i x_i\right)^2}{\sum_{i=1}^{n} x_i + \sum_{i\neq j} 2b_{ij} x_i x_j} \right] \tag{3}$$

Based on the decision tree algorithm, the Gini coefficient and entropy are plotted [11]. From these plots, we can visually see the features which are having more importance towards the class label (phishing or legitimate). From the correlation feature selection method, we will get to know how close two variables are having a linear relationship between them.

The reason for choosing these three feature selection methods is because the Gini index and entropy come under the decision tree and correlation, on the other hand,

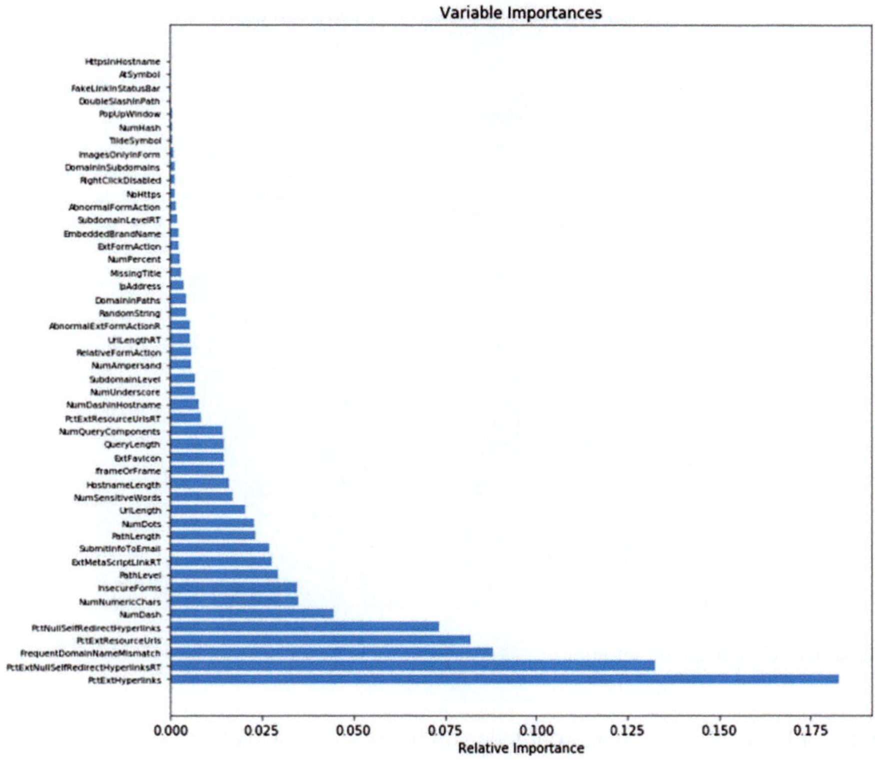

Fig. 3 Entropy feature selection

shows us how the subset is being formed and how efficient they are towards the labels [12].

From these three feature selection methods, the Gini index feature selection method is chosen for building the neural network model.

4.3 Multilayer Perceptron

MLP is basically a deep learning method. It is a feedforward neural network which generates some set of outputs from a set of inputs [13]. MLP works both as machine learning and deep learning also. Basically, in deep learning, a neuron will be connected to only some of the next layer's neurons, but in MLP each neuron is connected to all the neurons in the next layer and it goes further.

Generally, it can be said that many input layers in MLP are connected as a directed graph between both the output layer and the input layer as shown in Fig. 5.

MLP used back-propagation for training the network [14]. As the connection between the layers is in a directed graph manner, the signal path goes through the

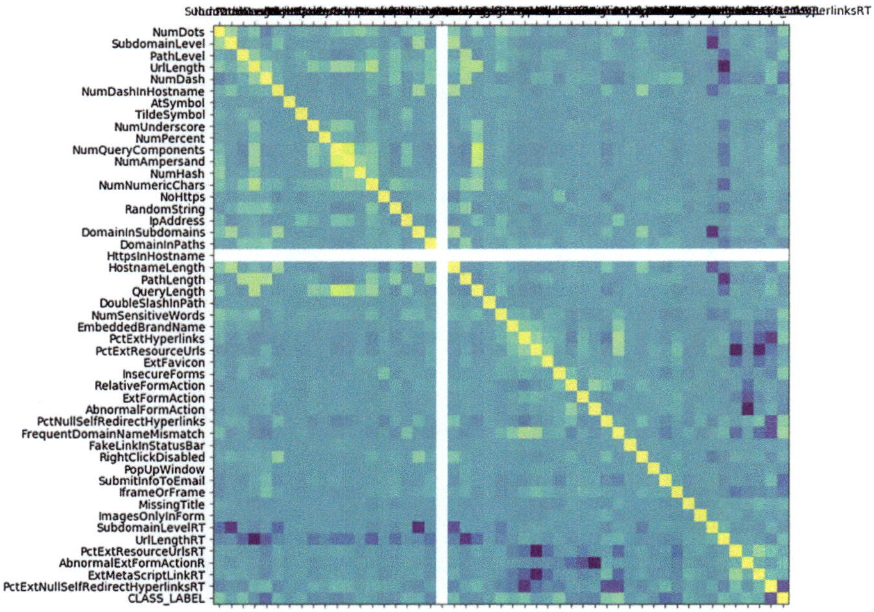

Fig. 4 Correlation feature selection

nodes only one way. Each neuron/node will be having an activation function. Nowadays, the ReLU activation function is the best-used one; the reason for using an activation function is, in a network, some neurons may affect by gradient decedent so to rectify that each layer consists of an activation function.

The reason for using MLP is to achieve nonlinear problems (unexpected test case). MLP acts as a supervised learning technique with the help of the back-propagation method.

For our problem, identifying the phishing website we need neural network technique to identify webpage which is not in our dataset and our model should identify it with high accuracy.

4.4 Model Performance Evaluation

This session addresses the topics of experimental settings and the results of machine learning algorithms and MLP. From the feature selection chosen method of the Gini index, the top five features are taken for model building because from Fig. 2 we can see that the top five features have more impact towards the class labels. The dataset is divided into three phases of training, validation and testing in the K-fold method. MLP model is set to three hidden layers. The model has many factors to set for the data to produce a good result. The main factors are set as activation

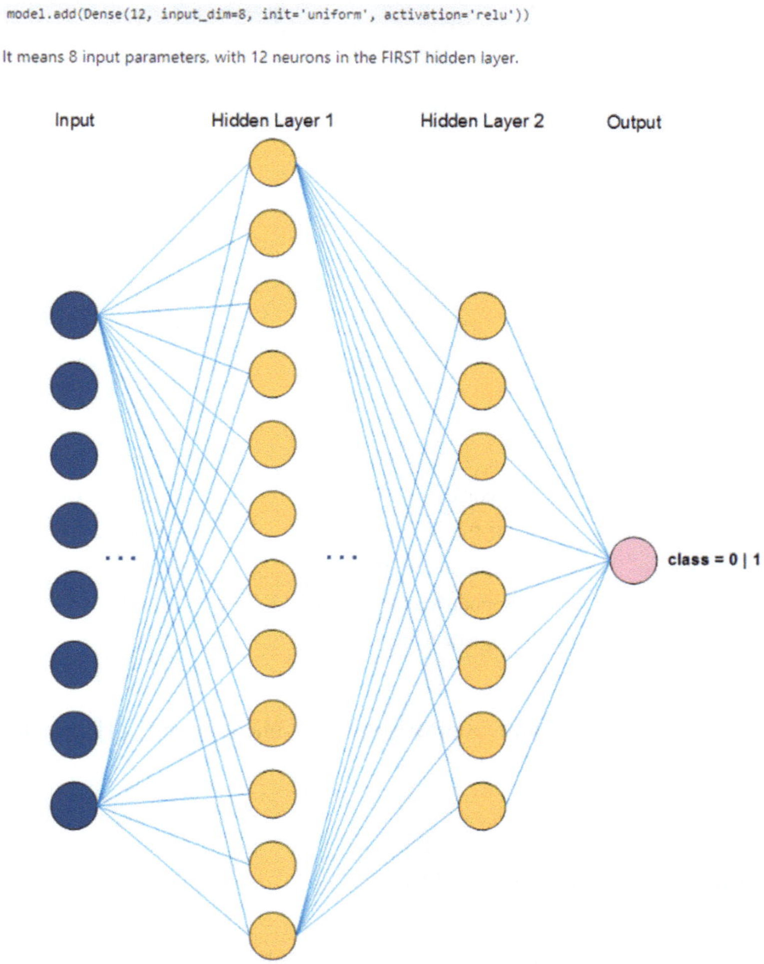

Fig. 5 MLP architecture

as ReLU, Kernel_initializer as uniform, loss function as Binary_crossentropy and optimizer as Adam optimizer. The model runs with batch_size of 10 and with 10 epochs (Iterations). The results are obtained by using Python code in the IPython console. MLP model of predicting the phished website is plotted in Fig. 6.

MLP model and other machine learning methods (random forest, support vector machine (SVM), logistic regression) performances are presented in Table 1. The main moto for identifying a phished website is to detect the phishes sites accurately with high accuracy and should also be performing good in terms of performance and speed, so the MLP model has achieved good results in all the criteria.

These results show the MLP has higher achievement compared to the machine learning methods. Furthermore, this study is also having high performance when

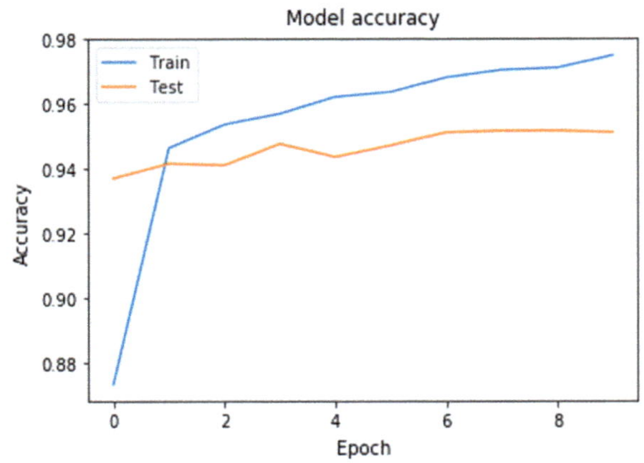

Fig. 6 MLP model accuracy

Table 1 Performance of machine learning methods

Methods	Accuracy (%)
MLP	96.80
Random forest	87.67
Support vector machine	91.60
Logistic regression	89.45

compared with the literature comparison of publication no. [4] which has a performance of 95.43%.

5 Conclusion and Future Work

The work proposes the use of multilayer perceptron in the detection of phishing webpage. The novelty of the work is that along with the protection of the phished webpage, the work proves that feature selection is also a major part of identifying the phishing webpage. The proposed work is built for the user who is doing online payments and for their security. The work is carried out in Python language. The result shows that using MLP in phishing detection proved to achieve a high accuracy of 96.80%. The future work is implementing feature extraction for these five features which are used for building the MLP model and detecting real-time phishing websites.

References

1. Saxe J, Harang R, Wild C, Sanders H (2018) A deep learning approach to fast, format-agnostic detection of malicious web content. In: 2018 IEEE security and privacy workshops (SPW).

IEEE, New York, pp 8–14
2. Lichman M (2017) UCI machine learning repository. University of California, School of Information and Computer Science, Irvine, CA (DATASET). [http://archive.ics.uci.edu/ml]
3. Aaron G, Manning R (2013) APWG Phishing Reports, APWG, 1 February 2013. (APWG). Available: http://www.antiphishing.org/resources/apwg-reports/. Accessed 10 Dec 2018
4. Sönmez Y, Tuncer T, Gökal H, Avcı E (2018) Phishing web sites features classification based on extreme learning machine. In: 2018 6th international symposium on digital forensic and security (ISDFS). IEEE, New York, pp 1–5
5. Mohammad RM, Thabtah F, McCluskey L (2014) Intelligent rule-based phishing websites classification. IET Inf Secur 8(3):153–160
6. Hadi WE, Aburub F, Alhawari S (2016) A new fast associative classification algorithm for detecting phishing websites. Appl Soft Comput 48:729–734
7. Mohammad R, McCluskey TL, Thabtah FA (2013) Predicting phishing websites using neural network trained with back-propagation. World Congress in Computer Science, Computer Engineering, and Applied Computing
8. Khonji M, Jones A, Iraqi Y (2013) An empirical evaluation for feature selection methods in phishing email classification. Int J Comput Syst Sci Eng 28(1):37–51
9. Hamid IRA, Abawajy J (2011) Phishing email feature selection approach. In: 2011 IEEE 10th international conference on trust, security and privacy in computing and communications. IEEE, New York, pp 916–921
10. Thabtah F, Abdelhamid N (2016) Deriving correlated sets of website features for phishing detection: a computational intelligence approach. J Inf Knowl Manage 15(04):1650042
11. Palaniswamy S, Tripathi S (2018) Emotion recognition from facial expressions using images with pose, illumination and age variation for human-computer/robot interaction. J ICT Res Appl 12(1):14–34
12. Kumar VK, Suja P, Tripathi S (2016) Emotion recognition from facial expressions for 4D videos using geometric approach. In: Advances in signal processing and intelligent recognition systems. Springer, Cham, pp 3–14
13. Vineetha KV, Kurup DG (2017) Direct demodulator for amplitude modulated signals using artificial neural network. In: International symposium on intelligent systems technologies and applications. Springer, Cham, pp 204–211
14. Purushothaman A, Vineetha KV, Kurup DG (2018) Fall detection system using artificial neural network. In: 2018 Second international conference on inventive communication and computational technologies (ICICCT). IEEE, New York, pp 1146–1149

A Self-trained Support Vector Machine Approach for Intrusion Detection

Santosh Kumar Sahu, Durga Prasad Mohapatra, and Sanjaya Kumar Panda

Abstract Intrusion refers to a set of attempts to compromise the confidentiality, integrity and availability (CIA) of the information system. Intrusion detection is the process of identifying such violations by analyzing the malicious attempts. Intrusion detection system is used to automate the intrusion detection process just in time or real-time and alert the system administrator for mitigating such efforts. Many researchers have been proposed several detection approaches in this context. In this paper, we adopt a semi-supervised learning-based support vector machine (SVM) approach for mitigating such malicious efforts. The proposed approach improves the learning process and the detection accuracy as compared to the standard SVM approach. Moreover, it requires less amount of labeled training data during the learning process. Our approach iteratively trains the labeled data, predicts the unlabeled data and further retrains the predicted instances. In this manner, it improves the training process and provides better result as compared to the standard SVM approach.

Keywords Support vector machine · Semi-supervised approach · Self-trained model · NSL-KDD · KDD corrected · GureKDD

1 Introduction

In general, intrusion is defined as a set of malicious attempts to compromise the security trait of the network or system. Here, an intruder performs an attack against an adversary or a victim [1–3]. The intruder has a malicious intention to carry out an

S. K. Sahu (✉) · D. P. Mohapatra
National Institute of Technology Rourkela, Rourkela, Odisha 769008, India
e-mail: santoshsahu@hotmail.co.in

D. P. Mohapatra
e-mail: durga@nitrkl.ac.in

S. K. Panda
National Institute of Technology Warangal, Warangal, Telangana 506004, India
e-mail: sanjayauce@gmail.com

A. K. Tripathy et al. (eds.), *Advances in Distributed Computing and Machine Learning*, Lecture Notes in Networks and Systems 127, https://doi.org/10.1007/978-981-15-4218-3_38

attack against the victim. There are various information gathering techniques adopted to collect information about the victim. The first one is vulnerability analysis, which discovers weakness of the target network or system. The intruder tries to gather all the information and tries to exploit the discovered vulnerabilities using the exploitation tools. The consequence of an attack compromises the security, and the intruder is able to gain access to the target system [4]. It in turn causes a heavy loss for an individual or organization. It is noteworthy to mention that such cases are reported to the cybercrime, and it increases exponentially day by day. For instance, India has witnessed 1852 malware attacks per minute on Windows machines [5] by several Windows malware attacking tools, such as ransomware, exploit, potential unwanted application (PUA), adware, cryptojacking, infector and worms. To mitigate such efforts, it is utmost essential to improve the detection approach. The approach needs to detect, drop or countermeasure such efforts and safeguard the information from the unauthorized access.

An intrusion detection system (IDS) [6–8] is a software or hardware that automates the intrusion detection process just in time or real-time. Ideally, an IDS should be able to detect the malicious network traffics or events on a computer usage. The detection approach is dynamic in nature and more powerful than a conventional firewall. For example, in case of a guessing attempt on login system, it would be able to detect the several failed attempts within a short duration of time and would alert to the administrator about the suspicious activity. However, in case of a firewall, it would fail to detect such activities due to predefined rules. Therefore, the objective of the IDS is to analyze the network connections or system calls and alert the suspicious outcomes. In this paper, a semi-supervised learning-based SVM approach is proposed to detect such malicious efforts more effectively as compared to the standard SVM approach.

The rest of this paper is structured as follows. Section 2 discusses the literature review. Section 3 presents the self-trained SVM approach. The experimental method is presented in Sect. 4 followed by results and discussion in Sect. 5. Finally, we conclude in Sect. 6.

2 Literature Survey

IDS is an active area of research on cybersecurity for more than two decades. Catania and Garino [9] have addressed the evolution of intrusion detection approaches and discussed several issues with the existing systems. Initially, researchers are working in signature-based IDS using Snort and Bro. Here, the known attack detection using the signature-based approach is outstanding, whereas it is not able to detect the unknown attacks. On the other hand, misuse-based systems reduce the demand of human intervention by applying data mining methods. In addition, several behavior-based detection approaches have been applied for intrusion detection.

Porras and Valdes [10] implemented a statistical-based technique for intrusion detection. Chen et al. [11] proposed an application of SVM and artificial neural network (ANN) for intrusion detection. Eskin et al. [12] proposed a hierarchical clustering technique, and Liao et al. [13] have elaborated the taxonomy and all-inclusive comparison of several IDS approaches. Catania and Garino [9] have also discussed the human interaction at the highest level that links with the other problems of IDS. It has been observed that several critical issues have been encountered during intrusion detection process, such as hyper-parameter tuning, identification of relevant traffic features, consumption of resources and lacking of public intrusion data. The anomaly-based detection approaches were reviewed by Patcha and Park [14]. The IDS fails to detect novel attacks due to lack of recent standard dataset, increase in false positive rate, high computational complexity and noise present in the available dataset. The prominent issues in IDS modeling are precise definition of normal behavior and incapable to process the encrypted network packets. Nowadays, it is easy to assess the victim and launch an attack very precisely without reconnaissance. Therefore, it becomes a challenging task to detect intrusions effectively and accurately. Moreover, misuse-based detection system is not enough to tackle the unknown attacks and reduce the false alarm. The anomaly- and signature-based approaches require an enormous amount of labeled data for modeling, which is an extremely difficult, time consuming and costly too. The process of transforming information into knowledge by using large labeled data (referred as data mining) includes different learning approaches like supervised, unsupervised and semi-supervised, which can also be applied for intrusion detection.

The main objectives of this paper are as follows.

- To study the performance of existing intrusion detection approaches
- To develop a novel intrusion detection approach, which overcomes the limitations of existing detection approaches in terms of accuracy and efficiency.

In the process of experimentation, we implement a network-based approach to achieve the above objectives. The traffic model generator and incident detector modules are extensively used for host-based intrusion detection.

3 A Self-trained Support Vector Machine Approach

The supervised and unsupervised learning approaches are the two most popular machine learning techniques [15]. The supervised approach requires huge amount of labeled instances, known as training set. During its learning process, the hyper parameters are tuned as per the training data. The classifiers, such as decision trees, SVM and neural network, come under supervised approaches. The unsupervised approach is based on the similarity measure, which finds the inherent structure of the unlabeled data. Clustering and outlier detection techniques come under unsupervised learning approaches.

The semi-supervised approach is the combination of supervised and unsupervised learning approaches. It uses both labeled and unlabeled data during its learning process. During model formation, supervised technique is used to initially train the model using labeled data and predict the unknown instances. Again, the predicted instances are used to retrain the model. This process is an iterative process and continues until the algorithm converges or errors reach to the threshold value.

3.1 Intrusion Detection Using Self-trained SVM

Chien et al. [16] have implemented a self-trained SVM model for the realization of transcription start sites. Maulik and Chakraborty [17] have used this approach to address the pixel classification problem of remote sensing imagery, and Li et al. [18] have implemented a speller system interface for an electroencephalogram (EEG)-based brain computer. In this work, we use it for testing and analyzing intrusions using a self-trained SVM.

3.2 Proposed Approach

The formulation of a standard support vector machine for a two-classification problem is given as follows.

$$\min \quad \frac{1}{2}\|\mathbf{w}\|^2 + C\sum_{i=1}^{N}\xi_i \tag{1}$$

with the following constraints

$$y_i(\mathbf{w}^T\mathbf{X}_i + b) \geq 1 - \xi_i \tag{2}$$

$$\xi_i \geq 0, i = 1, 2, 3, \ldots, N \tag{3}$$

where $\mathbf{X}_i \in \mathcal{R}^n$ is the feature vector for the ith training example. The outcome of each instance $y_i \in -1, 1$ is the class label of X_i, i = 1, 2, 3,..., N and $C > 0$ is the regularization constant. The self-trained SVM algorithm is given in Algorithm 1. The final trained SVM is treated as the final model for binary classification on intrusion dataset. Li et al. [18] have described the proof of convergence of the algorithm.

4 Experimental Method

The proposed approach is implemented using LIBSVM routine of SVM model. The library for SVM (LIBSVM) was developed by Chang and Lin [19], which supports Python, MATLAB, C, C++ and JAVA languages to deal with regression and classification problems. The object code was invoked virtually using different API calls. A

Algorithm 1 Self-Trained SVM Approach

Require:

F_l, F_U and σ_0
F_l: Set of training labeled instances $\mathbf{X}_i$, i = 1, 2, 3,..., N of N samples and the classes are denoted as $y_0(1)$, $y_0(2)$, $y_0(3)$,..., $y_0(N)$
F_U: The set of training instances without class label
σ_0: Threshold value for convergence

Ensure: Trained Model

```
Train a SVM Model using F_l as input labeled data and use the learned model to classify the class label of F_U
j = 2
while TRUE do
    F_N = F_l + F_U where classes of F_U classified by the previous trained model
    Retrain the model using F_N sample data and again classify next set of F_U
    Evaluate the objective function f(w^(j), ξ(j)) = 1/2 ||w^j||^2 + C Σ_{i=1}^{N1+N2} ξ_i
    if f(w^(j), ξ(j)) − f(w^(j−1), ξ(j − 1)) < σ_0 then
        Break
    end if
    j = j + 1
end while
```

radial basis function Kernel ($\exp(-\gamma * |u - v|^2)$) was used to implement cost-based SVM. The hyper parameters (c and γ) were determined either by a grid search or by the learning algorithm as discussed by Li et al. [18]. The comprehensive simulation approach is depicted in Fig. 1. The intrusion datasets are summarily depicted in Table 1. The classes of the datasets are divided into two classes, i.e., normal and attack by reducing from multi-class to binary class.

5 Results and Discussion

In the training phase of the model, the labeled and unlabeled instances are used as a ratio of 1:10. First, 500 labeled records/samples/data are used for training the model, and it predicts another 500 unlabeled data. Again, a total of 1000 records is used for training the model and it predicts next 500 unlabeled data. The model iteratively trains and predicts all the records of the different datasets. The result is shown in Fig. 2.

In another simulation, we consider 5000 labeled samples and 25,000 unlabeled samples for training, which is given in Fig. 3. As per the results obtained, self-trained process converges pretty quickly in all the cases. The detection accuracy depends on the number of iterations in the dataset, which improves the learning process. It has been observed that the detection accuracy becomes double for 5000 labeled samples as compared to 500 samples after 6 iterations as given in Fig. 3. From these results,

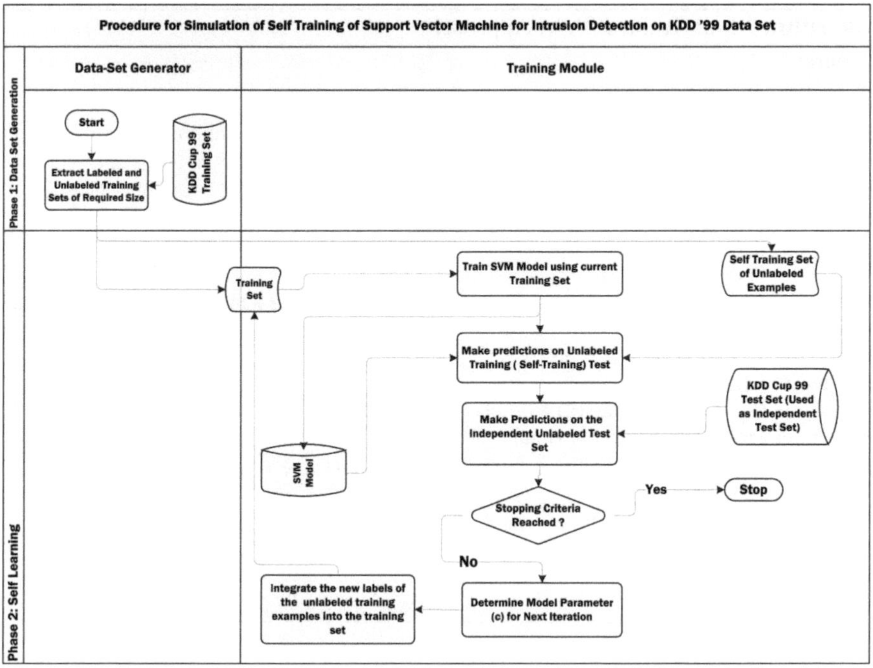

Fig. 1 Simulation of self-trained SVM approach for intrusion detection on the intrusion dataset

Table 1 Summary of KDDCup99 intrusion datasets

Dataset	Year	Classes	Normal	DoS	Probing	U2R	R2L	Total
KDD Full [20]	1999	24	972,781	3,883,370	41,102	052	01,126	4,898,431
KDD 10 [20]	1999	24	097,278	0,391,458	04,107	052	01,126	0,494,021
KDD corrected [20]	1999	38	060,593	0,229,269	04,925	070	16,172	0,311,029
NSL-KDD [21]	2007	38	087,832	0,054,598	02,133	052	00,999	0,145,586
GureKDD [22]	2008	28	174,873	0,001,124	00,000	958	01,124	0,178,810

we inferred that a large number of training data are required to improve the detection accuracy.

The accuracy is 75.5% for 500 samples, whereas it is improved to 86% for 5000 samples. As a result, the proposed approach can be used for the reduction of labeled training samples for intrusion detection. The reduction of the labeled samples is around 90% using our approach. We also compared the standard SVM with proposed SVM approach and given in Fig. 4.

The self-trained SVM model is applied NSL-KDD, GureKDD and KDD Corrected datasets. Figures 9 and 10 show the area under curve (AUC) of class 1 and class −1, respectively, on GureKDD dataset. For NSL-KDD dataset, Figs. 5 and 6 are visualizing the AUC. Similarly, Figs. 7 and 8 depict the AUC using KDD corrected dataset. Table 2 shows the detailed performance evaluation parameters, which

Fig. 2 Performance of proposed approach with 500 labeled samples and 5000 unlabeled samples

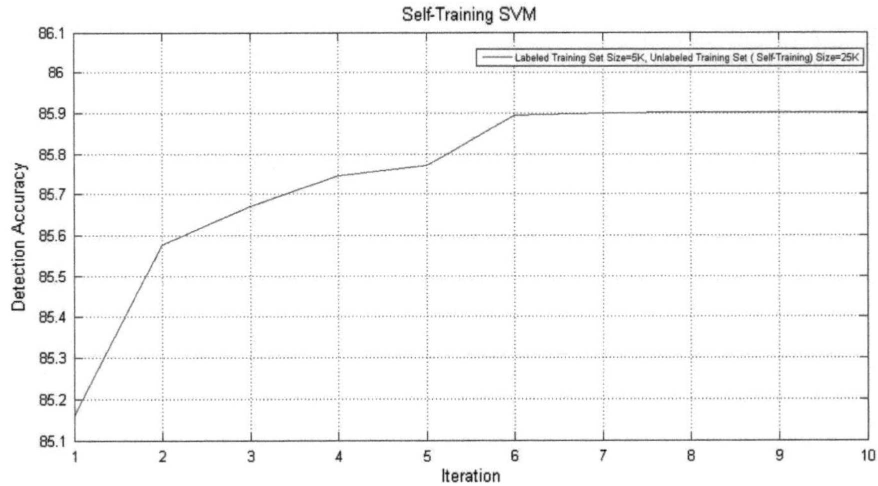

Fig. 3 Performance of proposed approach with 5000 labeled samples and 25,000 unlabeled samples

are based on the confusion matrix. Note that balanced accuracy is new criteria for evaluating the performance on the imbalanced dataset. It is calculated using false positive, false negative, precision and recall values.

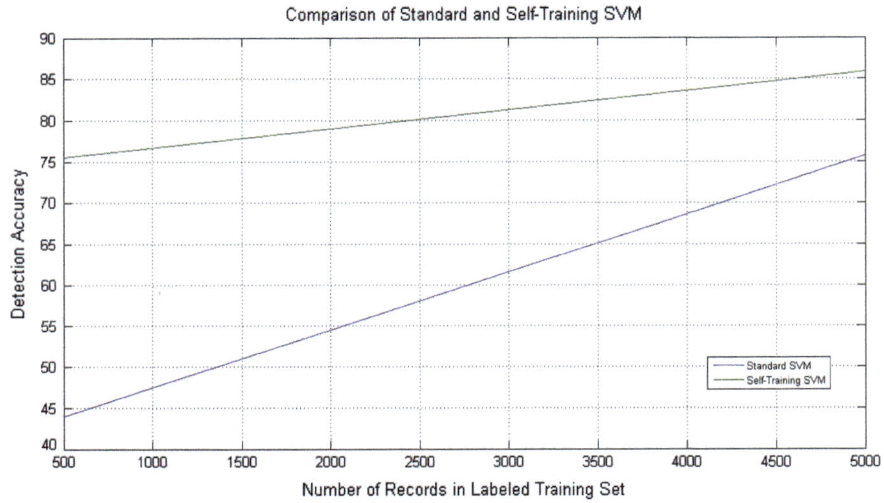

Fig. 4 Comparison of Standard SVM with proposed self-trained SVM

Fig. 5 Area under curve of class 1 in NSL-KDD dataset

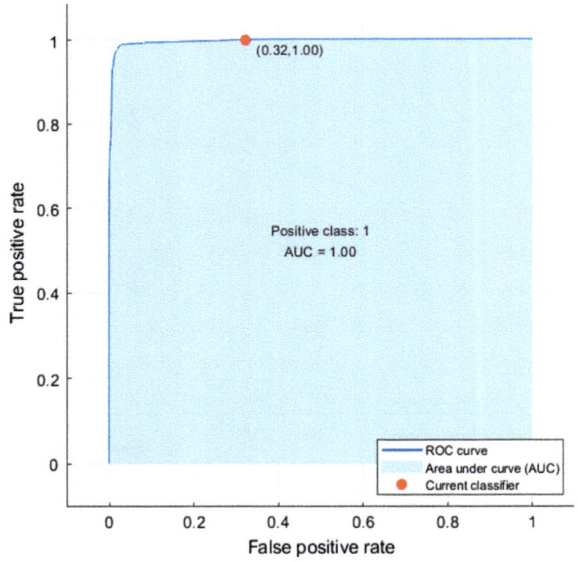

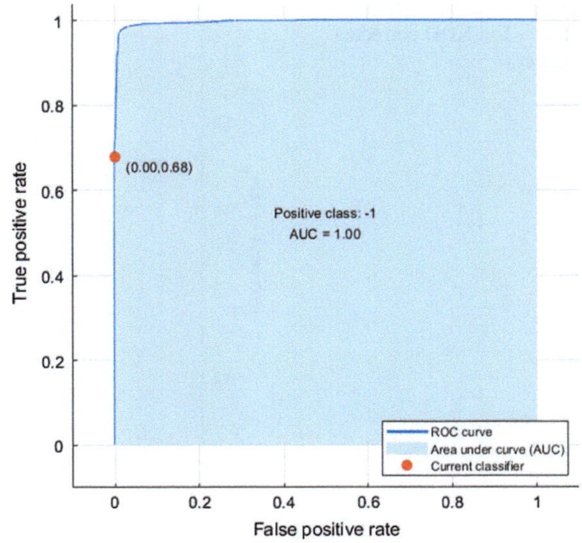

Fig. 6 curve of class -1 in NSL-KDD dataset

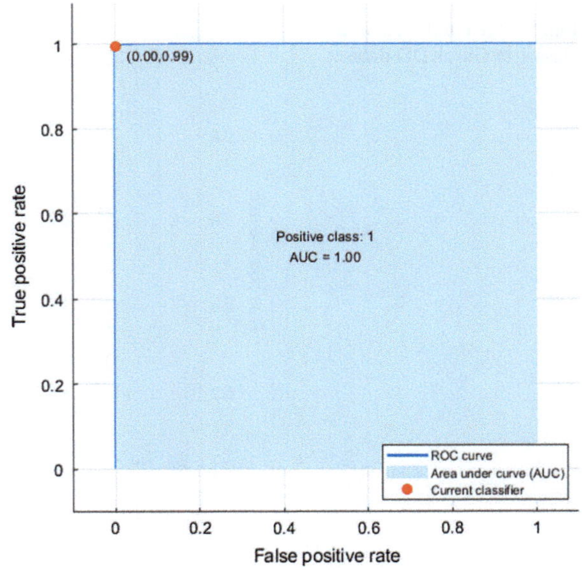

Fig. 7 Area under curve of class 1 in KDD corrected dataset

Fig. 8 Area under curve of class −1 in KDD corrected dataset

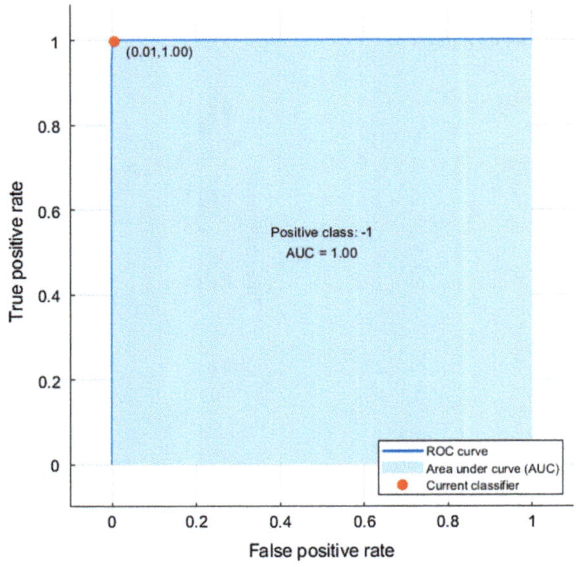

Fig. 9 Area under curve of class 1 in GureKDD dataset

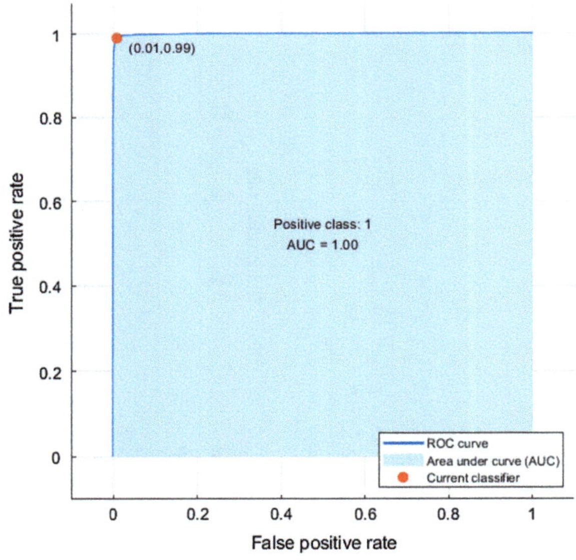

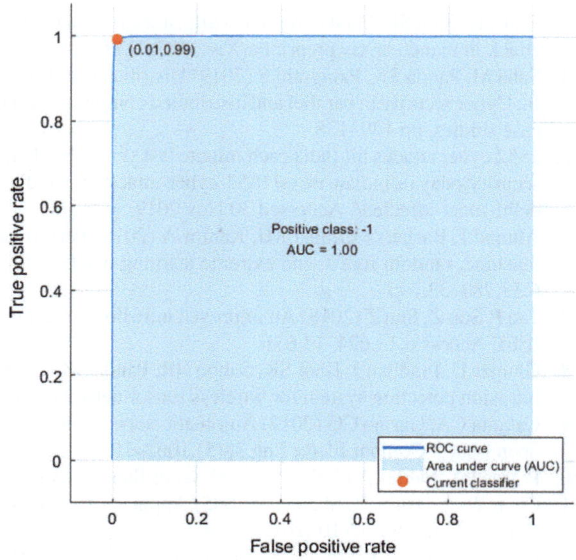

Fig. 10 Area under curve of class −1 in GureKDD dataset

Table 2 Performance assessment of proposed approach on different dataset

Dataset	TP	FP	FN	TN	Acc	Precision	Recall	FNR	TNR	FPR	F1	Bal_Acc
GureKDD	02,615	073	1241	156,975	99.18	0.97	0.68	0.320	0.99	0.06	0.81	82.55
NSL-KDD	11,716	075	0027	013,374	99.60	0.99	0.99	0.002	0.99	0.74	0.99	99.57
KDD corrected	29,138	450	0240	047,463	99.11	0.98	0.99	0.008	0.99	0.65	0.99	98.83

6 Conclusion

In this paper, we have presented a semi-supervised learning-based SVM approach and evaluated it using confusion matrix and area under curve. The KDD corrected, NSL-KDD and GureKDD datasets have been used for IDS assessment. As per the obtained results, the proposed approach has improved the learning process by reducing the training error. The detection accuracy and the balanced accuracy of the proposed approach outperforms on the intrusion datasets.

References

1. Allen J, Christie A, Fithen W, McHugh J, Pickel J (2000) State of the practice of intrusion detection technologies. Technical report, Carnegie-Mellon Software Engineering Institute, Pittsburgh, PA
2. Sahoo KS, Panda SK, Sahoo S, Sahoo B, Dash R (2019) Toward secure software-defined networks against distributed denial of service attack. J Supercomput 75(8):4829–4874

3. Rout JK, Bhoi SK, Panda SK (2014) Sftp: a secure and fault-tolerant paradigm against blackhole attack in manet. arXiv preprint arXiv:1403.0338
4. Saha M, Panda SK, Panigrahi S (2019) Distributed computing security: issues and challenges. In: Cyber security in parallel and distributed computing: concepts, techniques, applications and case studies, pp 129–138
5. 1852 cyber attacks hit India each minute last year; Mumbai, Delhi most affected. https://www.securitytoday.in/indian-news/1852-cyber-attacks-hit-india-each-minute-last-year-mumbai-delhi-most-affected/. Accessed 30 Nov 2019
6. Ahmad I, Basheri M, Iqbal MJ, Rahim A (2018) Performance comparison of support vector machine, random forest, and extreme learning machine for intrusion detection. IEEE Access 6:33,789–33,795
7. Tao P, Sun Z, Sun Z (2018) An improved intrusion detection algorithm based on GA and SVM. IEEE Access 6:13,624–13,631
8. Ghugar U, Pradhan J, Bhoi SK, Sahoo RR, Panda SK (2018) Pl-ids: physical layer trust based intrusion detection system for wireless sensor networks. In J Inf Technol 10(4):489–494
9. Catania CA, Garino CG (2012) Automatic network intrusion detection: current techniques and open issues. Comput Electr Eng 38(5):1062–1072
10. Porras PA, Valdes A (2001) Network surveillance. US Patent 6,321,338
11. Chen W-H, Hsu S-H, Shen H-P(2005) Application of SVM and ANN for intrusion detection. Comput Oper Res 32(10):2617–2634
12. Eskin E, Arnold A, Prerau M, Portnoy L, Stolfo S (2002) A geometric framework for unsupervised anomaly detection. In: Applications of data mining in computer security. Springer, Berlin, pp 77–101
13. Liao H-J, Richard Lin C-H, Lin Y-C,Tung K-Y (2013) Intrusion detection system: a comprehensive review. J Network Comput Appl 36(1):16–24
14. Patcha A, Park J-M (2007) An overview of anomaly detection techniques: existing solutions and latest technological trends. Comput Networks 51(12):3448–3470
15. Tan P-N (2018) Introduction to data mining. Pearson Education India, Delhi
16. Chien C-H, Sun Y-M, Chang W-C, Chiang-Hsieh P-Y, Lee T-Y, Tsai W-C, Horng J-T, Tsou A-P, Huang H-D (2011) Identifying transcriptional start sites of human micrornas based on high-throughput sequencing data. Nucleic Acids Res 39(21):9345–9356
17. Maulik U, Chakraborty D (2011) A self-trained ensemble with semisupervised SVM: An application to pixel classification of remote sensing imagery. Pattern Recogn 44(3):615–623
18. Li Y, Guan C, Li H, Chin Z (2008) A self-training semi-supervised SVM algorithm and its application in an EEG-based brain computer interface speller system. Pattern Recogn Lett 29(9):1285–1294
19. Chang C-C, Lin C-J (2011) LIBSVM: a library for support vector machines. ACM Trans Intell Syst Technol (TIST) 2(3):27
20. KDD cup 1999 data. http://kdd.ics.uci.edu/databases/kddcup99/kddcup99.html. Accessed 30 Nov 2019
21. NSL-KDD dataset. http://nsl.cs.unb.ca/NSL-KDD/. Accessed 30 Nov 2019
22. Gurekddcup dataset. http://www.sc.ehu.es/acwaldap/. Accessed 30 Nov 2019

Secure and Robust Blind Watermarking of Digital Images Using Logistic Map

Grace Karuna Purti, Dilip Kumar Yadav, and P. V. S. S. R. Chandra Mouli

Abstract Advances in the digital media ease the users to transmit/share digital data in the form of audio files, digital images and digital videos. However, the risk involved in such transmission or sharing is also increasing. The necessity to protect the copyrights of digital data and thereby identification of the owner is essential. Digital watermarking is the best solution for ownership identification and copyright protection. This paper proposes a novel blind watermarking approach based on logistic map. The watermark is embedded into a cover image using singular value decomposition (SVD). The experimental results and comparative analysis prove the efficacy of our method. For objective evaluation, peak signal to noise ratio (PSNR) and normalized cross-correlation (NCC) have been used to estimate the imperceptibility and robustness of our watermarking process.

Keywords Blind watermarking · Secure & robust watermarking · Logistic map · Singular value decomposition

1 Introdution

Illegal access, distribution and manipulation of digital content can be accomplished easily with the help of the developments in digital technology without any loss of data. Digital watermarking guarantees security of digital images. It hides the information

---

G. K. Purti · D. K. Yadav
Department of Computer Applications, National Institute of Technology Jamshedpur, Jamshedpur, Jharkhand, India
e-mail: grace.gothic@gmail.com

D. K. Yadav
e-mail: dkyadav.ca@nitjsr.ac.in

P. V. S. S. R. Chandra Mouli (✉)
Department of Computer Science, Central University of Tamil Nadu, Thiruvarur, Tamil Nadu, India
e-mail: chandramouli@cutn.ac.in

A. K. Tripathy et al. (eds.), *Advances in Distributed Computing and Machine Learning*, Lecture Notes in Networks and Systems 127, https://doi.org/10.1007/978-981-15-4218-3_39

and provides authentication [6]. A digital watermark gets embedded into a digital image without any imperceptibility [8].

There are two types of digital watermarking techniques—blind and non-blind. Watermark extraction can be possible without the cover data in blind watermarking, whereas cover data is mandatory in non-blind watermarking [5]. In general, a digital watermark can be a logo or an image.

The degree of robustness depends on the type of embedding domain. The data embedding capacity decides the imperceptibility of watermark [7].

During embedding, watermark is inserted into cover data, and hence, it should be robust to withstand attacks. Fidelity deal with imperceptibility [1].

Hsu et al. [2] used 2D-discrete cosine transform (DCT) for blind image watermarking. Local processing of cover data has been done by considering 8×8 non-overlapping blocks of cover data.

Liu et al. [3] proposed a RSA-based asymmetric encryption algorithm. Their work guarantees the security of the hidden data and possess good robustness and high computational efficiency. Logistic map is used to scramble the watermark. Then, RSA encryption algorithm was applied. Over the cover data, discrete wavelet transform (DWT) was applied. LL band was chosen for watermark embedding. The LL band was further decomposed using singular value decomposition (SVD). Finally, the watermark was embedded into the orthogonal matrix U obtained from SVD.

The motivation for this work is based on the work of Liu et al. [3]. The major differences between the proposed one and Liu et al. work is that the proposed is a semi-blind approach, whereas their approach was invisible; local block processing on cover data is performed in our approach whereas a global embedding scheme has been chosen in Liu et al.'s work. The reason for converting into a semi-blind approach is that the dependency and the essential requirement of having the original cover data in hand to extract watermark is not necessary. Thus, the semi-blind approach is more closer to real-world scenarios. The method is semi-blind because it needs a key value as an additional input to the extraction algorithm.

2 Preliminaries

2.1 *Logistic Map*

The logistic map is a polynomial map of degree 2, through which a one-dimensional sequence is generated in random, and it shows a complex chaotic behavior [4]. Mathematically, it is a nonlinear dynamical equation. Equation (1) specifies a simple logistic map.

$$x_{k+1} = r * x_k * (1 - x_k), \quad k = 0, 1, \ldots, n \tag{1}$$

The value of $x_k \in (-1, 1)$ whereas $0 < x_0 < 1$. The randomness of the sequence is maintained by the r value where the value of r ranges between the interval $(0, 4]$.

Logistic map is generated using iteratively for a fixed value of n. In this work, the value of n is chosen as the number of pixels in the given watermark image.

The sequence generated is a one-dimensional chaotic sequence and is normalized. The normalized sequence is bit XOR-ed with the the watermark, and thus, the watermark gets scrambled. The secret key $[X[0], r]$ has to be preserved for watermark extraction.

2.2 Singular Value Decomposition

SVD factorizes a given matrix into a product of three matrices. It is defined in Eq. (2).

$$I = L * S * R^T \tag{2}$$

where I is an image of size $M \times N$. L is a left singular matrix of size $M \times M$. R is a right singular matrix of size $N \times N$. S is a singular matrix of size $M \times N$.

3 Proposed Work

The proposed work is described here. The size of host image (H) and watermark (W) are $256 \times 256 \times 3$ and 64×64, respectively.

3.1 Watermark Embedding

Figure 1 depicts the process of embedding. The steps in the algorithm are as follows:

1. The first step is read the RGB image and convert it into YCbCr space.
2. The second step is read the watermark image and convert it into binary image by thresholding.
3. Generate the chaotic sequence using Eq. (1).
4. Scramble the watermark with the chaotic sequence by bit XOR-ing the watermark with the chaotic sequence.
5. Divide the Y into blocks of size 4×4.
6. Apply SVD on each 4×4 block of Y to get L, S and R. The size of L is also 4×4 in which the indices are labeled as $L_1, L_2, \ldots, L_{16}$ in columnar fashion.
7. Embed watermark bits into the L_2 and L_3 of L using the following equation where $L_{avg} = (L_2 + L_3)/2$.

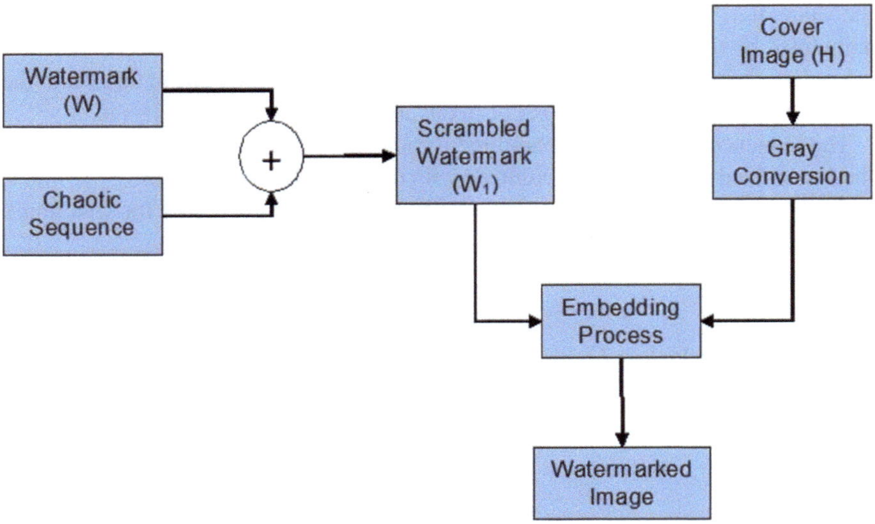

Fig. 1 Watermark embedding process diagram

$$L_2^{'} = \begin{cases} \text{sign}(L_2) * (L_{\text{avg}} + \alpha), & \text{if } w = 1 \\ \text{sign}(L_2) * (L_{\text{avg}} - \alpha), & \text{if } w = 0 \end{cases}$$

$$L_3^{'} = \begin{matrix} \text{sign}(L_3) * (L_{\text{avg}} - \alpha), & \text{if } w = 1 \\ \text{sign}(L_3) * (L_{\text{avg}} + \alpha), & \text{if } w = 0 \end{matrix}$$

8. Apply inverse SVD to get $Y^{'}$ using $Y^{'} = L^{'} * S * R^T$
9. The modified $Y^{'}$ combined with Cb and Cr is used to convert the watermarked image from YCbCr space to RGB space.

3.2 *Watermark Extraction*

Figure 2 depicts the extraction process. The steps in the algorithm are as follows:

1. The embedded image is converted to YCbCr space. The Y component is separated and is divided into blocks of size 4 × 4.
2. Apply SVD on each block.
3. Using the following equation, extract the scrambled watermark bits using

$$W^{'} = \begin{matrix} 0, & \text{if } L_2^{'} > L_3^{'} \\ 1, & \text{if } L_2^{'} \leq L_3^{'} \end{matrix}$$

4. $W^{'}$ is XOR-ed with the chaotic sequence to extract the extracted watermark.

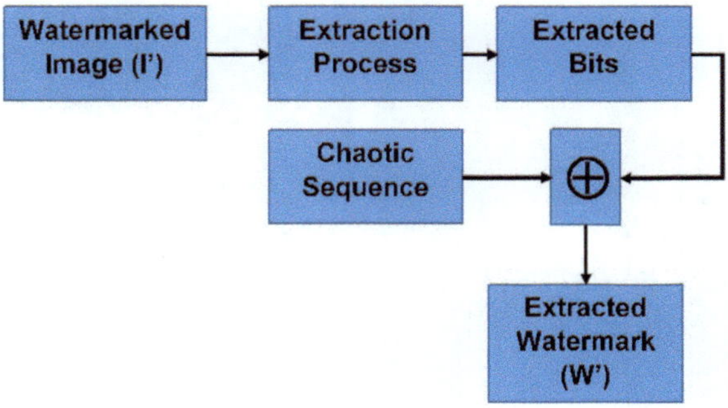

Fig. 2 Watermark extraction process diagram

4 Results and Analysis

4.1 Results

In this section, the experimental results and analysis of results are presented. Figure 3 shows the sample cover images with their names. Figure 4 shows the watermark used and the corresponding scrambled version. The watermarked images obtained after embedding phase are shown in Fig. 5.

4.2 Analysis

Two evaluation criteria have been used to analyze the results of the proposed method—peak signal to noise ratio (PSNR) and normalized cross- correlation (NCC).

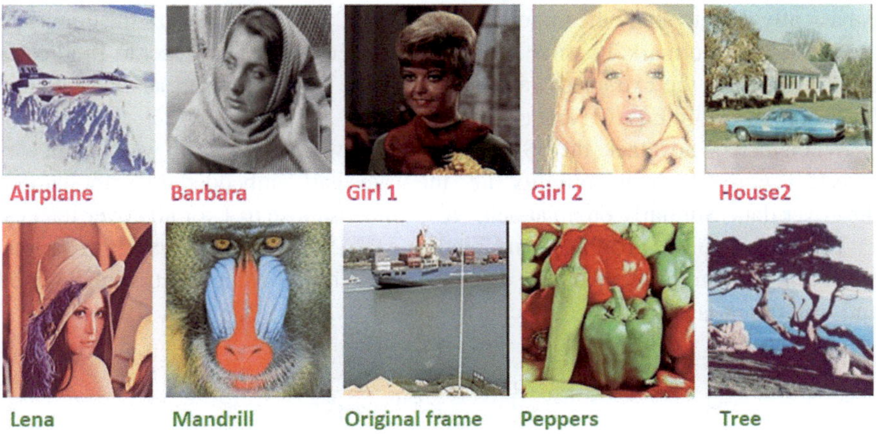

Fig. 3 Cover images with their names

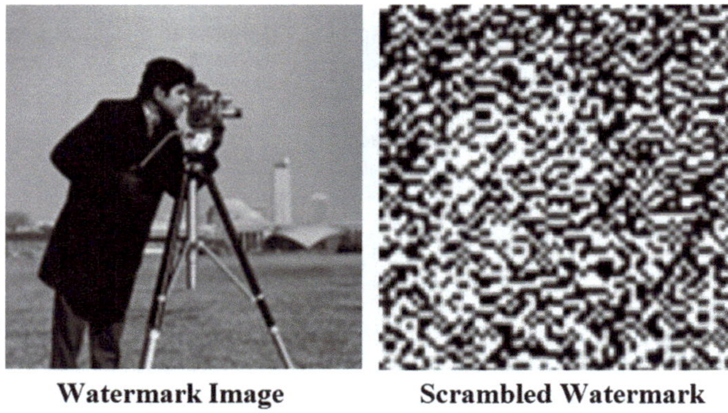

Fig. 4 Watermark image and scrambled watermark obtained after XOR-ing with the chaotic sequence

Fig. 5 Watermarked images

PSNR is the regular metric used between the cover data and the watermarked data. It is basically used to study the imperceptibility. NCC is used to analyze the quality of the extracted watermark. Its value should lie between 0 to 1. The closer to 1 indicates the similarity of the extracted watermark with the original watermark. PSNR and NCC are defined in Eqs. (3) and (4), respectively.

When there is no attack, the PSNR value lies between 30 to 40 db which indicates a perfect score for imperceptibility; i.e., the watermark embedded does not hamper the cover data. Similarly on no attach, NCC value is close to 1 for all cover images.

The other columns in Tables 1 and 2 indicate the PSNR and NCC values of various cover images over different attacks. In all the cases, the imperceptibility is maintained and watermark is extracted significantly.

$$\text{PSNR} = 20\log_{10}\frac{\text{MAX}_I}{\sqrt{\text{MSE}}} \tag{3}$$

Table 1 PSNR values between cover image and watermarked image

Image attack	No attack	Salt & pepper noise	Gaussian noise	Crop attack
Airplane	33.9508	11.8984	19.8929	11.8675
Barbara	37.2391	12.1438	20.0897	11.0522
Girl 1	39.3841	11.6761	20.4838	14.3654
Girl 2	36.9153	11.3088	20.6803	16.3804
House 2	34.6699	12.1213	19.9495	12.2922
Lena	37.4854	12.4133	20.0026	12.3388
Mandrill	35.1304	12.6158	19.8810	12.9762
Original frame	33.9862	12.3842	20.0337	13.5485
Peppers	36.6243	12.2453	20.0337	10.6383
Tree	33.2250	11.9612	19.9543	9.4155

Table 2 NCC between watermark and extracted watermark

Image attack	No attack	Salt & pepper noise	Gaussian noise	Crop attack
Airplane	0.9998	0.6607	0.6869	0.5404
Barbara	0.9965	0.6419	0.6525	0.5201
Girl 1	0.9860	0.6271	0.5985	0.5109
Girl 2	1.0000	0.6734	0.6983	0.5268
House 2	1.0000	0.6456	0.6647	0.5140
Lena	1.0000	0.6472	0.6403	0.5225
Mandrill	1.0000	0.6370	0.6368	0.5314
Original frame	0.9998	0.6463	0.6582	0.5215
peppers	0.9976	0.6500	0.6230	0.5224
Tree	1.0000	0.6417	0.6522	0.5102

$$\text{MSE} = \frac{\sum_{x=0}^{M-1} \sum_{y=0}^{N-1} (I(x, y) - I1(x, y))^2}{MN}$$

$$\text{NCC} = \frac{\sum_{a=0}^{M-1} \sum_{b=0}^{N-1} (W(a, b) \times W1(a, b))}{\sqrt{\sum_{a=0}^{M-1} \sum_{b=0}^{N-1} W(a, b)} \times \sqrt{\sum_{a=0}^{M-1} \sum_{b=0}^{N-1} W1(a, b)}} \tag{4}$$

5 Conclusion

In this paper, a novel blind watermarking scheme has been proposed using logistic map. The cover image is divided into small blocks, and each block is decomposed using SVD. The watermark is scrambled using logistic map. The U matrix of each block is targeted to insert the watermark bits. It is observed that there is high correlation between the second and third values of the first column in the U matrix for every block. At these positions, watermark bits are inserted. The PSNR and NCC values are the metrics employed to evaluate the proposed method. The values of both these metrics are in acceptable range when there is no attack. Subjecting the watermarked image to various attacks, the robustness is tested and evaluated again using the same metrics. The results show that the proposed method is secured and robust enough to identify the ownership and protect the copyright of digital images.

References

1. Cox I, Miller M, Bloom J, Fridrich J, Kalker T (2007) Digital watermarking and steganography. Morgan Kaufmann, Burlington
2. Hsu LY, Hu HT (2015) Blind image watermarking via exploitation of inter-block prediction and visibility threshold in DCT domain. J Vis Commun Image Representation 32:130–143
3. Liu Y, Tang S, Liu R, Zhang L, Ma Z (2018) Secure and robust digital image watermarking scheme using logistic and RSA encryption. Exp Syst Appl 97:95–105
4. Pareek NK, Patidar V, Sud KK (2006) Image encryption using chaotic logistic map. Image Vis Comput 24(9):926–934
5. Su Q, Niu Y, Wang G, Jia S, Yue J (2014) Color image blind watermarking scheme based on QR decomposition. Sign Process 94:219–235
6. Vaidya SP, Chandra Mouli PVSSR (2018) Adaptive, robust and blind digital watermarking using Bhattacharyya distance and bit manipulation. Multimedia Tools Appl 77(5):5609–5635
7. Verma VS, Jha RK (2015) An overview of robust digital image watermarking. IETE Tech Rev 32(6):479–496
8. Wang X, Niu P, Yang H, Wang C, Wang Al (2014) A new robust color image watermarking using local quaternion exponent moments. Inf Sci 277:731–754

Identifying Forensic Interesting Files in Digital Forensic Corpora by Applying Topic Modelling

D. Paul Joseph and Jasmine Norman

Abstract The cyber forensics is an emerging area, where the culprits in a cyber-attack are identified. To perform an investigation, investigator needs to identify the device, backup the data and perform analysis. Therefore, as the cybercrimes increase, so the seized devices and its data also increase, and due to the massive amount of data, the investigations are delayed significantly. Till today many of the forensic investigators use regular expressions and keyword search to find the evidences, which is a traditional approach. In traditional analysis, when the query is given, only exact searches that are matched to particular query are shown while disregarding the other results. Therefore, the main disadvantage with this is that, some sensitive files may not be shown while queried, and also additionally, all the data must be indexed before performing the query which takes huge manual effort as well as time. To overcome this, this research proposes two-tier forensic framework that introduced topical modelling to identify the latent topics and words. Existing approaches used latent semantic indexing (LSI) that has synonymy problem. To overcome this, this research introduces latent semantic analysis (LSA) to digital forensics field and applies it on author's corpora which contain 29.8 million files. Interestingly, this research yielded satisfactory results in terms of time and in finding uninteresting as well as interesting files. This paper also gives fair comparison among forensic search techniques in digital corpora and proves that the proposed methodology performance outstands.

Keywords Disc forensics · Uninteresting files · Interesting files · Latent semantic analysis · Topical modelling

D. P. Joseph (✉) · J. Norman
School of Information Technology and Engineering, VIT, Vellore, India
e-mail: pauljoseph.d@vit.ac.in

J. Norman
e-mail: jasmine@vit.ac.in

A. K. Tripathy et al. (eds.), *Advances in Distributed Computing and Machine Learning*, Lecture Notes in Networks and Systems 127, https://doi.org/10.1007/978-981-15-4218-3_40

1 Introduction

Cybercrimes have been incorporated in day-to-day human's life. As the word digitalization extended to all the areas, the attacks on the digitalized world also are increased. Internet not only connecting one person to another, but also interconnecting one person to attacker anonymously. Since the birth of internet, so many advancements have been taken place to combat the security threats. Even though many security protocols and software are developed, the advanced threats and malwares peep into digital network resulting in cybercrimes like data theft, cyber-stalking, cyber-warfare. Generally, when an attacker targets stand-alone system or a layman or an organization, cybercrime wing seizes the digital devices and submits to forensic investigators in a process of finding the culprits as well as the source, medium, and the intensity of the attack. Digital forensics comes into role, when the investigator has been asked to give the implications. Digital forensics, hereafter DF, is a branch of forensic science that compasses *identifying* the digital device-*searching* for the evidence in device-*seizing* of the device and device-*preserving* of the device [1, 2]. According to [3, 4], DF includes various domains like database forensics, disc forensics, network forensics, mobile forensics, memory forensics, multimedia forensics and cloud forensics. Typically, DF is a seven-stage process [5] but piled up into four stages [6, 7]. Authors have concentrated on disc forensics domain, and forensic disc analysis is performed on the author's corpora that contain 29.8 million files of different types.

2 Background

Once the cybercrimes are enumerated, the devices are seized and referred for the analysis. Survey according to FBI [8] reveals that in the year 2012 alone, 5.9 TB of data has been analysed, and further more alarming that, the rate of cybercrime cases pending is 61.9% in India [9]. The crucial reason behind this is the massive amount of digital data stored in the personal systems as well as enterprise systems. So far, numerous forensic investigation agencies perform analysis by keyword search and regular expression search after the data is indexed. The main hindrance in this is that, only the terms that are matched exactly to the query will be shown and rest are veiled. Another downside of this approach is semantically (polysemy) related words will not be shown while the standard search is performed. For example, if the investigator wants to search for "*flower*", standard keyword gives the result where the exact word "*flower*" is matched. The words which are semantically related to this could not be shown in the results. To overcome this approach, topical modelling is introduced [10]. Various modelling techniques have been developed, and one such model is latent semantic analysis [11]. Latent semantic analysis (LSA) is a statistical model that is developed to uncover the latent topics in a set of documents along with the semantically equivalent words [12]. LSA works on the distributional hypothesis mainly used for information retrieval in the corpus [13]; i.e. the words that are nearer to the meaning will occur more than once in the same text. For example, bat and ball appear oftentimes in the document that talks about cricket game; and lion, tiger and other animals will appear more times in the document that refers to the topic animals. With the context of information retrieval, the application of natural language processing over the neural networks is often termed as latent semantic analysis (LSA). Many articles quote that LSA or LSI is similar but differs in their usage and context.[1, 2,3]

3 Methodology

To overcome the drawbacks mentioned above, this research proposes two-tier forensic framework that serves twofold purpose. Firstly, it detects and eliminates the uninteresting files in forensic investigations by proposed algorithms. Secondly, it identifies interesting files in the resultant corpus with the help of machine learning algorithm as shown in Fig. 1. To classify a file in interesting category, primarily latent semantic analysis (LSA) is used along with few data pre-processing techniques.

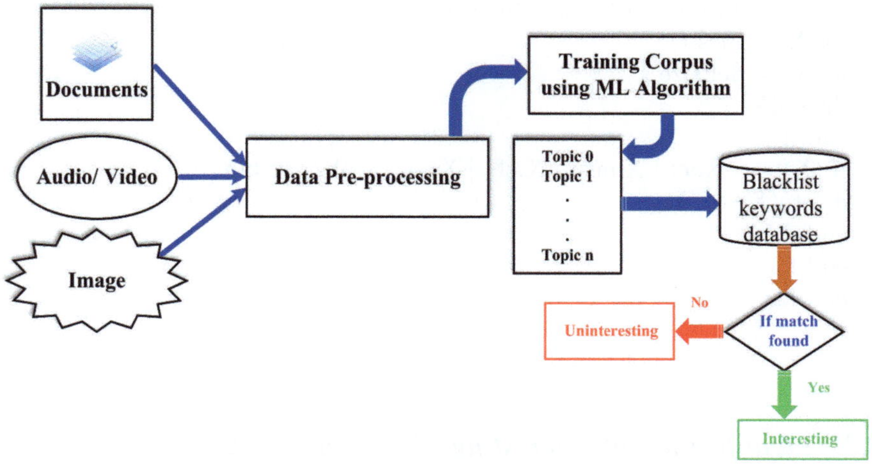

Fig. 1 Proposed architecture to identify the interesting files

[1] https://en.wikipedia.org/wiki/Latent_semantic_analysis.

[2] https://edutechwiki.unige.ch/en/Latent_semantic_analysis_and_indexing.

[3] https://www.scholarpedia.org/article/Latent_semantic_analysis.

3.1 Identification of Uninteresting Files

Once the acquisition process starts, as this research concentrates on disc forensics, only discs with supported formats are loaded into the framework. In the preliminary phase, the files which are irrelevant to forensic investigations can be removed [14]. These files can be system files, software files and auxiliary files. Identification and removal of uninteresting files are given in Algorithm 1. After the preliminary reduction phase, on the remaining files, data cleaning and data analysis are performed, in which LSA is used to identify the interesting files using semantic approach.

Algorithm 1: Deleting system files that meet threshold criteria

```
Input: X (application files) (exe | asm | dll | sys | cpl | cab | chm | icl)
Output: Y (application file) (size > threshold value)
n ⇐ 0
del_size_1 ← 1*1024
del_size_2 ← 512
for n ← 0 to max(n) //max(n) is end sector in disk
for X ← files ()
if (X) = del_size_1
delete(X)
end if
if (X) == X.icl | X.msu | X.cfg | X.so | X.pkg | X.bin
delete(X)
end if
end for
end for
```

3.2 Identification of Interesting Files Using LSA

In this section, an architecture is proposed to identify the forensic interesting files.

In Fig. 1, documents or multimedia data are given as input to the pre-processing task as this research is exclusively for textual data in forensic corpus. The following sections reveal how data is pre-processed for LSA.

Data Pre-processing

Raw data in the forensic corpus contains much inconsistency, as it includes redundant files, abbreviations, stop words, diacritics and sparse terms. Before training the data, all these inconsistencies must be removed and each word should be treated as a single token and so the data pre-processing methods are used. In existing work, data pre-processing methods consumed much time, and with this concern, this research has optimized pre-processing techniques within meantime. Few methods used in this process are as follows.

Removal of stop words

In documents, there will be many unwanted words that delay the investigations. For example, words like *the, hers, but, again, is, am, i* fall under this category. To remove this, authors have used spaCy package [15] in python. In the corpus, total of 1780 stop words are identified and removed after multiple iterations. The main advantage in removing the stop words is that it increases classification accuracy as fewer tokens will be left. Furthermore, in this research, customized stop words are added with respect to digital forensic terms, which reduced the corpus to much extent.

For example, *before the removal of stop words*: cyber-security is the emerging area as it involves many cyber-threats like cyber-assaults, cyber-espionage, cyber-stalking (total 17 words).

After removal: cyber-security emerging area involves cyber-threats cyber-assaults, cyber-espionage, cyber-stalking (total 10 words).

Removal of white spaces

While iterating through the corpus, it is found that many white spaces exist in ending and leading sentences. Since this research used python language primarily for data pre-processing and training the data, the pre-defined packages made this task done easily. By passing input to strip() function in loop, all the white spaces are removed.

Tokenization

Tokenization is the process of splitting the words in a document to individual tokens. Once the stop words are removed, all the individual words are treated as single tokens and stored in a python list in order to construct document-term matrix. In existing works, NLTK and genism are much used, which consumed a lot of time. To overcome this, this research collectively used spaCy and OpenNMT package which tokenized efficiently within $O(n)$ time complexity.

Subtree matching for extracting relations

Since it is arduous to build generalized patterns, knowing dependency sentence structures to enhance rule-based techniques for extracting information is necessary. In general, there exist many correlated words in forensic corpus [16], and therefore, it is very difficult to correlate manually, which is the main cause for delayed investigations. To overcome this, this research used subtree matching using spaCy toolkit to extract the different kinds of relations existing among the entities and objects. The resultant relations are shown as dependency graphs, by which investigators can easily identify the relationships among different persons involved in a cybercrime.

3.3 Latent Semantic Analysis

LSA, a statistical model technique in the natural language processing is developed to find the semantic relations among the documents and terms. This works by using the mathematical technique called singular value decomposition (SVD),[4] which is used for dimensionality reduction [17]. SVD is a least squares method, and hence, it constructs the words to the document in a matrix format known as word-document matrix. Words or objects that belong multiple dimensions are reduced to either single or two dimensional in space; i.e. SVD (matrix decomposition) aims to make subsequent matrix calculations simpler by reducing matrix into constituent parts[5]

$$M = PQR^T \tag{1}$$

where M the original matrix of $m \times n$ size is decomposed into P, Q and R matrices. $P(U)$ and R(V-transpose matrix) are orthogonal matrices of $m \times n$, and $Q(S)$ is a diagonal matrix of $m \times n$.

Constructing document-term matrix

Once the pre-processing steps are completed, then a $m \times n$ matrix is constructed such that there a m documents with n unique tokens. Initially, authors performed on small corpus that contains 17,398 tokens out of which 11,898 unique tokens are observed. Later LSA is extended to original corpus that contains 29.8 million files, which took 4 h in constructing matrix on distributed environment. For distributed setup, authors have used python's gensim pyro server with 8 logical servers on i5 processor of speed 2.53 GHz. The same corpus is trained on stand-alone system, which took 48 h for the documents to be processed. *tfidf* is calculated as per Eq. (2).[6]

$$tfidf_{i,j} = \left(W_{i,j}/W^*{}_j\right) * \log(X/X_i) \tag{2}$$

where $W_{i,j}$ = the number of times word i appears in document j (the original cell count).

$W^*{}_j$ = the number of total words in document j (just add the counts in column j).

X is the number of documents (columns) and X_i is the number of documents in which word i appears (the number of nonzero columns in row i).

Computing SVD

As explained earlier, SVD is applied on the term-document matrix to reduce the dimension to k (*number of topics*) *as stated below* (PQR^T). Each row P_x is a document-term matrix, which is the vector representation of document of length k (*number of topics*) and *Rx* is term-topic matrix which is the vector representation of terms in the document A.

---

[4]https://www.gnu.org/software/gsl/manual/html_node/Singular-Value-Decomposition.html.

[5]https://blog.statsbot.co/singular-value-decomposition-tutorial-52c695315254.

[6]https://pythonhosted.org/Pyro4/nameserver.html.

Algorithm 2: Calculating SVD ($M = PQR^T$)

Input: Matrix P where P=M, Matrix R (identity matrix)
repeat
for all i <j
compute submatrix of $P^P = (\alpha\beta\beta\delta)$
where $\alpha = \sum_{x=1}^{n} P_{xi}^2$, $\beta = \sum_{x=1}^{n} P_{xj}^2$ and $\gamma = \sum_{x=1}^{n} P_{xi}P_{xj}$
compute jacobi rotation that diagonalize $(\alpha\beta\beta\delta)$
update columns of matrix P
for x =1 $\leftarrow n$
$t = P_{xi}, whereP_{xi} = (t * s) – (u * P_{xj})$
$P_{x*j} = (u * s) + (t * P_{xj})$
end for
update V matrix of rsv (right singular vector)
for x = 1 $\leftarrow$ n
$s = R_{xi}$
$R_{xi} = (t * s) – (u * R_{xj}), R_{xj} = (u * s) + (t * R_{xj})$
end for
end for

4 Results and Discussion

Once all the steps are applied, the corpus is trained with LSA algorithm with 100 passes and 500 topics. Later, topic count is increased based on the usage. Since the output is in *n* topics, each topic word against document is compared with author's database that contains blacklisted keywords as described in the proposed methodology section. If any of the word matches, then that file is treated as interesting file, else flagged as uninteresting category. LSA technique with SVD is applied on the author's corpus which consists of 29.8 million files. Corpus contains 9.86% of textual files, 8.62% of word document files, 11.59% of pdf files and rest includes multimedia files, operating system files and application files. This technique is applied on pdf, text and word document files. Since there are thousands of files to be analysed, prioritizing which file to be analysed will be the major task for investigator. To meet this requirement, LSA is applied on the corpus after pre-processing techniques, which yields the documents based on the topics. This research also integrated keyword search along with LSA, such that no interesting files are skipped. For keyword search, authors have used blacklisted keywords identified by the National Security Agency Homeland (NSA). Not only the topics with their similarity measures are displayed, but also indicated whether that file is interesting or uninteresting to the investigator. For example, if the investigator searches for the word malware, only the files with word malware are returned in traditional search. Whereas after applying LSA on the corpus, files with the words like volatile, WannaCry, botnet, operating_system, malware, TOR are also retrieved as these words are closely related to the word "malware". Table 1 gives the comparison among the different approaches used in digital forensics. Even though LSI and LSA are similar, their representations with regard to different contexts yielded in different values as shown in Table 1.

Precession = Total relevant documents/Total retrieved documents and Recall = Documents retrieved relevant/total possible relevant documents. Higher the precession and recall rate, higher chances that maximum amount of data is retrieved. Since it is difficult to give all the results, only results are given for two keywords. Figure 2 represents the results for keyword K1, and Fig. 3 represents the results for keyword K2 as given below. From these figures, one can understand that LSA with the proposed methodology is proven to be the best searching algorithm in forensic investigations as it contains best precision and recall values.

5 Conclusion

In this research, traditional searching techniques in forensic investigations are given in detail. To overcome the disadvantages, this research introduced forensic framework that serves twofold purposes. Firstly, to identify uninteresting files based on threshold values. Secondly, to identify forensic interesting files using LSA. Existing techniques were used on limited set of documents, but this research has used the same approach with efficient pre-processing techniques and reduced the corpora to 25.5 million, which yielded satisfactory results by identifying 4.3 million files as forensically uninteresting. The overall performance with regard to time and classification of interesting files is highly satisfactory. Furthermore, this research draws comparison among traditional search methods and proves that the proposed methodology is the best approach in forensic investigations to detect uninteresting as well as interesting files along with accurate search results followed by keyword hits. The scope of further research based on the present study is to implement efficient topical modelling on forensic corpus, as synonymity occurs.

Table 1 Comparative analysis of various search techniques in forensic investigations

Searching type	Execution time (seconds)		Hit ratio		Auxiliary search	Precession		Recall		Strengths	Drawbacks
	K1	K2	K1	K2		K1	K2	K1	K2		
Keyword search	9.2	5.02	0.71	0.68	Boolean operators can be used	0.112	0.088	0.102	0.055	Used to find user's behaviour insight	(1) Stop words are included in search (2) Increased risk of keywords
Indexed search	4.51	8.46	0.64	0.69	Boolean operators and wildcards are used	0.110	0.098	0.101	0.095	Quick retrieval of query Faster access	Takes extra space on primary and secondary memory
GREP search	3.39	4.48	0.51	0.55	Usage of wildcards for pattern matching	0.47	0.59	0.39	0.47	Powerful in finding the expressions Easily configurable, but requires logical thinking	Memory overheads are caused
LSI	6.5	5.9	0.37	0.33	Keyword, indexed and grep search can be integrated after results	0.67	0.71	0.59	0.64	Semantic words Extracts conceptual content	Polysemy and synonymy
LSA	8.2	6.2	0.27	0.21	Keyword, indexed and grep search can be integrated after results	0.75	0.81	0.71	0.76	Semantic words and overcomes polysemy	Synonymy

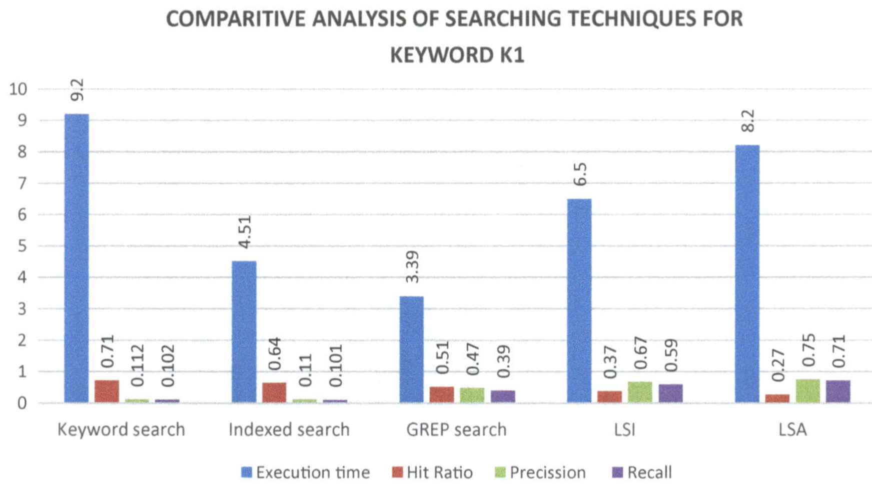

Fig. 2 Keyword versus indexed versus grep versus LSI versus LSA search for K1

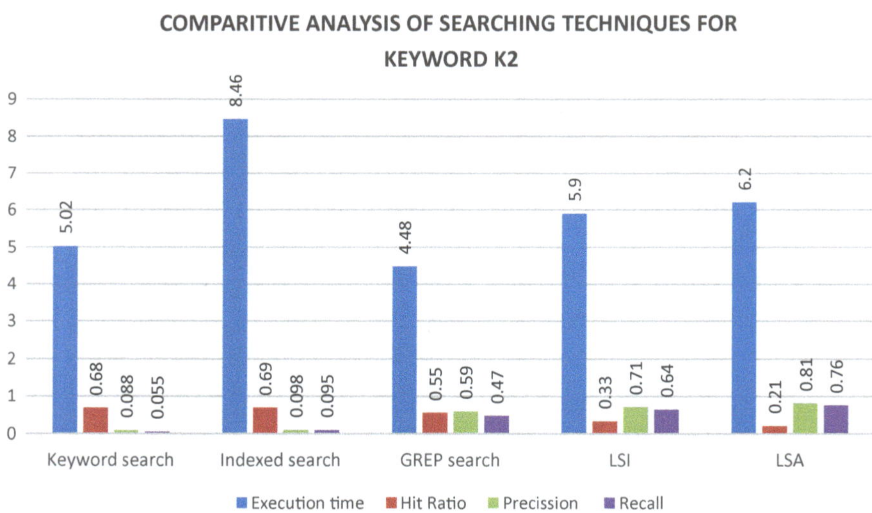

Fig. 3 Keyword versus indexed versus grep versus LSI versus LSA search for K2

References

1. Raghavan S (2013) Digital forensic research: current state of the art. CSI Trans ICT 1(1):91–114. https://doi.org/10.1007/s40012-012-0008-7
2. Beebe N (2009) Digital forensic research: the good, the bad and the unaddressed. In: Advances in digital forensics V, pp 17–36
3. Rogers MK, Seigfried K (2004) The future of computer forensics: a needs analysis survey. Comput Secur 23(1):12–16

4. Joseph P, Norman J (2019) An analysis of digital forensics in cyber security. In: First international conference on artificial intelligence and cognitive computing, vol 815, pp 0–7
5. Bem D, Feld F, Huebner E, Bem O (2008) Computer forensics—past, present and future. J Inf Sci Technol 5(3):43–59
6. Peterson G (2015) Digital Forensics XI. In: Peterson G, Shenoi S (eds) Advances in digital forensics XI 11th. Springer, Orlando, pp 74–89
7. Amari K (2009) Techniques and tools for recovering and analyzing data from volatile memory. Boston
8. Regional Computer Forensics Laboratory (2016) FBI Fiscal annual report. Mexico. Retrieved from https://abc.xyz/investor/pdf/2016_google_annual_report.pdf
9. Pratap Singh S (2016) Crime in India 2016. New Delhi, India. Retrieved from http://ncrb.gov.in/StatPublications/CII/CII2016/pdfs/NEWPDFs/CrimeinIndia-2016CompletePDF291117.pdf
10. Papadimitriou H, Berkeley UC (1998) Latent semantic indexing: analysis. In: Proceedings of the seventeenth ACM SIGACT-SIGMOD-SIGART symposium on principles of database systems, pp 159–168
11. Olmos R, León JA, Jorge-Botana G, Escudero I (2009) An introduction to latent semantic analysis. Behav Res Methods 41(3):944–950
12. Landauer TK, Foltz PW, Laham D (2009) An introduction to latent semantic analysis. Discourse Process 25(2–3):259–284
13. Landauer TK, Dumais ST, A solution to Plato's problem: the latent semantic analysis theory of acquisition, induction, and representation of knowledge
14. Joseph P, Norman J (2019) Forensic corpus data reduction techniques for faster analysis by eliminating tedious files. Inf Secur J 28(4–5):136–147. https://doi.org/10.1080/19393555.2019.1689319
15. Bird S, Loper E, Klein E (2009) Natural language processing with python. O'Reilly Media Inc.
16. Garfinkel SL (2006) Forensic feature extraction and cross-drive analysis. Digit Investig 3:71–81
17. Trefethern L, Bau D III (1997) Numerical linear algebra, vol 102. Soceity for Industrial and Applied Mathematics, Philadelphia

Predicting American Sign Language from Hand Gestures Using Image Processing and Deep Learning

Sowmya Saraswathi and Kakelli Anil Kumar

Abstract The use of computer vision along with machine learning techniques resulted in a breakthrough in medical and social fields. One such intervention is the sign language detection, enabling communication for differently-abled people. Hand gestures are the common method used for this. This paper deals with using image processing techniques like RGB to binary image conversion, skin detection, edge detection to extract the important features of the gesture and provide them as inputs to Convolution Neural Networks (CNN) to improve the accuracy in predicting American Sign Language alphabets (ASL). Using the preprocessed image data set of the hand gesture in our proposed mechanism has predicted the alphabets with higher accuracy up to 99% as compared to the model trained with original data set of images.

Keywords Image processing · Edge detection · Convolution neural nets · American sign language

1 Introduction

Computer vision, where image processing is used helps computers "see" and understand the content of digital images which in turn aids in the automation of tasks like gesture recognition, cancer detection, etc. is a trending topic among researchers. Our proposed work deals with recognition of ASL alphabets using image processing and deep learning. Figure 1 represents 26 alphabets of American Sign Language [1].

S. Saraswathi (✉) · K. A. Kumar
School of Computer Science and Engineering, Vellore Institute of Technology, Vellore, TN, India
e-mail: reddysowmya999@gmail.com

A. K. Tripathy et al. (eds.), *Advances in Distributed Computing and Machine Learning*, Lecture Notes in Networks and Systems 127, https://doi.org/10.1007/978-981-15-4218-3_41

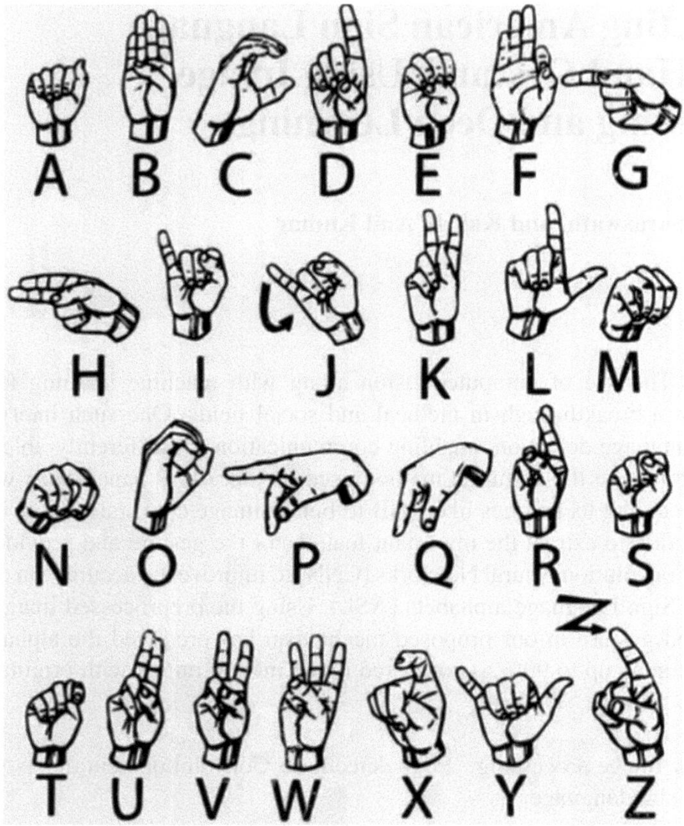

Fig. 1 American sign language

2 Related Work

The earlier research work done on hand gesture included applying image processing techniques like Applying Gabor transformation, feature extraction and classifications [2], a vision-based approach where the accuracy is increased using background subtraction method [3], Speed Up robust features (SURF) algorithm is used for extracting and detecting the features [4]. The following steps: recognizing the shape of the hand, tracing of the detected hand and converting the data into the required result were done for dynamic gesture recognition [5]. Gjorgji et al. [3] have discussed various hand gestures technologies. Two main techniques here were glove based and vision-based. For glove approach, the results depend on the sensitivity of the sensor and other hardware. For vision-based approach the unnecessary details in the background need to be eliminated for better results.

After extracting the features after image processing, they have used machine learning to classify the images based on those features. The system proposed by Patil et al. [6] used a basic neural network implemented using Haar Cascade Classifier as the Haar-like features can make the classification easier. AdaBoost learning algorithm was used for better accuracy where weak classifiers are combined into a comprehensive prediction. Garcia et al. [7] used Googlenet architecture trained on ILSVRC2012 dataset for CNN classifier. They used the concept of transfer learning to fine-tune the already trained model to fit their data set. The dataset considered by them had a horizontal flip of images also to consider the signs made with both hands. They created a web application that captures the images of hand gestures and sends it to the server to predict the output. A problem arises here when similar letters occur, the classifier is unable to differentiate between them or heavily prefers one of the two in a pair. Due to the lack of variation in their datasets, the accuracy obtained on the training dataset could not be produced during the testing of the web application. Hussain et al. [5] have proposed a technique to recognize the six static and eight dynamic hand gestures commands to control a computer. To trace these dynamic hand gestures, using HSV (Hue, Saturation, Value) skin colour algorithm for each frame, the palm is segmented. Thereafter, VGG16 CNN and AlexNet architectures were used to recognize the shape of the hand gestures and resulted with predicted accuracy up to 93.09% and 76.96% respectively.

Zengeler et al. [8] have taken depth data using time-of-flight sensors and applied machine learning models for hand gesture recognition. Segmentation and tracking of hand gestures form the sensor data followed by the extraction of 3D trajectory features. Using Principal Components Analysis (PCA) to extract only the palm and fingers as the input to a recurrent LSTM, which aggregates activation across the temporal dimension of the sequence. The accuracy increases with increasing the number of frames. After a decent training time, a classification accuracy of 99.8% was reported by SVM classifier. Using image processing and pattern matching, Pramada et al. [9] proposed a method for image processing phase which converted data set from RBG to Gray Scale followed by binary conversion. The generated coordinates of image are stored. If a new image is given as the input, the coordinates of the image are generated to compare with the stored coordinates. If the equivalence is observed then return the text or audio result. Based on research investigation of few earlier works in this field, it was concluded that Convolutional Neural Networks and Long Short-Term Memory are better in terms of accuracy.

Other non-machine learning techniques used [10, 11], the method proposed by Shivashankara et al. [11] for hand gesture recognition converts input image to HSV (Hue, Saturation, Value) components. Convert HSV to YCbCr components thereafter to Gray Scale and finally black and white image. Erosion and Dilation have been performed on the image in binary form (opening of image) to extract all the regions of interest followed by applying a filter to reduce salt and pepper noise. Next the Centroid (Central mass of the region) of the pre-processed image is obtained followed by Area (actual number of pixels in the region) and Perimeter (the distance around the boundary of the region) of the image. Finally, the boundary and the number of peaks is determined to recognize the gesture.

Rai et al. [12] takes the hand gestures as input and gives output in the form of speech and text and additional virtual mouse is used so that it is easier to use than physical mouse. The limitation is gestures need to be given in white background and in bright light conditions. Captured images are recognized using image processing techniques like Image Enhancement and Segmentation, colour filtering and skin segmentation and noise reduction technique. Contour analysis techniques are used to extract essential features and these obtained features are stored and used in comparison with the test images. These test images are compared with the stored images. If the gesture is recognized, the output is produced in the form of text and speech otherwise an error message will be resulted.

Ravikiran et al. [13] have proposed a gesture recognition algorithm with 5% error. They used canny edge detection where edges are determined based on the sharp brightness change. Unwanted edges identified in the above steps are clipped. The final step boundary tracing where for every five pixels traced the fifth pixel is made blue to aid in the fingertip detection. The method proposed by Sahoo et al. [14] is developed using three major folds: i) Pre-processing techniques: consists of illumination compensation, segmentation, filtering, hand region detection and image resize. YCbCr colour model is used for skin region extraction based on a threshold value. The features of the object and sign patterns information are extracted using Discrete Wavelet Transform (DWT) where the coefficients of the DWT are selected using Fisher-Ratio techniques. SVM classifier is used, which classifies the hand gesture.

Neelam et al. [15] proposed a real-time vision based can recognize 35 hand gestures present in Indian Sign Language (ISL) and American Sign Language (ASL) at a faster rate using SIFT algorithm with good accuracy. The implemented this with GUI where gestures are classified based on key points. Input RGB images were processed to Gray Scale to minimize errors. The accuracy of their system was 92–96%.

3 Research Framework

The dataset consists of colour images of hand gestures taken in both dimly lit plain background and well-lit background. The training data consists of the images with the plain background. There are at least 200 images for each alphabet. This dataset is preprocessed and given as input into the model [16]. In preprocessing phase, the images are cropped and resized to a smaller size. Canny edge detection algorithm is used to extract the edges from the images of original data set. The features are extracted from these images are used to train CNN model for testing accuracy as per the Fig. 2.

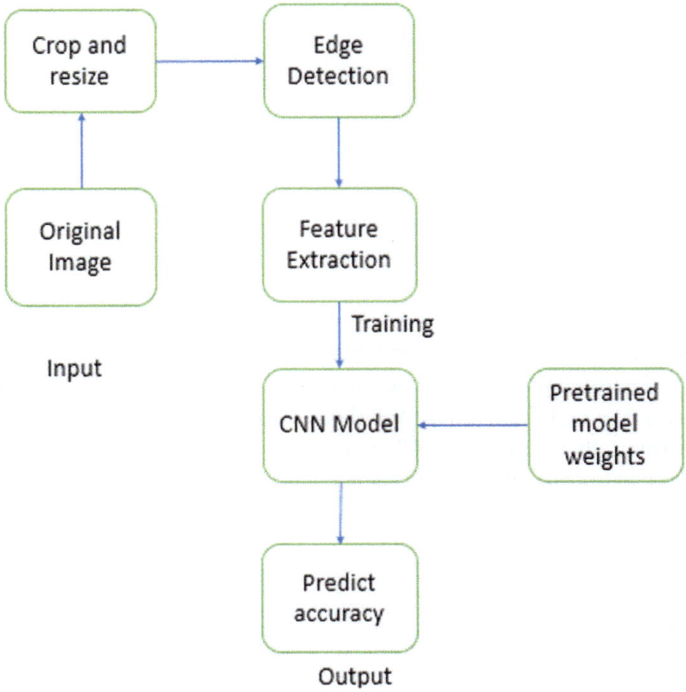

Fig. 2 Methodology of sign language recognition

4 Implementation

There are five key modules of implementation of sign language recognition are mentioned below.

4.1 Image Resizing

The image is cropped to remove the unnecessary background. This helps us to reduce the number of pixels we work with so that the processing is faster.

4.2 Image Processing and Feature Extraction

The dataset has set of images whose contrast is not uniform hence adjust their contrast values to apply edge detection algorithm. The Canny Edge Detection Algorithm (CEDA) runs in five steps.

- Smoothing/Pre-processing
- Gradient Calculation
- Non-maximum suppression
- Double thresholding
- Edge detection by hysteresis.

4.3 Neural Network

Convolution neural networks is a deep learning algorithm which helps the computers to understand and memorize the features of an image to classify a new image. The weights are generated from ImageNet dataset using VGG16 CNN model. Using transfer learning, these pretrained CNN model is further trained with preprocessed images. For each of the images and their features are extracted and given as input for classification.

4.4 Testing

After sufficient number of epochs, the test images are given as input to the model for testing its accuracy. From the dataset of RGB (Red, Green, Blue) images, the model has achieved an accuracy of 91%, and using Canny edge images the accuracy is achieved 99%. Figure 3a–d shows the results of accuracy vs epoch using RGB data

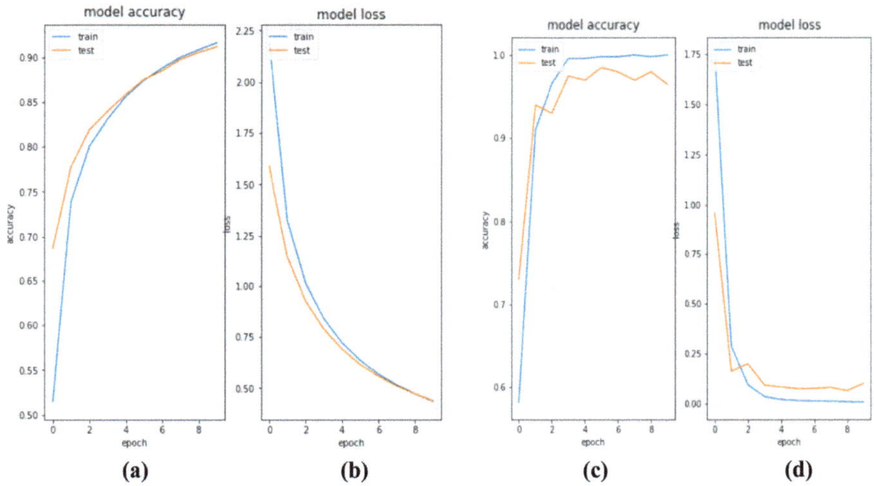

Fig. 3 Accuracy and model loss w.r.to epoch using RGB and Canny edge datasets

set, model loss v/s epochs using RGB data set, accuracy versus epoch using Canny Edge data set, and model loss v/s epochs using Canny Edge data set respectively.

5 Pseudocode

```
Input: ASL Hand gesture image dataset
Result: Predict the alphabet represented in the image
Initialization:
For each image in (images) do

    Image<-Canny_edge_detection(image)
    Image<-Resize (image)
    Image<-setContrast(image)
    Image<-RGBtoGRAY(image)
    Train_CNN(image)

Print(accuracy).
```

6 Results

The results of the proposed sign language recognition mechanism are visualized in Figs. 2 and 3, showing an increase in the accuracy for the processed images using Canny edge as compared to using original RGB images. The confusion matrix for the model trained with processed images is shown in Fig. 4.

7 Conclusion

We proposed a new mechanism of American Sign Language hand gesture recognition for physically disabled people using image processing and deep learning techniques. The edges of the hand are extracted using Canny edge detection algorithm. The image is resized and processed to reduce the processing speed. Training the Convolutional Neural Networks (CNNs) to predict the gestures have resulted in 99% accuracy. The performance of the proposed mechanism varies with respect to input image to CNN for training. We are using different data sets to extend our work with the emerging image processing techniques and other CNNs (ResNet, xception etc.) to identify the significant changes in terms of accuracy.

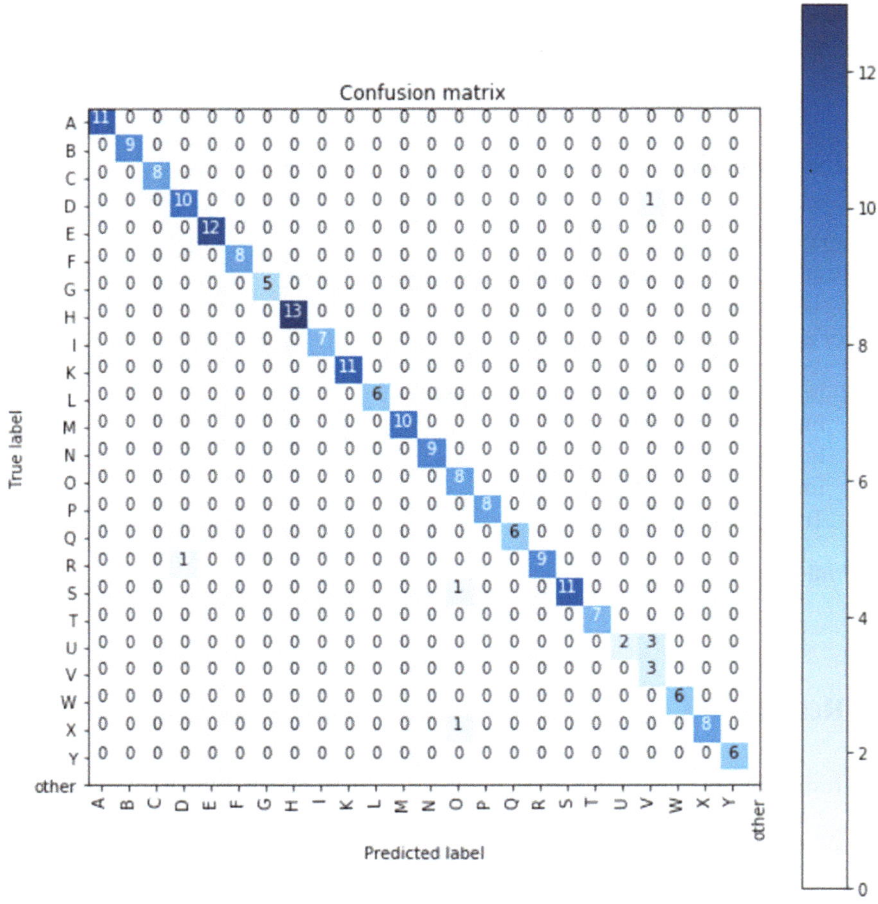

Fig. 4 Confusion matrix for model trained by Canny edge images

References

1. https://society6.com/product/asl-american-sign-language1722170_poster
2. Neethu PS, Suguna R, Divya Sathish (2018) Real time hand gesture recognition system. TAGA J 14:782–792
3. Strezoski G, Stojanovski D, Dimitrovski I, Madjarov G (2016) Hand gesture recognition using deep convolutional neural networks. In: ICT innovations (2016)
4. Kour KP, Lini Mathew (2018) Sign language recognition using image processing. Int J Adv Res Comput Sci Softw Eng 7(8)
5. Hussain S, Saxena R, Han X, Khan JA, Shin H (2017) Hand gesture recognition using deep learning. In: International SoC design conference (ISOCC)
6. Patil O, Vijay Gaikwad (2018) Hand gesture recognition using Haar-like features and cascade classifier. Int J Adv Res Comput Sci Manage Stud 6(2)
7. Garcia B, Viesca SA (2016) Real-time American sign language recognition with convolutional neural networks. In: Convolutional neural networks for visual recognition at Stanford University

8. Zengeler N, Kopinski T, Ha U (2018) Hand gesture recognition in automotive human-machine interaction using depth cameras. Sensors 19:59
9. Pramada S, Pranita DSN, Samiksha N, Vaidya AS (2013) Intelligent sign language recognition using image processing. IOSR J Eng (IOSRJEN) 3(2):45–51
10. Divyashree BA, Manjushree K (2019) Image processing techniques for hand gesture and sign recognition. Int Res J Eng Technol (IRJET) 6(1)
11. Shivashankara S, Srinath S (2018) American sign language recognition system: an optimal approach. Int J Image Graph Signal Process 10
12. Rai P, Alva A, Mahale GK, Shetty JS, Manjushree AN (2018) Gesture recognition system. Int J Comput Sci Mob Comput 7(5):164–175
13. Ravikiran J, Mahesh K, Mahishi S, Dheeraj R, Sudheender S, Pujari NV (2009) Finger detection for sign language recognition. In: Proceedings of the international multi conference of engineers and computer scientists, vol 1
14. Sahoo JP, SamitAri I, Ghosh DK (2017) Hand gesture recognition using DWT and F-ratio based feature descriptor, vol 12. The Institute of Engineering and Technology (IET)
15. Neelam GK, Ingle M (2014) Real time detection and recognition of Indian and American sign language using sift. Int J Electron Commun Eng Technol (IJECET) 5(5)
16. Gautam AK, Kaushik A (2017) American sign language recognition system using image processing method. Int J Comput Sci Eng 9

Skin Cancer Classification Using Convolution Neural Networks

Subasish Mohapatra, N. V. S. Abhishek, Dibyajit Bardhan, Anisha Ankita Ghosh, and Subhadarshini Mohanty

Abstract The incidence of skin cancers has been increasing over the past decades at an alarming rate. Right now, somewhere in the range of 2 and 3 million non-melanoma skin diseases and 132,000 melanoma skin malignant growths happen all-inclusive every year. One in each three diseases analyzed is skin malignant growth. As ozone levels keep being drained, the air loses increasingly more of its defensive channel capacity, and progressively sun-oriented UV radiation arrives at the Earth's surface. It is assessed that a 10 percent decline in ozone levels will bring about an extra 300,000 skin malignant growth cases. Hence, skin cancer today poses a serious threat to mankind. One of the major reasons for skin cancer fatalities is the absence of an early diagnosis. When detected early, skin cancer survival exceeds 95%. Therefore, to facilitate the early diagnosis of skin cancer, we propose this solution—a machine learning model trained on HAM10000 dataset (*a large collection of multi-source dermatoscopic images of common pigmented skin lesions*) using convolutional neural networks (CNN), to classify a given skin lesion image into various cancerous (or non-cancerous) skin conditions. This model incorporated into an online platform will enable doctors and laboratory technologists to know the three highest probability diagnoses for a given skin lesion. Hence, the model will be of immense help in quickly identifying high priority patients and speeding up the procedural workflow.

Keywords Convolutional neural networks (CNN) · Machine learning · Skin lesion

1 Introduction

In the current modern times, computer-aided diagnosis (CAD) systems have inevitably become the norm to analyze medical images so as to mine some significant insights into them. One such recent breakthrough application includes the use of CAD in the form of a mammogram for the successful detection of breast cancers in the United States (US). CAD has become an important cornerstone in

S. Mohapatra (✉) · N. V. S. Abhishek · D. Bardhan · A. A. Ghosh · S. Mohanty
CET, Bhubaneswar, India
e-mail: subasish.mohapatra@gmail.com

A. K. Tripathy et al. (eds.), *Advances in Distributed Computing and Machine Learning*, Lecture Notes in Networks and Systems 127, https://doi.org/10.1007/978-981-15-4218-3_42

medical research especially in diagnostic and imaging radiology [1–3]. When used to its optimum potential, the CAD is pivotal in early detection and diagnosis, thereby preventing human casualties. The treatment of cancer depends largely on its early detection during its early detection. In case of cancer, an early detection is integral as it is most often a complex mixture of numerous illnesses. In 2012, over 14 million new cases and 8.2 million deaths because of malignant cancer were reported worldwide. This clearly states how deadly a disease cancer is and how its early detection is vital.

Our subject of focus is skin cancer. The cases of skin cancer occur in skin which is subject to exposure to intense or continuous sunlight. Nevertheless, the probability of cancer occurring in other parts of the body is never taken out of the picture [4]. Usually, skin cancer starts forming on the topmost skin layer, known as the epidermis. Hence, it is quite visible. It clearly means that the CAD systems could be employed to analyze the images of skin lesion without exploiting any other bits of information to showcase preliminary or initial-stage diagnosis [5]. The deadliest form of skin cancer is melanoma, among the ones found in humans. It causes pigmented marks or signs on the skin. One of the reasons of melanoma is abnormality in melanocytes or melanin-producing cells. These cells impart coloration to the skin. Certain risk factors that encompass melanoma include sunburn history, poor or weakened immunity system, fair complexioned skin, heredity, extreme exposure to UV light, and the excessive use of tanning beds. According to [6], an approximate 91,270 new kinds of melanomas have been predicted to be identified in 2019, while 9320 human deaths due to melanoma have been reported.

It has been observed that melanoma has found a place in the most commonly occurring cancers in young women. If melanoma is not properly diagnosed at an early stage, it starts to grow and continues to spread, and the journey begins with the outermost skin layer, slowly finding its way into the deeper layers, where there is finally a connection with the lymph vessels and the blood [7, 8]. In cases of skin cancer, as it has been previously mentioned, the chances of warding off death casualties are high in cases of early detection as compared to later stages. However, skin cancer detection at an early stage is not easy and is usually an expensive process.

Generally, the multifaceted nature is a direct result of the way that skin injuries look exceptionally commonplace to each other, subsequently making it incredibly hard to break down whether the injury is threatening or amiable. Take the instance of a typical mole. It is similar shading as the skin, similar to dark colored, dark or tan, with an unmistakable fringe or layout isolating it from the contiguous skin scene. The state of the moles is round or oval and sizes under 0.25 in. (roughly 6 mm) in distance across. The size is generally as large as a pencil eraser. To distinguish strange mole qualities that could show skin disease or melanoma, a quest for moles with unpredictable fringes, shapes, hues, and moles with more than 6 mm measurement is finished.

To study the soreness of the skin and make a classification on whether it is malignant melanoma [9], or benign, multiple techniques are at the disposal. Like genetic algorithms, CNNs, artificial neural networks (ANNs), support vector machines (SVMs), the ABCDE rule.

These systems have all been proposed and in this manner confirmed as savvy, exceptionally productive, and less excruciating than customary medicinal procedures. In any case, in numerous PC vision issues, it becomes verifiable that both CNNs and profound learning are the method of decision.

Hence, how early diagnosis can be made possible with the help of technological intervention is our primary focus in this paper. Our proposed machine learning model trained on HAM10000 dataset, using convolutional neural networks (CNN) [10, 11], will classify a given skin lesion image into various cancerous (or non-cancerous) skin conditions. This model when incorporated into an online platform will become an effective tool at enabling oncologists and various other laboratory technologists to be cognizant of the three highest probability diagnoses for a given skin lesion. Hence, the model will be immensely helpful in quickly identifying high priority patients and speeding up the procedural processes.

2 Backgrounds

Skin disease is the wild development of unusual cells in the epidermis, the furthest skin layer, brought about by unrepaired DNA harm that triggers transformations. These changes lead the skin cells to increase quickly and structure dangerous tumors. The two primary drivers of skin malignancy are the sun's hurtful bright (UV) beams and the utilization of UV tanning machines. Fortunately if skin malignant growth is gotten early, your dermatologist can treat it with practically zero scarring and high chances of dispensing with it entirely. Regularly, the specialist may even recognize the development at a precancerous stage, before it has become an all-out skin malignancy or entered underneath the outside of the skin.

But imagine this. You wake up and find a frightening mark on your skin so you go to the doctor's office to get it checked up. They say it is fine so you go home and do not worry about it for a couple months, but then you have a throbbing pain from that spot—it looks ugly and menacing now. It has developed into a malignant tumor as a result of your doctor's misdiagnosis. The prevalence of misdiagnosis is scary. A study has shown that over one in 20 American adults have been misdiagnosed in that past and over half of these are harmful. A lot of skin lesions can be pretty much harmless but others can be life threatening. It is super important that these tumors are discovered right away, and this is when it is the easiest to treat them. Recently, we have seen AI become an integral part of our daily lives. We find it used in our text recommendations, ad personalization, virtual assistants, and so much more! With AI making such huge strides in these fields, there is unlimited potential for it to have an impact on health care through deep learning. With this in mind, we set out to make an end-to-end solution to classify skin lesions using deep learning.

In recent years, we have seen deep learning-based classification system surpasses humans in big datasets like ImageNet. In our proposed model, we have used CNN to achieve CAD through the help of deep learning. Our model is a complete end-to-end solution to classify skin lesions into malignant and benign categories.

3 Objective and Motivation

The objective of this project is to come up with a machine learning model which can predict the chance of occurrence of various types of cancer based on an image of the skin lesion. The ML model will be trained on HAM10000 [12] dataset using convolution neural networks (CNN). The model will classify a given skin lesion image into various cancerous (or non-cancerous) skin conditions. This model incorporated into an online platform will enable doctors and laboratory technologists to know the three highest probability diagnoses for a given skin lesion. Hence, the model will be of immense help in quickly identifying high priority patients and speeding up the procedural workflow.

The various benefits of the proposal to its stakeholders can be summarized as further. One of the major reasons for skin cancer fatalities is the absence of an early diagnosis. When detected early, skin cancer survival exceeds 95%. Therefore, this ML model can be used to facilitate the early diagnosis of skin cancer. Secondly, the healthcare industry—the healthcare professionals can make use of this predictive model to identify high priority patients and accordingly proceed with due procedure. In the case of low-risk predictions, the industry can go for a periodic checkup based on the indicative nature of the prediction score. Lastly, the govt. can also benefit from our work. Based on the predictions of the model over various subjects, govt. can go for effective policy planning measures. In case there is an abnormally high rate of cancer prediction from particular demography, govt. intervention becomes the need of the hour. These fruitful actions can be facilitated through the project. Different researchers have proposed various deep learning techniques for understanding and solving the skin cancer classification problem. We aim to develop a machine learning model which would perform better than these approaches.

4 Proposed Model

The dataset includes a collection of all the important diagnostic categories of pigmented lesions: actinic keratosis and intraepithelial carcinoma/Bowen's disease (**akiec**), basal cell carcinoma (**bcc**), benign keratosis-like lesions (solar lentigines/seborrheic keratoses and lichen planus like keratosis, **bkl**), dermatofibroma (**df**), melanoma (**mel**), melanocytic nevi (**nv**), and vascular lesions (angiomas, angiokeratomas, pyogenic granulomas and hemorrhage, **vasc**). More than 50% of lesions are confirmed through histopathology, and the ground truth for the rest of the cases is either the follow-up examination, expert consensus, or confirmation by in vivo confocal microscopy.

We propose a machine learning model, which after getting trained was able to classify a given skin lesion image into one of the given seven categories. CNNs were used for this image classification task because after the advent of powerful GPUs and ample sources of image data, CNNs have given better results than traditional image

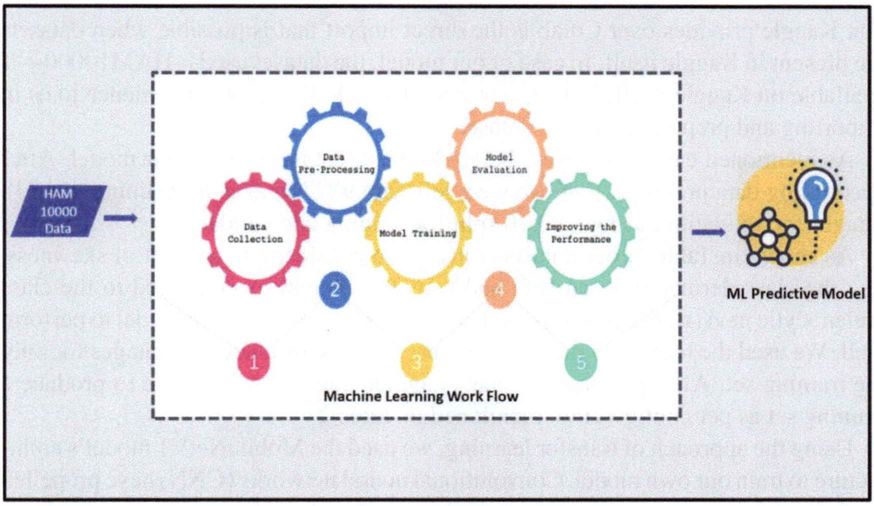

Fig. 1 Schematic diagram of the system

classification algorithms. This model was then published on a Web application. So, our solution can be broken down into two specific tasks, i.e., training a machine learning model which gave us satisfactory results and developing a Web application which provided an interactive platform for the users to use the machine learning model (Fig. 1).

5 Simulation Environment

The simulation environment thereby used is Kaggle environment. Kaggle based on a number of parameters provides a jump in speed. It gives free access to NVidia Tesla P100 GPUs in pieces. This benchmark shows that empowering a GPU to your Kernel brings about a 12.5× speedup during preparing of a profound learning model. The accurate accelerate differs dependent on various components including model design, bunch size, input pipeline intricacy, and so forth. All things are considered, and the GPU opens up incredible potential outcomes in Kaggle pieces. The flexibility on offer is tremendous. Many Jupyter Notebook keyboard shortcuts transfer exactly to the Kaggle environment. It allows a total quota of 30 h per week of free GPU access, which was found quite adequate for the purposes of training our machine learning model. Kaggle kernels come with all the essential Python libraries. Other online platforms, such as Google Colaboratory, also provide these kinds of facilities. But we have availed Kaggle services for a couple of reasons. Firstly, saving notebooks is easier in Kaggle as compared to Google Colab. Kaggle currently uses an advanced GPU chip—an upgraded Nvidia Tesla P100. Earlier Kaggle was powered by a K80, whereas Colab is still employing the NVidia Tesla K80. One of the major advantages that Kaggle provides over Colab is the direct import that is possible when datasets are present in Kaggle itself. In case of our model, the dataset used—HAM10000—is available on Kaggle itself. Hence, it provided a high degree of convenience to us in importing and preprocessing our dataset.

As mentioned earlier, we used the HAM10000 dataset to train our model. After performing data preprocessing, we were left with 9077 images for training and 938 images for validation, using a 90:10 split for training and validation.

As shown in Table 1, our dataset contained a considerable amount of skewness, i.e., the class dermatofibroma (**df**) had way less images as compared to the class melanocytic nevi (**nv**). This skewness needed to be removed for the model to perform well. We used the technique of image augmentation to produce new images for only the training set. After performing image augmentation, we were able to produce a training set as per configuration mentioned in Table 2.

Using the approach of transfer learning, we used the MobileNetV1 model's architecture to train our own model. Convolutional neural networks (**CNN**) have propelled a number of computer vision projects in the past decade and continue to attract huge interest from researchers, who look to modify and apply them for various use cases. We chose MobileNetV1 because it has a lightweight architecture, which makes it ideal for Web deployment.

Our model training specifications are as shown in Table 3.

Testing our trained model on the validation set produced the results as shown in Table 3.

Table 1 Total no. of images in training and validation sets for each class

Class	Training set	Validation set
nv	5954	751
bkl	1074	75
mel	1024	39
bcc	484	30
akiec	301	26
vasc	131	11
df	109	6

Table 2 Total no. of images in training and validation sets for each class after image augmentation (Class **nv** was not augmented)

Class	Training set	Validation set
nv	5954	751
bkl	5920	75
mel	5920	39
bcc	5858	30
akiec	5217	26
vasc	5290	11
df	4410	6

Table 3 Model training specifications

Specification	Metric used
Architecture	MobileNetV1
Optimizer	Adam(lr = 0.01)
Loss function	Categorical crossentropy
Model evaluation metrics	Categorical accuracy, top 2 accuracy, top 3 accuracy
Number of epochs	30

Table 4 Model performance on the validation set (Rounded to the nearest two decimal points)

Metric	Last epoch	Best epoch
Loss	0.72	0.56
Categorical accuracy	0.80	0.84
Top 2 accuracy	0.92	0.92
Top 3 accuracy	0.95	0.96

After evaluating the performance, the classification report, shown in Table 5, was generated (Table 4; Figs. 2, 3, 4 and 5).

The trained model was then saved in HDF5 file format.

Tensorflowjs's command line tool was used to convert the model.h5 file to JSON format, as shown below.

```
tensorflowjs_converter --input_format=keras
/tmp/model.h5 /tmp/tfjs_model
```

This command line statement will convert a model at /tmp/model.h5 and output a model.json file along with binary weight files to the tmp/tfjs_model/ directory.

Using Tensorflowjs's in-built methods, this model.json file was used directly by the Web application that was developed, to classify the image of a skin lesion, uploaded by a user, into three of the most likely classes of the given seven classes, ordered in descending order of their confidence.

Table 5 Classification report of our machine learning model

Class	Precision	Recall	$F1$-Score	Support
nv	0.91	0.97	0.94	751
bkl	0.86	0.08	0.15	75
mel	0.28	0.28	0.28	39
bcc	0.59	0.97	0.73	30
akiec	0.46	0.50	0.48	26
vasc	1.00	0.45	0.62	11
df	0.25	0.50	0.33	6
Macro average	0.62	0.54	0.51	938
Weighted average	0.86	0.85	0.82	938

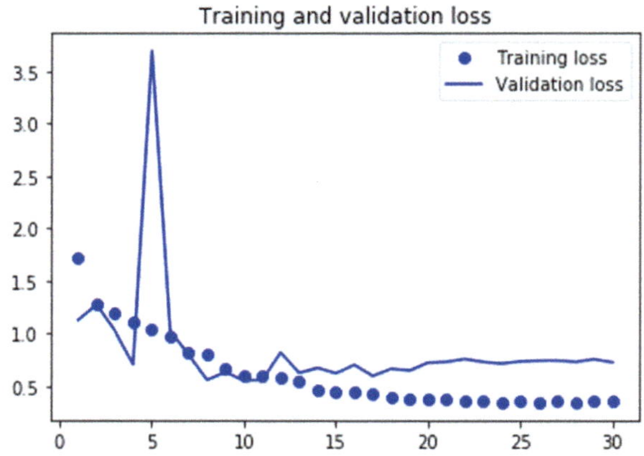

Fig. 2 Comparison plot between training loss and validation loss

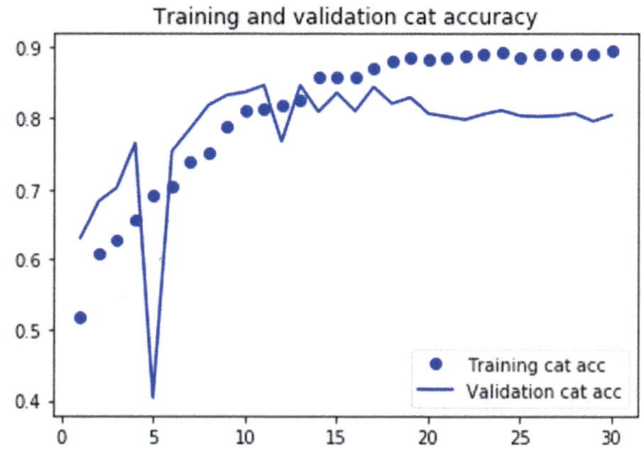

Fig. 3 Comparison plot between training categorical accuracy and validation categorical accuracy

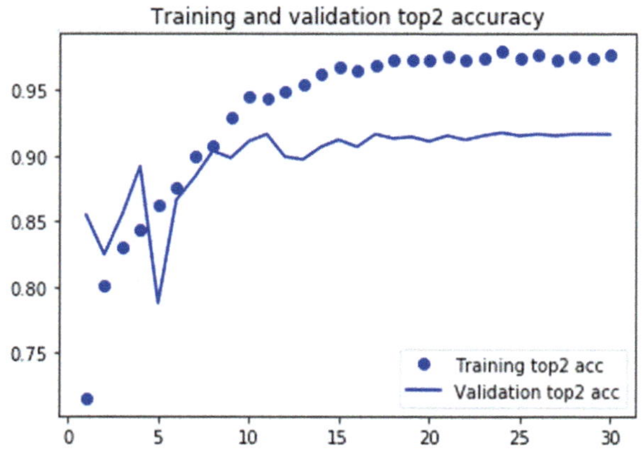

Fig. 4 Comparison plot between training top 2 accuracy and validation top 2 accuracy

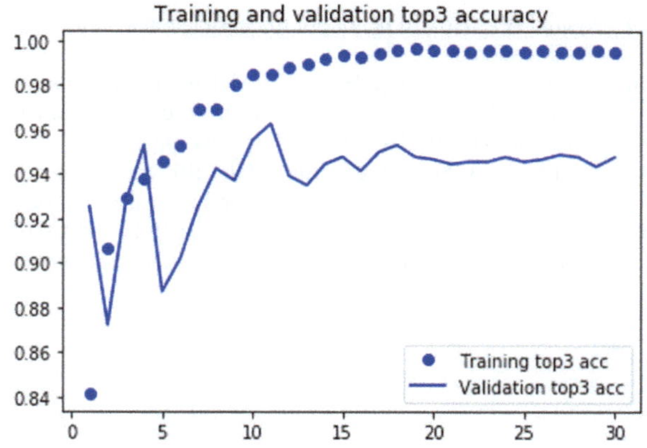

Fig. 5 Comparison plot between training top 3 accuracy and validation top 3 accuracy

6 Conclusion and Future Use

Through this model, we have made an endeavor to combat the malignancy of skin cancer, one that needs a timely diagnosis. With anticipation of an increase in the instances of skin cancer due to increased UV radiation penetration as a result of a highly depleted ozone layer, being on the horizon, this timely technological intervention would help oncologists and medical fraternity in delivering their best work and save millions around the globe through an early-stage diagnosis. The future scope and plan are to use more number of real, clinical datasets to achieve a wider and more effective implementation.

References

1. Masood A, Al-Jumaily AA (2013) Computer aided diagnostic support system for skin cancer: a review of techniques and algorithms. Int J Biomed Imaging
2. Doi K (2007) Computer-aided diagnosis in medical imaging: historical review, current status and future potential. Comput Med Imaging Graph 31(4–5):198–211
3. Mehta P, Shah B (2016) Review on techniques and steps of computer aided skin cancer diagnosis. Proc Comput Sci 85:309–316
4. Abdel-Zaher AhM, Eldeib AM (2016) Breast cancer classification using deep belief networks. Expert Syst Appl 46:139–144
5. Pathan SK, Prabhu G, Siddalingaswamy PC (2018) Techniques and algorithms for computer aided diagnosis of pigmented skin lesions—a review. Biomed Signal Process Control 39:237–262
6. The Skin Cancer Foundation (2018) Skin cancer information. Available: https://www.skincancer.org/skincancer-information. Accessed 25 Dec 2018
7. Hameed N, Ruskin A, Abu Hassan K, Hossain MA (2016) A comprehensive survey on image-based computer aided diagnosis systems for skin cancer. In: 2016 10th international conference on software, knowledge, information management & applications (SKIMA). IEEE, New York, pp 205–214
8. Esteva A, Kuprel B, Novoa RA, Ko J, Swetter SM, Blau HM, Thrun S (2017) Dermatologist-level classification of skin cancer with deep neural networks. Nature 542(7639):115

9. Nylund A (2016) To be, or not to be Melanoma: convolutional neural networks in skin lesion classification
10. Ali AA, Al-Marzouqi H (2017) Melanoma detection using regular convolutional neural networks. In: 2017 International conference on electrical and computing technologies and applications (ICECTA). IEEE, New York, pp 1–5
11. Krizhevsky A, Sutskever I, Hinton GE (2012) Imagenet classification with deep convolutional neural networks. In: Advances in neural information processing systems, pp 1097–1105
12. Tschandl P, Rosendahl C, Kittler H (2018) The HAM10000 dataset, a large collection of multi-source dermatoscopic images of common pigmented skin lesions. Sci Data 5:180161

An Analysis of Machine Learning Approach for Detecting Automated Spammer in Twitter

C. Vanmathi and R. Mangayarkarasi

Abstract With the evolution of social media, information sharing in online is turning into ubiquitous each and every day. Various facts are propagating thru online social networks along with each the wonderful and negative. In this paper, our essential focal point is to resolve terrible or incorrect data such as rumors. Rumor is a serious problem in online social media. The wrong rumors can create many problems in society; hence, it is important to block these rumors. First in this paper, we have compared Naive Bayes algorithm and logistic regression for deciding which algorithm is better for spam classification. Then, we have proposed a model for detecting spammers in twitters and automatically blocking those rumors by hindering an assured subset of the nodes. In order to decrease the influence of the rumors, nodes are allotted an acceptance time threshold. If a user posts the rumor more than a particular threshold, the user gets automatically blocked. We have created a Java-based Web application using tomcat SQL server, JavaScript html and CSS, like Facebook where a user can search for friends, accept friend requests, chat with them, change profile pictures and post messages. In this paper, we used a supervised learning technique known as Naive Bayes Algorithm for blocking of the users. The analysis of users blocked per month can be also done which will help in study of users and rumor details.

Keywords Twitter · Spammer detection · Machine learning

1 Introduction

The exchange of the information is done using the nodes. The nodes also contain hosts such as PCs, mobile phones, servers and the networking hardware. The two devices defined to be connected if the transfer or exchange of the information can be done. In today's world with the rising of social networks like Twitter, Facebook, Snapchats, sharing of the information with each other becomes easier. The people can make friends easily. It requires highly skilled network management professionals to

C. Vanmathi (✉) · R. Mangayarkarasi
School of Information Technology and Engineering, VIT University, Vellore, India
e-mail: vanmathi.c@vit.ac.in

A. K. Tripathy et al. (eds.), *Advances in Distributed Computing and Machine Learning*, Lecture Notes in Networks and Systems 127, https://doi.org/10.1007/978-981-15-4218-3_43

manage these social networks to run reliably and securely. These social networks help in sharing of the innovative ideas, share the day-to-day updates and develop their own skills. On the other hand, they have become a channel to spread wrong and malicious content or news. Some of the examples are some people may post the message of upcoming earthquake or any other natural disaster which may create big chaos in our society. Hence, it is important to detect these kinds of rumor and stop spreading these kinds of rumor. However, during the extreme cases such as terrorist attacks, it is very important to block these kinds of rumors which can help in preventing the negative influence in society. Most of the aforementioned works studied the problem of getting the most out of the influence of nice records via social networks. Fast approximation methods were also proposed to have an impact on the maximization problem. In contrast, the terrible affect minimization hassle has received a terrible lot less attention, but still there have been regular efforts on designing first-class strategies for blocking cruel rumors and minimizing the negative influence.

2 Literature Survey

The author discusses about detecting spammers through a hybrid approach based on characterization of users and mutual interactions such as the interaction with the followers. They have used about nineteen different features, and through these features, learning is done through classifiers such as Bayesian network, decision tree and random forest using real dataset. The discrimination of features is done and found out the least effective is metadata features and the most effective is community and interaction-based features. In this paper, the author has proposed a new feature which can help in differentiating spam accounts using analyzing features such as behavioral, content entropy, bait techniques and profile spammers [1]. The algorithms used supervised learning and generated the models tool CATS and detected the false detection rates [2]. Survey of different spam detection methods such as account-based, tweet-based, graph-based and hybrid method using the combination of the other methods. The advantages and disadvantages of these methods are discussed. The false positive and false negative values were also discussed [3]. The aim of paper is to detect sarcasm in Twitter using behavioral traits, model called SCUBA. The theories were employed using user's past tweets. The construction of behavioral modeling framework was built, and efficiency was demonstrated. The characterization of different forms of sarcasm is done.

The author's main aim was to create a large-scale critical dataset for detection of spams in Twitter [4]. The two algorithms were introduced for capturing spammer's behavior such as sharing of URLs that are malicious and contents which are exploited. The efficiency and effectiveness of the approach were found out [5]. A new approach was introduced for detecting bullying and aggressive behaviors on Twitter using machine learning classification algorithm is done using different features such as text, network and user-based attributes. Despoina et al. [6] developed a machine learning system with some orthogonal features to analyze the characteristics of spam in social networks. The statistical analysis such as F-measure is conducted, and analysis on trending topics is done [7]. Study on detection of fake news on the social media using psychological and social theories is done. They divided the process in two phases and matrices are evaluated using large datasets. The phases are divided as characterization and detection using data mining techniques [8]. Data in this paper are collected using Twitter streaming APIs. The term frequency analysis and relative normalized term frequency such as unigrams, bigrams and trigrams are done, and then the F-measure is obtained.

A hybrid approach of different classifier is used, and analysis of tweets using Google safe browsing API for security is done for intelligent spam detection [9, 10]. Different supervised and unsupervised techniques such as classification and clustering algorithms evaluated using different matrices and found that clustering provides better accuracy than other algorithms. The comparison of effectiveness and performance of different models such as hidden Markov model and random forest using different metrics such as F-measure and performance metrices were done using datasets and concluded that HMM is better than random forest [11]. The detection was done in two ways: automation and honeypot [12]. The performance of algorithms was analyzed using ROC curve on different algorithms like random boost and logit boost and is applied on two different random datasets. They concluded that the combination of logit boost and random boost provides better accuracy [13]. The author has discussed about SMS spam filtering and different difficulties faced by the developers. The different matrices such F-measure, accuracy and false positive rates are found out.

The author has used personal characteristics such as geographic location and IP address for spam detection using content analysis [14]. This method identifies who provides reviews for easy identifications of spam activities [15]. For overcoming spam post problems, the author has introduced new algorithm called ALO-Boosting which is the combination of Antlion Optimization (ALO) and boosting. Then, the author has compared different algorithms such as SVM, KNN and bootstrap aggregating with different matrices and concluded ALO-Boosting is giving better performance [16]. The author has taken a survey on different image spam filtering techniques such as OCR-based techniques which includes fuzzy OCR and Bayes OCR and concluded that fuzzy OCR is better than Bayes OCR [17]. The author has used a hybrid detection for spam detection which includes genetic algorithms (GA) and random weights network (RWN). It is named as Auto-GA-RWN for post spam detections and concluded that Auto-GA-RWN shows higher stability. Hossam et al. [18] focused to remove unwanted tweets. The different classification models were trained like maximum entropy model, extremely randomized trees, random forest model and C-support vector classification, multilayer perceptron and gradient boosting. The results are analyzed using Scikit learn [19]. The author has deliberated the edge server placement problem edge computing environments for mobile applications. The different approaches in terms of access delay and workload balancing are applied to smart cities.

In most of the previous works, the main focus of the people was to maximize the positive influence in society. Due to which the negative information influence was ignored due to which large chaos was created in society. Kimura et al. studied the matter of minimizing the propagation of malicious rumors by blocking a restricted range of links in a very social network. They provided two completely different definitions of contamination degree and planned corresponding optimization algorithms. Fan et al. investigated the smallest amount price rumor block drawback in social networks. They introduced the conception of "protectors" and checked out to pick a borderline range of them to limit the dangerous influence of rumors by triggering a protection cascade against the rumor cascade. However, square measure has limitations [20]. In the existing system, the rumors which were passed, it takes a large amount of time to know about it and delete the rumor. Hence, it is important to know about the rumor and block the users before it creates chaos in society.

3 Proposed System

Rumor is a serious problem in online social media. The wrong rumors can create many problems in society; hence, it is important to block these rumors. The main aim is to block the user automatically after giving three notifications. In this paper, we propose a model for detecting spammers in Twitters and automatically those rumors by blocking a certain subset of nodes. In order to decrease the influence of the rumors, each node is assigned a tolerance time threshold. If a user posts the rumor more than a particular threshold, the user gets automatically blocked. We have created a Java-based Web application using tomcat SQL server, JavaScript html and CSS, like Facebook where a user can search for friends, accept friend requests, chat with them, change profile pictures and post messages. In this paper, we used the supervised machine learning algorithm and dynamic blocking algorithm based on different nodes selection schemes and the maximum likelihood principle. The analysis of users blocked per month can be also done which will help in the study of users and rumor details.

3.1 Comparative Work

Naive Bayes
It is one of the methods for text classification. It is an intuitive model in finding probabilities. It helps us to calculate a probability of an event depending on the prior events. It involves two models using feature selector and without feature selectors. Naïve Bayes has very strong assumption on the shape of the data distribution. So, this algorithm is chosen for comparison.

Computing the probability of the message which contains a given spam word. If we first select the features, then the two text files ham keywords and spam keywords will be formed and we create a common data structure. Subsequently, the prior probabilities are evaluated, and frequencies of spam and ham words are computed. Finally the classification of spam and ham messages or posts is done.

Logistic Regression

It is the best model to handle text classification problems. It is based on binary output representation and involves probabilities of both dependent and independent variables. Logistic function can be represented as logistic function or also known as sigmoid function. The logistic regression output is between the values 0 and 1. The variable z represents the exposure to some set of risk factors, while $f(z)$ represents the probability of a particular outcome, given that set of risk factors. The variable z is a measure of the total contribution of all the risk factors used in the model and is known as the logit [21]

$$f(z) = \frac{1}{1 + e^{-z}}. \tag{1}$$

The logical regression algorithm best suits for training dataset and does not require more computational resources and input features to be scaled. The output of the algorithm is well calibrated and predicated probabilities. This algorithm works better if we remove the unrelated attributes to the output variables. Regression best suits for feature engineering (Figs. 1, 2 and 3).

Metrics

The metrics used for this analysis are accuracy, precision, recall and F-measure.

Datasets

For training, we have selected about 400 ham messages and about 100 spam messages. For testing of messages, we have selected about 200 messages. The datasets we have selected are in the form of text files (Fig. 4).

4 Results

Metrics	Logistic regression	Naive Bayes
Accuracy	0.885	0.975
Precision	0.8148148148148148	0.8888888888888888
Recall	0.55	1.0
F-measure	0.6567164179104479	0.9411764705882353

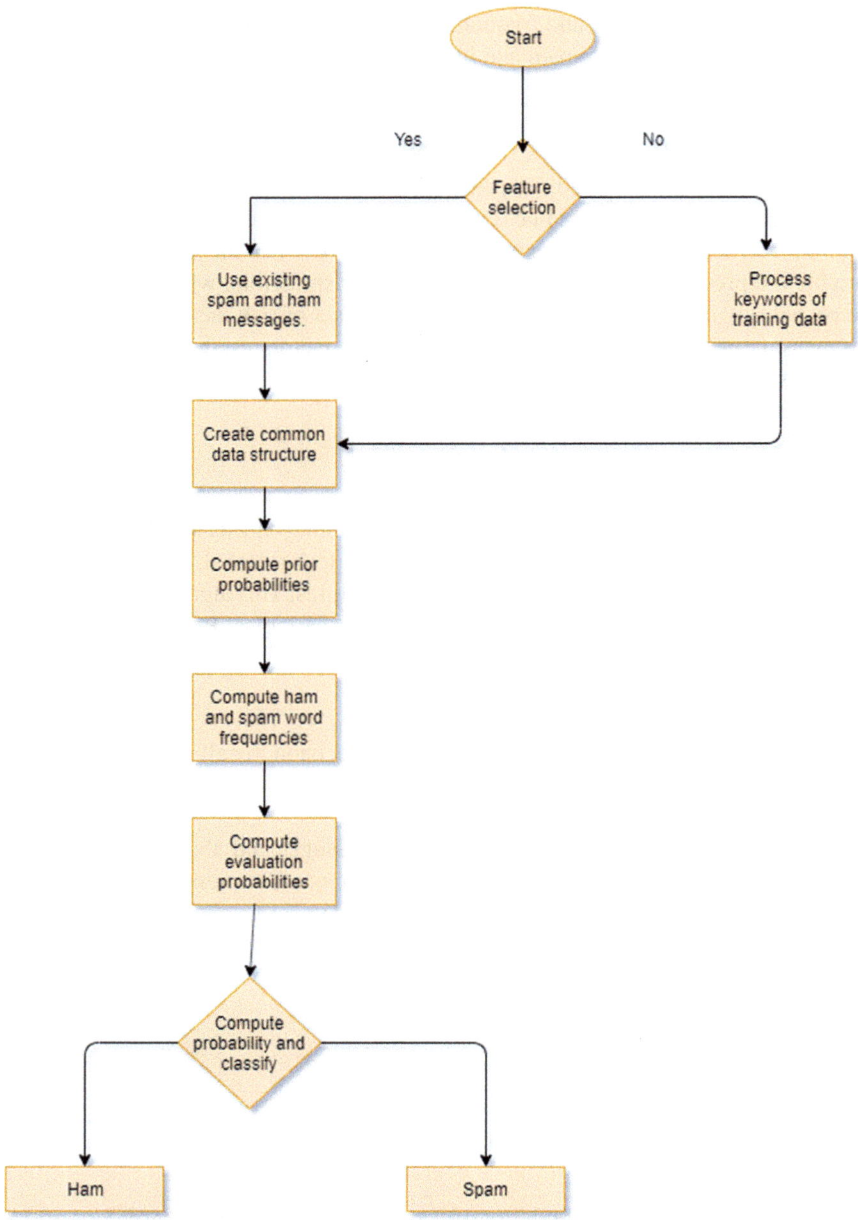

Fig. 1 Flow diagram of the system

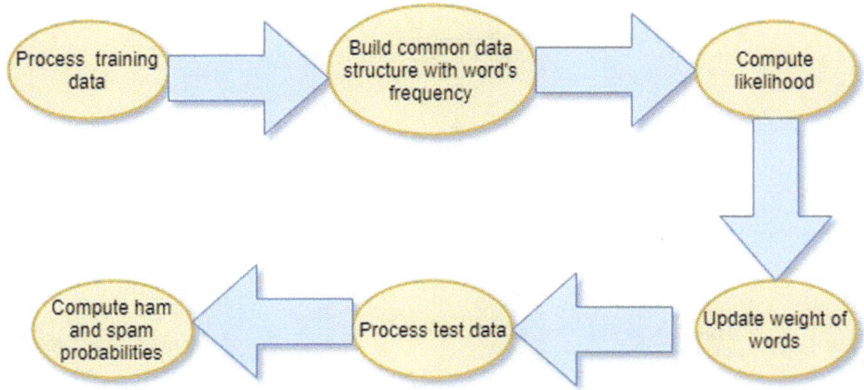

Fig. 2 Steps in spammer detection

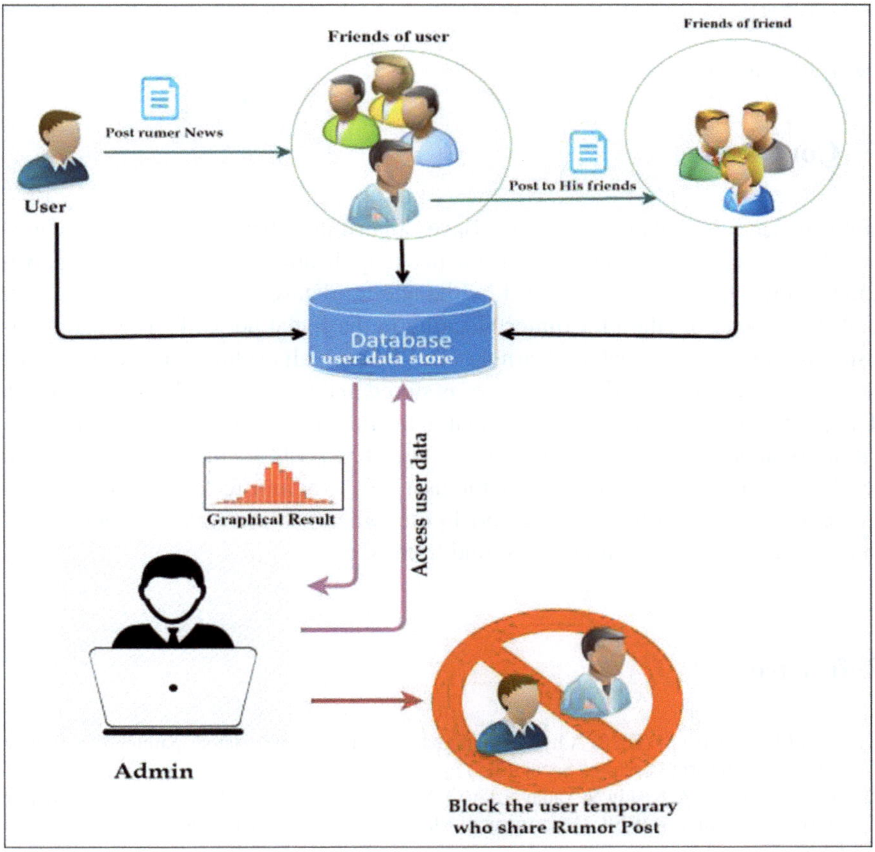

Fig. 3 System architecture

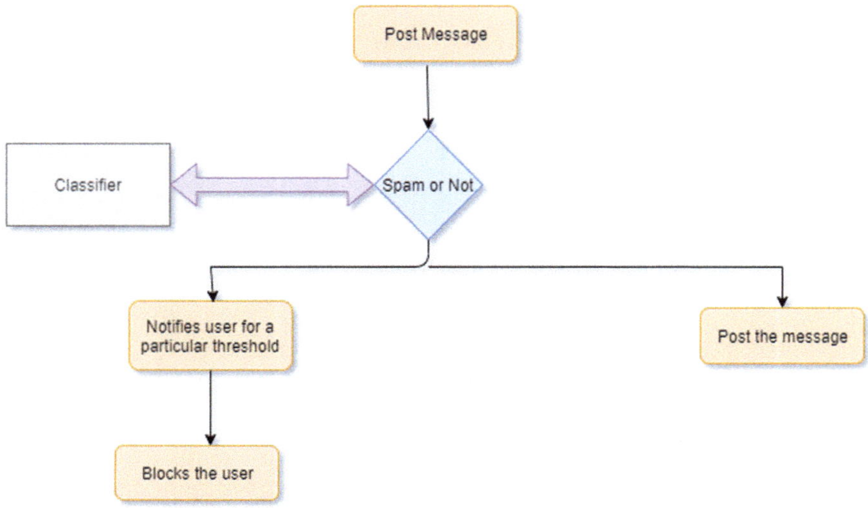

Fig. 4 Blocking user

5 Conclusion

In this paper, it is analyzed that Naive Bayes algorithm is better than logistic regression. Hence, using Navis Bayes algorithm in our application for rumor blocking gives good results. The data are provided by the admin (dataset). In this paper, we have analyzed a better method for rumor blocking in the social network. The analysis is tested for supervised machine learning classification algorithm to detect the spam messages. The list of words or messages is stored in the database which is restricted. If any user tries to post any rumor content in the timeline, the user will get the notification for about three times afterward, and also if he posts the message the user will be blocked. The admin can unblock the users if the admin feels. We have created an application similar to Facebook and Twitter where the user can post their views, share information, chat with friends, find friends, etc.

References

1. Fazil M, Abulaish M (2018) A hybrid approach for detecting automated spammers in Twitter. IEEE Trans Inform Forensics Secur 13(11)
2. Amleshwaram AA, Reddy N, Yadav S, Gu G, Yang C (2013) CATS: characterizing automation of Twitter spammers. IEEE, Department of Electrical and Computer Engineering Texas A&M University
3. Talha A, Kara R (2017) A survey of spam detection methods on Twitter. Int J Adv Comput Sci Appl (IJACSA) 8(3)
4. Rajadesingan A, Zafarani R, Liu H (2015) Sarcasm detection on Twitter: a behavioral modeling approach. In: WSDM'15, Shanghai, China, 2–6 Feb 2015

5. Concone F, Lo Re G, Morana M, Ruocco C (2019) Twitter spam account detection by effective labelling. In: CEUR-WS.org/Vol-2315
6. Chatzakouy D, Kourtellisz N, Blackburnz J, De Cristofaro E, Stringhini G, Vakali A (2017) Mean birds: detecting aggression and bullying on Twitter, 12 May 2017
7. Martinez-Romo J, Araujo L (2012) Detecting malicious tweets in trending topics using a statistical analysis of language. Elsevier, Amsterdam
8. Shuy K, Slivaz A, Wangy S, Tang J, Liuy H (2017) Fake news detection on social media: a data mining perspective, 3 Sept 2017
9. Benhardus J, Kalita J (2013) Streaming trend detection in Twitter. Int J Web Based Commun 9(1)
10. Vishwarupe V, Bedekar M, Pande M, Hiwale A (2018) Intelligent Twitter spam detection: a hybrid approach, Jan 2018
11. Fernandes MA, Patel P, Marwala T (2015) Automated detection of human users in Twitter. Procedia Comput Sci 53:224–231
12. Washhaa M, Qaroushb A, Mezghania M, Sedesa F (2017) A topic-based hidden Markov model for real-time spam tweets filtering
13. DeBarr D, Wechsler H (2012) Spam detection using random boost, 24 Mar 2012
14. Delany SJ, Buckley M, Greene D (2012) SMS spam filtering: methods and data. Expert Syst Appl 39
15. Bajaj S, Garg N, Singh SK (2017) A novel user-based spam review detection. Procedia Comput Sci 122
16. Naem AA, Ghali NI, Saleh AA (2018) Antlion optimization and boosting classifier for spam email detection. Future Comput Inform J 3
17. Biggio B, Fumera G, Pillai I, Roli F (2011) A survey and experimental evaluation of image spam filtering techniques. Pattern Recogn Lett 32
18. Farisa H, Al-Zoubia AM, Heidarib AA, Aljaraha I, Mafarja M, Hassonaha MA, Fujitad H (2019) An intelligent system for spam detection and identification of the most relevant features based on evolutionary Random Weight Networks. Inform Fusion 48
19. Inuwa-Dutse I, Liptrott M, Korkontzelos I (2018) Detection of spam-posting accounts on Twitter. Neurocomputing 315
20. Wang B, Chen G, Fu L, Song L, Wang X, Liu X (2016) DRIMUX: dynamic rumor influence minimization with user experience in social networks
21. Arkalgud N (2018) Logistic regression for spam filtering, 14 Feb 2008

A Fast Mode of Tweets Polarity Detection

V. P. Lijo and Hari Seetha

Abstract Polarity detection is an emerging area of research in text mining. Polarity detection is observing and identifying the sentiment inclination of text, whether it is positive or negative. In this paper, a fast mode of supervised learning for polarity detection on tweets is proposed, that is using datasets available in public. The feature selection strategy ensures reduced dimensionality. The low dimension data processing on Apache Spark supports scalability for large datasets. The experiment shows that the method is supporting high scalability and efficiency.

Keywords Sentiment analysis · Polarity detection · Big data · Twitter analysis

1 Introduction

Nowadays, people use twitter wide to get world news about political, environmental, recent technical advancements, etc. The social network's data growth is exponential. Its richness of sentiment and other aspects such as political and economic attract researchers to find insights into support applications that are using that information as to their essence of business. Sentiment analysis is dealing many aspects of information in the text such as identifying opinion existence, classification as positive, negative or neutral; categorize data based on the type of sentiment such as joy, sadness, anger, etc. Polarity detection is the process of classifying the given texts based on its polarity such as positive, negative and neutral.

The volume of data provides an opportunity to improve the performance of the algorithms in terms of accuracy, recall, precision, etc., by utilizing a large number of instances. Processing large data in a single machine is limiting with scalability. A

V. P. Lijo (✉)
Vellore Institute of Technology, Vellore 632014, India
e-mail: lijo.vp@vit.ac.in

H. Seetha
VIT-AP, Amaravati, AP, India
e-mail: hariseetha@gmail.com

A. K. Tripathy et al. (eds.), *Advances in Distributed Computing and Machine Learning*, Lecture Notes in Networks and Systems 127, https://doi.org/10.1007/978-981-15-4218-3_44

distributed or parallel computing environment can overcome this limitation. Most of the existing algorithms for sentiment analysis are designed for sequential processing. So it is a need to develop new algorithms or modifies existing algorithms to run on a distributed or parallel computing environment. Sentiment classification is the process of identifying the sentiment inclination (opinions) of a text and segregates them in different classes. The machine learning, lexicon-based approach and hybrid approach are the main categories of sentiment classification. The lexicon is a precompiled collection of known sentiment terms. The completeness and soundness of lexicons improve the performance of lexicon-based approaches. The machine learning approach follows either supervised learning or unsupervised learning strategies. Above all, the supervised machine learning approach SVM is very popular due to its generalization characteristics, and it works well in even unbalanced data. The SVM training (collecting a set of support vectors (SVs) and other related parameters) is considered as computationally costly. Numerous ways are available in the literature to improve the performance of the SVM by improving its training time.

Sequential minimal optimization (SMO) is a quick training strategy depended on certain heuristics yet consecutive in nature [12]. Priyadarshini [13] proposed a system with SVM and MapReduce for sentiment classification. Retraining is needed for building classifier, and it is a tedious task and does not contribute much to accuracy. The mixture of different strategies, for example, lexicon-based, machine learning and so forth, is valuable to improve the effectiveness of the sentiment analysis [9].

This research paper proposes a distributed model that is implemented in Apache Spark that transforms our tasks into MapReduce jobs. We modify the sequential minimal optimization (SMO) [12] for typically distributed sentiment classification. We investigate the effects of various parameters on the algorithm performance and computation cost, such as the number of instances, number of computing nodes by performing a substantial experimental evaluation.

The remaining sections of the paper are arranged as follows: In Sect. 2, details of SMO and Spark are discussed. Section 2 gives insights into the proposed method, Sect. 3 discusses experiments and results, and Sect. 4 gives a conclusion.

2 The Sentiment Analysis

The main components in system are data preprocessing, various classifiers and the SPARK framework, which are explained in detail in the following subsections.

2.1 Data Preprocessing

Stop-words removal [14], Stemming and lemmatization Part-of-Speech Tagging (POS), Handling Negations [7] and Tokenization into N-grams [6] are some of the important DPP steps.

2.2 Sequential Minimal Optimization (SMO)

Platt's [12] SMO is a fast training strategy for SVM, and it gains this speed up by reducing workloads in each iteration. SMO keeps decomposition at maximum by selecting each working set with two elements (minimum). The working set follows the condition:

$$\sum_{i=1}^{n} A_i Yi = 0 \tag{1}$$

where A_i is Lagrange multiplier and Y_i is class name. This helps to get analytical solutions for subproblems. To build a linear SVM, SMO keeps only one weight vector instead of the training examples which correspond to nonzero Lagrange multiplier values [1].

2.3 Apache Spark

Apache Spark is useful for processing large-scale data. It is a general-purpose open-source distributed computing engine for analyzing big data. It works in a cluster and distributes data among nodes in the cluster and processes the data chunks in parallel. Spark has one coordinator (Driver) and many distributed workers (executers). The driver is a Java process, and it communicates with a large number of executors. Figure 1 shows the Spark architecture.

The fast mode of operation is achieved by doing all the process from data collection to the classifier training in parallel, using SMO and Spark. The linear SVM expects the output from each as partial weight vector Wlocal for the local partition and threshold value b, where n number of instances.

Fig. 1 Spark architecture

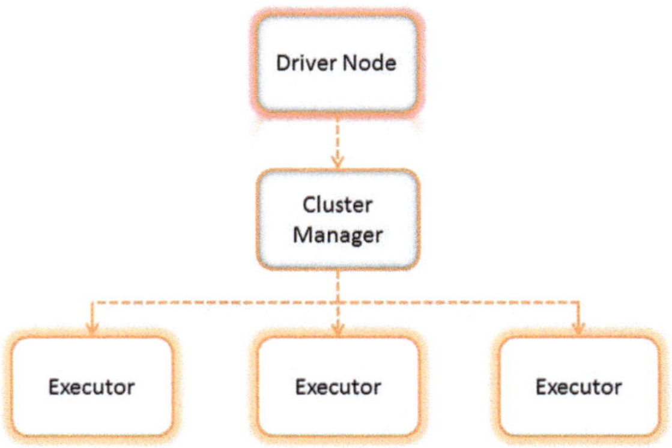

$$\text{Wlocal} = \sum_{i=1}^{n} y_i a_i x_i \tag{2}$$

$$\text{Wglobal} = \sum_{i=1}^{n} \text{Wlocal}_i \tag{3}$$

$$b_{\text{avg}} = \left(\sum_{i=1}^{n} b_i \right) / n \tag{4}$$

$$\text{SVM}_{\text{output}} = \text{Wglobal} * x_i - b_{\text{avg}} \tag{5}$$

The global weight vector is derived from local weight vectors. The reducer will do the task. The Wglobal and b are calculated as in Eqs. (3) and (4), respectively, for n number of partitions. Equation (5) shows the $\text{SVM}_{\text{output}}$ in general. The architecture of the proposed model is illustrated in Fig. 2. Distributed processing is carried out through the processes such as preprocessing and training.

The individual subsets directly supply to SMO in each node to get local results as weights: Wlocal and bias value b. Then, a global classifier is formed from local results. The final training set is using as input for classifiers and gets a trained classifier

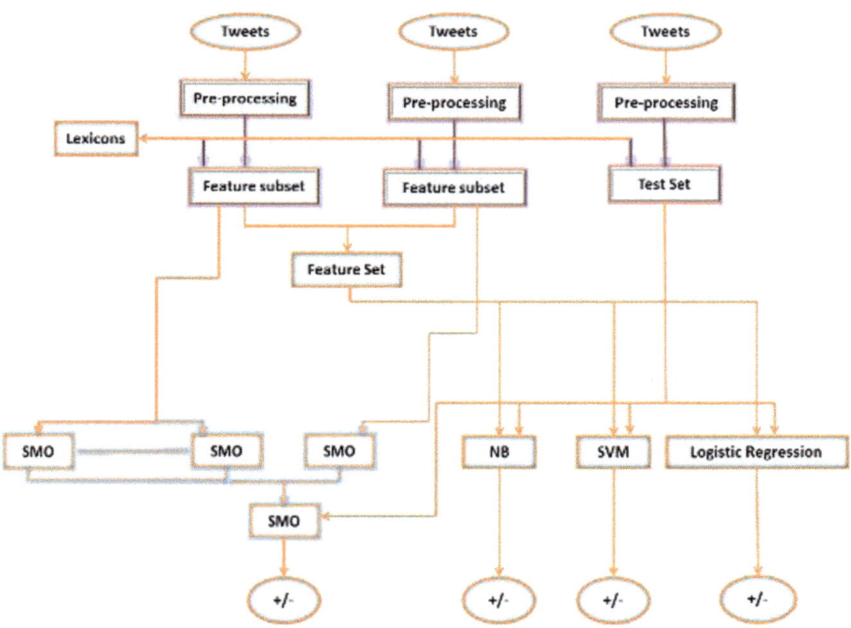

Fig. 2 The proposed model of sentiment classification

model. Here, Spark framework helps to execute each classifier in a different node. That will help to reduce the training time of multiple classifiers. Each SMO executes in a different node or each classifier executes in a different node. The SMO has chosen due to its fast training strategy.

In linear SVM, the weight vector and threshold value are the output of each map function. But the alpha array and threshold value *b* are the outputs for nonlinear SVM. The global alpha array is the result of the reducer, which joints all the maps' local alpha arrays. The reducer takes the average of *b* as in the case of linear SVM.

2.4 Lexicons and Methods

The literature gives large number of lexicons and methods for text polarity analysis. Popular lexicons are OpinionFinder [16], AFINN [11], NRC– (emotion [10], hashtag [10]), Bing Liu's Opinion [3], SentiWordNet [2], Sentiment140 [8], etc., SenticNet [5] SentiStrength [15] and Sentiment140 [8], are well-known methods for sentiment polarity calculation.

2.5 Features

The features considered in this work, lexicon-based [4] features and text-based features (presence of negation), are discussed in this subsection. Table 1 shows the lexicon-based features which are retrieved using the above-mentioned lexicons. For a comparison, the classifiers are also trained with 500 unigram features and results are collected.

2.6 Twitter Sentiment Analysis

We have used a supervised learning strategy for sentiment polarity classification in this work. The features are extracted from each tweet and represented them as vectors. Manually and automatically annotated data sets are used for training and evaluation. The analysis task is limited to sentiment polarity detection where tweets are classified as positive and negative.

Tweet Sentiment Representation The lexicon-based features used in [4] are used with changes for polarity detection. Besides, the presence of negation is noticed as binary data. This feature selection helps to reduce dimensionality significantly, and it overcomes the problem of sparse data in feature vectors. The collected features are ranked using information gain (IG) and considered only the first 15 features for each dataset.

Table 1 Lexicon-based features

Features	Source	Description
OFP	OpinionFinder	# of matching +ve words
OFN		# of matching −ve words
BLP	Bing Liu Lexicon	# of matching +ve words
BLN		# of matching −ve words
NRCP	NRC-emotion	# of matching +ve words
NRCN		# of matching −ve words
SWP	SentiWordNet	Score sum of +ve matching words
SWN		Score sum of −ve matching words
AFP	AFINN	Score sum of +ve matching words
AFN		Score sum of −ve matching words
SNP	SenticNet	Score sum of +ve matching concepts
SNN		Score sum of −ve matching concepts
NRChashP	NRC Hashtag lexicon	Score sum of +ve matching words
NRChashN		Score sum of −ve matching words
S140LexP	Sentiment140 lexicon	Score sum of +ve matching words
S140LexN		Score sum of −ve matching words
S140	Sentiment140 Method	Output label (−ve, N, +ve)
SSPOL	SentiStrength	Output label (−ve, N, +ve)
SSP		Strength output for the +ve class
SSN		Strength output for the −ve class
NEG	Negation	1 if Negation, 0 otherwise

3 Experiments and Results

In this session, we discuss the experiments and results. The training and testing of datasets are done using three datasets named SemEval, Sentiment140 and Twitter-Airline; they give training and testing data sets for polarity and subjectivity detection. Each tweet is tagged as positive, negative or neutral. For polarity detection, only positive and negative tweets are considered and discarded neutral. Feature analysis is performed over the features mentioned in Table 1. We analyze how well each feature split the data sets concerning polarity detection. Here, we consider multiple resources, and they gave better performance than individual resources. The principle behind this is the required feature may absent in one or more resources, but they may be compensated from other resources.

We have used the supervised classifiers for tweet level polarity detection. We compare the performance of the SMOS classifier with NB, SVM and logistic regression classifiers. The tenfold cross-validation is used in the experiments.

The experiments are carried out on 4 Hp computers with Ubuntu 16 OS, RAM- 8 GB and processor clock 2.3 GHz. The statistics of data sets SemEval, Sentiment140

Table 2 Statistics of datasets

Data sets	Positive	Negative	Neutral
SemEval	1458	3639	4585
Sentiment140	248,576	799,999	0
Twitter-Airline	2363	9178	3099

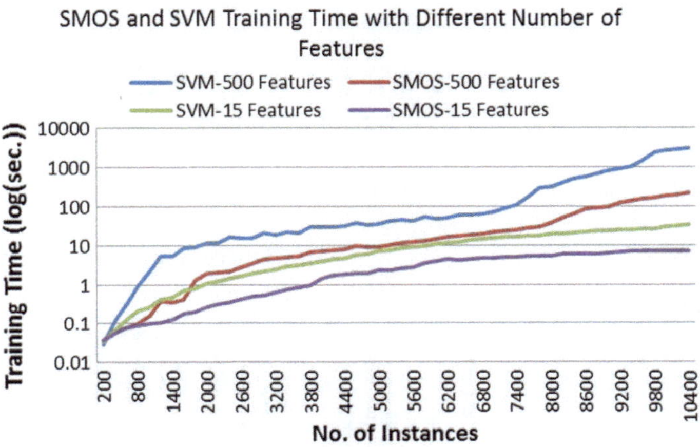

Fig. 3 Comparison of training time of SMOS and SVM classifier

and Twitter-Airline is shown in Table 2. Those data sets are publically available, which is ensuring the repeatability of experiments. Figure 3 shows the training time of SMOS and SVM with two different numbers of features. One is 500 unigram features, and the other is 15 lexicon-based features. There is a significant difference in the computation time of both cases.

SVM was consumed more than two hours to be trained with 10,000 instances of unigram features, but SMOS is used 300 s for the same set of instances. The accuracy achieved using the selected 15 features is comparable with the accuracy got with unigram features. But the time taken by SMOS for the same number of instances with selected 15 features is only 10 s. The positive impact of dimensionality reduction is supported by the performance of the Naïve Bayes and logistic regression. The scalability test is performed with a different number of instances from the Sentiment140 data set. The test results show that the SMOS operation is feasible for even large data sets, with less time complexity. Figure 4 shows the SMOS training time of various numbers (14, 15, 16 and 17) of features. Based on information gain, 15 features are used for classification. The highest accuracy of classification is gained with 15 and 16 features, so in favor of computation time, 15 features are used. The classification with more than 16 features does not contribute much to accuracy. Figure 5 shows the F1 and accuracy measures of SMOS with 14, 15, 16 and 17 features. Table 3 gives details about the performance of the classifier. This shows the classifiers' accu-

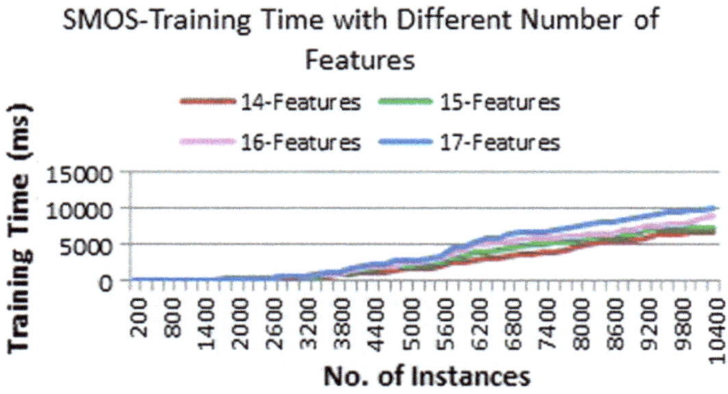

Fig. 4 SMOS—Computation with different number of features

Fig. 5 SMOS—Classification accuracy with different number of features

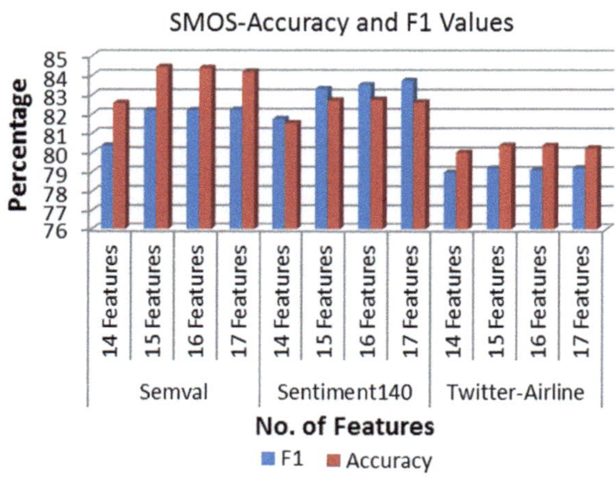

racy and F1 measure. Our classifier gives good accuracy with the less computational cost. When compared to other classifiers, the SMOS has some reduction in accuracy and F1 points, fall from 84.78 to 84.45 and 83.86 to 82.21, respectively, in case of SemEval, and the best accuracy is given by logistic regression. The SMOS processed the datasets with less computational cost with negligible accuracy loss. The proposed method with the reduced dimension of data shows better results also with naive Bayes (NB), SVM and logistic regression.

4 Conclusions

In this paper, we proposed a distributed classification method for sentiment analysis. The extensive experiments show that this method outperforms the sequential approaches for polarity detection concerning the computation cost without compro-

Table 3 Classification performance: tenfold cross-validation with 15 features

Data sets	Methods	F1	Accuracy
SemEval	NB	81.22	82.74
	SVM	83.67	85.21
	Logistic regression	83.86	84.78
Sentiment140	NB	77.89	76.89
	Logistic regression	75.91	77.51
Twitter-Airline	SVM	82.16	82.32
	NB	78.56	80.47
	Logistic regression	80.92	81.04
SemEval	SMOS	82.21	84.45
Sentiment140	SMOS	83.32	82.73
Twitter-Airline	SMOS	79.21	80.39

mising accuracy. The Spark and the modified SMO algorithm support scalability for processing large data sets. In this work, main focus is on reducing computational cost. As future work, take measures to improve the accuracy of sentiment classification.

References

1. Alham NK, Li M, Liu Y, Hammoud S (2011) A mapreduce-based distributed SVM algorithm for automatic image annotation. Comput Math Appl 62(7):2801–2811
2. Baccianella S, Esuli A, Sebastiani F (2010) Sentiwordnet 3.0: an enhanced lexical resource for sentiment analysis and opinion mining. In: LREC, vol 10, pp 2200–2204
3. Bing L (2012) Sentiment analysis: a fascinating problem. Morgan and Claypool Publishers, pp 7–143
4. Bravo-Marquez F, Mendoza M, Poblete B (2014) Meta-level sentiment models for big social data analysis. Knowl-Based Syst 69:86–99
5. Cambria E, Havasi C, Hussain A (2012) Senticnet 2: a semantic and affective resource for opinion mining and sentiment analysis. In: Twenty-Fifth international FLAIRS conference
6. Caropreso MF, Matwin S, Sebastiani F (2001) A learner-independent evaluation of the usefulness of statistical phrases for automated text categorization. Text Databases Doc Manage: Theory Pract 5478:78–102
7. Councill IG, McDonald R, Velikovich L (2010) What's great and what's not: learning to classify the scope of negation for improved sentiment analysis. In: Proceedings of the workshop on negation and speculation in natural language processing. Association for Computational Linguistics, pp 51–59
8. Go A, Bhayani R, Huang L (2009) Twitter sentiment classification using distant supervision. CS224N Project Report, Stanford 1(12):2009
9. Kolchyna O, Souza TT, Treleaven P, Aste T (2015) Twitter sentiment analysis: Lexicon method, machine learning method and their combination. arXiv preprint arXiv:1507.00955
10. Mohammad SM, Kiritchenko S, Zhu X (2013) NRC-Canada: building the state-of-the-art in sentiment analysis of tweets. arXiv preprint arXiv:1308.6242
11. Nielsen FÅ (2011) A new anew: evaluation of a word list for sentiment analysis in microblogs. arXiv preprint arXiv:1103.2903

12. Platt J (1998) Sequential minimal optimization: a fast algorithm for training support vector machines
13. Priyadarshini A et al (2015) A map reduce based support vector machine for big data classification. Int J Database Theory Appl 8(5):77–98
14. Salton G, McGill MJ (1983) Introduction to modern information retrieval. McGraw-Hill, New York
15. Thelwall M, Buckley K, Paltoglou G (2012) Sentiment strength detection for the social web. J Am Soc Inform Sci Technol 63(1):163–173
16. Wilson T, Wiebe J, Hoffmann P (2005) Recognizing contextual polarity in phrase-level sentiment analysis. In: Proceedings of human language technology conference and conference on empirical methods in natural language processing

One-Dimensional Chaotic Function for Financial Applications Using Soft Computing Techniques

A. Alli, J. Vijay Fidelis, S. Deepa, and E. Karthikeyan

Abstract Cash forecasting is essential as a delay in process could lead to dissatisfaction of customers in financial organization. To maintain the actual cash requirement without any financial loss is difficult. Particle swarm optimization (PSO) enhanced the efficiency to converge the dataset and it will be easier to fall into local optima might prove fatal. Hence, chaotic system has been introduced to significantly enhance the interpretation toward cash management model for any financial organization. In the proposed study, one-dimensional chaotic-particle swarm optimization (CPSO) functions were used to improve the search strategy in the given problem space to yield the optimal values. The experimental analysis was made using chaotic sequence which shows the effectiveness of our proposed approach. Hence, the proposed methodology has the ability to increase the convergence rate without premature convergence.

Keywords Cash forecasting · PSO · Chaotic sequence · Strategy · Convergence

1 Introduction

The process of maintaining the optimal cash balance in any financial organization is very decisive. In order to achieve the flawless system, banks must be able to identify the cash needs. Cash forecasting is essential to perform the day-to-day transaction without any shortage of funds. If the system fails to determine the cash needs in advance, it may lead to financial losses for the organization. The conventional methods practiced such as time series and Holtz method, Winter's method, and simple ANN were used in the financial system [1]. The different cash forecasting approach used by the researchers was a factor analysis, expert system, and neural networks [2]. Many attempts were made by the researchers to use ANN for prediction of global

A. Alli · J. Vijay Fidelis (✉) · S. Deepa
Department of Computer Applications, Presidency College, Bengaluru, Karnataka 560094, India
e-mail: veefeed@gmail.com

E. Karthikeyan
Department of Computer Application, Government Arts College, Udumalpet, Tirupur, Coimbatore 642126, India

A. K. Tripathy et al. (eds.), *Advances in Distributed Computing and Machine Learning*, Lecture Notes in Networks and Systems 127, https://doi.org/10.1007/978-981-15-4218-3_45

solar rRadiation (GSR) [3] as well as in the manufacturing process to solve linear and nonlinear problems [4].

The drawback with the traditional cash forecasting approximation was that the system may be unsuccessful to forecast the cash requirement in the dynamically changing environment. Studies were conducted for pecunirian institution such as ANN based cash forecasting using back propagation for short-term data [2]. An evolutionary computation technique called particle swarm optimization (PSO) based on swarm intelligence was initially proposed by Kennedy and Eberhart [5]. PSO has the features towards exploration and exploitation abilities to find the global best values. It exhibits some evolutionary computation for a population using random values [5]. Cash management model using PSO for short term data was optimized [5, 6]. Optimization of cash management using computational intelligence for both short term data and long term data was tested in various experiments carried forward [7].

Researchers proposed optimization techniques in various evolutionary techniques based on genetic algorithm (GA) as including PSO [7]. The limitation of GA was the convergence properties of this algorithm which causes the system to get stuck in local optimum and premature convergence. PSO has no limitations as what is faced by any other evolutionary algorithms (EA). As compared with GA, PSO has inclusion of memory as an enhancement. In PSO, individual objects which drift away from local optima are drawn back in tandem with them, and knowledge of good solutions is retained by all particles [5].

Some researchers have used PSO-based ANN, simulated annealing methods such that, to enhance the optimal performance of PSO [8]. Subsequently, the ANN–PSO, a hybrid evolutionary approach was used to forecast the cash requirement for financial organization [8]. To overcome the limitations of conventional PSO, multi-best and multi-swarm optimization have been proposed to improve the search strategy [9] as well as to obtain the best global optimal by tracking the velocity of the particle during iterations [10]. A recent study has been deployed to enhance the performance of PSO by inculcating chaotic series. CPSO has been used for the aforementioned applications but the same can be applied in finance application to better the interpretations of traditional PSO. Avoiding the premature convergence, there is a need to enhance the interpretation of PSO using chaotic mapping functions.

2 Background—Chaotic PSO

Chaos can be defined as a system with various features such as nonlinear, ergodicity, deterministic, and the initial values are sensitive [11]. Chaos can be applied in engineering, information sciences, and biological applications. Subsequently, the chaos was attracted by many researchers due to its effective search strategy in the problem space to avoid trapping in local minima [11].

The chaotic system has the ability to generate different sequence using various functions. The sequence generated by the chaotic series replaces a pseudorandom number used in PSO to avoid the results being trapped in local optima. Many experimental studies were made to analyze the benefits of using chaotic values instead of pseudorandom numbers. Hybrid genetic algorithm using chaotic logistic mapping function is used for image encryption which enhances encryption process of the image [7, 11]. Chaotic algorithm was effectively used to transmit the images in a secured manner using DNA sequence [12]. Chaos synchronization in both continuous and discrete systems was implemented using nonlinear differential equations [13].

The PSO was used by many research scholars to solve the nonlinear problems [14]. The limitations of conventional PSO were identified to improve the performance of PSO with dynamic adaptation was proposed [14]. In traditional PSO the parameters $r1$ and $r2$ alone cannot ensure the optimization because it uses pseudorandom number. Hence, the performance of standard PSO can be improved by using the chaotic sequence generated by chaos theory [15]. Therefore, the proposed methodology provides PSO with chaotic mapping to improve the global convergence.

3 One-Dimensional Chaotic Function for Financial Dataset

The logistic system, a one-dimensional mapping function which generates the different random values between 0 and 1 is defined using (1).

$$x_{n+1} = 4.0 * x_n * (1 - x_n) \tag{1}$$

where x_{n+1} is the new chaotic sequence and x_n is the current value. The initial value $x_n = 0.01$. Since it is a one-dimensional function, the newly generated sequence was assigned to chaos1 = chaos2 = x_n, 'n' ranges between 1 and 500. The chaos values were used in velocity update Eqs. (2) and (3).

To illustrate the dynamic behavior of the system generated for 500 iterations using the logistic function are shown in the following Fig. 1.

$$\begin{aligned} v_i(t+1) = {} & w * v_i(t) + c1 * \text{chaos1}(p_{\text{best}}(i, j) - x_{\text{i}}(t)) \\ & + c2 * \text{chaos2}(g_{\text{best}}(\text{j}) - x_i(t)) \end{aligned} \tag{2}$$

$$x_i^{t+1} = x_i^t + v_i^{t+1} \tag{3}$$

The chaotic sequence generated in CPSO influences the performance of the algorithm. In traditional PSO, exploring the global optimal solution which is on larger space, the accuracy was not as optimal as required due to rough search. Due to smaller scale of dataset involved might improve the accuracy but algorithm might be confined in local optimization. Hence, the paper presents an alternate method facilitating the algorithm to search for the optimal solution in wide space with the minimum number of iterations.

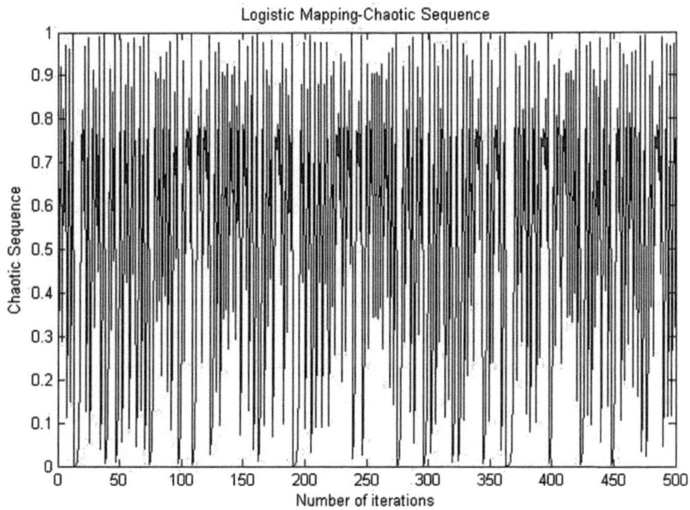

Fig. 1 To illustrate the dynamic behavior of the system generated for 500 iterations using the logistic function is in this sample caption

4 Results and Discussion

The CPSO deployed was to maintain the optimal cash. The evolutionary computing technique using four different chaos was developed to improve the efficiency of the system. The simulation has been made from two different sets of data which was short-term data confining to a specific bank branch located in an urban India.

The parameter selection plays a vital role in estimating the cash requirement for a particular day. The ANN and the traditional PSO parameters will not be adequate to find out the best optimal solution since it uses pseudorandom numbers [8]. Hence, there is need to use the chaotic functions in our cash management model to improve the search strategy.

To analyze the efficiency of CPSO, the forementioned four different chaotic mapping functions were used to generate different sets of chaotic sequence which replaces the pseudorandom numbers. The chaos1 and chaos2 are the chaos numbers. The logistic mapping by default provides the chaotic values between 0 and 1. The chaotic values generated by other chaotic functions used in CPSO were normalized to improve the accuracy of the results.

The experimental-based analysis was made for 50 independent runs keeping common parameters selected to bring down the impact of pseudorandom numbers. Total numbers of swarms were taken as 50 and numbers of hidden neurons to train the network was set to 9. The social and cognitive values were $C1 = C2 = 2$ to share the knowledge between the particles as well as to control the flow of particles to reach the optimal value. The convergence rate of CPSO could be enhanced by setting the initial inertia weight as 0.9. It was slowly reduced to 0.4 during the runs. The new velocity and position will be updated using the modified Eqs. (2) and (3) to identify the global best in the search space.

The convergence curve of CPSO used 6-9-1 pattern with the training parameters by varying the different chaotic sequence with the maximum of 500 iterations. The experimental analysis was made for both short-term and long-term data. The convergence point in the problem space guaranteed that the performance CPSO using logistic mapping is better than the traditional PSO for long-term data. It is clearly depicted in Fig. 2. The premature convergence can be avoided as well as the computational complexity can be reduced.

On an experimental basis, it was clearly examined that it has the capability toward finding an optimal solution including larger problem space with lesser number of repetitions. Therefore, it was concluded that CPSO is better than traditional PSO. The average MAE, MAPE, and MSE for one-dimensional chaotic methods were estimated for both short-term and long-term datasets which are tabulated as in Table 1.

Estimation of values was enough to keep up the optimal cash forecasting. The computational time for the CPSO using the logistic function was 321.31 s.

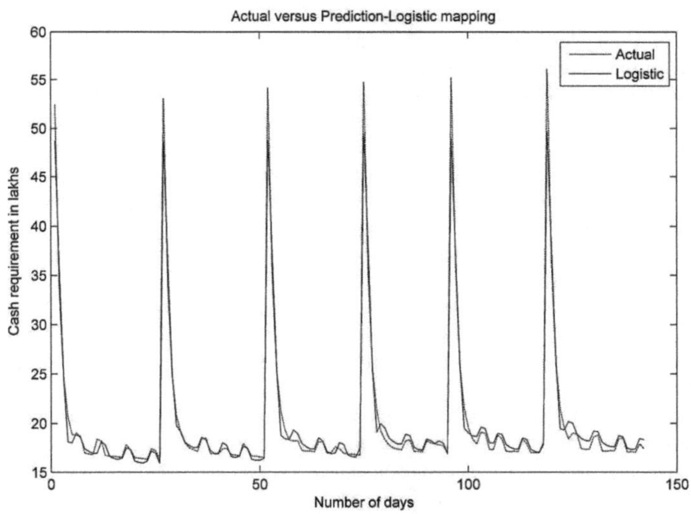

Fig. 2 To illustrate the performance of logistic mapping function

Table 1 Performance of chaotic functions—short-term data

Chaotic function	MAE	MAPE	MSE
Logistic-short term	1.5966	13.1305	4.1202
Logistic-long term	1.0345	4.0821	3.3102
Actual PSO	2.5672	15.2341	6.5673

5 Conclusion

In this article, a unique novel method is proposed to improve the (PSO) using chaotic maps CPSO for global optimization. We had used the various characteristics of the chaotic systems, such as sensitivity to initial value, randomness to improve the performance of the PSO. The proposed methodology uses different chaotic sequence of logistic map sequence. Therefore, the sequence generated from different chaotic systems was used by CPSO to avoid the premature convergence. The estimated values show that the logistic chaotic system provides better results for the used data sample.

References

1. Roodposhti FR, Heybati F, Musavi SR (2010) A comparison of classic time series models and artificial neural networks in anticipation of cash requirements of banks: A case study in Iran. In: Academic and Business Research Institute international conference, Orlando, USA
2. Kumar PremChand, Walia Ekta (2006) Cash forecasting: an application of artificial neural networks in finance. Int J Comput Sci Appl 3(1):61–77
3. Rehman S, Mohandes M (2008) Artificial neural network estimation of global solar radiation using air temperature and relative humidity. Energy Policy 36:571–576
4. Shi X, Schilling P, Boyd D (2004) Applying artificial neural networks and virtual experimental design to quality improvement of two industrial processes, vol 42. Taylor and Francis Group, London, pp 101–118
5. Kennedy J, Eberhart R (1995) Particle swarm optimization. In: Proceedings of the IEEE international conference on neural networks, Perth, Australia. IEEE Service Center, Piscataway, NJ, pp 1942–1948
6. Yi D, Ge X (2005) An improved PSO-based ANN with simulated annealing technique. Neurocomputing 63:527–533
7. Yang X, Yuan J, Yuan J, Mao H (2007) A modified particle swarm optimizer with dynamic adaptation. Appl Math Comput 189:1205–1213
8. Li J, Xiao X (2008) Multi-swarm and multi-best particle swarm optimization algorithm. IEEE, pp 6281–6286
9. Sun F, Ye T (2010) Transmission line image segmentation based GA and PSO hybrid algorithm. In: International conference on computational and sciences (ICCIS), pp 677–680
10. Liu L, Zhang Q, Wei X (2012) A RGB image encryption algorithm based on DNA encoding and chaos map. Comput Electr Eng (CEE) 38(5):1240–1248
11. Faeighi SKM, Mashhadi D (2014) Baleanu, Sampled-data nonlinear observer design for chaos synchronization: A Lyapunov-based approach. Commun Nonlinear Sci Numer Simul 19:2444–2453
12. Hosseinpourfard R, Javidi MM (2015) Chaotic PSO using the Lorenz system: an efficient approach for optimizing nonlinear problems. Çankaya Univ J Sci Eng 12:40–59
13. Liu BX, Wu Y, Xu X, Hu N, Cheng X (2014) Chaos adaptive improved particle swarm algorithm for solving multi-objective optimization. TELKOMNIKA Indones J Electr Eng 12:703–710
14. Abdullah A, Enayatifa R, Lee M (2012) A hybrid genetic algorithm and chaotic function model for image encryption. J Electron Commun 806–816
15. Altay EV, Alatas B (2019) Bird swarm algorithms with chaotic mapping. Springer Nature B.V.
16. Li Z, Hou J, Wang K, Liu Q (2013) PSO algorithm improvement based on particle tracing analysis. J Pattern Recogn Intell Syst 18–24
17. Tavazoei MS, Haeri M (2007) Comparison of different one-dimensional maps as chaotic search pattern in chaos optimization algorithm. J Appl Math Comput 187:1076–1085

Modified Convolutional Neural Network of Tamil Character Recognition

C. Vinotheni, S. Lakshmana Pandian, and G. Lakshmi

Abstract The handwritten Tamil character recognition in offline mode is challenging tasks as there are virtually different people who have different styles of writing the same characters. Deep convolution neural networks are playing a virtual role nowadays in recognizing handwritten character by automatically learning discriminative features from high dimensionality of input data. This work presents a modified convolution neural network (M-CNN) architecture to achieve a faster convergence rate and also to get the highest recognition accuracy. The M-CNN on different aspects along with layers design, activation function, loss function and optimization is discussed. Systematic experiments on isolated handwritten Tamil character dataset collected from various schools by ourselves. For these collected datasets, the proposed system recognized the characters with 97.07%.

Keywords Tamil handwritten recognition character · Modified convolutional neural network · Activation function

1 Introduction

A trendy move toward a paperless world with modern marvels in computer and communication technologies such as fax machines, word processors and email is still having numerous occurrences of prevalence in handwriting communications. In real time, the roles of handwritten recognition (HWR) is mainly applied for historical documents like palm scripts, handwritten checks in banks, medical handwritten reports and rooting mails in post office [1, 2]. The challenging part of HWR will

C. Vinotheni (✉) · S. Lakshmana Pandian · G. Lakshmi
Pondicherry Engineering College, Pondicherry, India
e-mail: vinotheni95@pec.edu

S. Lakshmana Pandian
e-mail: lpandian@pec.edu

G. Lakshmi
e-mail: lakshmipriya42012@pec.edu

A. K. Tripathy et al. (eds.), *Advances in Distributed Computing and Machine Learning*, Lecture Notes in Networks and Systems 127, https://doi.org/10.1007/978-981-15-4218-3_46

have to deal with many of individual writing styles because every particular author has a unique writing style, overlapping of characters, spaces in between characters have different size, and even single author handwritings might also vary at exclusive times. HWR is divided into two types, namely online and offline [3]. In online recognition, the users' handwriting is recognized by automatically converting the text into the digital form, when the user is writing. A digital ink concept is used, the input text is obtained from the digital pen and tablet as specialized hardware. Using a sensor, the tip of the pen's motions is measured as stokes which are converted into a list of coordinates. In offline recognition, the input is the image as users handwritten which is trained initially with the set of scanned images as characters. Later in testing, the input is a new character image which is given, and once it is recognizing accurately then accuracy is calculated by the difference between the actual text present in test data and the output text recognized by the proposed system.

Among all, HWR is a unique area where numerous experiments with deep neural networks are being conducted still. The machine learning techniques in traditional [4] have been used for handwritten character recognition (HCR) for a period of time with the following steps as preprocessing, segmentation, extracting the features and classification. In image processing, one of the areas is character recognition where characters are recognized by their characters image and converted to the machine-readable format. Basically, the deep learning network composed of layer yields input, several hidden layers and wound up with layer yields output. Every hidden layer comprises multiple neurons in the deep network with suitable weights for computation. The training parameters are incredibly large due to many hidden layers. A huge set of examples are needed to anticipate overfitting. The CNN as a convolutional neural network [5] is highly understood magnificence in deep mastering. CNN in neural system network is a mutlilayer feed forward which automatically extracts the discriminative features multiple times from high dimensionality of input data and also has the ability to learn high-dimensional complex with nonlinear mappings. Motivated by CNN ability to perform stages of classification of character where no explicit extraction of feature is required in a single pipeline. In prior work, they applied different algorithms with small datasets handwritten images only, and it is difficult for testing all the variations in handwriting. A newly developed architecture is a modified convolutional neural network (M-CNN) for this work to set a benchmark of recognizing 156 classes in Tamil characters from a real-time dataset.

1.1 Tamil Script

Tamil is an ancient South Indian morphologically rich language which is widely spoken in Southern India like Tamil Nadu, Sri Lanka and also in Malaysia [6]. Tamil script consists of vowels(uyirletter) as 12, consonants(meyyeletter) as 18, a special character(ayudhaezuthu) as 1 and 216 compound characters which is a merging of vowels as well as consonants shown in Table 1. Thus, the aggregate number of 247 characters is in Tamil. From Sanskrit, they borrowed five consonants that merged

Table 1 Combination of every single consonants with 12 vowel yields 216 compound characters. Illustrated with a single consonant

Every Single Consonants from க் ங் ச்......ன்	க்											
+												
Vowels (12)	அ	ஆ	இ	ஈ	உ	ஊ	எ	ஏ	ஐ	ஒ	ஓ	ஔ
Compound Characters	க	கா	கி	கீ	கு	கூ	கெ	கே	கை	கொ	கோ	கௌ

with 12 vowels and created further compound characters as 60, reflecting a sum of characters as 307. The universal standard is the standardized character encoding scheme for written as well as text. For each Unicode character is 2 bytes in nature. The Unicode range [7] of Tamil is U+0B80 to U+0BFF. The complete Tamil character dataset has 156 distinct characters from 126 Tamil and 30 Sanskrit compound characters. For example, சௌ (Sow) can be combined by 'செ', 'ள' distinct characters. Since Tamil script characters sets are huge and their letters have more angles and modifiers, despite the fact that recognition is more complicated in Tamil, the richness in content of Tirukkural, greatness in Kamba Ramayana, Silapathigaram and palm leaf scripts attracts the researchers in this field.

It organizes the remaining following portion of the work as follows: Sect. 2 addresses prior similar recognition of character research. The planned word for M-CNN architecture is explained with Sect. 3. Experimental outcomes are addressed with Sect. 4 and cited yields Sect. 5.

2 Previous Work

Before to this research, the issue of character recognition has been examined with different proposed methods [8] in numerous languages. Difficulties still exist in the recognition of normal and abnormal writing, similar shaped characters, slanting characters, joined characters, overlapping characters, curves and so on during recognition process. In current years, deep learning had accomplished massive achievement in CNN which automatically extracts the discriminative features many times in this character recognition field in languages like Chinese [3, 9–11], Arabic [12, 13] , Bangla [14], Devanagari [15] English [16], Georgian [17] and attracts the researchers to use.

Tamil [6] built up a CNN model with nine layers and achieved 97.7% recognition accuracy with dataset of 156 classes from hpl-Tamil-iso-char produced by HP laboratories India. In [7], the scanned input image is segmented as paragraphs using the technique of spatial space detection, paragraphs into lines and words utilizing vertical histogram and horizontal histogram, respectively, utilizing horizontal histogram the words are divided character image glyphs. The features that are extracted given to fuzzy neural network, support vector machine, radial basis network and self-organizing and classified utilizing supervised learning algorithm. The recognition rate for training achieved 100% and for testing 97% with speed of 0.1 s. Due to writing ambiguities, the classification is difficult in similar characters with writing ambiguous and mainly for Sanskrit characters. In [18], input image is resized to required pixels have worked on small characters dataset with multilayer CNN and achieved 90.19% recognition accuracy. Misclassified characters are due to the distinctive writing styles and font size problem. In [19], it built a recognition system in handwritten of characters on IWFHR-10 dataset utilizing convolutional neural network for characters in Tamil. The library ConvNetJS studies the features via improved traditional CNN architecture utilizing pooling with stochastic and probabilistic weighted and normalization with local contrast. Utilizing stochastic gradient descent, the training is accomplished with a learning rate as adaptive and carried out the classification as softmax in classifier layer got a recognition accuracy of 94.4%. In this paper, [20] used VGG 16 convolution neural network with transfer learning algorithm from dataset https://www.kaggle.com/seenivasanseeni/tamilhandwrittencharacters of 156 Tamil characters and achieved 94.52% accuracy with small amount of samples. For Devanagari, [16] used deep neural network with ISIDCHAR and V2DMDCHAR dataset for six different deep convolutional neural networks as DCNN architectures and Devanagari character recognition accuracy of 96.02% with adaptive gradient methods. In consideration of both the online as well as offline characters, Chinese [3] established another standard to recognize from 1CDAR-2013 dataset with 11-layer ConvNet and achieved 97.37% accuracy. In [13], it proposed the CNN model from one dataset OIHACDB-28 from scratch and achieved 97.32% Arabic character recognition accuracy and using transfer learning algorithm achieved 100% accuracy for their model. In [10], the network input is traditional gradient maps and acquired exactness as 96.95%. In additional layer as adaptation, the improved recognition exactness has been further increased 97.37% for setting another record to offline HCR. In [14], it proposed deep belief network as DBN design with dropout regularization methods with error detection rates as 2.73 and 2.27% for an Arabic character recognition, respectively. In [21], a promising idea in the application of CNN for isolating the features as well as classifier like support vector machine (SVM) with radial basis kernel which is supplanted in the CNN last layer has been developed for the identification of digits with MNIST dataset.

3 Proposed Approach

3.1 *Scanning*

A handwritten isolated characters document is positioned over the scanner. A scanner software is invoked that scans the document and it saves as TIF, JPG or GIF accessed for further purpose. To save the different dimensions of 350 images for each character, a set of 156 different folders are created.

3.2 *Preprocessing*

This is the first step of processing the scanned image with the input of raw pixel values. To keep the dataset in a standardized manner, the dataset is reduced to a uniform size of width and height of 28 pixels by cropped manner [14] because of the dataset composition nature with the learning model utilized, which is convenient for M-CNN conversion of an image of grayscale into binary image called as digitization (Fig. 1).

4 Convolutional Neural Network

The CNN architecture includes convolutional layer, activation function, pooling layer, optimization, loss function and fully connected layer as displayed in Fig. 2. The four key success ideas of CNN are multiple layers, local connection shared weight and pooling [5] . These are two major parts; convolution layer as the first part and second part as pooling layers alternatively in sequence which can also be added or deleted. Another part is sequence of fully connected layers. For M-CNN, the input

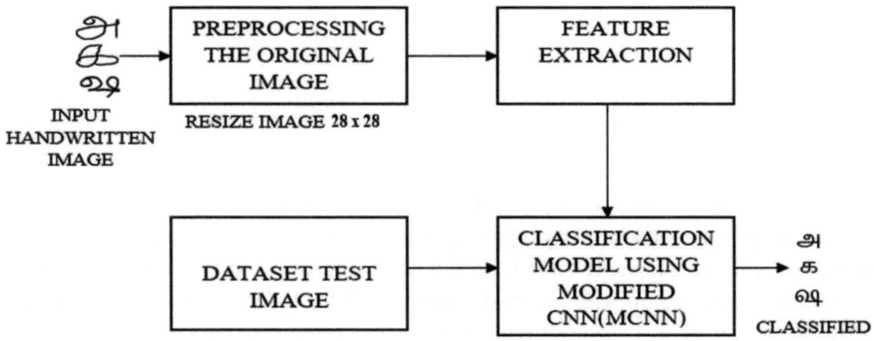

Fig. 1 Proposed architecture

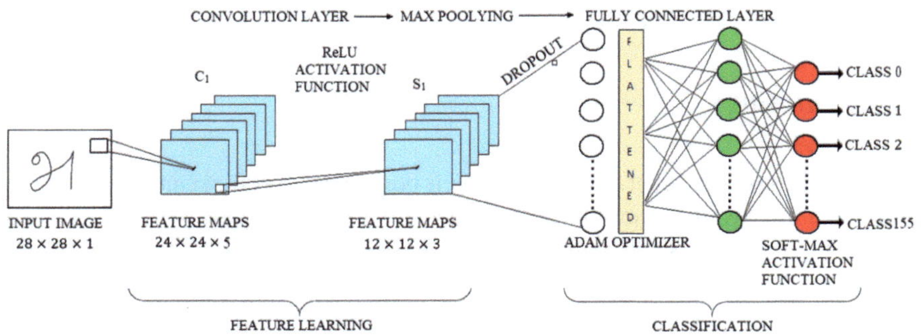

Fig. 2 Modified convolution neural network (M-CNN) architecture

image should be identical in height as well as width which implies that prior to the image is fed, a standardization procedure must pursue to change over the image as $h \times w \times c$ converted to $m \times m \times c$ size.

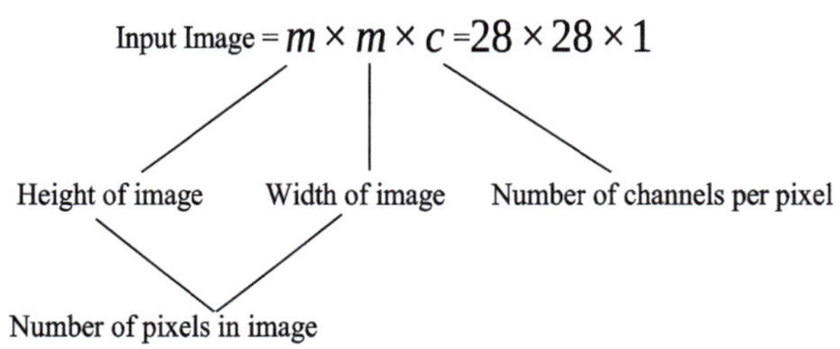

The input character image has raw pixel values of $m \times m \times c$ for grayscale or color intensity image in which $m \times m$ has the image height and width, respectively, if the grayscale image $C = 1$ and for colored intensity (RGB) image $C = 3$. The foremost advantage of CNN is the usage of shared weights which reduces parameter number and improves performance in convolution layers. All the inputs share the same filter in the layer which means every feature map units are weighted the same at different locations by the same features (Fig. 2).

4.1 Convolutional Layer

This is the foremost constructing block in CNN with filters for convolving with the input as image for each and every region appeared in it for producing the feature maps during the training phase. It differs from a neural network so that neurons (pixels) are no longer hooked up with the next weight and bias layer, however whole image splits into $n \times n$ matrix as small regions with the weights as well as bias known as kernels/filters [19]. For yielding a single feature in the entire image, the same filter is utilized to decrease the number of parameters for this convolution approach. The salient features of convolution layers are edges, corners and end points. The modified CNN takes the input image with two-dimensional normalized image of size as $m \times m \times c$ with a dot product of kernels (filters) range in dimension of $n \times n \times p$ where n refers the kernels height as well as width and p refers the number of channels in kernel in which both produce feature maps k of size in Eq. (1). The mathematical operation on convolution layer yields

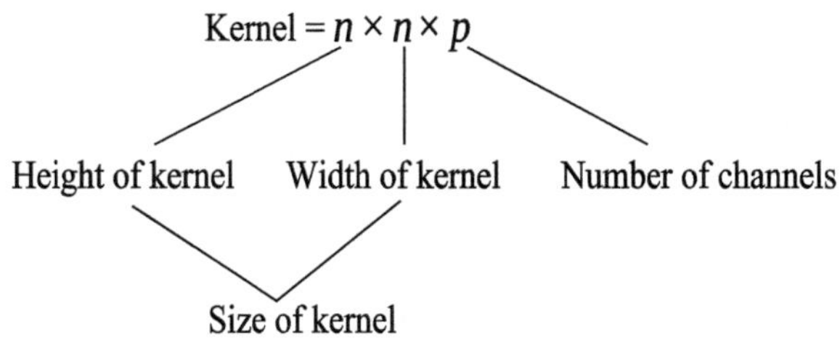

$$\begin{aligned}\text{Feature map of } C1 &= (m - n + 1) * (m - n + 1)\\ &= (24 - 5 + 1) * (24 - 5 + 1)\\ &= (20 * 20)\,. \end{aligned} \tag{1}$$

4.2 *Activation Function*

The convolution layer is often followed by activation functions. One of the nonlinear activation functions used in our proposed work is rectified linear units (ReLU). It helps the training process to speed up and found more effective result than other activation functions [15] like LReLU, PReLU and Swish.

4.3 *Polling Layer*

The convolution layer accomplished via layer as pooling/sub sampling helps to reduce the resolutions in the feature maps by removing distortion effects. Since numerous filters mean a lot of unknown weight used in convolution layer increases the stack of the images by huge volume, this layer reduces the data easily by applying mean, max or min operation. It divides images of newly generated into number of blocks and sub samples to produce single value for each block. In literature

[7, 12, 13, 15, 17, 19], they used max pooling nodes than others because it takes the maximum of the block that they are sub samples. A fixed characteristic is performed on inputs yet the parameter produced is none on this layer. All the max pooling algorithm is carried out $n \times n$ window. For our approach from the earlier convolutional layer, an 2×2 area of region is linked with units on this layer as pooling. Therefore, the feature maps size is reduced to convolutional layer by half as displayed in Eq. 2. The feature extracts can be calculated from the combination of convolution layers and pooling layers. The mathematical operation on max pooling layer yields

$$\begin{aligned}\text{Feature map of } S1 &= \left(\frac{m-n+1}{2}\right) * \left(\frac{m-n+1}{2}\right)\\ &= \left(\frac{20}{2}\right) * \left(\frac{20}{2}\right)\\ &= 10 * 10.\end{aligned} \tag{2}$$

4.4 Fully connected Layer

After the resultant series of convolutional as well as pooling layers with the flattened output of features which are forwarded into the layer as dense/ fully connected. It contains multiple neurons; each neuron gets the input that connects to all the neurons with the prior layer for executing the features piercing into recognizable objects of each class on this layer. In addition to training, there is a classifier on the dense layer in which each of the neuron uses the activation function as softmax to summarize the dot product to produce the effects of inputs and weights from the previous layer to output the results.

4.5 Transfer Learning

In this learning, the machine learning approach with a pre-trained neural network with some set of datasets is frequently utilized for objects detection, image recognition and many other applications [13, 22] . According to concerned application, the pre-trained network will once again be trained for desired datasets once more and proceed to start the testing stage.

5 Experimental Outcomes

5.1 *Datasets*

The required handwritten Tamil character dataset is developed by our own from the scratch to create a dataset and collected the documents written by different individuals from different fields. The dataset comprises 156 distinct isolated characters of Tamil which are stored in different 156 folders, and the image is of different dimensions. Basically, the Tamil characters are taken from Southern India that are all written by native Tamil writers. The characters are of different size and not at the center of the image which means it can be at any position in the image. To keep the dataset standardized, scanning and preprocessing were needed for grayscale image and resizing, respectively. The dataset contains 350 samples for each class with a total of 54,600 samples.

5.2 *Experimental Setup*

A serious of experiments were performed to compare the CNN and M-CNN convergence rates and to set a new benchmark for 156 distinct Tamil characters to be recognized with 350 samples for each class. The entire training testing experiment is executed in Ubuntu operating system of Intel-Core i7 processor CPU and 8 GB RAM. These models are written by Python, and Keras [21] was utilized as deep learning framework with dependencies which includes TensorFlow [23] as backend and NumPy. Thus, HDF5(h5py) also used to tackle storage problems with the dataset.

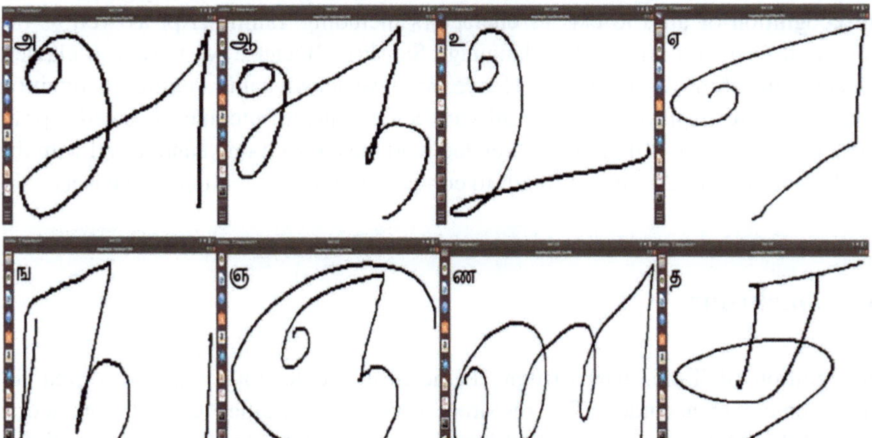

Fig. 3 Samples of the recognized characters from test image

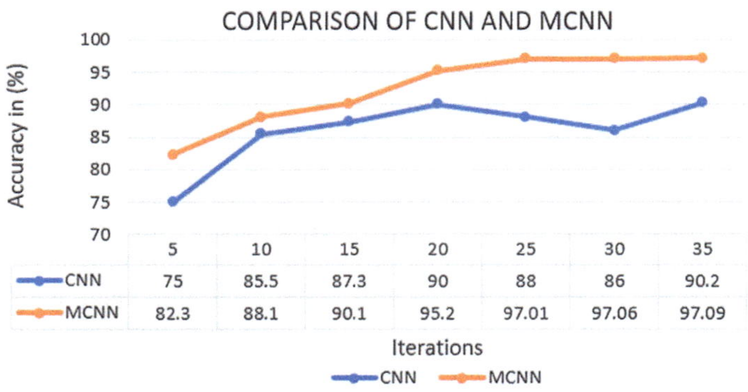

Fig. 4 Comparison of CNN and M-CNN

5.3 *Results and Discussion*

M-CNN is a deep learning neural network that modeled with components such as layers, activation function, optimizer function and learning rate. In our proposed model, it gives a best accuracy result to predict Tamil character by using the "n" number of iterations such as initially from 5 and incrementally as 10, 15, 25, 30 and 35, when compared with the existing model. After running all the iterations, the best accuracy was obtained at iteration 25 and continuously increasing in accuracy until it reaches iteration from 5 to 25. The iteration sets from 5 to 20 indicate a steady increase in precision till equilibrium at iteration 30 is reached. The sample images of recognized characters are shown in Fig. 3. The comparison of CNN and M-CNN recognition rate of percentages with several iterations is proven in Fig. 4. The test dataset was sent to the model of training in which the accuracy gave 97.09% in recognition of all 156 classes characters including Tamil script as well as the borrowed characters from other language Sanskrit. Because we have 156 classes, we could not show for all classes. Hence, we have shown some samples recognition images that are displayed. These results are better than the outcomes revealed in prior work [7, 10, 16, 18–20]. Since the related work has been trained and tested with the small datasets [18, 20], it is difficult to compare the results with large datasets.

6 Conclusion

Recognition of Tamil handwritten character is the leading researches area for improvement in accuracy. The research work uses modified convolution neural network(M-CNN) to set a standard for handwritten character recognition with the single sequence of convolution, pooling and fully connected layer with a new dataset which obtained 97.07% accuracy rate on test images. The reason of using deep learning was to take gain of CNN to handle high dimension of input and share their weights. Our dataset comprises 54,600 images as 156 distinct classes in Tamil isolated characters which are identically visual or written in a similar manner by most people. In future, another plan is to work with Tamil handwritten words and sentences recognition using deep learning architectures.

References

1. Thadchanamoorthy S, Kodikara ND, Premaretne HL, Pal U, Kimura F (2013) Tamil handwritten city name database development and recognition for postal automation. In: IEEE-ICDAR, pp 793–797
2. Feng B, Ren M, Zhang X, Suen C (2014) Automatic recognition of serial numbers in bank notes. Patt Recogn 47:2621–2634
3. Zhang X-Y, Bengio Y, Liu C-L (2017) Online and offline handwritten chinese character recognition: a comprehensive study and new benchmark. Patt Recogn 61:348–60
4. Ashay Singh and Ankur Singh Bist (2019) A wide scale survey on handwritten character recognition using machine learning. Int J Comput Sci Eng 7(6):124–134
5. Gu J, Wang Z, Kuen J, Ma L, Shahroudy A, Shuai B, Liu T et al (2018) Recent advances in convolutional neural networks. 77:354–377
6. Raghupathy KB, Chandrasekaran S (2019) Benchmarking on offline Handwritten Tamil Character Recognition using convolutional neural networks. J King Saud Univ Comput Informa Sci
7. Sureshkumar C, Ravichandran T (2010) Handwritten Tamil character recognition and conversion using neural network. Int J Comput Sci Eng 02(07):2261–2267
8. Vinotheni C, Lakshmana Pandian L (2019) A state of art approaches on handwriting recognition models. In: IEEE fifth international proceedings on science technology engineering and mathematics, vol 1, pp 98–103
9. Yang W, Jin L, Tao D, Xie Z, Feng Z (2016) A new training method to enhance deep convolutional neural networks for large-scale unconstrained handwritten Chinese character recognition. Patt Recogn 58:190–203
10. Zhang Y, Liang S, Nie S, Liu W, Peng S (2018) Robust offline handwritten character recognition through exploring writer-independent features under the guidance of printed data. Patt Recogn Lett 106:20–26
11. Li Z, Teng N, Jin M, Lu H (2016) Building efficient CNN architec-ture for offline handwritten Chinese character recognition. Int J Doc Anal Recogn (IJDAR) 21(4):233–240
12. Boufenar C, Kerboua A, Batouche M (2018) Investigation on deep learning for off-line handwritten Arabic character recognition. Cogn Syst Res 50:180–195
13. El-Sawy A, Loey M, Hazem EB (2017) Arabic handwritten characters recognition using convolutional neural network (2017) . WSEAS Trans Comput Res 5:11–19
14. Roy S, Das N, Kundu M, Nasipuri M (2017) Handwritten isolated Bangla compound character recognition: a new benchmark using a novel deep learning approach. Patt Recogn Lett 90:15–21
15. Jangid M, Srivastava S (2018) Handwritten Devanagari character recognition using layer-wise training of deep convolutional neural networks and adaptive gradient methods. J Imag 4(2):1–14
16. Raymond P, Petroski F, Such SP, Brockler F, Singh V, Hutkowski P (2018) Intelligent character recognition using fully convolutional neural net-works. Patt Recogn 88:604–613
17. Davit S, Magda T, Levan S, Irakli K, Shota A, Sandro J (2018) On Georgian handwritten character recognition. IFAC-PapersOnLine 51(30):161–165
18. Deepa M, Deepa R, Meena R, Nandhini R (2019) Tamil handwritten text recognition using convolutional neural networks. Int J Eng Sci Comput 9(3):20986–20988
19. Vijayaraghavan, Prashanth, and Misha Sra: Handwritten Tamil recognition using a convolutional neural network. MIT Media Lab, (2014)

20. Niu X-X, Suen CY (2012) A novel hybrid CNN–SVM classifier for recognizing handwritten digits. Patt Recogn 45(4):1318–1325
21. Sahare P, Dhok SB (2018) Multilingual character segmentation and recognition schemes for Indian document images. IEEE Access 6:10603–10617
22. Pragathi MA, Priyadarshini K, Saveetha S, Shavar Banu A, Mohammed Aarif KO (2019) Handwritten Tamil Character Recognition using deep learning. In: IEEE international proceedings on vision towards emerging trends in communication and networking (ViTECoN), pp 1–5
23. Abadi M, Agarwal A, Barham P, Brevdo E, Chen Z, Citro C, Corrado GS et al (2016) Tensorflow: large-scale machine learning on heterogeneous distributed systems. arXiv preprint arXiv:1603.04467

Epileptic Seizure Detection from Multivariate Temporal Data Using Gated Recurrent Unit

Saranya Devi Jeyabalan and Nancy Jane Yesudhas

Abstract The advancement in sensor and satellite technologies, biomedical informatics, climate informatics, and health care has led to the emergence of multivariate temporal data. Multivariate temporal data contains multiple time series with complex temporal behaviors. Mining Knowledge from such complex data remains challenging area of research. This paper primarily focuses on developing temporal decision support system in medical (TDSSM) to detect epileptic seizure from multivariate temporal data. This work uses deep neural networks to build temporal classification model for epileptic seizure detection. The temporal model for epilepsy seizure detection is constructed by training the gated recurrent unit using multivariate temporal dependencies of time series observations acquired from the EEG recording of 500 individuals.

Keywords Epileptic seizure · Temporal classification model · Gated recurrent unit

1 Introduction

Today's world is driven by collection of data from various domains, namely medical data, sensor data, geological data and many more. These data are often multivariate and evolves over time. Knowledge mining from such data becomes challenging area of research as these multivariate temporal data are enormous size, subject to missing value, outliers, etc. This paper aims in developing temporal decision support system in medical (TDSSM) for real-time epileptic seizure detection. Epilepsy is a nervous illness characterized by frequent and unprovoked seizure, affects almost 60 million people worldwide. Epileptic seizure disorder causes sudden disturbances in brain

---

S. D. Jeyabalan (✉) · N. J. Yesudhas
Madras Institute of Technology, Anna University, Chennai, India
e-mail: saran12.arc@gmail.com

N. J. Yesudhas
e-mail: nancyjaney@gmail.com

A. K. Tripathy et al. (eds.), *Advances in Distributed Computing and Machine Learning*,
Lecture Notes in Networks and Systems 127, https://doi.org/10.1007/978-981-15-4218-3_47

function producing loss of consciousness or awareness which may lead to critical condition [1].

Epileptic seizure is diagnosed by monitoring patient's scalp EEG for several days. This process is time-consuming and expensive. Further, several studies have suggested various algorithms to automatically detect seizures using EEG signals. Gotman developed automatic system to detect seizure in EEG [2]. Osorio et al. introduced Finite – Impulse Response (FIR) Filter for real-time seizure detection [3]. Seizure detection using nonlinear methods has been reported in several studies [4–10].

Well ahead, epileptic seizure is detected using several classifiers, namely linear discriminant analysis, artificial neural network (ANN), K-nearest neighbor (KNN), support vector machine (SVM), and fuzzy [11–17]. Real-time recognition of seizure for instant clinical analysis is highly important to enhance the clinical decision. This paper presents deep neural network-based approaches for real-time recognition of epileptic seizure from multivariate temporal data. Deep learning networks perform automatic pattern extraction from the input without any prior knowledge. Thereby, deep learning is capable of extracting features from high-dimensional data. Deep learning has yielded excellent results in recent advancement and applications.

2 Related Works

Maryann et al. presented genetic search process for selecting electroencephalogram features and electrode locations for epileptic seizure identification. The genetic search technique was applied to available preseizure records and baseline data. The authors demonstrated the proposed method on multiday recordings found from four patients surrounded with intracranial electrodes. The results revealed good specificity for epileptic seizure prediction that helps the neurology for treating the epilepsy [18].

Firpi et al. designed genetic programming-based artificial features (GPAF) algorithm to overcome the drawbacks of conventional methods for feature gathering to detect epilepsy seizure. The authors evaluated GPAF algorithm on seven patients. In addition, the author presented genetic detector for clinical, subclinical seizure detection, and also discovered seizures that occurred during experiments. The experimental result indicates that seizures were detected with low false negative rate [19].

Chisci et al. predicted real-time epileptic seizure using support vector machine from the online analysis of EEG signals. The authors carried out experiments on the Freiburg data sets. The experimental results exhibited that proposed algorithm provides a good combination of sensitivity and specificity for feasible epileptic seizure prediction. The experimental outcomes on the Freiburg data sets revealed good sensitivity for real-time seizure prediction using SVM classifier created on Kalman filter [20].

Zandi et al. proposed wavelet packet transform system for real-time epileptic seizure detection using scalp EEG. The authors developed normal index called common seizure index (CSI). CSI is obtained for every iteration of each EEG channel established on rhythmicity and relative energy of EEG. The algorithm was verified on scalp EEG recordings attained from 14 patients. The experimental results revealed high sensitivity for epileptic seizure discovery [21].

Raghu and Shiram presented an optimized configuration of multilayer perceptron for recognition of IEEG epileptic seizure. This study experiments the presented algorithm with EEG database attained from university of Bonn. Three epileptic seizure stages were considered, namely pre-ictal (set c), post-ictal (set D), and epileptic (set E) from intracranial EEG (iEEG). The proposed algorithm yielded good classification accuracy rate [22].

Raghu et al. presented minimum variance modified fuzzy entropy (MVMFzEn) technique for real-time prediction of seizure from EEG recordings. The goal of this paper was to set up automated biomarker for screening and discovery of seizure in real-time EEG recordings. The authors experimented the proposed algorithm on EEG recording obtained from 22 males and 16 females. The experimental results revealed good specificity and sensitivity for automated detection of epileptic seizure in real time [23].

3 Data Set Description

The data acquired for recognition of epileptic seizure is obtained from 500 individuals each cointaining 100 single channel EEG segments with the length of 23.6s. These segments were designated from the EEG recording systems generated continuously. The EEG recording generated from five participants with good health contain five sets represented from (A–E), set A and B consists of EEG recordings taken from volunteers when flexible with an eyes open (A) and eyes closed (B). Set C and D encloses activity dignified during seizure-free intervals. Set E comprises activity seizure activity [24].

4 System Architecture

The temporal decision support system in medical (TDSSM) using gated recurrent neural network (GRU) is developed in this paper. The TDSSM supports the physician in real-time diagnosis of seizure and provides aid in clinical decision. The system architecture for such detection is drafted in Fig. 1. Deep learning has attained favorable outcomes in various research problems speech recognition, image classification, information retrieval, and so on [25–27]. In this study, the deep neural network GRU is used for epileptic seizure detection.

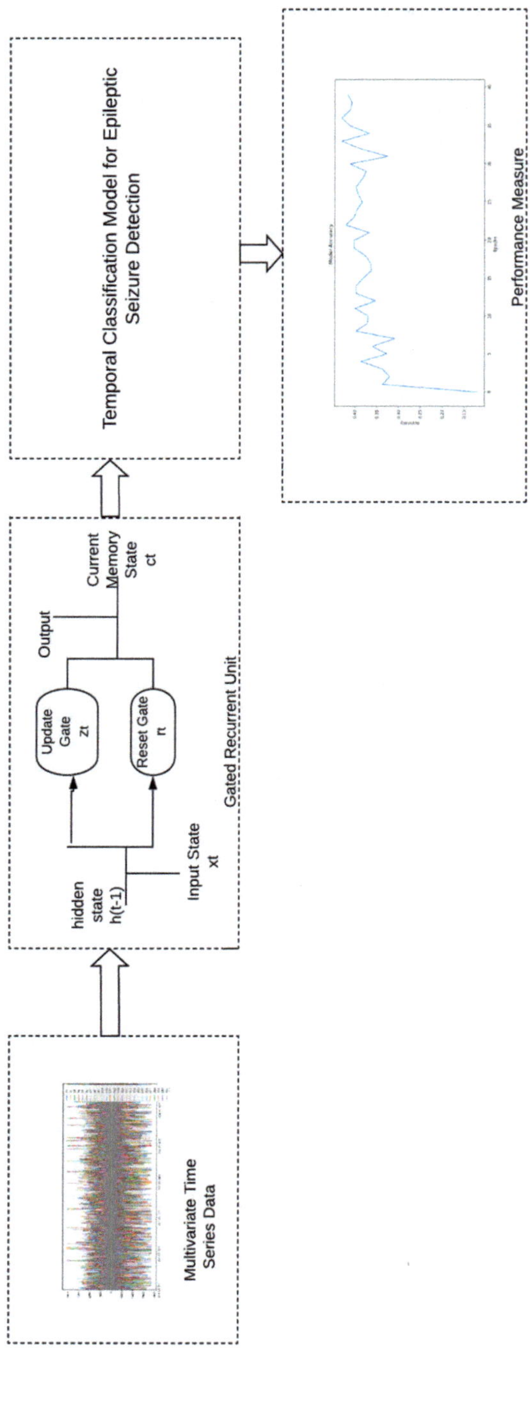

Fig. 1 System architecture for epileptic seizure detection using GRU

4.1 Gated Recurrent Unit (GRU)

The simple recurrent network (RNN) is hard to learn dependencies from long temporal sequential data sets because of the gradients of exploding and vanishing gradients. GRU is a particular model of RNN with gating mechanism which is suitable for modeling temporal dependencies. GRU was proposed by Cho et al. that simplifies the long short-term memory (LSTM) with two gates reset gate rt and update gate zt [28]. The GRU structure is shown in Fig. 5.

4.2 Temporal Classification Model for Epileptic Seizure Detection

The multivariate attributes with temporal sequence of observations obtained from 500 individuals of epilepsy patients are fed as the input to the GRU. The GRU network fully reveals its memory content at each time stamp and balances output between the previous memory state and new memory state.

The reset gate r_t of GRU is computed by Eq. (1)

$$h_t = \sigma(w_r x_t + u_r h_{(t-1)}) + b_r \tag{1}$$

where σ is the sigmoid function, x_t is the input state at current time stamp t, $h_{(t-1)}$ is the hidden state value of previous output of the GRU at time stamp $(t-1)$, w_t is the feed-forward connections, u_r is the recurrent matrices, and b_r is the bias.

The update gate z_r of GRU is computed by Eq. (2)

$$z_t = \sigma(w_z x_t + u_z h_{(t-1)}) + b_z \tag{2}$$

where the parameters of the update gate of GRU is same as the reset gate.

The current memory state c_t is processed with a hyperbolic tangent function that uses the reset gate r_t to store the relevant information from the past, and it is computed using Eq. (3)

$$c_t = \tanh(w_c x_t + u(r_t * h_{(t-1)})) + b_c \tag{3}$$

where tanh is the hyperbolic tangent function and * denotes element wise multiplication

The output of the GRU is computed using Eq. (4)

$$h_t = z_t * h_{(t-1)} + (1 - z_t) + z_t \tag{4}$$

Hence, GRU are capable to store and filter the data by means of reset gate and update gates. GRU excludes the vanishing gradient problem since it retains the new input that arrives every single time and preserves the appropriate information and moves it to succeeding time stamp. Thereby, GRU can perform well in modeling temporal sequential data sets.

5 Experimental Results

In this paper, the multivariate temporal data sets attained from epilepsy patients are used for experimentation. Figure 2 shows the graphical plot of multivariate temporal data obtained from 500 individuals of epilepsy patients with 23.6 s duration. The input data are promulgated into the GRU for building temporal classification model. The GRU model is trained with 80% of the data sets, and the remaining 20% of the data sets were used for model testing. The performance measurement of the constructed temporal model is valued in terms of classification accuracy and root mean square error (RMSE). RMSE measures the variance between the results projected by the model and the actual results observed from the data sets. A series of experiments were conducted with different combinations of network, namely LSTM, simple RNN, multilayer perceptron (MLP), and support vector machine (SVM).

Table 1 depicts the classification accuracy of different network learning techniques. The performance of the classification model was evaluated with precision, recall, and F1 score. Precision measures the actual positive result out of total positive results predicted by the model. Recall measures the actual positive result out of total positive results. Figure 4 shows the graphical plot of RMSE value obtained by temporal classification model built using GRU for epileptic seizure detection. The

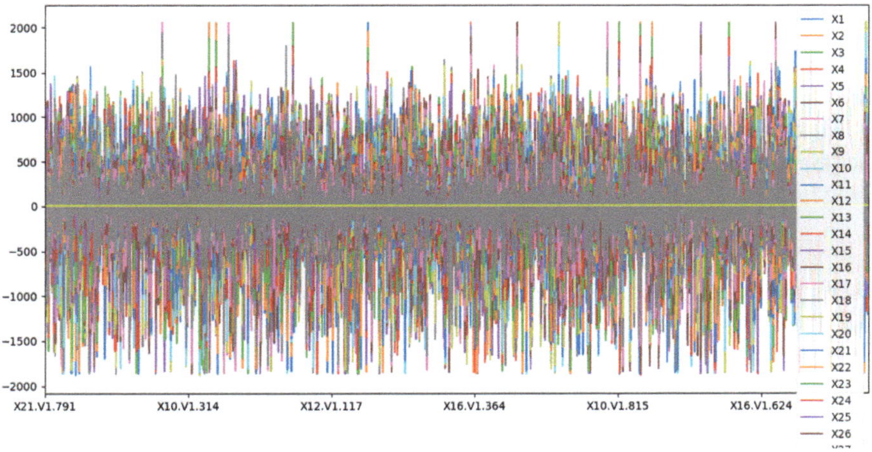

Fig. 2 Time series plot of data sets obtained from epilepsy patients

Table 1 Comparison of performance measures of GRU model with different network learning techniques

Network training techniques	Accuracy (%)	Precision	Recall	F1-score
GRU [28]	91	0.9130	1.000	0.9545
LSTM [29]	88	0.9261	1.000	0.9563
RNN [30]	83	0.9004	0.9731	0.9373
MLP [31]	80	0.8062	1.000	0.8951
SVM [32]	72	0.8571	0.9176	0.8862

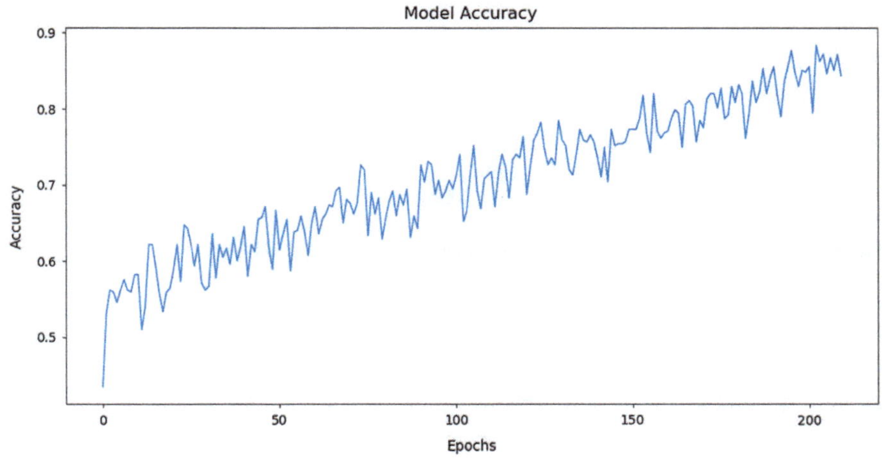

Fig. 3 Accuracy graph plot of temporal classification model by GRU for epileptic seizure detection

RMSE value of the model decreases as the epochs increases as shown in Fig. 4. The classification accuracy of the temporal model built using GRU for epileptic seizure detection is illustrated by the graph as shown in Fig. 3, and in this graph, as the number of epochs increases, the accuracy of the model increases. Thus, the classification accuracy of GRU is higher as depicted in Table 1, because of its gating mechanisms GRU is capable of training long duration temporal dependencies in the data sets.

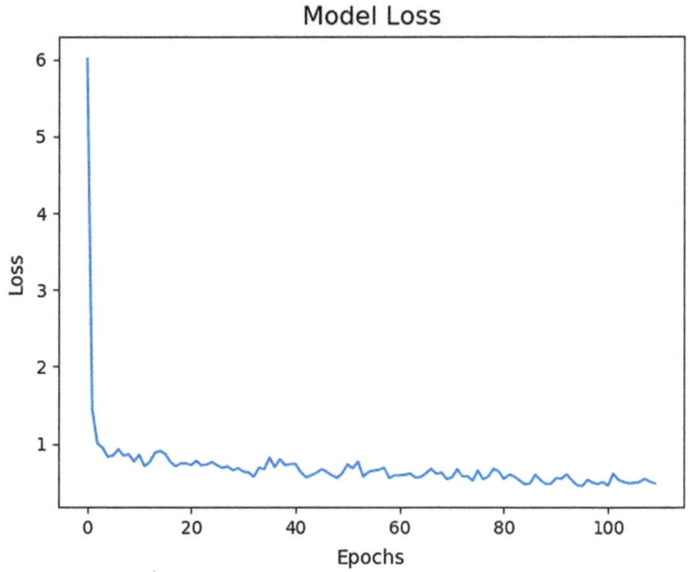

Fig. 4 RMSE value graph plot of GRU for epileptic seizure detection

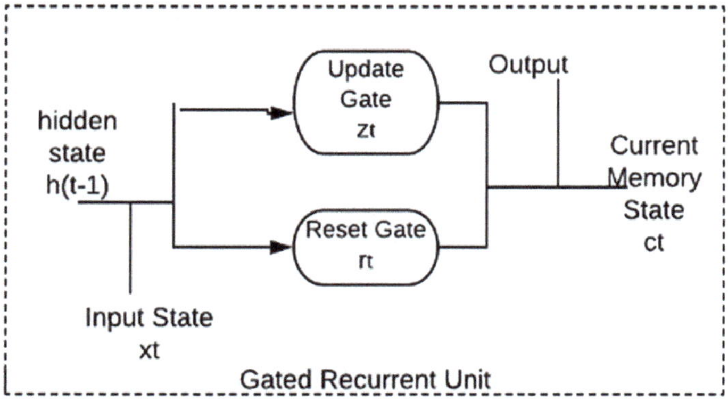

Fig. 5 Structure of GRU

6 Conclusion

The data sets acquired for epilepsy seizure detection are multivariate temporal sequence. This work presents GRU-based temporal classification model that is used to develop temporal decision support system in medical (TDSSM). This TDSSM enhances the real-time diagnosis of epileptic seizure that assists the physician for real-time medical decision making. This work is experimented with multivariate temporal data sets acquired from 500 individuals each containing multichannel EEG signal segmentation of 23.6 s duration. The experimental outcome shows that the classification accuracy of constructed GRU-based temporal model for epileptic seizure detection is comparatively higher than LSTM, RNN, and other non-gated neural networks (MLP and SVM). Hence, GRU outperforms LSTM, RNN and other neural networks as it uses less training parameters and less memory, and by this way, it trains the long temporal sequential data sets faster with less computational cost.

References

1. Epilepsy facts and figures (1999) Epilepsy Foundation of America, Landover
2. Gotman J (1982) Automatic recognition of epileptic seizures in the EEG. Electroencephalography Clin Neurophysio 54:530–540
3. Osorio I, Frei M, Wilkinson S (1998) Real-time automated detection and quantitative analysis of seizures and short-term prediction of clinical onset. Epilepsia 39:615–627
4. Lehnertz K, Elger CE (1995) Spatio-temporal dynamics of the primary epileptogenic area in temporal lobe epilepsy characterized by neuronal complexity loss. Electoencephalogr Clin Neurophysiol 95:108–117
5. Lehnertz K, Elger CE (1998) Can epileptic seizures be predicted? Evidence from nonlinear time series analysis of brain electrical activity. Phys Rev Lett 80:5019–5022
6. Le Van Quyen M, Martinerie J, Baulac M, Valera F (1999) Anticipating epilepticseizuresinrealtimebyanon-linearanalysisofsimilaritybetween EEG recordings. NeuroReport 10:2149–2155
7. Iasemidis LD, Shiau DS, Chaovalitwongse W, Sackellares J, Pardalos PM, Principe JC, Carney PR, Prasad A, Veeramani B, Tsakalis K (2003) Adaptive epileptic seizure prediction system. IEEE Trans Biomed Eng 50(5):616–627
8. Sackellares JC, Shiau DS, Principe JC, Yang MC, Dance LK, Suharitdamrong W, Chaovalitwongse W, Pardalos PM, Iasemidis LD (2006) Predictability analysis for an automated seizure prediction algorithm. Expert Syst Appl 23(6):509–520
9. Adeli H, Ghosh-Dastidar S, Dadmehr N (2007) Awavelet-chaosmethodology for analysis of EEGs and EEG subbands to detect seizure and epilepsy. IEEE Trans Biomed Eng 54(2):205–211
10. Sabesan S, Chakravarthy N, Tsakalis K, Pardalos P, Iasemidis L (2009) Measuring resetting of brain dynamics at epileptic seizures: application of global optimization and spatial synchronization techniques. Comb J Optim 17(1):74–97
11. Meier R, Dittrich H, Schulze-Bonhage A, Aertsen A (2008) Detecting epileptic seizures in long-term human EEG: a new approach to automatic online and real-time detection and classication of polymorphic seizure patterns. Clin J Neurophysiol 25(3):119–131
12. Islam MK et al (2016) A wavelet-based artifact reduction from scalp EEG forepilepticseizure-detection. IEEE J Biomed Health Informat 20(5):1321–1332
13. Zhou W et al (2013) Epilepticseizure detection using lacunarity and Bayesian linear discriminant analysis in intracranial EEG. IEEE Trans Biomed Eng 60(12):3375–3381
14. Srinivasan V et al (2005) Articial neural network based epileptic detection using time-domain and frequency-domain features. Med J Syst 29(6):647–660
15. Sriraam N, Raghu S (2017) Classication of focal and non-focal epileptic seizures using multi-features and SVM classier. Med J Syst 41(160):1–14
16. Aarabi A et al (2009) A fuzzy rule-based system for epilepticseizure detection in intracranial EEG. Clin Neurophysiol 120:1648–1657
17. Srinivasan V et al (2007) Approximate entropy-based epileptic EEG detection using articial neural networks. IEEE Trans Inf Technol Biomed 11(3):288–295
18. Maryann D, Alessandro, Esteller R, Vachtsevanos G, Hinson A, Echauz J, Litt B (2003) Epileptic seizure prediction using hybrid feature selection over multiple intracranial EEG electrode contacts: a report of four patients. IEEE Trans Biomed Eng 50(5)

19. Firpi H, Goodman ED, Echauz J (2007) Epileptic seizure detection using genetically programmed articial features. IEEE Trans Biomed Eng 54(2)
20. Chisci L, Mavino A, Perferi G, Sciandrone M, Anile C, Colicchio G, Fuggetta F (2010) Real-time epileptic seizure prediction using AR models and support vector machines. IEEE Trans Biomed Eng 57(5)
21. Zandi AS, Javidan M, Dumont A, Tafreshi R (2010) Automated real-time epileptic seizure detection in scalp EEG recordings using an algorithm based on wavelet packet transform. IEEE Trans Biomed Eng 57(7)
22. Raghu S, Shiram N (2017) Optimal configuration of multilayer perceptron neural network classifier for recognition of interancial epileptic seizure. Expert Syst Appl
23. Raghu S, Sriraam N, Pradeep Kumar G, Hegde AS (2018) A novel approach for real-time recognition of epileptic seizures using minimum variance modified fuzzy entropy. IEEE Transactions on Biomedical Engineering. 65(11), (2018)
24. Georgiy R, John B, Martha J, Richard N (2010) Patient-specific early seizure detection from scalp electroencephalogram. J Clin Neurophysiol 27(3)
25. Palangi H, Deng L, Shen Y, Gao J, He X, Chen J, Song X, Ward R (2016) Deep sentence embedding using the long short-term memory networks: analysis and application to information retrieval. IEEE/ACM Trans Audio Speech Lang Process 24(4):694–707
26. Graves A, Mohamed AR, Hinton G (2013) Speech recognition with deep recurrent neural networks. In: IEEE international conference on acoustics, speech and signal processing (ICASSP), pp 6645–6649
27. Krizhevsky A, Sutskever I Hinton GE (2012) Imagenet classication with deep convolutional neural networks. In: Advances in neural information processing systems, pp 1097–1105
28. Cho K, Van Merrienboer B, Gulcehre C, Bougares F, Schwenk H, Yoshua B (2014) Learning phrase representations using rnn encoder-decoder for statistical machine translation. arXiv preprint arXiv:1406.1078
29. Abdelhameed AM, Hisham G, Magdy B (2018) Deep convolutional bidirectional LSTM recurrent neural network for epileptic seizure detection. In: IEEE international new circuits and systems conference (NEWCAS)
30. Mohammad AN, Homayoun MN (2010) Analysis and classification of EEG signals using spectral analysis and recurrent neural networks. In: 17th Iranian conference of biomedical engineering (ICBME)
31. Samiee K, Kovacs P, Gabbouj M (2015) Epileptic seizure classification of EEG time-series using rational discrete short-time fourier transform. IEEE Trans Biomed Eng 62(2):541–552
32. Han-Tai S, Vladimir C, Jieun L, Brandon V, Edward E, Benjamin H, Gregory A (2017) SVM-based system for prediction of epileptic seizures from iEEG signal. IEEE Trans Biomed Eng 62(2):1011–1022

A Comparative Study of Genetic Algorithm and Neural Network Computing Techniques over Feature Selection

R. Rathi and D. P. Acharjya

Abstract Internet made a big revolution in the real world and thus poses so many challenges to the researchers by generating an enormous amount of data. The data generated contains an enormous amount of unwanted information. Before processing with such a dataset, the important features present in the dataset must be retrieved. The feature selection process is important because the performance of a model built for the purpose of classification, prediction or clustering depends mainly on the number of relevant features present in the dataset. In this proposed work, the real coded genetic algorithm is used to find the important features by considering the fuzzy rough degree of dependency as its fitness function for finding out optimum features for agricultural dataset, iris dataset and Pima Indian diabetes dataset. The experimental results show that the proposed work produces relevant features by maintaining classification accuracy.

Keywords Real coded genetic algorithm · Fuzzy rough · Crossover · Mutation

1 Introduction

There has been a huge rise in the data being generated. All such huge data may not be relevant for all the applications. Right data for the right application should be retrieved efficiently so that the model developed with the data yields unbelievable results. In the field of biology, medicine, engineering and agriculture, huge amount of data is retrieved, and it is practically impossible to work with huge data which contains redundant or irrelevant data. So attribute reduction or feature selection poses

Supported by organization x.

R. Rathi (✉) · D. P. Acharjya (✉)
VIT University, Vellore, India
e-mail: rathi.r@vit.ac.in

D. P. Acharjya
e-mail: dpacharjya@gmail.com

A. K. Tripathy et al. (eds.), *Advances in Distributed Computing and Machine Learning*, Lecture Notes in Networks and Systems 127, https://doi.org/10.1007/978-981-15-4218-3_48

many challenging tasks for the researchers in the current research. An efficient feature selection method is that the exact meaning should be preserved even after feature selection. The significant goal of feature selection is that it expands effectiveness, diminishes computational cost, lessens storage spaces and yields better exactness.

Rough set's core and reduct are an important and useful concept to remove the superfluous attributes from the input information system [1]. Many research works were carried out using rough set theory to remove the superfluous attributes. Basabi and Goutam Chakraborthy carried out their research work to remove the irrelevant data using rough set and genetic algorithm [2]. The authors found the optimal solution with the discriminant index as the fitness function to genetic algorithm. Fuzzy rough set feature selection approach is proposed by Jensen [3]. A data reduction or feature selection methodology merges vagueness of fuzzy set and indiscernability concept of rough set. Richard carried out the research work by hybridizing ant colony optimization and fuzzy rough concept to yield better features. Richard Jensen also carried out the research work with the fuzzy rough-assisted technique for feature selection [4]. The authors have also proposed new approaches to fuzzy rough feature selection techniques [5]. Many feature reduction techniques which include discernability matrix-based approach, fuzzy rough positive region-based approach, fuzzy rough boundary region-based approach are proposed in that research work. Modified fuzzy rough-based approach for feature selection is proposed which focuses on improving the performance of fuzzy rough set attribute reduction algorithm [6]. Fuzzy rough feature selection method consumes most of the time for finding the reduct set. To overcome this drawback, a model is proposed called accelerator which is used to redesign the fuzzy rough feature selection technique [7]. Further, the feature selection technique is developed by combining correlation feature selection with fuzzy rough feature selection for improving the overall quality of feature selection process [8]. A hybrid model is proposed based on genetic algorithm for feature selection in which uncertainty present in datasets is not dealt with [9]. Further, a model is developed for feature selection based on guided genetic algorithm to minimize the number of cost function evaluation [10].

From the literature survey, it is observed that improving the performance of fuzzy rough feature selection is carried out by many researchers but they failed to concentrate on finding out the optimum reduct set. With this idea, the research work is carried out. The rest of the paper is organized as follows. Section 2 discusses about the foundation of comparative methods used for finding classification accuracy, Sect. 3 discusses about the proposed work carried out with real coded genetic algorithm and fuzzy rough set, Sect. 4 focuses on the brief note on various information systems, Sect. 5 focuses on experimental results, Sect. 6 focuses on comparative work done with different feature selection methods, and Sect. 7 ends with the conclusion.

2 Foundations of Comparative Methods

An efficient feature selection method is that the meaning should be preserved even after feature selection, i.e., the classification accuracy should remain the same or the classification accuracy should improve after the removal of irrelevant features from the dataset. The features are selected by the genetic fuzzy rough set (GFRS) approach, and the classification accuracy is checked before and after the feature selection using neural networks (NN), genetic algorithm (GA), Naive Bayes (NB) and C4.5 algorithm.

2.1 *Artificial Neural Network*

Artificial neural network is a computational model constructed on a biological neural network that consists of interconnected artificial neurons. In addition, some neurons are arranged to take inputs from the outside environment. Here, a backpropagation neural network is used to measure the accuracy of the features selected. Backpropagation is an efficient method for training multilayer artificial neural network for a given set of input patterns with known classification [11]. The network has many inputs but only a single output. The input to the jth node from the ith node is denoted as a_i. The strength of the input signal is determined by the adaptive coefficients known as weights. The weight on the network from the ith node to the jth node is denoted as v_{ij}. Each node calculates a net input value y_{in} defined as

$$y_{in} = t_i + \sum v_{ij} a_i \tag{1}$$

Additionally, the net input is transformed to an activation value, using an activation function. Sigmoid activation function is used here since it is most widely used. The output value y_{io} of the ith neuron is determined by the equation

$$y_{io} = g(y_{in}) \tag{2}$$

2.2 *Bayesian Classification*

Intelligent decision making depends on hidden information which is hidden with plenty of databases. Data analysis can be carried out with classification and prediction that can help provide us with a better understanding of the high-dimensional data. In general, classification is used to predict future data trends. However, classification also predicts categorical labels. Here, we discuss the fundamental concepts of Bayesian classification that can predict class membership probabilities. Bayesian classification is derived from Bayes theorem [12]. Let S be a data tuple. In Bayesian terms, V is considered as evidence. As usual, it is described by measurements made on a set of q-attributes. Let R be some hypothesis, such that the data tuple S belongs to some specified class C. For classification, we determine $P(R/S)$, the probability that the hypothesis R holds given the evidence S. Bayes theorem is defined as follows:

$$P(R/S) = \frac{P(S/R)P(R)}{P(S)} \tag{3}$$

2.3 C4.5 Decision Tree

C4.5 decision tree algorithm is derived from ID3 decision tree algorithm. C4.5 algorithm [13] can be used to handle continuous data and missing data efficiently. The algorithm initially starts with training data. In every node of the decision tree, the algorithm chooses an attribute which most efficiently divides the data into subsets of data. Subsets of attributes are retrieved by normalized information gain. The highest normalized information gain attribute is used to make the decisions.

C4.5 is used to handle unknown values in an attribute set by calculating the probabilities of different classes. The new gain will be defined as follows

$$\text{Gain}(p) = F(\text{Info}(T) - \text{Info}(p, T)) \tag{4}$$

where

$$\text{Info}(p, T) = \sum_{j=1}^{n}(p_j \times \text{Entropy}(p_j)) \tag{5}$$

$\text{Info}(T) = \text{Entropy}(T)$ where F is the number of known value samples present in the database to the total number of samples in a set of attribute data, and p_j represents the set of possible values for the attribute T. To handle the continuous data value, C4.5 algorithm calculates gain as discussed in [14].

2.4 Genetic Algorithm

Genetic algorithms are global optimization and search-based technique following the principle of natural selection and natural genetic [15]. It creates an environment where populations of data compete, and finally the fittest survive. The most promising part of the genetic algorithm is the fitness function. The steps of the genetic algorithm are as follows:

1. Start the algorithm with first generation by generating random population of n-chromosomes, where chromosomes are individuals in each generation. Addi-

tionally, the parameters of genetic algorithm like initial population, crossover probability and mutation probability are initialized.
2. Evaluate each chromosome with fitness function in the population.
3. Select two parent chromosomes for the next generation from the current population according to their fitness.
4. Crossover swaps the two individuals with a crossover probability and forms new offspring (children) by selecting a random point.
5. Mutation is performed with a mutation probability by selecting a random point and changing the chromosome values from 1 to 0 and 0 to 1 to get a new offspring.

3 Genetic Fuzzy Rough Hybridization for Feature Selection

Genetic fuzzy rough feature selection (GFRS) model is proposed in this research work. The significance of hybridizing the genetic algorithm with fuzzy rough is that the proposed technique will splendidly work for managing uncertainty present in the dataset utilizing approximations, and it likewise manages to work with continuous data. By hybridizing genetic algorithm, best optimized feature selection results will be obtained. Inorder to apply the GFRS algorithm, the input information system should be converted to a membership value by a fuzzification process. We utilize the concept of real coded genetic algorithm and embed the feature selection problem within it. Genetic algorithm starts its execution with the set of initial population. Each and every individual chromosome forms the population, and every individual feature or gene forms a chromosome. Randomness is used to select the chromosome for feature selection. For example, if attributes 1, 2 and 3 are selected at random, then the chromosome is like 0.4 0.2 0.1 0 0 0.5 [5]. Like such, the population is formed, and the fitness function is used to evaluate the population . Fuzzy rough degree of dependency is used as the fitness function [3]. Based upon this fitness function, the fittest individuals are selected with the help of roulette wheel selection procedure. The selected chromosome will undergo a transformation with crossover and mutation operator. Crossover is the exchange of properties of two parents chromosomes to produce the offspring or child. Single point crossover is used here in which a fragmentation point is chosen by random on parent 1 and parent 2. Mutation is used to maintain diversity of the population. Elitism is the optional operator used in genetic algorithm to yield the best fittest chromosome from two different generations. The iterating nature of genetic algorithm should be stopped meaningfully so that it achieves a best optimal feature set. If the highest dependency value for the features does not change for generations, then the combination of attributes is concluded as the best features.

4 Brief Note on Various Information Systems

This section provides brief introduction to various information systems considered for comparative analysis.

4.1 IRIS Data

Iris dataset is collected from UCI repository. Iris dataset is a small and standard dataset which consists of four different features, three different classes and 150 total data. The four different features are (1) sepal length, (2) sepal width, (3) petal length and (4) petal width. The different classes include Iris setosa, Iris versicolor and Iris virginica.

4.2 Pima Indian Diabetes Dataset

This dataset is originally from the National Institute of Diabetes and Digestive and Kidney Diseases. Pima dataset has a total of 768 objects with eight features. The eight features are (1) number of times pregnant, (2) plasma glucose concentration, (3) diastolic blood pressure, (4) tricep skin fold thickness, (5) serum insulin, (6) body mass index, (7) diabetes pedigree function and (8) age in years.

4.3 Agricultural Dataset

Agricultural dataset is collected for Tiruvannamalai district from soil testing laboratory and Krishi Vigyan Kendra [16]. A total of 4156 objects with 14 features and nine different classes are considered with agricultural dataset. Different features are (1) soil EC, (2) soil pH, (3) nitrogen, (4) phosphorous, (5) potassium, (6) ferrous, (7) zinc, (8) manganese, (9) copper, (10) water PH, (11) water electrical conductivity, (12) calcium, (13)magnesium and (14) sodium [17]. The different classes are brinjal, snake gourd, chilies, capsicum, lady's finger, tomato, little gourd and tapioca.

5 Experimental Results

To show the efficiency of the proposed model, the features selected for different datasets as discussed above are tested for classification accuracy, precision, recall and cross validation error (CV error). The features selected for IRIS data with the proposed model are 2,3 and 4. The classification results for iris dataset with all the features and the selected features with the proposed model are shown in Table 1. The proposed model selects seven features as the important features for Pima Indian diabetes dataset by removing one feature which is skin thickness (4). The classification results for Pima dataset without and with features selected are shown in Table 2. The features selected for the agricultural dataset by applying the proposed model are 2, 3, 4, 5, 6, 8, 12 and 13. The classification results obtained are shown in Table 3.

From the above experimental results, it can be noted that the proposed feature selection method has efficiently selected the relevant features. The classification power should remain the same even after the irrelevant features are removed. It can be clearly understood from the experimental results that the classification accuracy, precision and recall remain the same or improve after the relevant features are selected.

It is also interesting to note from Figs. 1 and 2 that the classification accuracy remains the same even after the best features are selected. The features extracted by the proposed model are proven to be the best features based upon the input received from domain experts as well as with the classification accuracy results.

Table 1 Iris dataset classification results without and with feature selection

Algorithms used	Accuracy achieved (%)	Precision	Recall	Features selected	CV error
NN	94	0.94	0.94	All	6
GA	96.67	0.96	0.96	All	3.33
C4.5	96	0.96	0.96	All	4
NB	96	0.96	0.96	All	4
NN	96.66	0.967	0.967	2,3,4	3.33
GA	97.7	0.96	0.96	2,3,4	3.33
C4.5	98	0.98	0.98	2,3,4	3
NB	96	0.96	0.96	2,3,4	4

Table 2 Pima dataset classification results without and with feature selection

Algorithms used	Accuracy achieved (%)	Precision	Recall	Features selected	CV error
NN	72.39	0.714	0.724	All	27.60
GA	75.26	0.747	0.753	All	24.73
C4.5	73.82	0.735	0.738	All	26.17
NB	76.30	0.759	0.763	All	23.69
NN	73.6	0.734	0.734	1,2,3,5,6,7,8	27.60
GA	76.2	0.77	0.753	1,2,3,5,6,7,8	24.73
C4.5	73.82	0.735	0.738	1,2,3,5,6,7,8	26.17
NB	76.30	0.759	0.763	1,2,3,5,6,7,8	23.69

Table 3 Agricultural dataset classification results without and with feature selection

Algorithms used	Accuracy achieved (%)	Precision	Recall	Features selected	CV error
NN	90.83	0.902	0.904	All	9
GA	93.35	0.93	0.93	All	6.6
C4.5	91.33	0.912	0.913	All	8.6
NB	91.96	0.91	0.91	All	16.29
NN	91.02	0.91	0.91	2,3,4,5,6,8,12,13	8.9
GA	93.71	0.93	0.93	2,3,4,5,6,8,12,13	6.2
C4.5	91.45	0.914	0.915	2,3,4,5,6,8,12,13	8.5
NB	92.2	0.92	0.92	2,3,4,5,6,8,12,13	7.7

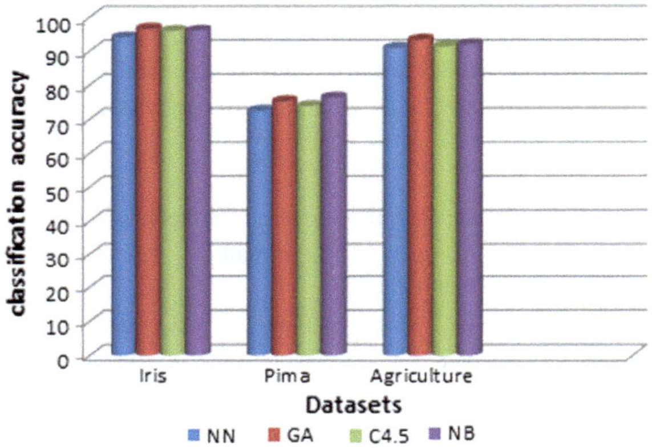

Fig. 1 Classification accuracy without feature selection

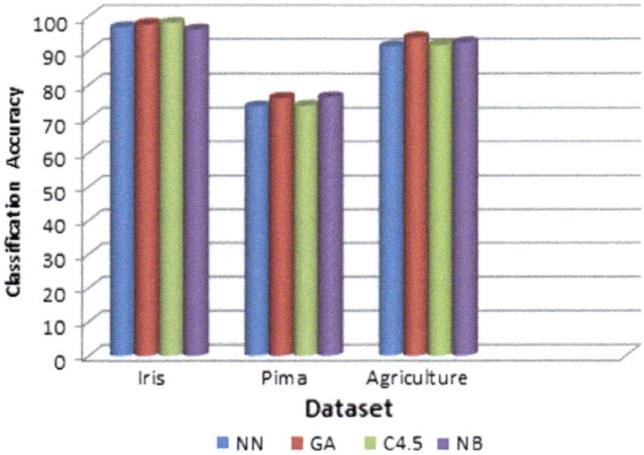

Fig. 2 Classification accuracy with feature selection

Fig. 3 Comparison of different feature selection methods

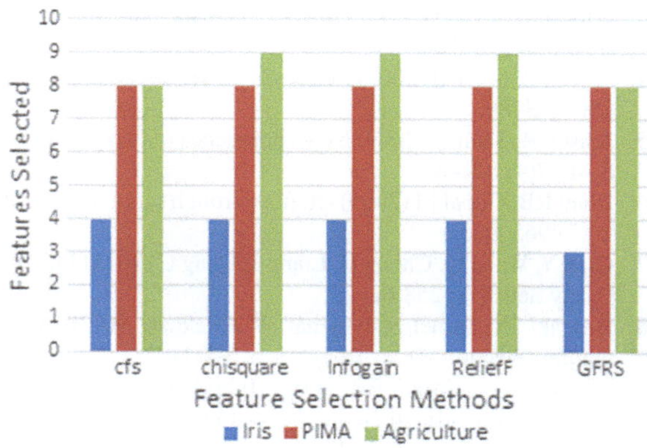

6 Comparative Analysis with Different Feature Selection Methods

To show the efficiency of the proposed model of feature selection, it is compared with some of the standard feature selection techniques. We have included CFS [18], chi-square [19], information gain [13] and ReliefF [20]. The results for CFS, chi-square, information gain and ReliefF are obtained by using the implementation of Waikato Environment for Knowledge Analysis (WEKA) [20]. It can be clearly understood from Fig. 3 that GFRS outperforms well when compared to other models and methods.

7 Conclusion

Feature selection technique's aim is to remove irrelevant features which helps in less time consumption for processing the irrelevant features and also improves the classification power. The proposed model GFRS is tested for its feature selection capability by considering three different datasets like Iris, Pima and Agriculture datasets, and its classification results are obtained. Also, it is compared with other feature selection techniques. From the results observed, it is concluded that GFRS outperforms well for all the datasets as well as the methods.

References

1. Pawlak Z (2012) Rough sets: theoretical aspects of reasoning about data, vol 9. Springer Science & Business Media, Berlin
2. Chakraborty G, Chakraborty B (2004) A rough-GA hybrid algorithm for rule extraction from large data. In: IEEE International conference on computational intelligence for measurement systems and applications (CIMSA 2004), pp 85–90

3. Jensen R, Shen Q (2005) Fuzzy-rough data reduction with ant colony optimization. Fuzzy Sets Syst 149(1):5–20
4. Jensen R, Shen Q (2007) Fuzzy-rough sets assisted attribute selection. IEEE Trans Fuzzy Syst 15(1):73–89
5. Jensen R, Shen Q (2009) New approaches to fuzzy-rough feature selection. IEEE Trans Fuzzy Syst 17(4):824–838
6. Bhatt RB, Gopal M (2005) On fuzzy-rough sets approach to feature selection. Patt Recogn Lett 26(7):965–975
7. Qian Y, Wang Q, Cheng H, Liang J, Dang C (2015) Fuzzy-rough feature selection accelerator. Fuzzy Sets Syst 258:61–78
8. Anaraki JR, Samet S, Banzhaf W, Eftekhari M (2016) A new fuzzy-rough hybrid merit to feature selection. Trans Rough Sets XX:1–23
9. Oh IS, Lee JS, Moon BR (2004) Hybrid genetic algorithms for feature selection. IEEE Trans Patt Anal Mach Intell 26(11):1424–1437
10. Jung M, Zscheischler J (2013) A guided hybrid genetic algorithm for feature selection with expensive cost functions. Procedia Comput Sci 18:2337–2346
11. Hecht-Nielsen R (1992) Theory of the backpropagation neural network. In: Neural networks for perception. Academic, London, pp 65–93
12. Acharjya DP, Roy D, Rahaman MA (2012) Prediction of missing associations using rough computing and Bayesian classification. Int J Intell Syst Appl 4(11):1–13
13. Quinlan JR (1986) Induction of decision trees. Mach Learn 1(1):81–106
14. Hssina B, Merbouha A, Ezzikouri H, Erritali M (2014) A comparative study of decision tree ID3 and C4. 5. Int J Adv Comput Sci Appl 4(2):13–19
15. Goldberg DE (2006) Genetic algorithms. Pearson Education India, New Delhi
16. Rathi R, Acharjya DP (2017) A rule based classification for agriculture vegetable production for Tiruvannamalai District using rough set and genetic algorithm. Int J Fuzzy Syst Appl
17. Rathi R, Acharjya DP (2018) A framework for prediction using rough set and real coded genetic algorithm. Arab J Sci Eng 43(8):4215–4227
18. Hall MA (2000) Correlation-based feature selection of discrete and numeric class machine learning
19. Liu H, Setiono R (1995) Chi2: feature selection and discretization of numeric attributes. In: Seventh international conference on tools with artificial intelligence. IEEE, pp 388–391
20. Kononenko I (1994) Estimating attributes: analysis and extensions of RELIEF. In: European conference on machine learning. Springer, Berlin, pp 171–182

A Comparative Study on Prediction of Dengue Fever Using Machine Learning Algorithm

Saif Mahmud Khan Dourjoy, Abu Mohammed Golam Rabbani Rafi, Zerin Nasrin Tumpa, and Mohd. Saifuzzaman

Abstract Dengue is the most common viral fever for the people. This is also known as life-threatening disease. Dengue has become more and more evident this year in Bangladesh. It has taken the lives of many in our country. And the number of dengue fever patients is increasing day by day. There are many people at risk from dengue. Early forecast of dengue can spare individual's life by cautioning them to take legitimate conclusion and care. But it is difficult to say in advance whether this will happen or not. The aim of this piece of research work is to analysis the symptoms of dengue fever and early prediction of the symptoms that can be seen in years ahead. For predicting the symptoms, two different machine learning algorithms have been used. Support vector machine (SVM) and random forest classifier algorithm have been used. Finally, the accuracy of these two has been evaluated and the confusion matrix has been shown. And then, we have talked about the algorithm which is better for our dataset.

Keywords Dengue fever · Symptoms · Support vector machine · Random forest classifier · Prediction · Dataset

S. M. K. Dourjoy (✉) · A. M. G. R. Rafi · Z. N. Tumpa · Mohd. Saifuzzaman
Department of Computer Science and Engineering, Daffodil International University, Dhaka, Bangladesh
e-mail: saif15-7098@diu.edu.bd

A. M. G. R. Rafi
e-mail: rabbani15-6719@diu.edu.bd

Z. N. Tumpa
e-mail: zerinnasrin@gmail.com

Mohd. Saifuzzaman
e-mail: Saifuzzaman.cse@diu.edu.bd

A. K. Tripathy et al. (eds.), *Advances in Distributed Computing and Machine Learning*, Lecture Notes in Networks and Systems 127, https://doi.org/10.1007/978-981-15-4218-3_49

1 Introduction

Dengue is viewed as one of the most significant arboviral diseases on the planet which is transmitted by Aedes mosquitoes, principally *Aedes* (Stegomyia) *aegypti* and furthermore *Ae. Albopictus*. This is also known as life-threatening disease nowadays. The disease causes influenza like ailment, and once in a while forms into a possibly deadly inconvenience or extreme dengue, for example, dengue haemorrhagic fever and dengue shock syndrome. Dengue is found in tropical and sub-tropical atmospheres around the world, for the most part in urban and semi-urban territories. Extreme dengue is a main source of genuine disease and passing among kids in some Asian and Latin American nations. There is no particular treatment for dengue but early prediction of dengue can protect person's life and they can drive for taking appropriate finding and care.

Dengue is a mosquito-borne disease found in tropical and sub-tropical locales around the globe. Lately, transmission has expanded prevalently in urban and semi-urban zones and has turned into a significant worldwide general well-being concern. Serious dengue was first acknowledged in the 1950s while dengue rises in the Philippines and Thailand [1]. Nowadays, serious dengue influences most Asian and Latin American nations and has turned into a main source of hospitalization and passing among children in these districts.

The occurrence of dengue has developed drastically around the globe in late decades. Over 2.5 billion individuals—over 40% of the total populace—are currently in danger from dengue [2]. But in Bangladesh, the number of dengue patients has been increasing yearly. Since January 1 till October 5, upwards of 89,616 individuals have been influenced by dengue [3]. So a large number of people have been affected by dengue this year. And many people have died of dengue in 2019. Dengue ailments are realized by four immovably related diseases of infections known as DEN-1, DEN-2, DEN-3, and DEN-4 [4]. The DEN-5 has been presented in 2013 [5]. Nearby are three class characterizations of dengue fever arranges that are [6]

- (DF) Dengue fever—In DF stage, body temperature increases day by day and patient will feel headache, pain in muscle, and itching.
- (DHF) Dengue hemorrhagic fever—In DHF stage, body temperature decreases and minor bleeding from patients nose, gums, skins, etc.
- (DSS) Dengue shock syndrome—In DSS stage, body temperature will be fluctuating and vomiting accompanied by the flow of minor blood.

Our aim is to develop a model for predicting symptoms of dengue using some machine learning algorithms and increase the awareness of the people of our country and reducing the number of deaths affected by dengue.

2 Related Works

One of the most consuming subject or episode of this current year is the monstrous breakout of the dengue fever. Dengue fever exhibits a huge and developing weight of infection to endemic nations, where kids are at specific hazard. Around the world, no viable enemy of viral treatment has been recognized; hence, vector control is a key for malady avoidance.

In order to predict the dengue fever, Iqbal et al. have used the dataset which is made with clinical data, genes, and climate factors. In their research, using this dataset and tree-based model, neural-based model, evolutionary-based classifier, ensemble classifier algorithms, they have predicted whether or not dengue will occur [5]. There are many research for predicting dengue fever but Jain et al. made a dataset using meteorological data, lag variable of surveillance of the disease, clinical data, socioeconomic data and used algorithms like generalized additive models, R-squared for prediction [7]. For predicting the dengue fever, Guo et al. used demographic data, clinical data, laboratory data, dengue cases, meteorological data, search and query data as data and support vector machine (SVR), gradient boosted regression model (GBM), step-down linear regression model, negative binomial regression (NBM), LASSO model, GAM model, and RMSE as algorithm [8].On the other hand, Rahim et al. predicted dengue fever using real data on dengue and weather data as data to make dataset and decision tree, support vector machine, random forest, CN2, Naïve Bayes as algorithm for prediction [9]. For the purpose of predicting dengue fever, Sanjudevi et al. made a dataset using dengue cases and used support vector machine (SVM), SMO, decision tree algorithms for detection and prediction [10]. Muhilthini et al. with the intention of predicting dengue fever used gradient boosting regression (GBR), mean square error algorithms over temperature, population, rain fall, urbanization, dengue cases data [11]. With the objective to predict dengue fever, Choi et al. took help from environment, temperature, rain fall, humidity data, and generalizes linear model, negative binomial regression (NBR) algorithms [12]. In order to predict dengue fever, Rajathi et al. used dataset which was consist of dengue cases and that were collected from health department and for predicting the dengue fever Naive Bayes, SMO, random forest classifier and many other algorithms have been used [13]. For predicting dengue fever, Suhaeri et al. took temperature, leukocytes, thrombocyte, and hemoglobin data for making dataset and ELM algorithms for prediction [14].

3 Methodologies

3.1 Data Collection

For our research work, data collection was the most important tasks for making our dataset. Data collection or information accumulation is the way toward social event and estimating data on factors of enthusiasm in a built up precise design that empowers one to respond to expressed research questions, test speculations, and assess results. Data have been collected by surveying and using Internet. We have surveyed more than 600 hundreds patients for collecting data and finding information. Symptoms of dengue fever have been used to make the dataset. Dengue fever symptoms have been predicted in this research for reducing the future damage of a patient. Early dengue symptoms prediction can save a patient life and they can go for better and early diagnosis (Fig. 1).

3.2 *Dataset*

Almost 1047 data have been collected and some data which is shown above. 20% data have been used for testing and 80% of data have been used for training (Fig. 2).

3.3 *Working Process*

Collected data have been prepared and cleaned up using data preprocessing techniques. Then data have been generalized, and finally, a output has been generated by algorithm when it has been given to the algorithm for prediction (Fig. 3).

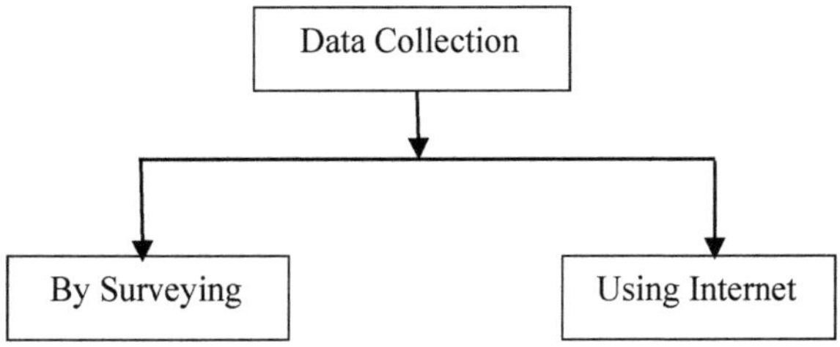

Fig. 1 Data collection procedure

Headache	Pain in th	Cough, so	Rashes all	Body pain	Aversion	Loose mo	Joint pain	Instability	Feeling di	Bleeding	Frequent	Rapid plat	Acute abd	Abnormal	Year
1	0	0	1	1	1	1	0	1	0	0	0	1	0	0	2016
1	0	1	0	0	0	0	0	0	0	0	0	0	0	1	2019
1	0	1	0	1	1	0	1	1	0	0	0	0	0	0	2015
0	0	0	0	0	0	0	0	0	0	0	0	1	0	0	2018
1	0	1	0	1	1	1	1	1	1	1	1	1	1	0	2017
1	0	0	0	1	0	1	1	0	0	0	0	0	1	0	2007
1	1	0	0	1	1	0	0	0	1	0	0	1	0	0	2013
0	0	1	0	0	0	0	0	0	0	0	0	0	0	0	2011
0	0	1	0	1	1	0	0	1	1	0	1	0	0	0	2009

Fig. 2 Dataset

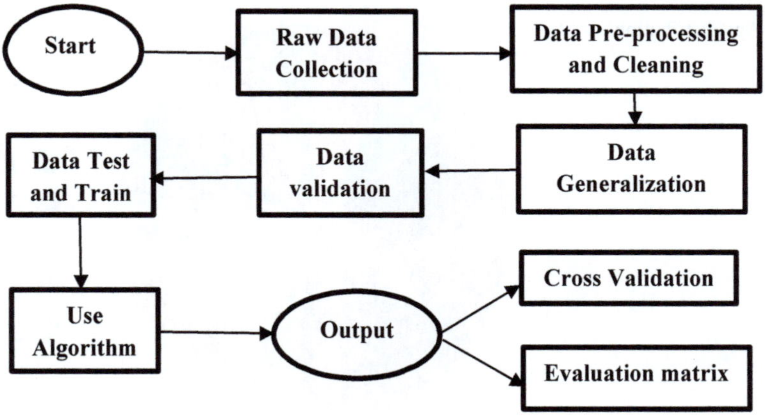

Fig. 3 Working procedure

4 Results and Discussions

4.1 Statistical Results

After surveying, some analytical statistic data have been found and these are shown below.

Gender Distribution

According to this figure, it turns out that male are in more danger than female. The pace of dengue fever is higher among men; however, more ladies are among the individuals who have died from the mosquito-borne viral ailment in Bangladesh [15]. After our surveying, we have found that 62.5% male, 36.9% female, and less than 1% other got attacked by dengue this year in our country. This number is too high comparing to the previous years. Many people of them already have died of dengue and female are more in number (Fig. 4).

Fig. 4 Gender distribution

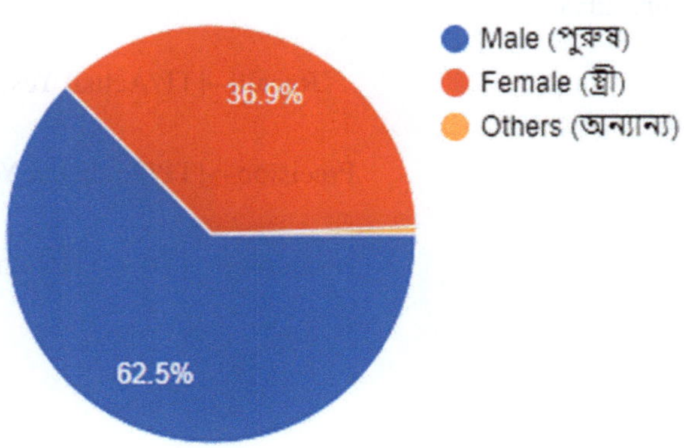

Fig. 5 Age distribution

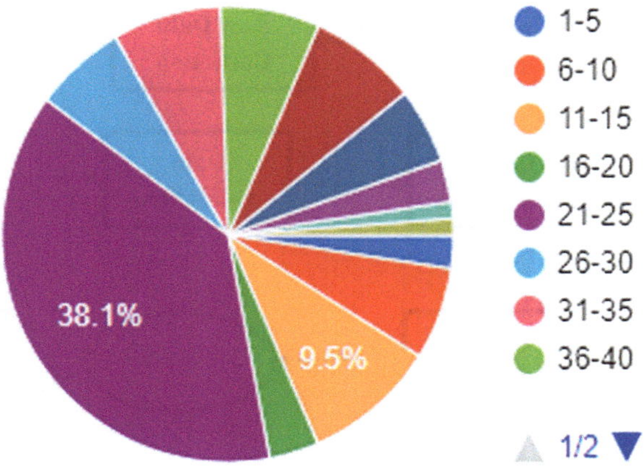

Age Distribution
According to this figure, it turns out that age among 21–25 are in a great danger. 38.1% from age 21–25 have attacked by dengue in our country according to our survey. Age among 6–10, 11–15, and 36–40 are also in great danger (Fig. 5).

Prevalent Symptoms
According to this figure, it turns out that headache will be the most prevalent symptoms among patients in next year. The percentage of headache will be 86.92% which is a huge number of percentage. After that, body pain will be the second in prevalent symptoms and it will be 76.92. After getting the results of our predicted values, it can be said that headache, body pain, aversion, feeling dizzy, joint pain, platelet depletion, and pain back of the eye will be the most seen symptoms in the next year (Fig. 6).

4.2 Evaluation Matrix

Formulas

Recall—[TP/Actual Yes]

Precision—[TP/Predicted Yes]

Accuracy—[TP + TN/Total]

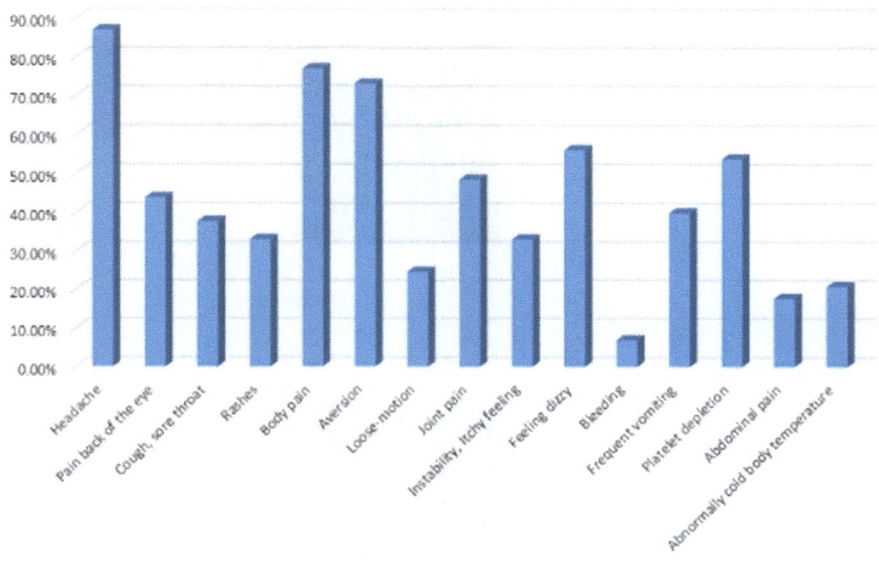

Fig. 6 Prevalent symptoms for 2020

In order to evaluate the value of recall, precision, and accuracy from a confusion matrix, above formulas are needed.

Confusion Matrix

A confusion matrix is a counter that is regularly used to portray the exhibition of an order model on a lot of test information for which the genuine qualities are recognized. It permits the perception of the exhibition of a calculation. Fifteen confusion matrices have been found for our model. Fifteen different confusion matrices have been found because 15 symptoms have been used as parameter in our research work.

Confusion Matrix for Feeling Dizzy

According to this figure, the number of actual yes and predicted yes are, respectively, 122 and 173. True positive and true negative are, respectively, 112 and 25 and the number of total is 208 (Fig. 7).

	Recall	Precision	Accuracy
Feeling dizzy	0.92	0.65	0.66

Confusion Matrix for Platelet Depletion

According to this figure, the number of actual yes and predicted yes are, respectively, 124 and 165. True positive and true negative are, respectively, 104 and 23 and the number of total is 208 (Fig. 8).

	Recall	Precision	Accuracy
Platelet depletion	0.84	0.63	0.61

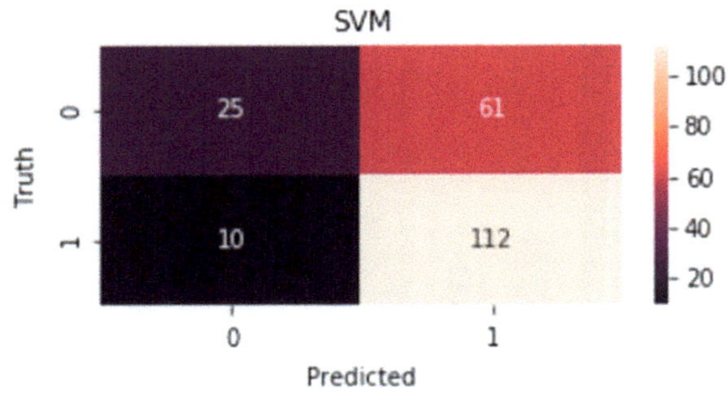

Fig. 7 Matrix of feeling dizzy

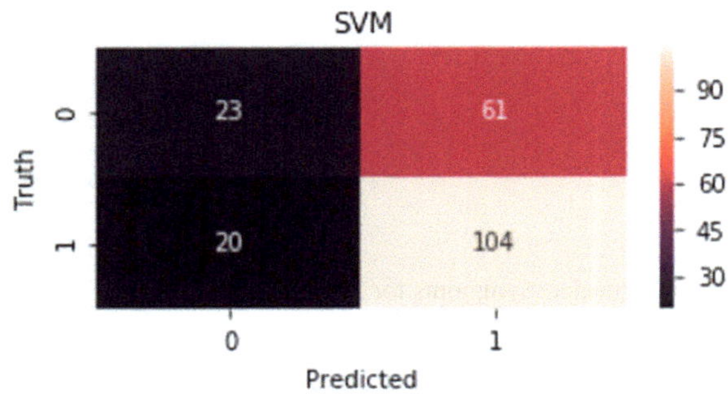

Fig. 8 Matrix of platelet depletion

4.3 Accuracy of Algorithms

Support vector machine (SVM) and random forest classifier have been used in our research work for predicting dengue fever and these two's accuracies have been evaluated.

	Accuracy
Support vector machine	0.6947073620306817
Random forest classifier	0.6810406668468509

By using two different algorithms, two different accuracy rates have been found. But the accuracy level is almost similar.

In order to predict dengue fever, there are many research using clinical data, genes, climate factors, laboratory data, dengue cases, meteorological data, temperature, population, rain fall, urbanization data and values and predicted whether or not dengue will occur. But research for predicting dengue fever symptoms using symptoms as parameter is much rare in research works. In our research work, symptoms have been used as parameter for predicting dengue fever symptoms which can be seen in the future years. The early prediction of dengue symptoms will help to prevent the damage of dengue and will save our lives.

5 Conclusions

The present work shows the significant jobs of female mosquito disease rates from the past season in dengue episodes. Joining this factor in the model essentially improves the predicting intensity of dengue hemorrhagic fever determining models. Dengue disease has risen as a significant well-being worry in Southeast Asia particularly Bangladesh and till a particular antibody or an antiviral operator become accessible, we need to depend on early distinguishing proof of hazard variables related with creating intricacies of DF; simply like the admonition includes just as the ones recognized in this research. A model for identification of dengue symptoms has been developed so that they can take quick and proper treatment for perfect and early overcome. While doing our research work with our dataset, two different accuracy rates have been found from these different algorithms. But the rates are almost similar. Though it can vary from dataset to dataset. But support vector machine gives simply better accuracy than random forest. So, I have a strong recommendation for you to use or try support vector machine in predicting research work. Furthermore, we know, ML plays roles in collaboration with IoT [16, 17], Security [18], Data Mining [19], and many more sector to process and analyses a huge amount of data and information where we use ML towards a common viral fever which will be a little contribution from us towards the world.

References

1. World Health Organization [Online]. Available: https://www.who.int/denguecontrol/disease/en/
2. Molla M (2019) Front page. The Daily Star, 27 July 2019 [Online]. Available https://www.thedailystar.net/frontpage/news/dengue-outbreak-south-asia-climate-change-the-culprit-1777585
3. Tribune D (2019) DGHS: 269 more hospitalized for dengue in 24 hrs. Dhaka Tribune, 5 Oct 2019 [Online]. Available https://www.dhakatribune.com/bangladesh/2019/10/05/dghs-269-more-hospitalized-for-dengue-in-24hrs
4. Mahmud F (2019) News/Bangladesh, Aljazeera.com, 7 Aug 2019 [Online]. Available https://www.aljazeera.com/news/2019/08/bangladesh-grappling-record-dengue-fever-outbreak-190807091210915.html
5. Iqbal NIM (2017) Machine learning for dengue outbreak prediction: an outlook. Int J Adv Res Comput Sci 8:93–102
6. Hadinegoro SRS (2012) The revised WHO dengue case classification: does the system need to be modified? Paediatrics Int Child Health 32(Suppl 1):33–38
7. Jain R, Sontisirikit S, Iamsirithaworn S, Prendinger P (2019) Prediction of dengue outbreaks based on disease surveillance, meteorological and socio-economic data. BMC Infect Dis 19(1)
8. Guo P, Liu T, Zhang Q, Wang L, Xiao J, Zhang Q, Luo G, Li Z, He J, Zhang Y, Ma W (2017) Developing a dengue forecast model using machine learning: a case study in China. PLOS Negl Trop Dis 11(10):e0005973
9. Rahim NF, Rahim FNF (2018) Dengue fatality prediction using data mining. J Fundam Appl Sci 9(65):671–683
10. Sanjudevi DSR (2019) Dengue fever prediction using classification techniques. Int Res J Eng Technol (IRJET) 6(02):558–563

11. Muhilthini P, Meenakshi B, Santhanalakshmi S (2018) Dengue possibility forecasting model using machine learning algorithms. Int Res J Eng Technol 5(3):1661–1665
12. Choi Y, Tang CS, McIver L, Hashizume M, Chan V, Abeyasinghe RR, Iddings S, Huy R (2016) Effects of weather factors on dengue fever incidence and implications for interventions in Cambodia. BMC Public Health 16(1)
13. Rajathi N, Kanagaraj S, Brahmanambika R, Manjubarkavi K (2018) Early detection of dengue using machine learning algorithms. Int J Pure Appl Math 118(18):3881–3886
14. Suhaeri S, Nawi NM, Fathurahman M (2018) Early detection of dengue disease using extreme learning machine. Int J Adv Sci Eng Inform Technol 8(5):2219–2224
15. Bdnews24.com (2019) Health, Bdnews24.com, 20 Aug 2019 [Online]. Available https://bdnews24.com/health/2019/08/20/women-dying-more-from-dengue-but-disease-afflicting-more-men-in-bangladesh
16. Saifuzzaman M, Khan AH, Moon NN, Nur FN (2017) Smart security for an organization based on IoT. Int J Comput Appl 165(10):33–38
17. Saifuzzaman M, Moon NN, Nur FN (2017) IoT based street lighting and traffic management system. In: 2017 IEEE Region 10 Humanitarian Technology Conference (R10-HTC), 21–23 Dec 2017
18. Shetu SF, Saifuzzaman M, Moon NN, Nur FN (2019) A Survey of Botnet in Cyber Security. In: 2019 2nd International Conference on Intelligent Communication and Computational Techniques (ICCT), Jaipur, India, 2019, pp 174–177
19. Shetu SF, Saifuzzaman M, Sultana S, Yousuf R, Moon NN (2020) Students performance prediction through education data mining depending on overall academic status and environment. In: 3rd International Conference on Innovative Computing and Communication (ICICC-2020) on 21–23rd February 2020, New Delhi, India

Does Machine Learning Algorithms Improve Forecasting Accuracy? Predicting Stock Market Index Using Ensemble Model

T. Viswanathan and Manu Stephen

Abstract Forecasting market performance and understanding the mechanism of price discovery is inherent to develop trading strategies. This paper examines the predictive power of machine learning algorithms in forecasting stock index. The study applies various machine learning algorithms and suggests the best model for forecasting stock index. We have developed an ensemble model to predict the daily closing index of Nifty 50 based on open, high, low and previous day's close. The ensemble model includes a mix of simple linear regression, gradient boosted tree, decision tree and random forest. The parameters of the models are tested for its accuracy train-test split under supervised learning. The predicting accuracy of machine learning algorithms is further refined using cross-validation techniques that include leaving one out cross-validation and k-fold cross-validation. We found that the ensemble model provides an accurate forecast of the stock market index for the short term. The outcome of the study would facilitate the investors and portfolio managers to use the appropriate model for forecasting and take an informed decision by considering the nature of stock market volatility.

Keywords Forecasting · Linear regression · Gradient boosted tree · Decision tree and random forest · Ensemble

1 Introduction

Forecasting market return and predicting volatility play a significant role in the spot and derivatives market. Extant literature shows earlier studies have used statistical and econometric tools for forecasting. However, the models face constraints to issues such as parameter optimisation, feature selection, overfitting and underfitting. Machine learning algorithms can overcome the limitations of the traditional model

T. Viswanathan (✉) · M. Stephen
Symbiosis Institute of Business Management, Bengaluru, India
e-mail: visu_prof@rediffmail.com

M. Stephen
e-mail: manu.stephen21@sibm.edu.in

A. K. Tripathy et al. (eds.), *Advances in Distributed Computing and Machine Learning*, Lecture Notes in Networks and Systems 127, https://doi.org/10.1007/978-981-15-4218-3_50

and improve forecasting accuracy. Further, the algorithms bring flexibility in dealing with big data and also improve forecasting accuracy. We test the forecasting accuracy of linear regression, gradient boosted tree, decision tree and random forest on predicting the market return of Nifty 50. For post testing the accuracy, we create an ensemble model using machine learning algorithms to predict Nifty's closing index (C) of the day based on open (O), high (H) and low price (L). These parameters jointly named as (OHLC) is a type of chart used to analyse the daily price movement of stock, currencies and commodities. Garman and Klass [1] and Alizadeh et al. [2] estimated weekly stock volatility using OHLC. Following the methodology, we develop an ensemble model to forecast stock index.

2 Literature Review

Adebiyi et al. [3] examined the forecasting performance of ARIMA and artificial neural networks model. The study found that neural networks are superior to ARIMA model in forecasting. Feuerriegel and Gordon applied text based model to examine the effect of new information on the short and long term stock index forecasting. The research found the text-based models succeed in reducing forecast errors. Fischer and Krauss [4] applied long- and short-term memory networks (LSTM) for predicting S&P 500 and found LSTM outperform random forest, a deep neural net and logistic regression classifier. Göçken et al. [5] used hybrid artificial neural network (ANN) models which consist in exploiting capabilities of harmony search (HS) and genetic algorithm (GA) are found to be the dominant model for stock market forecasting.

3 Objective

To create an efficient machine learning model that can predict the stock index value for the next day. The prediction is for a short duration since the financial market is volatile. So, long-term prediction is not a reliable solution. The model created should be flexible to even predict other index values like the next day Low, High, Open or Close depending on the requirement. Also, the model intended should be a generalised model that would work across different stock exchanges and that would be able to imbibe weights for different parameters as per the users need so that it could be used as an efficient tool in the investor's hands.

4 Methodology

The methodology used in the research is detailed below. The research was performed using various regression models. The historical values of NSE 50 index have been used for the initial model creation and validation. The model was created using machine learning libraries in Pyspark. Jupyter notebook interpreter was used for running the code.

4.1 Input Data

The data used for the research is the last 5 years daily values of NSE50 index, collected from yahoo finance. A sample from the input data is provided in Appendix A. It shows the Date, previous day Open price, previous day Close price, previous day Low price, previous day High price, previous day Volume and the Closing price.

4.2 Modelling Using Machine Learning

The input data was pre-processed to remove null values. They were then converted into double data type, and correlation testing was performed using Pyspark packages to find the association of present day close with previous day Open, High, Volume, Low and Close. There is a significant association between four of these variables, as mentioned in Table 1.

The Volume has a low association. So, Open, High, Low and Close were taken as the independent variable and present day close (label) has a dependent variable for the model. Also, the adjusted R square was measured to check if there is any inflation in the model due (Table 2).

Table 1 Correlation values of daily price movement

S. No.	Comparison variables with label	Correlation value
1.	Open	0.9968
2.	High	0.9977
3.	Low	0.9976
4.	Close	0.9982
5.	Volume	0.2114

Table 2 Computation of adjusted R square

Rsquare	Adjusted *R* square	Standard error of the estimate	*R* square change	*F* change	df1	df1	Sig *F* change
0.998	0.997	81.222	0.997	87765.68	4	1216	0.000

4.3 Ensemble Model for Stock Price Prediction

The feature for the model was constructed, combining previous day Open, High, Volume, Low and Close using Vector Assembler. It was then passed to individual linear regression, random forest, decision tree and GBT models. The model summary and efficiency were analyzed and RMSE value was compared. Then, these individual results were passed as a feature to an ensemble model created using linear regression and the final stock index prediction was obtained.

5 Results and Analysis

5.1 Descriptive Statistics

The final data for the modelling was obtained after pre-processing the input file. The rows containing null values were removed to make the data more efficient for modelling (Table 3; Fig. 1).

Table 3 Descriptive statistics of dependent and independent variables

Summary statistics					
Statistics	Previous open	Previous high	Previous low	Previous close	Closing price
Count	1221	1221	1221	1221	1221
Mean	9594.017	9634.105	9535.877	9584.972	9588.045
Standard deviation	1380.664	1383.127	1374.752	1379.055	1380.19
Min	7023.65	7034.2	6825.8	6970.6	6970.6
Max	12,146.2	12,158.8	12,099.95	12,151.15	12,151.15

Fig. 1 Line graph between date and closing stock price

5.2 Linear Regression

All the assumptions for regression, like normality, linearity, homoscedasticity were satisfied. The multicollinearity was disregarded for the research since it was not causing higher inflation and for stock price prediction model, these correlated variables have a major importances so avoiding them reduces the flexibility and significance of the intended model. The data was passed to a linear regression model and RMSE score of 88.509 was obtained (Figs. 2 and 3).

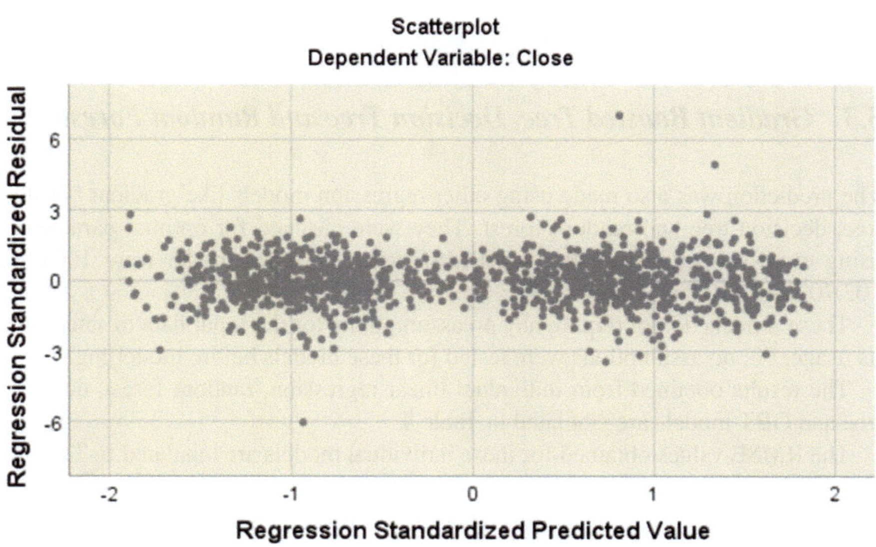

Fig. 2 Testing homoscedasticity using SPSS

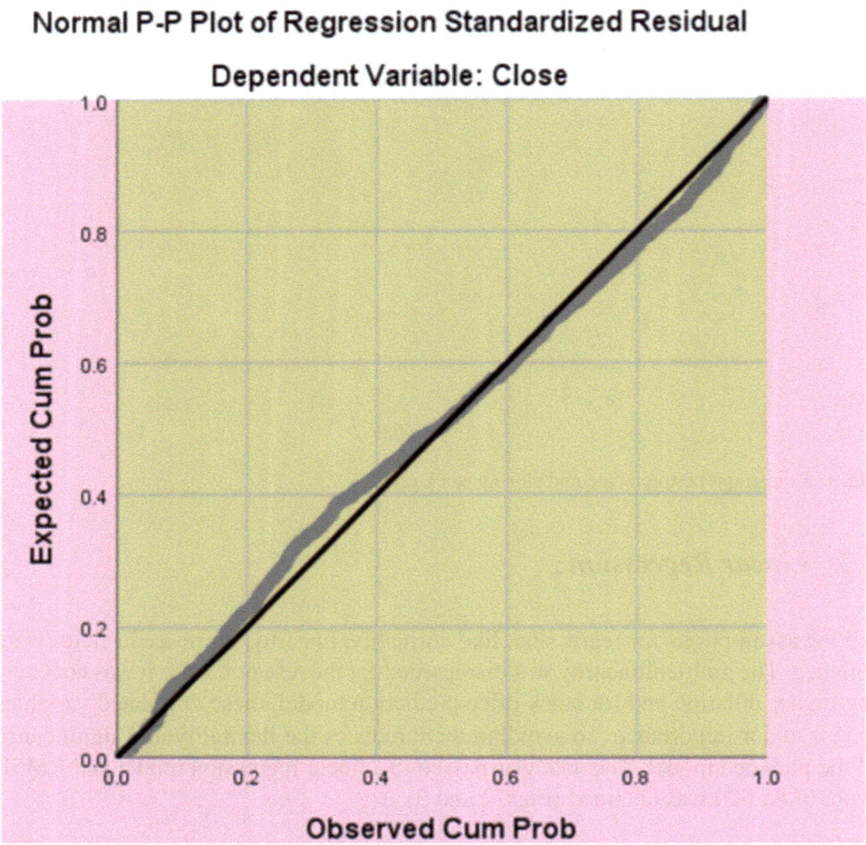

Fig. 3 Testing normality using SPSS

5.3 Gradient Boosted Tree, Decision Tree and Random Forest

The prediction was also made using other regression models like gradient boosted tree, decision tree and random forest. They were checked for optimal parameters using cross-validation and the best RMSE values in each model were 106.086, 102.709 and 110.849, respectively.

These models do not require any preassumptions for the input data to satisfy for its usage. So, no assumptions were tested for these models before modelling.

The results obtained from individual linear regression, random forest, decision tree and GBT models are tabulated in Table 4.

The RMSE values obtained for these individual models are tabulated in Table 5.

Table 4 Eight days forecast of Nifty index

Date	Gradient boosted tree prediction	Decision tree prediction	Random forest prediction	Linear regression prediction	Closing price (original)
01-06-2017	9781.939709	9758.506641	9668.222618	9626.086007	9616.1
09-01-2018	10,622.63191	10,588.04673	10,625.18912	10,607.22736	10,637
10-04-2017	9100.054395	9100.84807	9226.395987	9215.80933	9181.45
18-01-2017	8393.87302	8448.738513	8412.496934	8412.426577	8417
19-09-2016	8853.657951	8841.414472	8761.978056	8791.608015	8808.4
22-03-2017	9113.091549	9100.84807	9226.395987	9123.277079	9030.45
26-03-2015	8475.624779	8448.738513	8517.58108	8550.20802	8342.15
30-06-2016	8199.545727	8198.54678	8201.519542	8193.208999	8287.75

Table 5 RMSE values for each individual models

S. No.	Individual models	RMSE value
1.	Linear regression	88.509
2.	Random forest	102.709
3.	Decision tree	110.849
4.	GBT	106.086

5.4 Results from Ensemble Model

The final model was derived taking the individual models' prediction values and adding it as a feature to an ensemble model, which uses linear regression. The model was tested under different settings to determine the best fit model (Table 6).

The RMSE value obtained in the final model after recursive parameter tuning was 42.529, which was a very low value providing reliable evidence for the efficiency of the model. Also, there was no significant difference in the RMSE values obtained during testing with training and validation data, proving concrete evidence that the derived model is free from under-fitting and over-fitting.

Table 6 Analysis on the final ensemble model

S. No.	Analysis of final model flexibility	RMSE values
1.	Testing with training data	36.24
2.	Testing with validation data	41.73
3.	Testing with test data	42.52
4.	Train-validation split model	56.82

6 Limitations

The model was created and validated using NSE50 data only. Other Index data are yet to be tested. Also, the modelling was performed in a standalone desktop deterring the high efficiency that could be obtained running the program in a server, with a huge chunk of input data. Different regression models were tested during the time of model creation, yet, parameter tuning was not utilised to its full potential.

7 Future Work

In future, a volatility model would be created, predicting the volatility movement in percentages instead of direct stock index prediction, so it could be rendered to get insight while using stop loss during investment decisions. Also, parameter tuning was not done to the full potential. Hence, sufficient analysis would be performed to find the optimal values for each parameter. For the model to be used efficiently in the financial market decision making, it should be tested across index data across various stock exchanges. Therefore, a comprehensive testing would be performed in the immediate future, testing a comprehensive set of input data with values from various Stock exchanges. The sentiment analysis of different indexes would also be taken as a factor in the upcoming models to improve the efficiency of the model.

8 Conclusion

An ensemble model was created to help wealth management banks, investment bankers and corporate investors, financial institutions and macro-economic policymakers to determine the stock index value for the next day with the current day movements. The model performed efficiently compared to the existing models with appreciable metrics. Also, the developed model is flexible to add any other macro-economic factors, seasonality and other shocks as factors while determining index value and also to give weight to some factors. Since investment decisions are costly and should be taken with a high amount of care and with lower risk, this model would be a good fit to be added in the kit.

References

1. Garman MB, Klass MJ (1980) On the estimation of security price volatilities from historical data. J Bus 53(1):67–78
2. Alizadeh S, Brandt MW, Diebold FX (2002) Range-based estimation of stochastic volatility models. J Finan 57(3):1047–1091

3. Adebiyi AA, Adewumi AO, Ayo CK (2014) Comparison of ARIMA and artificial neural networks models for stock price prediction. J Appl Math 2014
4. Fischer T, Krauss C (2018) Deep learning with long short-term memory networks for financial market predictions. Eur J Oper Res 270(2):654–669
5. Göçken M, Özçalıcı M, Boru A, Dosdoğru AT (2016) Integrating metaheuristics and artificial neural networks for improved stock price prediction. Expert Syst Appl 44:320–331
6. Feuerriegel S, Gordon J (2018) Long-term stock index forecasting based on text mining of regulatory disclosures. Decis Support Syst 112:88–97

Correction to: Power Allocation-Based QoS Guarantees in Millimeter-Wave-Enabled Vehicular Communications

Satyabrata Swain, Jyoti Prakash Sahoo, and Asis Kumar Tripathy

Correction to: Chapter 4 in: A. K. Tripathy et al. (eds.), *Advances in Distributed Computing and Machine Learning*, Lecture Notes in Networks and Systems 127, https://doi.org/10.1007/978-981-15-4218-3_4

In the original version of the book, the following belated corrections has been incorporated: In chapter "Power Allocation-Based QoS Guarantees in Millimeter-Wave-Enabled Vehicular Communications", the reference has been changed from "16. Mason F, Giordani M, Chiariotti F, ZanellaA, Zorzi Z (2019) An adaptive broadcasting strategy for efficient dynamic mapping in vehicular networks. IEEE Trans Wirel Commun (TWC)" to "16. Mason F., Giordani M., Chiariotti F., Zanella A., Zorzi M. An Adaptive Broadcasting Strategy for Efficient Dynamic Mapping in Vehicular Networks. IEEE Transactions on Wireless Communications, vol. 19, no. 8, pp. 5605-5620. Aug. 2020". The correction chapter and the book has been updated with the changes.

The updated version of this chapter can be found at
https://doi.org/10.1007/978-981-15-4218-3_4

A. K. Tripathy et al. (eds.), *Advances in Distributed Computing and Machine Learning*,
Lecture Notes in Networks and Systems 127, https://doi.org/10.1007/978-981-15-4218-3_51

Author Index

A. K. Tripathy et al. (eds.), *Advances in Distributed Computing and Machine Learning*, Lecture Notes in Networks and Systems 127, https://doi.org/10.1007/978-981-15-4218-3

MIX
Papier aus verantwortungsvollen Quellen
Paper from responsible sources
FSC® C105338

If you have any concerns about our products,
you can contact us on
ProductSafety@springernature.com

In case Publisher is established outside the EU,
the EU authorized representative is:
Springer Nature Customer Service Center GmbH
Europaplatz 3, 69115 Heidelberg, Germany

Printed by Libri Plureos GmbH
in Hamburg, Germany